Environmental Science and Engineering
Subseries: Environmental Science

Series Editors: R. Allan • U. Förstner • W. Salomons

For further volumes:
http://www.springer.com/series/7487

Teja Tscharntke ·
Christoph Leuschner ·
Edzo Veldkamp · Heiko Faust ·
Edi Guhardja · Arifuddin Bidin
Editors

Tropical Rainforests and Agroforests under Global Change

Ecological and Socio-economic Valuations

Springer

Editors

Prof. Dr. Teja Tscharntke
Georg-August University
Dept. Crop Sciences
Agroecology
Waldweg 26
37073 Göttingen
Germany
ttschar@gwdg.de

Prof. Dr. Christoph Leuschner
Georg-August University
Albrecht-von-Haller-Institute of Plant Sciences
Plant Ecology
Untere Karspüle 2
37073 Göttingen
Germany
cleusch@gwdg.de

Prof. Dr. Edzo Veldkamp
Georg-August University
Buesgen Institute
Soil Science and Forest Nutrition
Buesgenweg 2
37077 Göttingen
Germany
eveldka@gwdg.de

Prof. Dr. Heiko Faust
Georg-August University
Institute of Geography
Human Geography
Goldschmidtstr. 5
37077 Göttingen
Germany
hfaust@gwdg.de

Prof. Dr. Edi Guhardja
Bogor Agricultural University
Dept. Biology
IPB Kampus Baranang Siang
Bogor 16144
Indonesia
storma-ipb@indo.net.id

Dr. Arifuddin Bidin
Tadulako University
Dept. Forest Management
Kampus Bumi Tadulako
Tondo Palu
Indonesia
arif_bidin@yahoo.com

ISSN 1863-5520
ISBN 978-3-642-00492-6 e-ISBN 978-3-642-00493-3
DOI 10.1007/978-3-642-00493-3
Springer Heidelberg Dordrecht London New York

Library of Congress Control Number: 2010920289

Cover design: Integra Software Services Pvt. Ltd., Pondicherry

Typesetting: Camera-ready by Dr. Stella Aspelmeier

Printed on acid-free paper

Springer is part of Springer Science+Business Media (www.springer.com)

Contents

Tropical rainforests and agroforests under global change: Ecological and socio-economic valuations – an introduction
Teja Tscharntke, Christoph Leuschner, Edzo Veldkamp, Heiko Faust, Edi Guhardja, Arifuddin Bidin 1

Part I Agroforestry management in an ecological and socio-economic context

Biodiversity patterns and trophic interactions in human-dominated tropical landscapes in Sulawesi (Indonesia): plants, arthropods and vertebrates
Yann Clough, Stefan Abrahamczyk, Marc-Oliver Adams, Alam Anshary, Nunik Ariyanti, Lydia Betz, Damayanti Buchori, Daniele Cicuzza, Kevin Darras, Dadang Dwi Putra, Brigitte Fiala, S. Robbert Gradstein, Michael Kessler, Alexandra-Maria Klein, Ramadhanil Pitopang, Bandung Sahari, Christoph Scherber, Christian H. Schulze, Shahabuddin, Simone Sporn, Kathrin Stenchly, Sri S. Tjitrosoedirdjo, Thomas C. Wanger, Maria Weist, Arno Wielgoss, Teja Tscharntke 15

The potential of land-use systems for maintaining tropical forest butterfly diversity
Christian H. Schulze, Stefan Schneeweihs, Konrad Fiedler 73

Insect pollinator communities under changing land-use in tropical landscapes: implications for agricultural management in Indonesia
Bandung Sahari, Akhmad Rizali, Damayanti Buchori 97

Structure and management of cocoa agroforestry systems in Central Sulawesi across an intensification gradient
Jana Juhrbandt, Thomas Duwe, Jan Barkmann, Gerhard Gerold, Rainer Marggraf 115

Land tenure rights, village institutions, and rainforest conversion in Central Sulawesi (Indonesia)
Jan Barkmann, Günter Burkard, Heiko Faust, Michael Fremerey, Sebastian Koch, Agus Lanini 141

Rural income dynamics in post-crisis Indonesia: evidence from Central Sulawesi
Jan Priebe, Robert Rudolf, Julian Weisbrod, Stephan Klasen, Iman Sugema, Nunung Nuryartono 161

Gender division of labor in agroforestry activities within households: a case of Wonogiri - Central Java - Indonesia
Herien Puspitawati, Ma'mun Sarma 177

The robustness of indicator based poverty assessment tools in changing environments - empirical evidence from Indonesia
Xenia van Edig, Stefan Schwarze, Manfred Zeller 191

Demography, development, and deforestation at the rainforest margin in Indonesia
Stephan Klasen, Heiko Faust, Michael Grimm, Stefan Schwarze 213

Part II Climate change effects on tropical rainforests and agroforests

Functional biodiversity and climate change along an altitudinal gradient in a tropical mountain rainforest
Jörg Bendix, Hermann Behling, Thorsten Peters, Michael Richter, Erwin Beck 239

Spatiotemporal trends of forest cover change in Southeast Asia
Stefan Erasmi, Muhammad Ardiansyah, Pavel Propastin, Alfredo Huete . 269

Comparison of tree water use characteristics in reforestation and agroforestry stands across the tropics
Diego Dierick, Norbert Kunert, Michael Köhler, Luitgard Schwendenmann, Dirk Hölscher 293

A comparison of throughfall rate and nutrient fluxes in rainforest and cacao plantation in Central Sulawesi, Indonesia
Carsten Gutzler, Stefan Koehler, Gerhard Gerold 309

Effects of "ENSO-events" and rainforest conversion on river discharge in Central Sulawesi (Indonesia)
Gerhard Gerold, Constanze Leemhuis 327

Adaptation to climate change in Indonesia - livelihood strategies of rural households in the face of ENSO related droughts
Norbert B. Binternagel, Jana Juhrbandt, Sebastian Koch, Mangku Purnomo, Stefan Schwarze, Jan Barkmann, Heiko Faust 351

Terrestrial herb communities of tropical submontane and tropical montane forests in Central Sulawesi, Indonesia
Daniele Cicuzza, Michael Kessler, Ramadhanil Pitopang, Sri S. Tjitrosoedirdjo, S. Robbert Gradstein 377

The hydraulic performance of tropical rainforest trees in their perhumid environment - is there evidence for drought vulnerability?
Alexandra Zach, Bernhard Schuldt, Viviana Horna, Soekisman Tjitrosemito, Christoph Leuschner 391

Part III Integrated concepts of land use in tropical landscapes

---

Principle and practice of the buffer zone in biosphere reserves: from global to local – general perspective from managers versus local perspective from villagers in Central Sulawesi, Indonesia
Marion Mehring, Susanne Stoll-Kleemann.......................... 413

Institutions for environmental service payment programmes - evidence of community resource management arrangements in Central Sulawesi, Indonesia
Christina Seeberg-Elverfeldt, Stefan Schwarze, Heiko Faust 431

Agricultural expansion in the Brazilian state of Mato Grosso; implications for C stocks and greenhouse gas emissions
Eleanor Milne, Carlos Eduardo P. Cerri, João Luis Nunes Carvalho.... 447

Contribution of agroforestry to biodiversity and livelihoods improvement in rural communities of Southern African regions
Kanungwe Felix Kalaba, Paxie Chirwa, Stephen Syampungani, Clifford Oluyede Ajayi 461

Human ecological dimensions in sustainable utilization and conservation of tropical mountain rain forests under global change in southern Ecuador
Perdita Pohle, Andrés Gerique, Martina Park, María Fernanda López Sandoval 477

Linkages between poverty and sustainable agricultural and rural development in the uplands of Southeast Asia
Manfred Zeller, Tina Beuchelt, Isabel Fischer, Franz Heidhues 511

Index of keywords 529

List of Contributors

Stefan Abrahamczyk
Institute of Systematic Botany, University of Zürich, Zollikerstrasse 107 CH-8008 Zürich, Switzerland

Marc-Oliver Adams
Agroecology, University of Göttingen, Waldweg 26, D-37073 Göttingen, Germany

Department of Animal Ecology and Tropical Biology (Zoology III), Biozentrum, University of Würzburg, Am Hubland, D-97074 Würzburg, Germany

Clifford Oluyede Ajayi
World Agroforestry Centre (ICRAF), P. O. Box 30798, Lilongwe 03, Malawi

Alam Anshary
Faculty of Agriculture, University of Tadulako, Palu, Central Sulawesi, Indonesia

Muhammad Ardiansyah
Bogor Agricultural University, Department of Soil Sciences and Land resources, Bogor, Indonesia

Nunik Ariyanti
Department of Biology, Faculty of Mathematics and Natural Sciences, Bogor Agricultural University, Jalan Padjajaran, 16144 Bogor, West Java, Indonesia

Jan Barkmann
Environmental & Resource Economics, Department of Agricultural Economics and Rural Development, Georg-August-University Göttingen, Platz der Göttinger Sieben 5, 37073 Göttingen, Germany

Erwin Beck
Dept. of Plant Physiology, University of Bayreuth, Universitaetsstr. 30, D 95440 Bayreuth, Germany

Hermann Behling
Department of Palynology and Climate Dynamics, Albrecht-von-Haller Institute of Plant Sciences, University of Goettingen, Untere Karspuele 2, D 37073 Goettingen, Germany

Jörg Bendix
Laboratory for Climatology and Remote Sensing, Faculty of

Geography University of Marburg, Deutschhausstr. 10, D 35032 Marburg, Germany

Lydia Betz
Agroecology, University of Göttingen, Waldweg 26, D-37073 Göttingen, Germany

Tina Beuchelt
Department of Agricultural Economics and Social Sciences in the Tropics and Subtropics, Section Rural Development Theory and Policy, Universität Hohenheim, (490a), D-70593 Stuttgart, Germany

Norbert B. Binternagel
Georg-August-University Göttingen, Faculty of Geosciences and Geography, Department of Human Geography, Goldschmidtstr. 5, 37077 Göttingen, Germany

Damayanti Buchori
Peka Indonesia Foundation (Indonesian Nature Conservation Foundation)-Wildlife Trust Alliance. Jl. Uranus Blok H No 1 Perum IPB Sindang Barang 2, Bogor, West Java-Indonesia

Department of Plant Protection, Bogor Agricultural University, Kampus IPB Dramaga, Bogor-West Java-Indonesia

Günter Burkard
Rural Sociology, Faculty of Organic Agricultural Sciences, University of Kassel, Steinstr. 19, 37213 Witzenhausen, Germany

João Luis Nunes Carvalho
Center for Nuclear energy in Agriculture (CENA), The University of São Paulo,
C.P. 96, C.E.P. 13.400-970, Piracicaba, SP, Brazil

Carlos Eduardo P. Cerri
Center for Nuclear energy in Agriculture (CENA), The University of São Paulo,
C.P. 96, C.E.P. 13.400-970, Piracicaba, SP, Brazil

Paxie Chirwa
Stellenbosch University, Department of Forest and Wood Science, Stellenbosch 7602, South Africa

Daniele Cicuzza
Institute of Systematic Botany, University of Zürich, Zollikerstrasse 107 CH-8008 Zürich, Switzerland

Albrecht-von-Haller-Institute of Plant Sciences, University of Göttingen, Untere Karspüle 2, D-37073 Göttingen, Germany

Yann Clough
Agroecology, University of Göttingen, Waldweg 26, D-37073 Göttingen, Germany

Kevin Darras
14 allée de Frênes F-01210 Versonnex, France

Agroecology, University of Göttingen, Waldweg 26, D-37073 Göttingen, Germany

Diego Dierick
Tropical Silviculture and Forest Ecology, Burckhardt Institute, University of Göttingen, Büsgenweg 1, D- 37077 Göttingen, Germany

Dadang Dwi Putra
Celebes Bird Club, Jl. Thamrin 63A, Palu, Central Sulawesi, Indonesia

Thomas Duwe
Department of Landscape Ecology, Institute of Geography, Georg-August University Göttingen, Goldschmidtstr. 5, 37077 Göttingen, Germany

Xenia van Edig
Department of Agricultural Economics and Rural Development, University of Göttingen, Germany

Stefan Erasmi
University of Göttingen, Institute of Geography, Goldschmidtstr. 5, 37077 Göttingen, Germany

Heiko Faust
University of Göttingen, Institute of Geography, Division of Human Geography, Goldschmidtstr. 5, 37077 Göttingen, Germany

Brigitte Fiala
Department of Animal Ecology and Tropical Biology (Zoology III), Biozentrum, University of Würzburg, Am Hubland, D-97074 Würzburg, Germany

Konrad Fiedler
Department of Population Ecology, Faculty of Life Sciences, University of Vienna, Rennweg 14, A-1030 Vienna, Austria

Isabel Fischer
Department of Agricultural Economics and Social Sciences in the Tropics and Subtropics, Section Rural Development Theory and Policy, Universität Hohenheim, (490a), D-70593 Stuttgart, Germany

Michael Fremerey
Rural Sociology, Faculty of Organic Agricultural Sciences, University of Kassel, Steinstr. 19, 37213 Witzenhausen, Germany

Andrés Gerique
Institute of Geography, Friedrich-Alexander-University Erlangen-Nürnberg, Kochstrasse 4/4, 91054 Erlangen, Germany

Gerhard Gerold
Department of Landscape Ecology, Institute of Geography, Georg-August University Göttingen, Goldschmidtstr. 5, 37077 Göttingen, Germany

S. Robbert Gradstein
Albrecht-von-Haller-Institute of Plant Sciences, University of Göttingen, Untere Karspüle 2, D-37073 Göttingen, Germany

Michael Grimm
International Institute of Social Studies, Erasmus University Rotterdam, Kortenaerkade 12, 2518AX The Hague, The Netherlands

Carsten Gutzler
Institute of Geography, Department of Landscape Ecology, University of Göttingen, Goldschmidtstr. 5, 37077 Göttingen, Germany

Franz Heidhues
Department of Agricultural Economics and Social Sciences in the Tropics and Subtropics, Section Rural Development Theory and Policy, Universität Hohenheim, (490a), D-70593 Stuttgart, Germany

Dirk Hölscher
Tropical Silviculture and Forest Ecology, Burckhardt Institute, University of Göttingen, Büsgenweg 1, D- 37077 Göttingen, Germany

Viviana Horna
Plant Ecology, Albrecht von Haller Institute for Plant Sciences, University of Göttingen, Untere Karspüle 2, 37073 Göttingen, Germany

Alfredo Huete
University of Arizona, Department of Soil, Water and Environmental Science, Tucson, AZ, USA

Jana Juhrbandt
Environmental and Resource Economics, Department of Agricultural Economics and Rural Development, Georg-August University Göttingen, Platz der Göttinger Sieben 5, D-37073 Göttingen, Germany

Kanungwe Felix Kalaba
Copperbelt University, School of Natural Resources, P.O. Box 21692, Kitwe, Zambia

Michael Kessler
Institute of Plant Sciences, University of Göttingen, Untere Karspüle 2, 37073 Göttingen, Germany

Institute of Systematic Botany, University of Zürich, Zollikerstrasse 107 CH-8008 Zürich, Switzerland

Stephan Klasen
University of Göttingen, Faculty of Economics, Platz der Göttinger Sieben 3, 37073 Göttingen, Germany

Alexandra-Maria Klein
Agroecology, University of Göttingen, Waldweg 26, D-37073 Göttingen, Germany

Sebastian Koch
Environmental & Resource Economics, Department of Agricultural Economics and Rural Development, Georg-August-University Göttingen, Platz der Göttinger Sieben 5, 37073 Göttingen, Germany

Department of Cultural and Social Geography, Institute of Geography, Georg-August-University Göttingen, Goldschmidtstr. 5, 37077 Göttingen, Germany

Michael Köhler
Tropical Silviculture and Forest Ecology, Burckhardt Institute, University of Göttingen, Büsgenweg 1, D- 37077 Göttingen, Germany

Stefan Koehler
Landscape Ecology and Land Evaluation, Faculty for Agricultural- and Environmental Sciences, University of Rostock, Justus-von-Liebig Weg 6, 18059 Rostock, Germany

Norbert Kunert
Tropical Silviculture and Forest Ecology, Burckhardt Institute, University of Göttingen, Büsgenweg 1, D- 37077 Göttingen, Germany

Agus Lanini
Faculty of Law, Universitas Tadulako, Kampus Bumi Tondo, Palu 94118, Indonesia

Constanze Leemhuis
Center for Development Research, University of Bonn, Walter-Flex-Str. 3, 53113 Bonn, Germany

Christoph Leuschner
Plant Ecology, Albrecht von Haller Institute for Plant Sciences, University of Göttingen, Untere Karspüle 2, 37073 Göttingen, Germany

María Fernanda López Sandoval
Escuela de Geografía, Facultad de Ciencia Humana, Pontificia Universidad Católica del Ecuador (PUCE), Quito

Rainer Marggraf
Environmental and Resource Economics, Department of Agricultural Economics and Rural Development, Georg-August University Göttingen, Platz der Göttinger Sieben 5, D-37073 Göttingen, Germany

Marion Mehring
Applied Geography and Sustainability Science, GoBi (Governance of Biodiversity) Research Group, Institute of Geography and Geology, Ernst-Moritz-Arndt Universität Greifswald, Jahnstr. 16, 17487 Greifswald, Germany

Eleanor Milne
The Macaulay Institute, Craigiebuckler, Aberdeen, AB158QH, UK

Colorado State University (NREL), Fort Collins, CO 80523-1499 USA

Nunung Nuryartono
Institut Pertanian Bogor, Indonesia, 16710 Bogor, International Center for Applied Finance and Economics, Kampus IPB Baranang Siang, Gedung Utama Lt.
II, Jalan Raya Paiajaran

Martina Park
Institute of Geography, Friedrich-Alexander-University Erlangen-Nürnberg, Kochstrasse 4/4, 91054 Erlangen, Germany

Thorsten Peters
Institute of Geography, University of Erlangen, Kochstr.4/4, D 91054 Erlangen, Germany

Ramadhanil Pitopang
Peka Indonesia Foundation (Indonesian Nature Conservation Foundation)-Wildlife Trust Alliance. Jl. Uranus Blok H No 1 Perum IPB Sindang Barang 2, Bogor, West Java, Indonesia

Department of Forest Managment and Herbarium Celebense, Tadulako University, Palu, Indonesia

Perdita Pohle
Institute of Geography, Friedrich-Alexander-University Erlangen-Nürnberg, Kochstrasse 4/4, 91054 Erlangen, Germany

Jan Priebe
University of Göttingen, Faculty of Economics, Platz der Göttinger Sieben 3, 37073 Göttingen, Germany

Pavel Propastin
University of Göttingen, Institute of Geography, Goldschmidtstr. 5, 37077 Göttingen, Germany

Mangku Purnomo
Brawijaya University of Malang, Faculty of Agriculture, Department of Socio-Economics, Jalan Veteran, 65145 Malang, Indonesia

Herien Puspitawati
Department of Family and Consumer Sciences, Faculty of Human Ecology, Bogor Agricultural University, Jalan Lingkar Akademik Kampus IPB, Darmaga, Bogor 16680, Jawa Barat, Indonesia

Michael Richter
Institute of Geography, University of Erlangen, Kochstr. 4/4, D 91054 Erlangen, Germany

Akhmad Rizali
Peka Indonesia Foundation (Indonesian Nature Conservation Foundation)-Wildlife Trust Alliance. Jl. Uranus Blok H No 1 Perum IPB Sindang Barang 2, Bogor, West Java-Indonesia

Department of Plant Protection, Bogor Agricultural University, Kampus IPB Dramaga, Bogor-West Java-Indonesia

Agroecology, University of Göttingen, Waldweg 26, 37073 Göettingen, Germany

Robert Rudolf
University of Göttingen, Faculty of Economics, Platz der Göttinger Sieben 3, 37073 Göttingen, Germany

Bandung Sahari
Peka Indonesia Foundation (Indonesian Nature Conservation Foundation)-Wildlife Trust Alliance. Jl. Uranus Blok H No 1 Perum IPB Sindang Barang 2, Bogor, West Java-Indonesia

Ma'mun Sarma
Department of Management, Faculty of Economics and Management, Bogor Agricultural University, Indonesia

Christoph Scherber
Agroecology, University of Göttingen, Waldweg 26, D-37073 Göttingen, Germany

Stefan Schneeweihs
Department of Population Ecology, Faculty of Life Sciences, University of Vienna, Rennweg 14, A-1030 Vienna, Austria

Bernhard Schuldt
Plant Ecology, Albrecht von Haller Institute for Plant Sciences, University of Göttingen, Untere Karspüle 2, 37073 Göttingen, Germany

Christian H. Schulze
Department of Population Ecology, Faculty of Life Sciences, University of Vienna, Rennweg 14, A-1030 Vienna, Austria

Stefan Schwarze
University of Göttingen, Department of Agricultural Economics and Rural Development, Platz der Göttinger Sieben 5, 37073 Göttingen

Luitgard Schwendenmann
Tropical Silviculture and Forest Ecology, Burckhardt Institute, University of Göttingen, Büsgenweg 1, D- 37077 Göttingen, Germany

Christina Seeberg-Elverfeldt
Natural Resources Management and Environment Department (NRD), FAO, Viale delle Terme di Caracalla, 00153 Rome, Italy

Shahabuddin
Faculty of Agriculture, University of Tadulako, Palu, Central Sulawesi, Indonesia

Simone Sporn
Albrecht-von-Haller-Institute of Plant Sciences, University of Göttingen, Untere Karspüle 2, D-37073 Göttingen, Germany

Kathrin Stenchly
Agroecology, University of Göttingen, Waldweg 26, D-37073 Göttingen, Germany

Susanne Stoll-Kleemann
Applied Geography and Sustainability Science, GoBi (Governance of Biodiversity) Research Group, Institute of Geography and Geology, Ernst-Moritz-Arndt Universität Greifswald, Jahnstr. 16, 17487 Greifswald, Germany

Iman Sugema
Institut Pertanian Bogor, Indonesia, 16710 Bogor, International Center for Applied Finance and Economics, Kampus IPB Baranang Siang, Gedung Utama Lt.
II, Jalan Raya Paiajaran

Stephen Syampungani
Copperbelt University, School of Natural Resources, P.O. Box 21692, Kitwe, Zambia

Soekisman Tjitrosemito
Faculty of Science and Mathematics, Institute Pertanian Bogor, Jl. Raya Pajajaran, Bogor, 16144 Indonesia

Sri S. Tjitrosoedirdjo
Department of Biology, Faculty of Mathematics and Natural Sciences, Bogor Agricultural University, Jalan Padjajaran, 16144 Bogor, West Java, Indonesia

Teja Tscharntke
Agroecology, University of Göttingen, Waldweg 26, D-37073 Göttingen, Germany

Thomas C. Wanger
Agroecology, University of Göttingen, Waldweg 26, D-37073 Göttingen, Germany

Environment Institute, School of Earth and Environmental Sciences, University of Adelaide, Australia

Julian Weisbrod
University of Göttingen, Faculty of Economics, Platz der Göttinger Sieben 3, 37073 Göttingen, Germany

Maria Weist
Agroecology, University of Göttingen, Waldweg 26, D-37073 Göttingen, Germany

Arno Wielgoss
Agroecology, University of Göttingen, Waldweg 26, D-37073 Göttingen, Germany

Department of Animal Ecology and Tropical Biology (Zoology III), Biozentrum, University of Würzburg, Am Hubland, D-97074 Würzburg, Germany

Alexandra Zach
Plant Ecology, Albrecht von Haller Institute for Plant Sciences, University of Göttingen, Untere Karspüle 2, 37073 Göttingen, Germany

Manfred Zeller
Department of Agricultural Economics and Social Sciences in the Tropics and Subtropics, Section Rural Development Theory and Policy, Universität Hohenheim, (490a), D-70593 Stuttgart, Germany

Tropical rainforests and agroforests under global change: Ecological and socio-economic valuations – an introduction

Teja Tscharntke[1], Christoph Leuschner[2], Edzo Veldkamp[3], Heiko Faust[4], Edi Guhardja[5], and Arifuddin Bidin[6]

[1] Agroecology, University of Göttingen, Waldweg 26, 37073 Göttingen, Germany
[2] Plant Ecology, Albrecht-von-Haller-Institute of Plant Sciences, University of Göttingen, Untere Karspüle 2, 37073 Göttingen, Germany
[3] Soil Science and Forest Nutrition, University of Göttingen, Büsgenweg 2, 37077 Göttingen,Germany
[4] Human Geography, Institute of Geography, University of Göttingen, Goldschmidtstr. 5, 37077 Göttingen, Germany
[5] Kampus IPB Baranang Siang, Ex-Aula Kantor Pusat, Institut Pertanian Bogor, JL Raya Pajajaran, Bogor 16144, Indonesia
[6] Universitas Tadulako, Kampus Bumi Tondo, Palu 94118, Indonesia (Sulteng)

Tropical rainforests are disappearing, causing irreversible species losses, especially of rare and specialised species (e.g. Owens and Bennett 2000, Acebey et al. 2003, Kessler et al. 2009, Maas et al. 2009). Despite an increasing recognition of the value of biodiversity and associated ecosystem services at national and international levels, rainforests continue to be seriously threatened by human-induced global change such as agricultural intensification and climate change. Tropical rain forests provide critical ecosystem services to the local, regional and global communities (e.g. water supply, store of biodiversity, large pool of carbon), while each year an estimated 13 million ha of forest is destroyed (FAO 2006), 5.6 to 8.8 Gt of carbon are emitted (Nabuurs et al. 2007) and an estimated 14,000 to 40,000 species disappear from tropical forests (Hughes et al. 1997). In general, ecosystem properties and human well-being are known to be profoundly influenced by changes in land use management (Robertson and Swinton 2005), while land use intensification and diversification have led to rapid changes in biogeochemical and hydrological processes (Vitousek et al. 1997). Understanding these processes needs an integrated scientific approach linking ecological, economic and social perspectives at different scales, from the household and village level to landscapes and regions (Olschewski et al. 2006, Steffan-Dewenter et al. 2007, Tscharntke et al. 2007).

Tropical forests are threatened by interacting forces that operate at different scales (Brook et al. 2008). At the global level, climate change processes, globalized markets for agricultural products (e.g. biofuels) and international

T. Tscharntke et al. (eds.), *Tropical Rainforests and Agroforests under Global Change*, Environmental Science and Engineering, 1–11, DOI 10.1007/978-3-642-00493-3_1,

conventions to protect climate and biodiversity have an impact on the spatial extend and quality of tropical forests. Policies in the areas of food and agriculture, forestry, nature conservation, rural development and extension work at the national and sub-national level, while at the local level policies implemented by local institutions (e.g., road construction and community based forest management) also impact the stability of forest areas (Maertens et al. 2006, Muller and Zeller 2002).

In the past few decades, intensification of agriculture through the use of high-yielding crops, fertilization, irrigation and pesticides has led to an unprecedented increase in food production. However, land conversion and intensification have also altered biotic interactions in ecosystems which can have serious local, regional and global environmental consequences that are expected to increase in the near future (Matson et al. 1997, Brook et al. 2008, Tscharntke et al. 2008). The intensive use of forest resources, expansion of agricultural activity into primary or near primary habitat, as well as land-use intensification are primary drivers for biodiversity loss (Sala et al. 2000). Some of the impacts of intensification are a doubling of nitrogen and phosphorus-driven eutrophication and pesticide use, which will lead to an extraordinary loss of ecosystem services and species extinction (Tilman et al. 2002). Especially in regions where agricultural intensification is a relatively recent process, like in the tropics, these problems are expected to occur (Robertson and Swinton 2005, Lee and Jetz 2008). For example, intensification of cacao agroforestry led to a tripling of emissions of the greenhouse gas N_2O (Veldkamp et al. 2008). Analyses of economic and ecological consequences of land use intensification showed that low-shade agroforestry provided the best available compromise between economic and ecological needs (Steffan-Dewenter et al. 2007). Certification schemes for shade-grown crops may provide a market-based mechanism to slow down the current intensification process.

Climate change is widely recognised as one of the most serious environmental threats facing humanity today. The Intergovernmental Panel on Climate Change (IPCC) concluded in its Fourth Assessment Report that warming of the earths climate is now indisputable, and that it is very likely that this is due to emissions of greenhouse gasses from human activities, particularly from the last half of the 20th century onwards. At present, human-induced climate change is already causing extremes in temperatures and precipitation which is likely to continue for centuries (Karl and Trenberth 2003). First evidence is appearing that changes in temperature extremes can affect tropical forest growth and probably forest carbon stocks (Clark et al. 2003, Phillips et al. 2008). Also dry season length has increased in various parts of the tropics, including Southeast Asia (Corlett and LaFrankie 1998) which led to a decline in annual precipitation in parts of the humid tropics (Bunker and Carson 2005). Several climate models predict a warming trend of 1.5 - 2.5 °C in annual mean temperature for large parts of the tropics, and some climate scenarios predict a more frequent occurrence of droughts of increasing severity induced by global warming (Timmermann et al. 1999). Such scenarios are of serious concern, not only for land use systems that depend on the regular supply of rain or irrigation water but also for the future development of natural rainforests as drought stress has been shown to affect tree growth and species composition in old-growth forests (Wright 1991, Walsh and Newbery 1999, Engelbrecht et al. 2007). A drought experiment conducted in a cacao agroforestry plantation showed that this plantation was surprisingly resilient to an induced drought of more than a year (Schwendenmann et al. 2009). However, droughts can have a strong impact on household incomes from agriculture, they strongly affect the vulnerability to poverty and thus have to be analyzed as important exogenous shocks to households, forcing them to adjust their behaviour and develop strategies to cope with these problems.

The stability of rainforest margins is a critical factor in the protection of tropical rainforests (Tscharntke et al. 2007). At present, however, rainforest margins in many parts of the tropics are far from stable, both in socio-economic and in ecological terms. For example, protected areas may attract, rather than repel, human settlement, which may be due to international donor investment in national conservation programs (Wittemeyer et al. 2008). An alternative hypothesis is that protected areas might be compromised if leakage takes place, that is, if impacts that would take place inside the restricted area are displaced to a nearby, undisturbed area (Ewers and Rodrigues 2008). The challenge of understanding (in-)stability of rainforest margins and of generating policy-relevant knowledge how to counteract destabilizing processes reaches across scientific disciplines and requires an integrated approach. As recommended by the Global Land Project Science Plan (GLP 2005) interactions between decision making, ecosystem services and global environmental change form important feedbacks from human activities at local, regional and global scales.

In this book, we focus on three research topics, each of which will integrate socio-economic and ecological research:

Part I: Agroforestry management in an ecological and socio-economic context
Part II: Climate change effects on tropical rainforests and agroforests
Part III: Integrated concepts of land use in tropical landscapes

Part I: Agroforestry management in an ecological and socio-economic context

The paper of Clough et al. focuses on functional biodiversity of a wide range of plant and animal groups, changing with patterns in land-use type. The rise of Indonesia to the third largest cocoa-producing country in the world is at the core of recent agricultural intensification processes determining landscape-wide abundance and distribution of species. Results show that generally shaded agroforestry and distance to nearest forest are major determinants of species richness. Functionally important groups such as insectivorous and seed-dispersing birds benefit from tall shade trees, shade-tree diversity and proximity to forest edge.

Schulze et al. summarize studies analyzing effects of forest disturbance and conversion on butterfly assemblages. Old-grown secondary forests turned out to be an important habitat type for forest butterflies in all major tropical regions, whereas agroforestry systems harbour only a small set of species. Butterflies species with narrow geographic ranges are particularly prone to local and global extinctions. The authors conclude that in all four major tropical regions the future of tropical butterflies depends on a strict protection of large forest blocks.

Insect pollinator communities under changing land use are the topic of Sahari et al. Increasing land-use intensity and larger distance to forests may reduce pollinator availability and pollination success of cultivated plants. Changes in pollinator community structure may be as important as decreased pollinator species richness, which is discussed at length by the authors.

Juhrbandt et al. analyze the structure and management of cocoa agroforests in Central Sulawesi. Agricultural intensification by removal of shade trees is ongoing in the region. The authors developed a structural intensification index that was positively related to cocoa yield. This strong economic incentive for cocoa farmers to intensify production threatens local biodiversity and ecosystem functioning. Soil nutrient status is mostly sufficient, while phytosanitary and soil amelioration management are often suboptimal, thereby jeopardizing quantity and quality of cocoa bean yield.

In contrast to the argument that security of land tenure rights fosters resource conservation, Barkmann et al. found in their study in Sulawesi that security of formal land titles attracts migrants who are aggressive buyers of land for cocoa production. In the end, a substantial share of the autochthonous population finds itself either landless or is forced to cut marginal forest land, often inside protected rainforests. Restricting land ownership to traditional forms of community land rights may be a way out.

Priebe et al. analyzed a new panel data set collected in Central Sulawesi and found a sharp increase in rural incomes after the Asian Financial Crisis of 1997/98. While traditional agriculture still constitutes the backbone of household incomes, the ability to alleviate poverty and to enjoy income growth has been associated with the households ability to diversify into the non-agricultural sector of the economy, to focus on higher value-added agricultural activities and its capability to invest into new production techniques.

Gender division of labour in Indonesian agroforestry is the topic of Puspitawati and Sarma. In general they found an imbalance of gender partnership in labour division, for example when it comes to cultivation, processing and marketing of cashew nuts. The community and gender empowerment includes the agreement among communities (both men and women) in the village to reformulate its regional planning, the increase of womens potential skills and knowledge, and the strengthening of farmer institutions.

Eradicating poverty is one of the highest priorities of development policies, and better-off households may less encroach rainforests and may also be less vulnerable to shocks by natural hazards, argue van Edig et al. in their chapter. Using data from household surveys in 2005 and 2007 from Sulawesi, they test the robustness of newly developed poverty assessment indicators over time. Almost 20% of the rural population of Central Sulawesi has been identified as being very poor with individuals living on less than $ 1 US per capita and day in purchasing power parities, contrasting with Indonesian average of 7.5%.

Klasen et al. link issues of long-term economic development, technological change, and conservation of the rainforest. They base their chapter on econometric analyses of a village survey in Central Sulawesi. They found that a key driver of economic development is demographic change, particularly immigration into new areas. This in turn affects the emergence of land rights that spur technological change and economic development. However, immigration and most technological change also promotes deforestation, while economic development itself appears to reduce deforestation.

Part II: Climate change effects on tropical rainforests and agroforests

Bendix et al. investigate possible consequences of climate change for a hotspot of biodiversity in the South Ecuadorian Andes. Changes of climate and vegetation during the Holocene are described showing possible fluctuations between paramo-like grassland and tropical mountain forest. To assess the effects of temperature shift along an altitudinal gradient, two model approaches were applied to distribution patterns of moths, the species-area and the energetic equivalence rule. Applying realistic scenarios of climate change, 31% of moth species losses can be predicted until the year 2100, while woody plants are much more resilient. In their outlook, they discuss a priority ranking of climate change vs. direct anthropogenic impacts for conservation.

The current state of tropical forest cover and its change have been identified as key variables in modelling and measuring human impact on ecosystems. Erasmi et al. focus on Southeast Asia, which exhibits highest rates of forest loss worldwide. Results on a national level are compared with an analysis at the regional level in Central Sulawesi and provide recommendations for future remote sensing based forest assessment in the tropics.

Dierick et al. focus on reforestation and agroforestry, which help mitigating climate change by carbon sequestration, while high water use of trees might be a problem. Data on tree sap flux and water use from Indonesia, Panama and the Philippines suggest high tree species-specific differences. Across sites, tree diameter explained 65% of the observed differences in tree water use, while remaining variability could be linked to species-identity. When water resources are limited or climate scenarios predict decreasing precipitation, species selection may become important to control tree water use in reforestation and agroforestry.

Gutzler et al found in Central Sulawesi that throughfall rates differed between cacao agroforestry (90%) and rainforest sites (81%). As expected, nutrient enrichment through canopy leaching was higher in the natural forest than on the cacao plots. Conversion of forest to cacao increases water input by 10%, while additional increases through reduced transpiration and higher runoffs can be expected.

Effect of ENSO related droughts and rainforest conversion in Central Sulawesi has been studied by Gerold et al. In a hydrological modelling approach, the authors found a strong relationship between deforestation rates and discharge variability. Main results of the scenarios of ENSO related droughts include that (i) precipitation leads to an increase of discharge variability, (ii) strong reductions of water yield may occur, (iii) the potential area of paddy rice cultivation is decreased, and (iv) annual crops show a higher increase in river discharge than perennial crops (cacao) with high increase in overland flow and flooding risks.

Adaptation to climate change can reduce social vulnerability, as shown by Binternagel et al. for ENSO-related droughts in Central Sulawesi. All interviewed households have been affected by extreme climate effects such as ENSO related droughts, which led to a decline in agricultural outputs. Most common adaptation strategies are reactive or include also, influenced by social and human capital, anticipatory strategies. For the successful adoption of innovations to cope with ENSO related droughts, membership in certain ethnic groups, local institutions and networks turned out to be most important.

Most botanical research in the tropics focuses on trees, although terrestrial herbs exhibit high richness and play important roles. Cicuzza et al. compared herb communities in tropical submontane and montane forest sites of Central Sulawesi and found 91 and 77 angiosperms, respectively. These high species numbers point to a previously underappreciated richness and functional role of plant assemblages. Ramadanil et al. (2008) quantified how composition of herb communities, including invasive species, changes between land use types.

Zach et al studied hydraulic properties of various tree species in a species-rich tropical forest of Central Sulawesi to address the question whether there is evidence for drought vulnerability in perhumid forests. The high observed plasticity with variation in tree height may be a direct response to changes in ambient conditions along the microclimate gradient from the forest understory to the canopy top. Large trees might be more resistant against drought events in the short run, but may be more vulnerable to prolonged drought periods.

Part III: Integrated concepts of landuse in tropical landscapes

Mehring and Stoll-Kleemann discuss principle and practice of buffer zone management around biosphere reserves. Based on a field study in Central Sulawesi and a global study among biosphere reserves, the authors compare the perspectives of villagers with those of reserve managers. While managers found human uses in the buffer zones to be sustainable, villagers miss basic management activities such as clear boundary demarcation and law enforcement as barriers to a more sustainable development. The authors interpret the completely different perception as an indicator of missing/bad communication between both, villagers and management.

Payments for Environmental Service (PES) schemes are discussed by Seeberg-Elverfeldt et al. as a possibility to promote conservation of natural resources, but pilot schemes are frequently small in size and face high transaction costs, leading to the exclusion of smallholders. Using the example of the institution of the community conservation agreements (CCA) in Central Sulawesi, they assess whether a community arrangement can provide the framework conditions to implement a PES project. Four points are necessary: an organisational structure representing the village households, participation of the resource users in the institutional implementation, monitoring and enforcement by the institution of the forest usage regulations and, finally, the institutions ability to administer funds. The authors discuss how to better inform and involve community members in the management of natural resource projects to increase compliance with regulations.

The effects of agricultural expansion on carbon stocks and greenhouse gas (GHG) emissions are the topic of Milne et al. The states of Rondonia and Mato Grosso in Brazil make up the worlds largest agricultural frontier and deforestation reached unprecedented rates. Interviews revealed that management practices employed by the small (<500 ha) farms differed markedly from those employed by medium (> 500 ha) and large (> 10,000 ha) farms, affecting carbon stocks and GHG emissions. In an outlook, the possibility of REDD being included in a successor to the Kyoto protocol is discussed.

Kalaba et al. analyze in a study from Southern African regions, how improved agroforestry systems provide benefits that contribute to rural livelihoods, improve socio-economic status and ecosystem functioning of land-use systems. Compared with subsistence agriculture, improved agroforestry systems provide added benefit by generating cash income from the marketing of diverse products and by promotion of biodiversity conservation.

Pohle et al. focus on different topics of the sustainable use and conservation of tropical mountain rainforests in southern Ecuador. Based on an ethnobotanical study, the cultivation of plant species in demand in home gardens have been identified as promising options for increasing household incomes. Although deforestation is prevalent in the research area, farmers highly value the economic importance of forests, e.g. as an agricultural reserve. Livelihood strategies were found to vary between communities depending on subsistence agriculture and those engaged in agro-pastoral activities. Land use and land tenure conflicts are severely dependent on state policies and land adjudication.

The paper of Zeller et al. deals with the disadvantaged upland areas in Southeast Asia and focuses on linkages between poverty and sustainable agricultural and rural development. Apart form the market approach and the population approach, governance issues appear to be particularly relevant for upland areas that are often politically and institutionally marginalized. The authors conclude with implications for rural and agricultural development policies.

Most papers of this book were presented at an international symposium on Tropical Rainforests and Agroforests under Global Change held in Bali in October 2008. The editors thank the coordinators of the Collaborative Research Centre STORMA (Stability of Rainforest Margins in Indonesia), Wolfram Lorenz, Melanie Grosse and Surya Tarigan, for their invaluable support, and many peers for their thoughtful reviews of the book chapters. Technical assistance during the editorial process was provided by Stella Aspelmeier. The financial support by the German Research Foundation (Deutsche Forschungsgemeinschaft) is gratefully acknowledged.

References

Acebey A, Gradstein SR, Krmer T (2003) Species richness and habitat diversification of corticolous bryophytes in submontane rain forest and fallows of Bolivia. Journal of Tropical Ecology 18:9-18

Brook BW, Sodhi NS and Bradshaw C.J.A. (2008) Synergies among extinction drivers under global change. Trend Ecol Evol 23: 453-460

Bunker DE, Carson WP (2005) Drought stress and tropical forest woody seedlings: effect on community structure and composition. Journal of Ecology 93:794-806

Clark DA, Piper SC, Keeling CD, Clark DB (2003) Tropical rain forest tree growth and atmospheric carbon dynamics linked to interannual temperature variation during 1984-2000. Proceedings of the National Academy of Sciences of the United States of America 100:5852-5857

Clough Y, Dwi Putra D, Pitopang R, Tscharntke T (2009) Local and landscape factors determine functional bird diversity in Indonesian cacao agroforestry. Biological Conservation 142: 1032-1041

Corlett RT, LaFrankie JV (1998) Potential impacts of climate change on tropical Asian forests through an influence on phenology. Climatic Change 39:439-453

Engelbrecht BMJ et al. (2007) Drought sensitivity shapes species distribution patterns in tropical forests. Nature 447:80-U82

Ewers RM and Rodrigues ASL (2008) Estimates of reserve effectiveness are confounded by leakage. Trends Ecol Evol 23: 113-116

FAO (2006) Global Forest Resources Assessment 2005: Progress towards sustainable forest management. In: FAO Forestry Paper. Food and Agriculture Organsiation, Rome, p 320

GLP (2005) Science plan and implementation strategy. In. IGBP Secretariat, Stockholm, p 64

Hughes JB, Daily GC, Ehrlich PR (1997): Population diversity: Its extent and extinction. Science 278:689-692

IPCC (2007) Climate Change 2007: the physical science basis. Contribution of working group I to the fourth assessment report of the intergovernmental panel on climate change. Cambridge University Press, Cambridge, UK and New York, USA

Karl TR, Trenberth KE (2003) Modern global climate change. Science 302:1719-1723

Kessler M, Abrahamczyk S, Bos M, Buchori D, Putra DD, Gradstein SR, Hoehn P, Kluge J, Orend F, Pitopang R, Saleh S, Schulze CH, Sporn SG, Steffan-Dewenter I, Tscharntke T (2009) Alpha and Beta Diversity of Plants and Animals along a Tropical Land-use Gradient. Ecological Applications (in press)

Lee TM and Jetz W (2008) Future battlegrounds for conservation under global change. Proc R Soc Lond B 275: 1261-1270

Maas B, Putra DD, Waltert M, Clough Y., Tscharntke T., Schulze C. (2009) Six years of habitat modification in a tropical rainforest margin of Indonesia do not affect bird diversity but endemic forest species. Biological Conservation 142: 2665-2671

Matson, PA, Parton WJ, Power AG, Swift MJ (1997) Agricultural intensification and ecosystem properties. Science 277:504-509

Maertens M, Zeller M, Birner R (2006) Sustainable agricultural intensification in forest frontier areas. In: Agricultural Economics 34(2), pp. 197-206

Muller D, Zeller M (2002) Land use dynamics in the central highlands of Vietnam: a spatial model combining village survey data with satellite imagery interpretation. Agricultural Economics 27:333-354

Nabuurs GJ et al. (2007) Forestry. In: Metz B, Davidson OR, Bosch PR, Dave R, Meyer LA (eds) Climate Change 2007: Mitigation. Contribution of Working Group III to the Fourth Assessment Report of the Intergovernmental Panel on Climate Change. Cambridge University Press, Cambridge, United Kingdom and New York, NY, USA

Olschewski R, Tscharntke T, Benítez P C, Schwarze S, Klein A M (2006) Economic valuation of pollination services and pest management comparing coffee landscapes in Ecuador and Indonesia. Ecology and Society 11 (1): 1-7

Owens IPF, Bennett PM (2000) Ecological basis of extinction risk in birds: habitat loss versus human persecution and introduced predators. Proc Nat Acad Sci USA 97:12144-12148

Phillips OL, Lewis SL, Baker TR, Chao KJ, Higuchi N (2008) The changing Amazon forest. Philosophical Transactions of the Royal Society B-Biological Sciences 363:1819-1827

Ramadanil, Tjitrosudirjo SS, and Setiadi, D. 2008. Structure and composition of understory plant assemblages of six land-use types in The Lore Lindu National Park, Central Sulawesi, Indonesia. Bangladesh Journal of Plant Taxonomy. 15(1): 1-12

Robertson GP, Swinton SM (2005) Reconciling agricultural productivity and environmental integrity: a grand challenge for agriculture. Frontiers in Ecology and the Environment 3:38-46

Sala OE, Chapin FS, Armesto JJ, Berlow E, Bloomfield J, Dirzo R, Huber-Sanwald E, Huenneke LF, Jackson RB, Kinzig A, Leemans R, Lodge DM, Mooney HA, Oesterheld M, Poff NL, Sykes MT, Walker BH, Walker M, Wall DH (2000) Biodiversity - global biodiversity scenarios for the year 2100. Science 287:1770-1774

Schwendenmann L, Veldkamp E, Moser G, Hlscher D, Khler M, Clough Y, Anas I, Djajakirana G, Erasmi S, Hertel D, Leitner D, Leuschner Ch, Michalzik B, Propastin P, Tjoa A, Tscharntke T, van Straaten O (2009) Effects of an experimental drought on the functioning of a cacao agroforestry system, Sulawesi, Indonesia. Global Change Biology, published online Jul 21 2009, doi: 10.1111/j.1365-2486.2009.02034.x

Steffan-Dewenter I, Kessler M, Barkmann J, Bos MM, Buchori D, Erasmi S, Faust H, Gerold G, Glenk K, Gradstein SR, Guhardja E, Harteveld M, Herteld D, Hohn P, Kappas M, Kohler S, Leuschner C, Maertens M, Marggraf R, Migge-Kleian S, Mogea J, Pitopang R, Schaefer M, Schwarze S, Sporn SG, Steingrebe A, Tjitrosoedirdjo SS, Tjitrosoemito S, Twele A, Weber R, Woltmann L, Zeller M, Tscharntke T (2007) Tradeoffs between income, biodiversity, and ecosystem functioning during tropical rainforest conversion and agroforestry intensification. Proceedings of the National Academy of Sciences of the United States of America 104, 4973-4978

Tilman D, Cassman KG, Matson PA, Naylor R, Polasky S (2002) Agricultural sustainability and intensive production practices. Nature 418:671-677

Timmermann A, Oberhuber J, Bacher A, Esch M, Latif M, Roeckner E (1999) Increased El Nino frequency in a climate model forced by future greenhouse warming. Nature 398:694-697

Tscharntke T, Leuschner C, Zeller M, Guhardja E, Bidin A (eds) (2007) The stability of tropical rainforest margins, linking ecological, economic and social constraints of land use and conservation. Springer Verlag, Berlin, pp 513

Tscharntke T, Sekercioglu CH, Dietsch TV, Sodhi NS, Hoehn PH, Tylianakis JM (2008) Landscape constraints on functional biodiversity of birds and insects in tropical agroecosystems. Ecology 89: 944-951

Veldkamp E Purbopuspito J Corre MD, Brumme R (2008) Land-use change effects on trace gases fluxes in the forest margins of Central Sulawesi, Indonesia. Journal of Geophysical Research 113: G02003, doi:10.1029/2007JG000-522

Vitousek PM, Mooney HA, Lubchenco J, Melillo JM (1997) Human domination of Earth's ecosystems. Science 277:494-499

Walsh RPD, Newbery DM (1999) The ecoclimatology of Danum, Sabah, in the context of the world's rainforest regions, with particular reference to dry periods and their impact. Philosophical Transactions of the Royal Society of London B 354:1869-1883

Wittemeyer G, Elsen P, Bean WT, Coleman A, Burton O and Brashares JS (2008) Accelarated human population growth at protected area edges. Science 321; 123-126

Wright SJ (1991) Seasonal drought and the phenology of understory shrubs in a tropical moist forest. Ecology 72:1643-1657

Part I

Agroforestry management in an ecological and socio-economic context

Biodiversity patterns and trophic interactions in human-dominated tropical landscapes in Sulawesi (Indonesia): plants, arthropods and vertebrates

Yann Clough[1*], Stefan Abrahamczyk[2], Marc-Oliver Adams[1,3], Alam Anshary[4], Nunik Ariyanti[5], Lydia Betz[1], Damayanti Buchori[6], Daniele Cicuzza[2,9], Kevin Darras[7,1], Dadang Dwi Putra[8], Brigitte Fiala[3], S. Robbert Gradstein[9], Michael Kessler[2], Alexandra-Maria Klein[1], Ramadhanil Pitopang[10], Bandung Sahari[11], Christoph Scherber[1], Christian H. Schulze[12], Shahabuddin[4], Simone Sporn[9], Kathrin Stenchly[1], Sri S. Tjitrosoedirdjo[5], Thomas C. Wanger[1,13], Maria Weist[1], Arno Wielgoss[1,3], and Teja Tscharntke[1]

[1] Agroecology, University of Göttingen, Waldweg 26, 37073 Göttingen, Germany
[2] Institute of Systematic Botany, University of Zürich, Zollikerstrasse 107 CH-8008 Zürich, Switzerland
[3] Department of Animal Ecology and Tropical Biology (Zoology III), Biozentrum, University of Würzburg, Am Hubland, D-97074 Würzburg, Germany
[4] Faculty of Agriculture, University of Tadulako, Palu, Central Sulawesi, Indonesia
[5] Department of Biology, Faculty of Mathematics and Natural Sciences, Bogor Agricultural University, Jalan Padjajaran, 16144 Bogor, West Java, Indonesia
[6] Department of Plant Protection, Bogor Agricultural University, Kampus IPB Darmaga, Bogor, West Java, Indonesia
[7] 14 allée de Frênes F-01210 Versonnex, France
[8] Celebes Bird Club, Jl. Thamrin 63A, Palu, Central Sulawesi, Indonesia
[9] Albrecht-von-Haller-Institute of Plant Sciences, University of Göttingen, Untere Karspüle 2, D-37073 Göttingen, Germany
[10] Herbarium Celebense (CEB), Tadulako University, Kampus Bumi Tadulako Palu, Central Sulawesi 94118, Indonesia
[11] Peka Indonesia Foundation (Indonesian Nature Conservation Foundation)-Wildlife Trust Alliance. Jl. Uranus Blok H No 1 Perum IPB Sindang Barang 2, Bogor, West Java, Indonesia
[12] Department of Population Ecology, Faculty Center of Biodiversity, University of Vienna, Rennweg 14, A-1030 Vienna, Austria
[13] Environment Institute, School of Earth and Environmental Sciences, University of Adelaide, Australia

*corresponding author: Y. Clough, email: yclough@gwdg.de

T. Tscharntke et al. (eds.), *Tropical Rainforests and Agroforests under Global Change*, Environmental Science and Engineering, 15–71, DOI 10.1007/978-3-642-00493-3_2,

Summary

The need to capture primary production in order to sustain and improve economic livelihoods has lead to increasing conversion of natural habitat and intensification of agricultural practices in many parts of the world including most tropical regions. Understanding how these processes affect ecosystems and their functioning, in particular in the high-diversity ecosystems of the tropics, has become a key issue in ecological research. In this chapter, our focus is on the agriculture-forest landscapes of Central Sulawesi, Indonesia, an island widely known for its endemic yet still poorly known flora and fauna. The rise of the region to one of the largest cacao producing areas in the world is at the core of recent land-use change and intensification processes. Covering plants (trees, rattan palms, herbs, bryophytes) and several invertebrate (ants, dung beetles, cacao insect herbivores, fruit-feeding butterflies, parasitic Hymenoptera, spiders) and vertebrate groups (amphibians, birds, murids, reptiles), we give an in-depth overview of the determinants of biodiversity in cacao landscapes, including both management and landscape-scale variables into our analyses. Results show that shaded agroforests host a rich community of species. By adopting a large-scale study design we showed that proximity of natural forest is a key predictor for species richness of plants, invertebrates and vertebrates alike. Endemics and forest specialists benefit most from indigenous shade tree cover and proximity to natural forest. Importantly, several functionally important groups such as insectivorous and seed-dispersing birds benefit from tall shade trees, shade tree diversity and proximity to forest edge, while parasitoid diversity is greatest close to natural forests. Available data on the effects of landuse change in cacao landscape of Central Sulawesi is increasing. Change in landscape configuration and management practices are being clearly reflected in the composition of species communities, with likely impacts on ecosystem services such as pest control and pollination. More knowledge is needed especially in terms of species interactions and ecosystem functioning, but also on how existing knowledge can contribute to effective conservation in human-dominated landscapes outside protected areas.

Keywords: agricultural intensification, agroforestry, amphibians, ants, arthropods, bees, biodiversity, birds, bryophytes, butterflies, cacao, cocoa, community structure, *Conopomorpha cramerella*, decomposition, dung beetles, forest distance, fungal disease, herbivores, herbivory, herbs, Hymenoptera, insects, land-use change, landscape ecology, lianas, mammals, Muridae, Nymphalidae, parasitoids, pollination, plants, predation, rattan palms, rats, reptiles, shade trees, spiders, *Theobroma cacao*, trees, trophic interactions, vertebrates

1 Introduction: Patterns of biodiversity and associated processes in changing landscapes

Human activity has led to the modification of increasingly large tracts of the terrestrial biosphere, with estimates ranging up to 40% of the total area (Foley et al. 2005). While the expansion and intensification of agricultural activity is to a large degree made necessary by the high rate of increase in the world population, negative impacts on ecosystems may be tremendous. The disappearance of (near-) primary habitat is often associated with irreversible species losses, especially of rare and/or specialised species (e.g. Acebey et al. 2003, Owens and Bennett 2000). The intensive use of forest resources, expansion of agricultural activity into primary or near primary habitat, as well as land-use intensification are primary drivers for biodiversity loss (Sala et al. 2000). Conservationists therefore seek to promote the creation of protected areas in which human activity is severely restricted, with a strong focus on those areas in the world containing the most, and the most unique biodiversity (e.g. Myers et al. 2000). Land-use change is not just about losing species, though (Kaimowitz and Sheil 2007). It may seriously undermine the capacity of ecosystems to *sustain food production, maintain freshwater and forest resources, regulate climate and air quality, and ameliorate infectious diseases*, to cite Foley et al. (2005). The most biodiverse regions tend to be in tropical regions where high poverty, especially in rural areas, is paralleled by a strong dependence on natural resources, and thus the toughest fights for better livelihoods and improved conservation essentially share the same battleground (Chazdon et al. 2008, Kaimowitz and Sheil 2007).

In comparison with natural habitats such as tropical forests, agricultural systems are much more simplified and harbour only a fraction of their biodiversity, especially at larger scales (Kessler et al. 2009). The diversity of species in agricultural habitats may have important ecological functions, some not or barely recognised, even though the species assemblages may be poor and/or modified when compared with natural habitats. Yet, the interactions between organisms within agricultural habitats and at the margin to natural habitats are potentially of crucial importance for sustained agricultural production and thus economic livelihoods of the people involved (Daily et al. 2000, Matson et al. 1997, Ricketts et al. 2004). There is an urgent need to shift more attention to the role of species in the functioning of these ecosystems to understand the impacts of current and future land-use changes.

Sulawesi is a key landmass within the Wallacea biogeographic region, one of the world's biodiversity and endemism hotspots which makes it extremely valuable in terms of conservation (Cannon et al. 2007, Myers et al. 2000, Stattersfield et al. 1998, Whitten et al. 2002). The loss of forest habitat and forest degradation on this equatorial island (FWI/WRI/GFW 2002; Cannon et al. 2007) reflect the situation found in several countries of Southeast Asia: deforestation is still happening, possibly even at increasing rates (Achard et al. 2002, Koh 2007, Sodhi et al. 2004 , Sodhi et al. 2006), with new forms of land-use gaining ground. While the progression of oil palm plantation has so far been less widespread on Sulawesi than in other parts of the archipelago (Sulawesi 0.12 Million ha, GAPKI 2006), the cacao boom has resulted in the conversion of large areas of primary and secondary forest as well as forest gardens. In Sulawesi, the total area planted with cacao is estimated to be around 912,000 ha, with a total production of 508,135 t/year (Direktorat Jenderal Perkebunan 2007). Thus, Sulawesi produces about 65% of Indonesian cocoa, the rest being spread rather evenly across the other provinces.

While the cocoa boom has taken place largely in a context of hands-off policy on behalf of government agencies (Akiyama and Nishio 1996), the current slump in production due to cacao tree aging and/or increased losses to pests and diseases such as cacao pod borer *Conopomorpha cramerella* (Snellen) and black pod disease *Phytophthora palmivora* (Butler) have prompted the Government of Indonesia to initiate a rehabilitation of the sector by replanting of 450,000 ha with resistant (unspecified, but most likely against cacao pod borer, black pod disease and vascular streak dieback *Oncobasidium theobromae*) varieties. The national aim is to boost national cocoa bean production to 1 million t/year by 2013 (Antaranews, 2008).

The ecology of human-disturbed habitats in Sulawesi has only been poorly investigated so far, and crops such as cocoa, which have only recently become dominant, do not feature in the only monograph focusing on the island's ecology (Whitten et al. 2002). Knowledge about the distribution of species and the nature of their interactions is essential for both biodiversity conservation and sustainable agricultural management.

This chapter summarises key findings of research conducted within the DFG-funded collaborative research center SFB-552 STORMA . Using published and unpublished results from a wide range of organism groups, it addresses the following key questions:

1. What determines biodiversity patterns in human-dominated landscapes of Central Sulawesi?
2. Which key trophic aboveground species interactions affect processes important for agricultural production?

2 Plants: trees, rattan palms, herbs, bryophytes

Unlike many groups of animals, plants are directly influenced by agricultural management practices, including logging, planting, weeding, and the application of fertilisers or herbicides. Almost by definition, the intensification of land use from natural forests across agroforestry systems to annual cultures represents a gradient of decreasing tree density and diversity. However, this decline is not linear and different aspects of plant community composition have to be considered, dependent on agricultural practices.

In Central Sulawesi, there are two main patterns of forest conversion. In slash-and-burn agriculture, forests are completely logged, usually leaving no or very few remnant trees. After burning, fields are then usually planted with annual crops (especially maize) for a few seasons, while perennial crops (mainly cacao, formerly also coffee) are established. This system leads to major losses of natural plant diversity and results in agricultural systems with only 1-3 tree species (the crop plus 1-2 shade trees) covered by an impoverished layer of epiphytes (bryophytes, lichens) and a more diverse layer of terrestrial herbs with dominance of pantropical weeds.

A second, more traditional way of forest conversion involves the gradual removal of native trees to establish agroforestry systems. Typically, some parts of the original forest tree layer are maintained while most of the understory is cleared. This sheltered, semi-shaded habitat with well-preserved soils is then used to plant cash crops such as cacao or coffee as well as a range of perennial and annual crops for local consumption. Over time, management of these agricultural systems is commonly intensified, ultimately resulting in species-poor plantations of similar structure to those established through slash-and-burn agriculture.

Tree diversity in submontane forests of Central Sulawesi (800-1200 m a.s.l.) typically ranges from 50 to 60 species per 0.25 ha or around 150 species per ha (Kessler et al. 2005, Gradstein et al. 2007). These values are unusually high for forests at this elevation in Southeast Asia, especially considering the isolated island position of Sulawesi. Cacao agroforestry systems with shade trees remaining from the original forest have around 12-30 tree species per 0.25 ha, whereas newly established or highly intensified cacao plantations harbour between 1 and 12 tree species, often with a predominance of introduced legume trees (*Erythrina falcata* Benth., *Glyricidia sepium* (Jacq.) Kunth ex Walp.) whose nitrogen-fixing capability and fast growth make them suitable shade trees in cacao plantations.

Secondary forests, which are rare in the study region because little land is allowed to lie fallow for longer periods, have fairly high levels of species richness (20-35 per 0.25 ha) but with markedly different taxonomic composition (Pitopang et al. 2004). Whereas Lauraceae, Meliaceae, and Euphorbiaceae are predominant in primary forests, Euphorbiaceae, Rubiaceae and Myristicaceae dominate in the agroforestry systems, and Euphorbiaceae, Urticaceae, and Ulmaceae in the secondary forests. In general, agroforestry systems differ from primary forests by a lower density of understory trees, while secondary forests have fewer species of commercial interest (timber trees). A special case is posed by newly created agroforestry systems with a fairly diverse cover of planted shade trees (up to 15 species per 0.25 ha), most of which are introduced species with specific uses (timber, fibres, fruit, sap). Because trees are the main structural components of forests, declines in tree density are paralleled by decreases in above- and below-ground biomass as well as in productivity (Leuschner et al. 2006, Harteveld et al. 2007, Hertel et al. 2007, Steffan-Dewenter et al. 2007).

Commercially, rattan palms (Calamoideae) , which can be processed into material for furniture-making, are the most important plant group beside timber trees. These liana-like palms occur with about 50 species in Central Sulawesi, including many as yet undescribed species of presumably restricted geographical distribution (Mogea 2002, Siebert 2005). Currently, only about four species are intensively harvested commercially, but additional species also have commercial potential, even if the quality of the cane is slightly inferior. Although rattan can be cultivated in coffee and cacao agroforests, there have so far been no large-scale attempts to establish rattan plantations in Sulawesi (Siebert 2000). Instead, most rattan palms are harvested in natural forests, with collectors extracting palms from up to 10 kilometres distance from the forest edge. This activity represents the deepest and most intensive incursion of commercial activities into the forests, except for the localised hunting of larger mammals. Although rattan collectors only extract selected individuals of commercially valuable size, collecting affects the reproduction of the palms (reducing the number of mature, flowering and fruiting plants) and increases vegetative (clonal) reproduction (Siebert 2002) and may also lead to shifts in species composition. The ecological implications of these shifts have not been explored, but could involve changes in fruit availability to frugivorous animals, and to changes in vegetation structure. It thus becomes clear that the transition from truly pristine to agricultural ecosystems is not restricted to the forest edge but rather extends several kilometres into the forest. Along with rattans, other lianas are also an important structural component of tropical forests (Putz 1984). Natural forests in Central Sulawesi comprise about 6-12 species of liana per 40 m^2 (Steffan-Dewenter et al. 2007). In agroforests, this diversity is maintained in the less intensively managed systems, albeit with a shift towards smaller, more herbaceous taxa of smaller diameters. In intensively managed systems liana richness drops to 0-5 species per 40 m^2.

In contrast to woody plants, the floristic composition and spatial variation of terrestrial herbs in tropical forests and agricultural systems are poorly known (Tuomisto et al. 2003). The number of herb species reported to date fluctuates from 48 species in Puerto Rico (Smith 1970) to 121 species in Brunei (Poulsen and Pendry 1995). In Central Sulawesi, the number of herbs in two natural forests at 1000 m and 1400 m is 204 and 171 species, respectively, which is very high compared to other tropical forests (Cicuzza et al., unpublished data). However, because of low light levels, herbs are rather thinly spread on the ground in closed-canopy natural forests. By contrast, the density and diversity of herbs increases notably in agroforestry systems, where the open canopy allows more light to reach the ground, (Tjitrosoedirdjo et al. 2004, Steffan-Dewenter et al. 2007). It is only in the very intensively managed plantations, where herbicides are commonly applied, that herbs become rare and only few species are found. No less than 176 species of terrestrial herbs have been recorded in cacao plantations in central Sulawesi. Only 16 of these are shared with natural forests, showing an almost complete turnover of species, as well as of families (Cicuzza et al. unpublished data). In the forests, the dominant families are Araceae, Orchidaceae, Zingiberaceae, and several fern groups, whereas plantations are dominated by members of the

Araceae, Asteraceae, and Poaceae. Also, whereas many species in the forests have restricted geographical distributions most of the species found in cacao plantations are widespread and several of them are pantropical weeds.

The natural forests of Central Sulawesi have a rich and diverse epiphytic bryophyte flora, with over 150 bryophyte species on just eight mature trees in submontane forest at 1000-1100 m (Sporn 2008), which are among the highest numbers recorded world-wide. Small understory trees harbour few additional species, adding only about 10% to the total number. Forest conversion leads to a loss of about 70% of forest bryophytes (Sporn 2008). The remaining moss species, most of them from the forest canopy, may re-establish in nearby cacao agroforests with a well-developed shade layer. The bryophyte flora of these cacao agroforests is to a large extent (ca. 85%) derived from the forest. Shaded cacao agroforests may thus play a role in conserving forest bryophyte biodiversity (Aryanti et al. 2008, Sporn 2008). Because dense bryophyte layers covering the trunks of the cacao trees are generally believed to reduce fruit set and cocoa production, this epiphyte layer is often removed by the farmers. Experimental studies have shown, however, that such an inhibition does not take place and that the removal is unnecessary for improving the productivity of cacao (Sporn et al. 2007).

In addition to agroforestry systems, there are a number of other agricultural systems in Central Sulawesi. Perhaps the most important and at the same time most intensively managed systems are rice paddies. These cover extensive areas in flat valley floors that have sufficient water supply and often support a wide range of more or less noxious weeds. Although specific studies from Sulawesi are missing, in other parts of Indonesia herbaceous weeds are common and diverse in rice paddies (Soerjani et al. 1987).

Finally, house gardens of 300-2500 m^2 in size are found around most houses in rural villages and support a varied flora of high social and economic as well as potential ecological value. In the Napu valley of Central Sulawesi, Kehlenbeck (2007) found 206 crop, 162 ornamental, and 58 weedy species in 50 home gardens spread across five villages. On average 30-50 crop species were found per garden, mainly for subsistence-oriented non-staple foods (fruits, vegetables, spices) and non-food items (medicine, fodder, fuel wood). Richness and composition of the home garden flora differed between homes and villages depending on the cultural and social background of the gardeners as well as on ecological conditions, especially soil fertility. Furthermore, Kehlenbeck (2007) detected shifts in species composition between 2001 and 2004, with increasing emphasis on cash crops in villages with good market access and overall decreasing diversity.

3 Cacao insects and herbivory

Herbivorous insects are megadiverse in the tropics (Lewinsohn and Roslin 2008). Studies conducted in forests show that herbivory also tends to be higher at tropical than at temperate latitudes (Coley and Aide 1991, Coley and Barone 1996), suggesting either weaker defences in tropical plants or higher herbivore pressure (Coley and Aide 1991). In agricultural systems, losses of primary productivity through herbivorous insects can be considered an ecosystem dis-service (Zhang et al. 2007). Reduction in diversity of plants in such systems is often found to coincide with a higher prevalence of pest insects compared to natural habitats, in some cases even leading up to devastating outbreaks of relevant pests (Tahvanainen and Root 1972, Root 1973, Wallner 1987). Herbivorous insects in agricultural landscapes in the Malay Archipelago have been the subject of two early monographs (Dammerman 1929, Kalshoven 1950). Much of what has been published in the last 30 years pertains to pests and pest control in rice, and has been covered by Matteson (2000) and Settle et al. (1996). In the following, we give a brief outline of herbivore identity and patterns of herbivory in cacao agroforestry of Sulawesi, referring to other agroecosystems when appropriate.

Despite having been introduced to Indonesia, *Theobroma cacao* supports surprisingly high numbers of herbivorous insect species in the region. Strong (1974) showed that the greater the area of cacao plantations in a given region, the richer the herbivore community. Sap-sucking families such as Miridae and Flatidae, as well as a number of Lepidoptera and Coleoptera families constitute the majority of cacao pest species worldwide. Among mirids, *Distantiella theobroma* Distant and *Sahlbergella singularis* Haglund are responsible for considerable damage in Ghana and West Africa, while *Helopeltis* spp. is a major pest in South-East Asia (ICCO 2009, Kalshoven 1950). Noteworthy among the Lepidoptera in this region is the cacao pod borer (*Conopomorpha cramerella* Snellen), which - after first being reported in 1841 - now infests all cacao-producing provinces in Indonesia (Adi Prawoto 1996, ICCO 2009). In addition, a number of Geometridae (e.g. *Hyposidra talaca* Walker and *Ectropis* spp.), Psychidae, Limacodidae, and stem- or bark-boring species are known to feed on cacao (Conway 1969, Kalshoven 1950). Coleopteran herbivores are represented largely by the Chrysomelidae and Scarabaeidae families (Kalshoven 1950, Adams pers. obs.). The negative impact resulting from infestation by the aforementioned insects can be twofold: in addition to damaging the cacao trees directly through reduction of the photosynthetically active surface or destruction of flowers and fruit stands, sap-sucking homopterans and heteropterans are suspected to act as vectors for viral and fungal pathogens, such as *Phytophthora* spp., the cause of black pod disease (Mitchell 2004).

The intensification of cacao cultivation encompasses a wide range of changes in the ecology of the plantation that may all - either directly or indirectly - affect the likelihood and prevalence of herbivore infestation. However, from this host of changes, the reduction in species diversity and overall number of shade trees and the increasing isolation of plantations from larger forest expanses due to the expansion of settlements arguably exert the greatest influence. Shade in itself may influence the attractiveness and nutritional qualities of a host for herbivores by affecting the plants metabolism, and thereby altering its chemical composition, as well as the concentration and distribution of allelochemicals (Crone and Jones 1999, Downum et al. 1991, Maiorana 1981, Mattson 1980, Mole et al. 1988, Schoonhoven et al. 1998). Transition from shaded to unshaded plantations entails microclimatic changes in the understory brought about by higher insolation and greater air movement (Beer et al. 1998, Siebert 2002), which may alter the attractiveness and/or suitability of a plantation for herbivores. Shade trees also increase the ecological complexity of cacao plantations. While the species composition of these tree communities changes and simplifies markedly in the transition from traditional, rustic plantations to planted shade cultivation (Kessler et al. 2005, Steffan-Dewenter et al. 2007), even the latter remain capable of supporting roughly similar numbers of - predominantly herbivorous - beetle species, but with a different species composition (Bos et al. 2007a). Although species that depend on intact forest habitats vanish early after the initial conversion to farmland, a second sharp drop in insect diversity is only apparent once the majority of shade trees has been removed (Bos et al. 2007b, Steffan-Dewenter et al. 2007). Although microclimatic changes and loss of ecological complexity tend to be closely interrelated, several studies point towards tree diversity as the factor possessing greater explanatory value (Jactel and Brockerhoff 2007, Riihimäki et al. 2005). The maintenance of relatively high levels of diversity is relevant in the light of the natural enemies hypothesis (Tahvanainen and Root 1972), which predicts that such complex habitats are capable of supporting a larger community of entomophagous insects by providing shelter and a greater variety of alternative prey species, in turn enabling these predators to effectively control herbivore populations and prevent pest outbreaks. Findings from Ghana, for example, suggest a connection between the greater number and diversity of ants found in shaded cacao and the lower prevalence of many pest species (Bigger 1981, but see section 4 of the present chapter). Likewise, studies in Sulawesi have demonstrated the expected shift of predator activity with the conversion from rustic to sun-grown cacao plantations, emphasizing that these changes had a greater impact on the entomophagous species compared to the herbivores (Klein et al. 2002a). We found the extent of leaf damage in 44 cacao plantations located in Central Sulawesi to be inversely related to shade cover (see Figure 1).

Proximity to larger expanses of forest or other relatively undisturbed habitats has also been shown to increase predation. Klein et al. (2006) have suggested that such habitats allow the build-up of larger populations in higher trophic levels, which then colonise or forage in adjacent plantations (Altieri and Schmidt 1986, Klein et al. 2006). Parasitism rates in Hymenoptera (Klein et al. 2006) and Lepidoptera (Adams, unpublished data) were indeed found to be higher in plantations located closer to tracts of intact forest (see section 5 of the present chapter).

We found higher populations of certain herbivores - namely, Chrysomelidae and Geometridae - closer to the natural forest, but no higher herbivore-related damage (Adams, unpublished data). Claims related to the detrimental effect

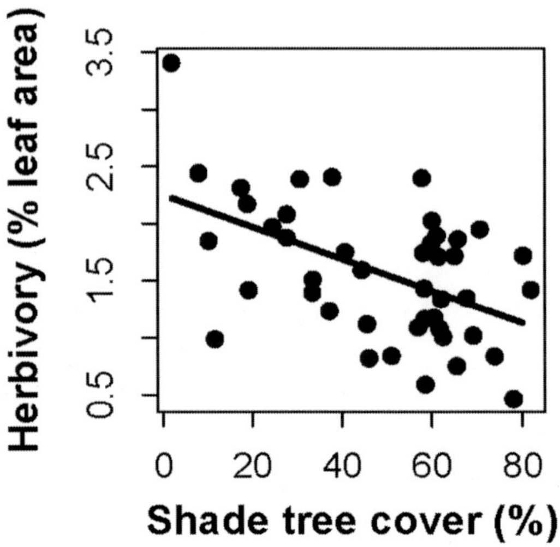

Fig. 1. Leaf herbivory decreases with increasing shade tree cover in smallholder cacao plantations of Central Sulawesi (Adams and Clough, unpublished data)

of shade trees and forest proximity on cacao health and herbivore pressure, which are often put forth by smallholders in favour of shade reduction (Ruf and Schroth 2004) are thus not supported by the present evidence as far as herbivory is concerned. While shade trees or secondary forest are indeed known to serve as habitat for potential pest species, the beneficial effects of these environments on natural enemies seem to outweigh negative effects (Conway 1969).

Beside shade removal, other aspects of intensification, for instance fertiliser and pesticide use, may affect herbivory. Changes in ground nutrient levels are reflected in the nutritional value of various plants and plant parts (Hartemink 2005, Schoonhoven et al. 1998), and herbivores are capable to detect these differences and choose accordingly, allowing them to feed more efficiently on more nutritious plant tissue (Schoonhoven et al. 1998). We did not find any effects of nitrogen fertilization on folivory (Adams, unpublished data), but damage by *Helopeltis* bugs was much increased in fertilised plots (Clough et al. unpublished data). Past experience in Malaysian cacao agroforests has indicated indiscriminate and wide-spread use of insecticides as important contributor to severe pest outbreaks. While reducing the target species populations, resurgence occurs shortly thereafter, likely due to the negative effect of the treatments on natural enemies of pests (Conway 1969, Wood 2002)

The removal of shade trees also allows light to penetrate deeper into the understory, potentially offering a competitive advantage to more aggressive weeds (Beer et al. 1998), which in turn require higher weeding frequencies and greater input of labour to keep in check. We detected a reduction in the prevalence of certain herbivores (Chrysomelidae) and in the extent of leaf damage when shorter weeding intervals were adopted (Adams, unpublished data). The mechanism underlying this pattern is unknown. One possible explanation is that the herbaceous layer serves as an additional source of food and shelter for generalist herbivores. Its frequent removal would thus lower the capacity of a given plantation to sustain large pest populations.

4 Ants

In the tropics, ants play are very numerous and play a key role in natural and agriculturally used habitats (Delabie et al. 2007). Ant communities are often impoverished in agricultural settings, as shown by Pfeiffer et al. (2008) in Southeast Asian oil palm plantations, for instance. Occasionally though, they have also been found to be very species-rich in human-dominated landscapes, in some cases with little species loss from forest to agricultural land-uses, as was shown for agroforests of Central Sulawesi by Bos et al.(2007a). Whether diverse ant communities can be maintained in cacao over time is an important issue which has not yet been investigated.

As efficient predators, ants may limit herbivore populations, including potential insect pests in cocoa agroecosystems. Observations were made for the black cocoa ant *Dolichoderus thoracicus* (Smith) (Hymenoptera: Formicidae) which seems to reduce the incidence of *Helpoletis sulawesii* and cocoa pod borer *Conopomorpha cramerella* in Southeast Asia (See and Khoo 1996, Khoo and Ho 1992). Because of these properties, the introduction of the black cocoa ant has been recommended as an integrated pest management strategy in orchards and cocoa plantations (Van Mele and Cuc 2001, Ho and Khoo 1997). The role of ants as vectors of fungal plant diseases is less well known, but has been recognised years ago by agricultural researchers. Several studies show that various ants, especially tree-nesting and tent-building species, can play a role in the dissemination of black pod disease (Evans 1973, Taylor and Griffin 1981, McGregor and Moxon 1985).

In a large-scale study on ants in cacao plantations in Central Sulawesi (Wielgoss et al. submitted), Indonesia, we found a single ant species of the genus *Philidris* nesting in 63% of the trees and 81% of the study plots. Where present, it ecologically dominates the ant fauna due to its high abundance and aggressive behaviour. This is the first time an ant species of the genus *Philidris* has been reported to be so widespread and dominant. From an agricultural point of view, *Philidris* sp. plays an ambiguous role: On the one hand, it is a highly efficient predator of the major insect pests of cacao in Southeast Asia, *Helopeltis* spp. and the cocoa pod borer, and protects the cacao pods significantly from damage. On the other hand, it also acts as an effective vector of the pathogenic fungus *Phytophthora palmivora*, as was apparent from field studies and experiments (Figure 2, Clough et al. in prep).

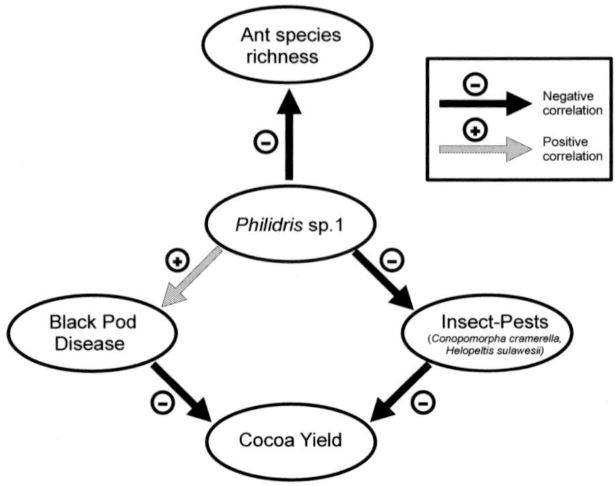

Fig. 2. Effect of the dolichoderine ant *Philidris* sp. 1. on components of the cacao agroecosystem in Central Sulawesi, Indonesia.

Overall, ant species richness was primarily reduced by *Philidris.* Furthermore, it was found to be diminished in more intensively managed plots with higher mean temperatures, while fertilisation and weed management as well as the proximity of forest edge did not exert significant influence.

The ubiquity of *Philidris* in cacao may be recent, because two studies conducted in our research area between 2001 and 2005 report *Philidris* as absent in one case, and as present but only very locally dominant in another (Hosang 2004, Bos et al. 2008).

Our results suggest that the presence of the dominant and aggressive ant species *Philidris* sp. results in a level of pest control in the tropical agroforestry systems at least equal to that on trees occupied by several less dominant ant species, These data, together with previous research on *Dolichoderus thoracicus* as a potential biocontrol agent, could lend support to the doubts of Gove (2007) on the presumption that higher predator species diversity automatically entails higher predation pressure. However, fluctuations in the population levels of other dominant ants such as *Anoplolepis gracilipes* (Bos et al. 2008, Clough et al. pers. obs.) in Central Sulawesi, along with the complementarity in behaviour and activity patterns within ant species communities (Philpott and Armbrecht 2006) suggest that temporal stability of ecosystem services such as predation provided by diverse assemblages may be higher than that of populations of dominant species.

5 Parasitoids

Fragmentation and the continuing loss of natural habitats are the number one threats for species persistence, as has been shown by a number of studies across the globe (Henle et al. 2004, Kruess and Tscharntke 1994, Andren 1997, Steffan-Dewenter and Tscharntke 1999, Kruess and Tscharntke 2000, Rogo and Odulaja 2001, Klein et al. 2002b). These species losses are due to the progression of agricultural habitats (Geist and Lambin 2002), which creates patches of natural forests. The increasing isolation of natural habitats may negatively affect species richness, abundance, and community structure which in turn affects many ecosystem services, such as pollination (Klein et al. 2002b, Kremen et al. 2002) and natural control (Klein et al. 2002a, 2006). Parasitic Hymenoptera are particularly sensitive to land-use changes and are likely to be vulnerable to local or even global extinction, because they are typically very specialised, have a limited dispersal ability and occupy a high trophic level in the food web (Shaw and Hochberg 2001). Since a large proportion of Hymenoptera live as parasitoids (LaSalle and Gauld 1993), the changes in their composition and population abundance might result in overall shifts of food web characteristics (Tylianakis et al. 2008). The questions at the core of this section are thus: 1) How does the loss of parasitoid assemblage affect the functioning of food webs in human-influenced ecosystems? 2) What is the impact of this diversity loss toward ecosystem stability and natural control of agricultural pests?

Parasitic Hymenoptera are among the most species-rich and biologically diverse animal taxa (Mason and Huber 1993; Naumann 1991; Quicke 1997; Whitfield 1998), and may be one of the insect groups playing the most valuable role in maintaining the diversity of natural communities (Quicke 1997). The majority of species develop as parasitoids of other insects (LaSalle and Gauld 1993) and play an important role in natural regulation of populations of herbivorous insects (Quicke 1997; Godfray 2004). Many species of parasitic Hymenoptera have been successfully used as biological control agents against agricultural pests (Noyes and Hayat 1984; Quicke 1997). Hawkins et al. (1997) reported that natural enemies (predators, parasitoids, or pathogens) represent an important and ubiquitous source of herbivore mortality. Among them, parasitoids cause higher mortality of herbivores than do either predators or pathogens (Hawkins et al. 1997, Snyder and Ives 2003). Since parasitic Hymenoptera represent a key factor in regulating natural insect populations, their loss can result in a serious destabilization of natural ecosystems.

5.1 Landscape change and diversity of parasitic Hymenoptera

In tropical landscapes, deforestation and fragmentation of habitats have caused the creation of ecological islands, whereby the natural habitats are functioning as a *source* for insects colonizing adjacent cultivated areas (*sinks*). Based on the theory of island biogeography (MacArthur and Wilson 1967), the immigration of species is a function of the distance from the forest, and agricultural areas closer to the remaining forest habitat should thus support higher insect species richness than those situated further away from the forest. This hypothesis was proven to be true by several studies done in the cacao agroforestry systems in Lore Lindu National Park. The relationship between increasing isolation of cacao agroforestry systems and species composition as well as species richness was recorded by Sahari (2004). The study identified that the number of species as well as the abundance of parasitic Hymenoptera decreased with increasing distance to the forest margin. The highest species richness was found inside the forest and at the forest margin (see Figure 3). There was a pronounced change in species composition between sites with distance to the forest (see Figure 4). Forest sites and cacao trees within forest sites have widely different species assemblages compared to cacao plantations far from the forests. These findings are reminiscent of the results of several studies conducted by Kruess and Tscharntke (1994, 2000) which revealed that parasitoids may be particularly sensitive to habitat fragmentation. Their studies indicated that species richness of parasitoids and parasitism decreased significantly within a few hundred meters from the source area.

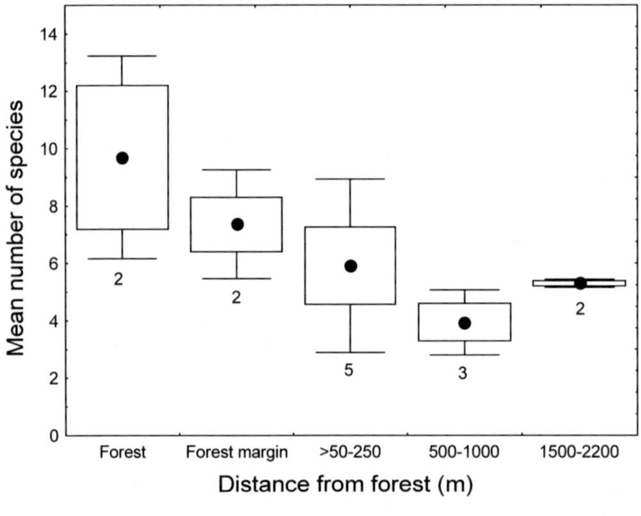

Fig. 3. Mean number of recorded parasitic Hymenoptera species ± SE (box) and SD (whisker) in cacao agroforestry systems located in different distances to the nearest forest: $F_{4,11} = 2.07$, $p = 0.153$, $n = 16$. Numbers below the whiskers indicate the number of sampled sites (Sahari 2004).

Why is there a difference in species richness with respect to distance? Is it related to the dispersal ability of the parasitic hymenopteran from the natu-

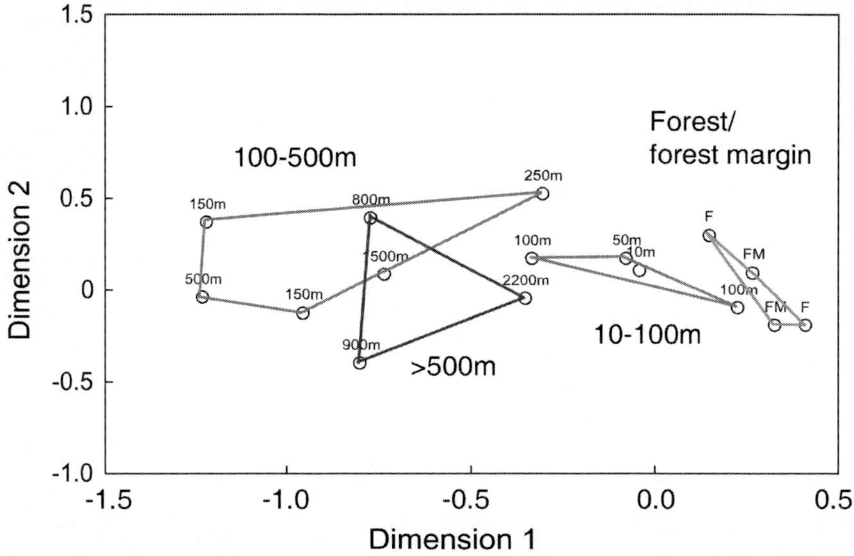

Fig. 4. Two-dimensional scaling plot based on Srensen indices for measuring similarity of species composition between single sampling sites. F: forest sites; FM: cacao plantations inside forest margin; Distances represent degree of isolation of cacao plantations from the forest margin. Connecting lines indicates defined groups of habitats. (Twelve cacao plantations at varying distances (10-2200m) to remaining forest (represented by Lore Lindu National Park), two cacao plantations inside the forest and two forest sites were selected. Selected forest trees had a size and shape similar to that of cacao trees. Hymenoptera were sampled from five cacao trees per cacao plantation and from five forest trees at each forest sites. For quantifying the similarity of different samples, we pooled the samples from the five trees sprayed per plantations. Srensen indices were used to quantify the similarity of species composition between sampled sites. Based on Srensen values we performed non-linear multidimensional scaling (MDS) which led to a two-dimensional projection of distances between the sites (Sahari 2004).

ral habitats, or to the presence and abundance of hosts in cacao plantations? Noyes (1989) proposed that the higher species richness inside and at the edge of the forest relative to cacao plantations located at a distance from the forest is most likely related to a more diverse assemblage of hosts for parasitic Hymenoptera. Marino and Landis (2000) mentioned the scarcity of adult food sources, appropriate microclimates or alternate hosts as important reasons for a reduced abundance, diversity and species richness in agricultural landscapes. The relationship between flowering plants and Hymenopteran parasitoids was also documented by Tooker and Hanks (2000a). Sperber et al. (2004) recorded a relationship between species richness of shade trees and the number of parasitic Hymenoptera species. This was also supported by Saaksjarvi et al. (2006), who identified a strong relationship between species composition of plants and hymenopteran parasitoids. The relationship between plant community structure and natural enemy diversity was also documented by Tooker and Hanks (2000b). Therefore, lower diversity of understory plants and trees in cacao plantations compared to forest sites at the margin of Lore Lindu National Park (Schulze et al. 2004) is certainly related to a significant decrease of herbivores acting as potential hosts for parasitic Hymenoptera. For this reason, a decrease in species richness can be expected also for the parasitic wasps. This is strongly supported by Langer and Hance (2004) who found that more intensive agriculture frequently reduces the availability of non-crop habitats where alternative hosts may be present. Marino et al. (2006) emphasize that generalist hymenopteran parasitoids were strongly associated with alternate lepidopteran hosts that feed on trees and shrubs, whereas oligophagous and specialist hymenopteran parasitoids were associated with lepidopteran alternate hosts that feed on ruderals and shrubs.

All these studies indicated that habitat transformation from forest to agricultural land is decreasing non-crop diversity that could act as host plant for alternate lepidopteran hosts. The modification of natural into human-dominated landscapes is changing the landscape complexity and trophic interactions between plant, herbivorous insect, and parasitoids. This is supported by Thies et al. (2003), Marino et al. (2006) and Tscharntke et al. (2007), who review evidence that biological control by native hymenopteran parasitoids strongly depends on landscape complexity. Cronin (2003) upheld this argument and showed that the landscape matrix influenced the movement, oviposition behavior, and spatial distribution of *Anagrus columbi* Perkins, a common egg parasitoid of the planthopper *Prokelisia crocea* Van Duzee.

5.2 Implications for biological control functions

The fact that tropical rainforest affects species composition and diversity of parasitoids in adjacent agroforestry emphasizes the need to develop a landscape perspective and a mosaic of natural and managed ecosystems for conservation programs that will sustain a diverse parasitoid communities and their potential importance in biological control.

Structurally complex landscapes enhance local diversity in agroecosystems, which may compensate for local high intensity management (Tscharntke et al. 2005). Sperber et al. (2004) proposed that the proper management of shade tree diversity plays a vital role in maintaining the sustainability of cacao agroforestry production systems in the tropics and, concurrently, will maintain high parasitoid diversity in these locations. Low species richness of shade trees may be partly responsible for the low levels of species richness found in cacao plantations in Central Sulawesi, (Schulze et al. 2004). Specific information on spatial dynamics of parasitic Hymenoptera and how they are affected by changing landscapes can thus provide important implications for maintaining a high extent of diversity by developing more effective conservation strategies. This information may also have important implications for landscape management, which should target the maintenance of species-rich parasitoid communities with a high potential to control outbreaks of pest insects. Since parasitic Hymenoptera are the most important biological control agents and are responsible for most of the economic and environmental benefits produced by biological control programs (Marchiori et al. 2007), the changing pattern of parasitoid communities is related to changes in ecological function. In essence, both management and landscape modification can entail tremendous effects on parasitoid community assemblages, and on ecosystem function and services.

6 Spiders

Spiders (Arachnida: Araneae) are considered to be a good indicator group to evaluate effects of land-use change on biodiversity (Marc et al. 1999). Still, the assessment of spider diversity losses through agricultural intensification and their possible effects on multilevel processes in tropical ecosystems is not trivial. Floren and Linsenmair (2005) define the approximate value for the loss of species following anthropogenic disturbances as the relation of overall species numbers to primary forest species numbers. However, the species communities in disturbed habitats are not simply an impoverished version of the less-disturbed primary forest (Chen and Tso 2004).

Spiders are abundant predators in natural as well as in cultivated landscapes, hence the interest in spiders as potential biological control agents. There are numerous studies on the specific arachnofauna in agroecosystems that describe differences in biodiversity between habitats or management types and their bioindicative potential within diverse agroecosystems. Most of these studies focus mainly on temperate zones e.g. apple (Marc and Canard 1997) and citrus orchards (Mansour et al. 1982) or cotton fields (Whitcomb et al. 1963). In the tropics, the spiders of paddy fields have received a fair share of attention (Okuma und Wongsiri 1973; Fagan et al. 1998) but there are only few published studies on spider diversity in Southeast Asia (e.g. Chen and Tso 2004, Russel-Smith and Stork 1994, Floren and Deelemann-Reinhold 2005, Tsai et al. 2006), or on spiders in cacao agroforests (e.g. Pérez-de La Cruz 2007).

In a study conducted between 2006 and 2008, we investigated the spider fauna in 43 cacao plots in Central Sulawesi within agroforestry sites along the margin of the Lore Lindu National Park. Our aim was to study the impact of distance to forest, shade tree management and cultural practices which may all radically alter population sizes and compositions of spiders (Mansour et al. 1983). We caught canopy dwelling spiders using branch eclectors, while the herb and litter layers were investigated by sweep netting, pitfall traps as well as litter sifting. In addition, we recorded all spider webs detected visually from ground-level to 4m height on 420 cacao trees. Every detected web was described by its web type, web size, height and their position within the cacao tree. The classification of the 5143 detected webs was based upon the reduction of the sometimes complexly structured webs into five simple web classes (orb web, tangle web, lattice web, sheet web and line web).

Up to now, we identified 20 spider families in total. Tetragnathidae, Clubionidae and Salticidae, each comprising about 15% of all individuals, were the most abundant families in cacao plantations across all strata. Families Heteropodidae (10%), Gnaphosidae as well as Theridiidae (both 8%) were also abundant in the investigated cacao plots. Salticidae, Heteropodidae as well as Gnaphosidae were caught in all strata, but density of individuals increased from litter to canopy for the first two families. Generally, half of the families were caught only in one of the three investigated strata, showing a clear preference for certain microhabitats (Figure 5).

The canopy-active spiders belonged to 16 families of which the Clubionidae (24.4%) and Salticidae (22.4%) were the dominant families. Within the Salticidae, the genus *Thorelliola* and the ant-mimics *Myrmarachne* spp. showed the highest number of individuals. Besides Gnaphosidae, the family Heteropodidae is well represented in Southeast Asia and both were also abundant families on tree trunks. The Gnaphosidae are often referred to as ground spiders, but in the tropics some genera are more likely to be found on shrubs and plants (Murphy and Murphy 2000). The predominance of free-hunting spider families is evident and is due to the selective catching method of branch eclectors. The number and diversity of spiders living in the canopy of cacao trees were positively correlated with the extent of the herb cover.

Within the herb layer, initial analyses showed a spider community that included 14 families of which the Tetragnathidae had a eudominant fraction of 55.7%. This family was represented by the genus *Leucauge* sp. and *Opadometa* sp., both of which are known from the tropics worldwide and prefer to build orb webs in shaded vegatation (Murphy and Murphy 2000). The variance in total number of spiders increased with elevation (R^2=0.33, p<0.01) and canopy openness (R^2=0.35, p<0.01) measured below the cacao canopy at a height of 130cm (R^2=0.70 (corrected), p<0.001). Sample size for all reported statistics is the number of cacao plots (n=43). Russell-Smith and Stork (1994) investigated the spider fauna of the canopy in primary forests in Sulawesi with regard to differences in spider composition along an altitudinal gradient. They found spider abundance to increase with elevation from 210 to 1150 m a.s.l. In cacao plantations, the family richness was positively correlated with the canopy openness. The subdominant Thomisidae are of topical interest because their prey spectrum contains *Helopeltis* spp. (Heteroptera: Miridae) (Simanjuntak 2001), a pest insect of cacao in Southeast Asia (see section 3 of the present chapter). Thomisidae showed preferences for cacao habitats with a reduced number of shade trees and a structurally developed understory. Chen and Tso (2004) as well as Floren and Deeleman-Reinhold (2005) observed a

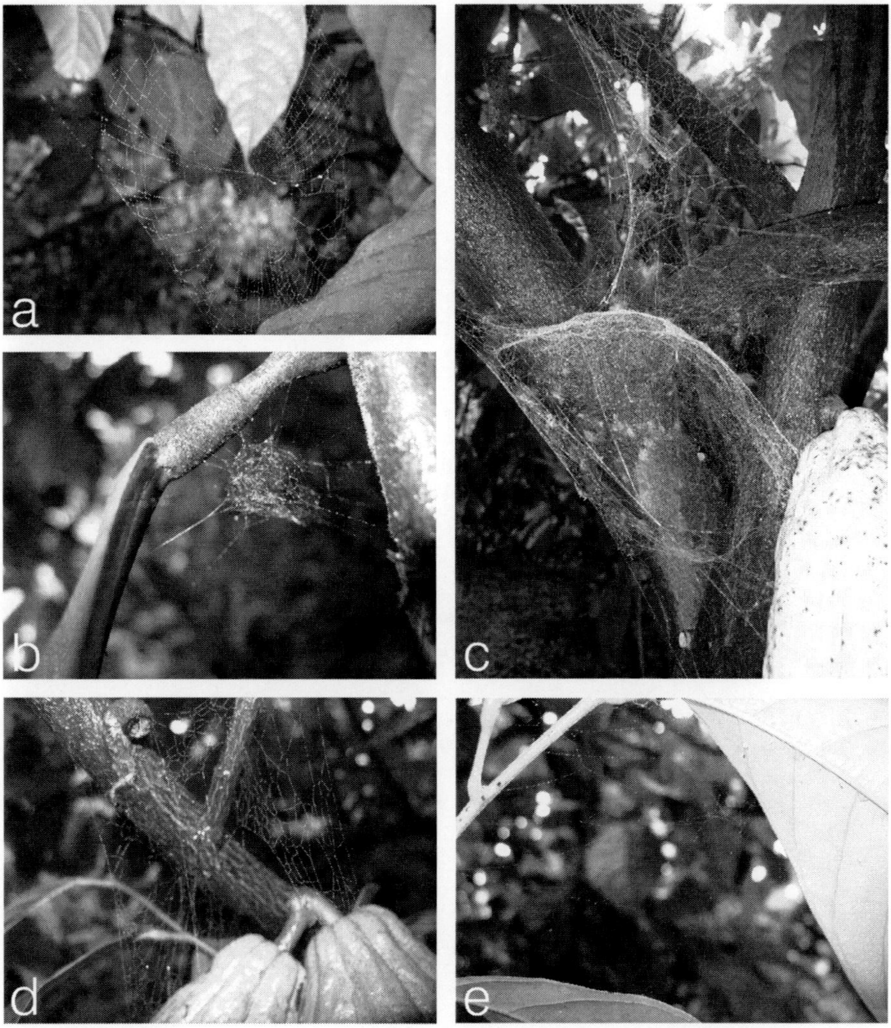

Fig. 5. Web constructions of the five spider web guilds that are associated with cacao trees (a) orb web (b) tangle web (c) sheet web (d) lattice web (e) line web (all photos by Kathrin Stenchly)

higher number of Thomisidae in disturbed forests (i.e. firewood plantations) compared to primary forest sites.

Finally, the litter stratum showed a high abundance of Theridiidae, Lycosidae, Linyphiidae and Salticidae. Various species of the Lycosidae can be characterised as wide ranging habitat generalists (Jocqué and Alderweireldt 2005) and were often found in plantations but less in tropical primary forests (c.f. Lo-Man-Hung et al. 2008). Hence the abundance of ground living spiders was positively influenced by distance to forest (R^2=0.36, p<0.01) and also by litter density (R^2=0.29, p<0.05).

The analysis of different web types that occur in cacao trees shows a high abundance of line webs which constituted 41.9% (2157) of all detected spider webs. Their biological control potential is likely to be small in contrast to the orb-web type which was the second most dominant web type with a total number of 1528 (29.7%). Interestingly, *Conopomorpha cramerella* incidence was reduced on trees with increased number of orb webs (R^2=0.47, p<0.01). Orb webs are built by the spider families Araneidae, Tetragnathidae and Uloboridae whereas several representatives of the Araneids have webs that are specially adapted to catch adult Lepidoptera (Maloney et al. 2003). Furthermore, 648 webs (12.6%) belonged to the lattice web type, while sheet webs showed a proportion of 9% and tangle webs 6.8%. Most webs were detected on branches, leaves and twigs. In conclusion, our results suggest that spiders may play an important ecological role in agroforests. Further studies are needed to fill the large gap in the knowledge about the role of spiders in agroforests, both in terms of biodiversity and ecological functions such as predation.

7 Bees and pollination

Over the past decade, the importance of pollinators as a key element of biodiversity supporting human livelihoods has been increasingly recognised in temperate and tropical regions. This is not surprising as most plants benefit from pollinators, encompassing organisms as contrasting as bees, flies, birds, bats and other mammals. For example, 90% of Angiosperms in tropical rainforests are now estimated to rely on animal pollination (Bawa 1990); most marketable horticultural products, increasingly delivered from developing, mainly tropical, regions (Aizen et al. 2008) benefit from pollinators (Klein et al. 2007).

The main drivers reducing terrestrial biodiversity are land-use change and intensification (Sala et al. 2000). The proportion and configuration of natural habitats in agricultural landscape matrices seem to be the major land-use variables promoting pollinator diversity and consequently the mutualistic interactions associated with the services pollinators deliver to plants (Brosi et al. 2008, Ricketts et al. 2008). The functional consequence of plant-pollinator interactions associated with natural habitat might then entail human benefits such as horticultural, wild plant species, and genetic diversity. In reality humans are still eliminating natural habitats like rainforest despite their ecological importance.

In the STORMA project in Central Sulawesi, Indonesia, case studies increased our knowledge of how rainforest habitats and land-use management affect bee communities and pollination services. One of the focal crops in this respect was coffee, which was more widely cultivated before the cacao boom reached this region. Klein et al. (2002a) used a land-use gradient –ranging from coffee agroforestry gardens established in the rainforest margin under the shade of highly diverse forest trees to agroforestry systems dominated by coffee under a single shade tree species – outside the rainforest to test the responsiveness of the coffee flower-visiting bee community to land-use intensification. The total number of species and individuals of social bees (honeybees and stingless bees) decreased, while individual numbers of solitary bee species of various bee families increased with land-use intensification. These results indicate that functional bee guilds respond differently to land-use intensification.

More detailed studies (e.g. Klein et al. 2003a, b) demonstrated that the vicinity of coffee plantings to the forest margin and variables associated with local management practices (shade density and overall plant diversity) structured flower-visiting communities. Increased distance of coffee plantings to the forest margin generally had an adverse effect on wood- and cavity-nesting bees; in this nesting group, the small bees were especially negatively affected by forest isolation (Klein et al. 2003b, 2008). Reduced shade mainly promoted the diverse group of soil-nesting bees relying on open ground in the soil (Klein et al. 2003b, see also Klein et al. in press). In experiments conducted in parallel to the flower-visitor observations, the pollination service to coffee plants in 24 sites was measured using data on initial fruit set of cross-, open-, and hand-pollinated plants (Klein et al. 2003c) and berry weight at harvest (Olschewski et al. 2007). Despite the functional responses of different nesting groups to forest vicinity and shade, overall pollinator species richness and abundance as well as coffee fruit set and berry weight decreased with increasing forest distance (Klein et al. 2003, Olschewski et al. 2006, 2007). Bee species richness was directly related to the pollination service, while functional group richness appeared to be more important than species richness and abundance (Klein et al. 2008).

In a follow-up study, these results were confirmed for another crop pollination system, pumpkin, planted in a land-use gradient (Hoehn et al. 2008). While in the coffee studies, the mechanisms behind the diversity-pollination service relationship could be only speculated upon, Hoehn et al. (2008) could explain how bee species richness enhances pollination services. They characterised the species-specific differences in spatio-temporal visitation of pumpkin flowers using a land-use gradient comprising high through medium to low bee species richness. At sites characterised by high bee-species richness, pumpkins reached a maximum in seed numbers (hand pollinated flowers served as reference) (319 ± 105 SD), whereas in the sites of lowest bee species richness, the number of seeds per pumpkin was significantly lower (179 ± 44 SD, Klein et al. 2009). Comparing the spatial (flower-visiting height of the plant) and the temporal (flower-visiting time of the day) activity of the flower-visiting species in both systems, it could be shown that different species occupied different spatial and temporal flower niches (Figure 6). Reductions in bee species richness, from 10 to three species resulted in non-occupied flower niches, both spatially and temporally at sites of lower species richness (Hoehn et al. 2008, Klein et al. 2009). Additionally, the ten species found in the high bee species sites were able to attend to the entire spectrum of spatio-temporal flower niches. This increased the probability that flowers at any height were visited and flower visitors seemed to be active during the whole receptive period of the flowers.

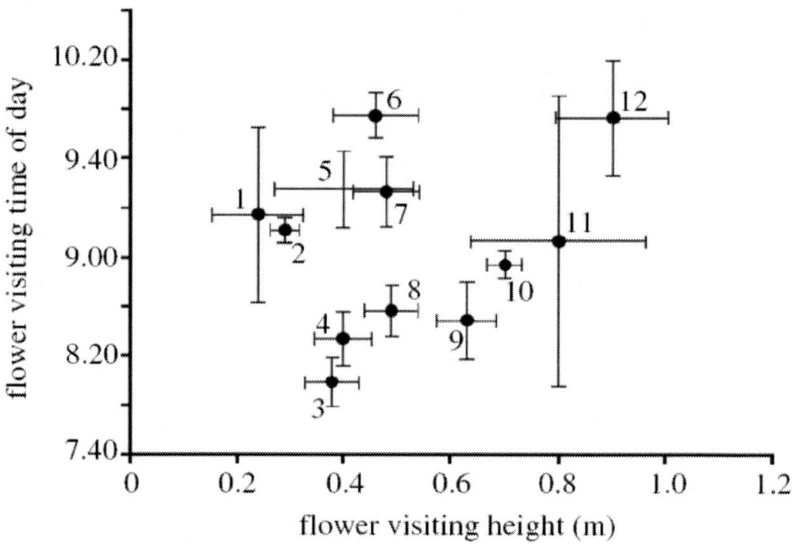

Fig. 6. Height and visiting time of flowers preferred by each specific bee species. Arithmetic means and s.e. are given. For mean values, standard error and significance levels, see Hoehn et al. 2008. Numbers represent species identity: 1, *Nomia concinna*; 2, *Lasioglossum* sp.; 3, *Apis cerana*; 4, *Xylocopa dejeani*; 5, *Nomia fulvata*; 6, *Ceratina. cognata*; 7, *Trigona* sp.; 8, *Amegilla* sp.; 9, *Xylocipa confusa*; 10, *Lasiolglossum halictoides*; 11, *Apis dorsata*; 12, *Xylocipa nobilis.* (Reproduced from Hoehn et al. 2008, with permission)

It is increasingly apparent that incorporating the spatial heterogeneity in the availability of flowers enhances the correlation between bee diversity and pollination services (Tylianakis et al. 2008).

The agricultural landscape at the forest margin in Indonesia is highly variable in the temporal availability of flower resources. Both the temporal variation in bee species richness of understory flower resources and spatial variation of coffee flower-visiting bee species richness were shown to increase with forest distance. The variation in bee species richness decreased the mean and increased the spatial variation of bee-pollinated coffee fruit set per agroforest. Hence, coffee grown near intact rainforests may fluctuate less in bee-pollinated fruit set among coffee plants than coffee plants in isolated coffee agroforests that receive pollination services of a less species rich pollinator community (Klein et al. in press).

Case studies from the STORMA project highlight the advantage of conserving rainforest in agricultural landscapes as habitat for bees and to stabilize crop pollination services. Yet, Olschewski et al. (2006) demonstrated that farmers will often not chose to conserve rainforest but rather tend to cut more rainforest to increase the land devoted to agriculture. The reasons are simple and economically driven: farmers are receiving more net revenues when cutting the forest to plant economically attractive crops than conserving the rainforest to increase pollination services in adjacent plantings of coffee and other crops.

8 Dung beetles and dung decomposition

Dung beetles (Coleoptera: Scarabaeidae) are particularly important in most terrestrial ecosystems since their dung burial activity maintains soil fertility (Omaliko 1984) and plant regeneration through seed dispersal activity (Andresen 2003). Therefore, a reduced dung beetle population most likely results in a cascading and long-term effect throughout the ecosystem.

Because biodiversity inventories cannot be complete given the range of expertise needed to cover all taxa, conservation planning is very often based on surrogates for which data are available and assumed effective for biodiversity conservation (Rodrigues and Brooks 2007). In contrast to many other taxa, dung beetles have been identified as one of the most cost-effective group for biodiversity surveys in tropical forests (Gardner et al. 2008, see also Kessler et al. 2009). In addition they have proven to be very suitable to assess effects of disturbances on tropical ecosystems (Nichols et al. 2007, Shahabuddin et al. 2008), for example because they are abundant, vary highly with respect to species traits, and respond rapidly to environmental change (Slade et al. 2007, Shahabuddin et al. 2008). Another advantage is that relatively complete species inventories and data on the abundance of individual species can be achieved rapidly with standardized methods (Larsen and Forsyth 2005).

With the continuing loss of tropical forests, agroforestry systems are gaining more attention due to their potential value for conserving tropical biodiversity. Agroforestry systems can maintain a high portion of tropical biodiversity (Rice and Greenberg 2000; Schulze et al. 2004a; McNeely and Schroth 2006).

Particularly, traditional agroforestry systems such as cacao and coffee cultivations established under a diverse layer of forest trees have been shown to be of high conservation value (Perfecto et al. 2003; Reitsma et al. 2001; Arellano et al. 2005; Harvey et al. 2006).

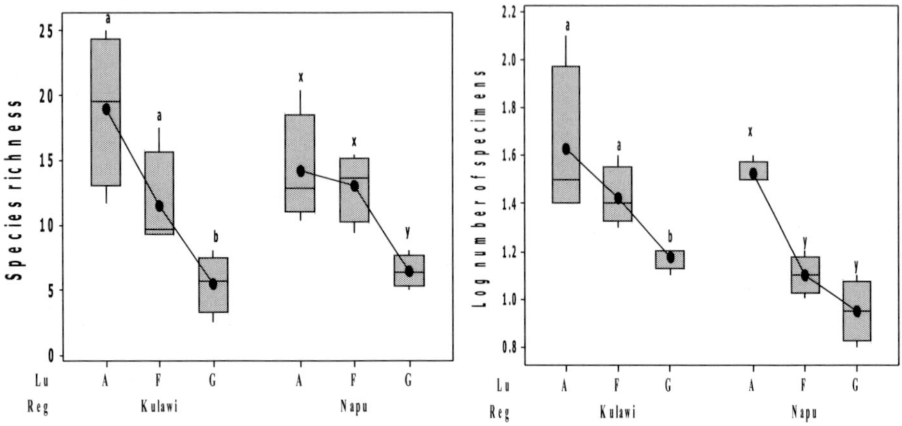

Fig. 7. Species richness (left) and abundance (right) of dung beetles across three sampled land-use type at Kulawi and Napu region of Lore Lindu National Park in Central Sulawesi. A, forest ; F, agroforestry system; G, open cultivated area (n=4 per land-use type). Box and whisker indicated the inter quartile range of median with it is maximum and minimum value. Different letters on top of the bars in the same region indicate significant differences between habitats (Tukey's HSD test; $\alpha = 0.05$). Dung beetle specimens were collected using pitfall traps baited with cattle dung (*Bos taurus*) as described in Shahabuddin (2008b)

Studies on the potential contribution of agroforestry systems to the maintenance of tropical forest dung beetles in Central Sulawesi (e.g. Shahabuddin et al. 2005, 2007, 2008) revealed that species richness and abundance decreased from natural forest to agroforestry systems and annual cultures (Figure 7) and species composition between those three land-use types were significantly different. (ANOSIM; global $R = 0.85$, $p < 0.001$, see also fig. 8). Of 28 species recorded, about 50 % were found at both forest and agroforestry sites and only about 18 % were recorded in all habitat types. These results indicate that agroforestry systems at the forest margin support a substantial proportion of dung beetle diversity. This is supported by Bos et al. (2007), who showed that compared with all studied insect groups (ants, lower canopy beetles, and wasps) dung beetles had the highest faunal overlap between forest and agroforestry sites at forest margins in Sulawesi. Of the 17 species recorded at the forest sites, a total of 13 (76.5 % of all forest species) were also found in agroforestry systems and all species were relatively common. In addition, additive partitioning of species richness showed that about 43 % of the dung beetle species richness could be explained by differences between land-use types (Shahabuddin 2007).

In an experimental field study (Shahabuddin et al. 2008), we found that dung decomposition, in which dung beetles play a key role, significantly declined from natural forest to open cultivated areas. The decline was most likely related to the decrease of the functional group of large-bodied tunnelers from forest to open cultivated area (Shahabuddin et al. 2008). This indicates that the ecosystem function provided by dung beetles was dependent on species identities and the functional groups involved. These results are in line with other studies, which reported that large-bodied beetles are the most sensitive to human disturbances such as habitat fragmentation (Andresen 2003) and modification of tropical forests (Slade et al. 2007, Navarrete & Halffter 2008).

Laboratory and field experiments (Horgan 2005, Shahabuddin et al. in prep.) demonstrated that dung decomposition is best predicted by mean body size and biomass of species but not richness of dung beetle assemblages. Slade et al. (2007) found that there were complementary effects among functional groups of dung beetles on dung removal. This points towards the importance of functional group diversity for completing dung removal activities. In general, the studies mentioned suggest that the functional consequences of conversion of forest to human- dominated land-use systems in terms of dung removal by beetles will depend on the specific responses of species and functional groups to landscape modification.

Although certain land-use systems like cacao agroforests can be characterised by a relatively rich dung beetle fauna, their importance for the maintenance of regional diversity has to be treated cautiously. The decreasing area of natural habitats, which may act as important ecological source, may also result in a mid- to long-term decrease of dung beetle richness in agroforests. The importance of remaining forests as source area is highlighted by the decrease of insect diversity in agroforestry systems with increasing distance to the nearest natural forest (Klein et al. 2006). For a substantial proportion of species, agroforestry systems at the forest margin may rather represent ecological sinks than habitats maintaining self-sustaining populations of forest species. Therefore, the true conservation value of land-use systems for insect diversity can only be assessed when considering the entire landscape matrix. Isolation from natural habitats may be of major importance for determining insect communities in tropical (as well as temperate) land-use systems (Tscharntke et al. 2005, Mendoza et al. 2005).

Even if high species richness is found in modified forest habitats, the loss of endemic species can be substantial (Schulze et al. 2004b, Fermon et al. 2005). The decrease of taxonomic diversity and biogeographically distinct species has to be considered for evaluating the conservation value of habitats (Hill 1999, Fermon et al. 2005). In other words, disturbed sites appear to have low complementarity, i.e. low beta diversity (Waltert et al. 2004, Tylianakis et al. 2005), even if they have high local alpha diversity. Sustaining high beta diversity and complementarity on a landscape scale (Kessler et al. 2009) is arguably

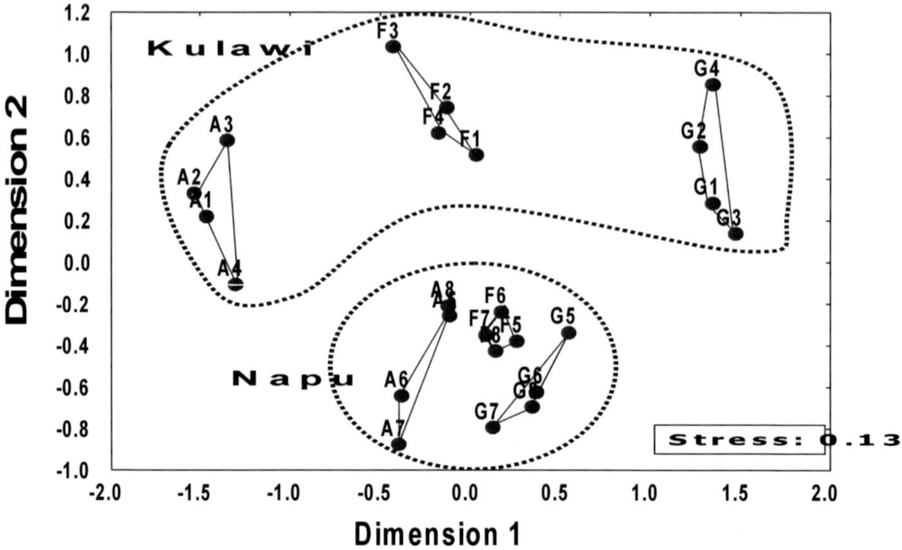

Fig. 8. Two-dimensional scaling plot based on Bray-Curtis indices quantifying the similarity of dung beetle communities sampled at single sites at Kulawi and Napu region in Central Sulawesi. Land-use types are indiated by different letters: A - natural forest, F - agroforestry system, G - open, cultivated area (Shahabuddin 2007).

more important for conservation nationally and globally than sustaining high site richness.

9 Fruit-feeding butterflies

Fruit-feeding butterflies of the Nymphalidae family are species-rich and easily monitored using fruit-baited live traps. This, together with the possibility of identifying most species visually, makes them an ideal group to study the response of biodiversity to landscape and land-use change. Recent research in the Lore Lindu national park area has focused on the butterfly communities in natural or mature forests, compared to disturbed (Fermon et al. 2005) and secondary (Veddeler et al. 2005) forests. Overall species richness was reported higher in disturbed forests and lower in secondary forests. However, endemic species richness was consistently higher in the undisturbed habitat.

In an unpublished study carried out by Darras in Central Sulawesi from March to May 2008, we linked taxon-wise abundance and overall diversity measures of butterfly communities in smallholder cacao plantations, situated along shade and landscape gradients, to environmental variables. Sampling was carried out in 43 cacao plots of the Kulawi and Palolo valleys using baited live-traps suspended around 1 m above the soil. We recorded around 2300 individuals from 35 species.

The species *Melanitis leda* Linné, *Mycalesis horsfieldi* Moore and *Mycalesis janardana* Moore, in order of decreasing abundance, accounted for 82% of the total number of individuals.

These species all belong to the Satyrinae subfamily, which forms the bulk of the baited specimens with 19 species. All other families (Charaxinae, Limenitinae, Nymphalinae, Morphinae, Apaturinae) were rarer and less well represented in terms of species.

Using generalised linear models, we analysed how environmental variables affected different population attributes. A forward selection approach was used to yield models explaining the greatest variance and using the maximum number of significant explaining variables.

Abundance of Charaxinae (25 specimens, 32% explained variance) as well as Nymphalinae (27 specimens, 21% explained variance) were positively related to plot temperature. This is consistent with previous results, as in the more sunny secondary forest habitats in the study of Veddeler et al. (2005), Charaxinae and Nymphalinae were more abundant too.

Lohora spp. butterflies are predominantly forest understory dwellers endemic of Sulawesi (Fermon et al. 2005, Vane-Wright and deJong 2003). Veddeler et al. (2005) report a significant increase in abundance as the shading degree of the forest increases, which is itself taken as a substitute for successional stage. Our findings are concordant, as the best model explaining *Lohora* spp. abundance showed a negative influence of both temperature and canopy openness (35 specimen for both valleys: 30% explained variance; 25 specimen in Palolo valley: 37% explained variance). The results of Fermon et al. (2005) also suggest that all *Lohora* (with one exception: *L. transiens* Fruhstorfer) are more rare in the more open, disturbed site. However, we did not determine *Lohora* individuals to the species level, so we cannot comment on the *L. transiens* exception.

Mycalesis spp. were more abundant in plots with high herb species richness and lower slope (998 specimen, 29 % explained variance). Veddeler et al. (2005) also show that *Mycalesis* abundance closely follows vascular plant richness. Apparent preference for even terrain may suggest association with lowland habitats.

For *Melanitis* spp., the best model revealed a negative influence of litter thickness and a positive influence of canopy openness (1138 specimen, 41% explained variance), the latter being in accord with preferences found in the other studies. The only remarkable exception is *M. velutina* Felder, which occurs more often in the undisturbed forest in both studies, so this species was excluded from the model.

A possible caveat of our sampling method is that canopy openness could have artificially increased the apparent abundance (and richness) as upper-dwelling butterflies come down to the understory (DeVries 1988) where our traps were located. On the other hand, such a bias was controlled for by

Fermon et al. (2005), who still found similar trends for *Melanitis* spp. and other taxa.

Community evenness was linked positively to shading and negatively to *Melanitis* abundance (both factors together explain 37% of the variance). Most even communities were found in forest-like habitats where understory was thick and the dominant genus *Melanitis* was rare.

Estimated species richness was not strongly affected by any of the variables included in the study (R^2 not above 10%).

Environmental determinants have taxon-specific impacts on abundance, which may be divergent. All studied taxa here showed a significant positive relationship between abundance and species richness. We can say that no single environmental measure determined the richness and abundance of all taxa. For this particular group, biodiversity studies were more informative when taxa or functional groups were considered separately. In conclusion, the studies conducted on fruit-feeding butterflies in Central Sulawesi demonstrate that landscape change and local characteristics of agroforests have a discernable impact on the composition of their communities. This also supports the necessity of having a variety of habitats under different environmental conditions to maintain higher overall species richness across the landscape.

10 Birds

10.1 The birds of Sulawesi: land-use and conservation

Sulawesi is home to 224 resident land bird species including eleven endemic genera and 91 (43%) endemic species (Coates et al. 1997, Stattersfield et al. 1998). Most of these species are forest inhabitants. Accordingly, many endemic forest species are strongly affected by deforestation. Much of the logged forests have been converted into agricultural land, including both annual crops and perennial plantations such as oil palm, copra, coffee, and cacao. Agroforests with considerable tree cover, relatively high tree species diversity and complex structure provide many ecosystem functions and processes also found in primary forests (Siebert 2002, Schroth et al. 2004, Steffan-Dewenter et al. 2007). They are often considered to be important alternative habitats for many rainforest species (Rice and Greenberg 2000, Schroth et al. 2004, Faria et al. 2007). These plantations provide habitats and dispersal pathways and assist in the conservation of the local species composition (Faria et al. 2007).

For bird species composition, the vegetation structure of the plantation appears to be most important (Waltert et al. 2004, Van Bael et al. 2007a, Abrahamczyk et al. 2008). Cacao agroforestry systems with a high proportion of large remnant forest trees provide a wide range of breeding niches and food resources for many species (Sodhi et al. 2005, Van Bael et al. 2007) and shelter a much higher number of forest specialists and endemics compared to plantations with planted, even-aged shade trees. Interestingly, agroforests with planted and often exotic shade trees harbour the same number of generalists as plantations with remnant forest trees but species composition differs strongly. In plantation with planted shade trees the proportion of widely distributed generalists with a low conservation priority like *Passer montanus* L. or *Lonchura malacca* L. is much higher. However, the presence of some forest specialists and the high number of generalists, including many endemics and the two near-threatened species *Loriculus exilis* Schlegel and *Ptilinopus subgularis* Meyer & Wiglesworth, show that even plantations without rainforest trees have a certain value for bird conservation.

Another important factor for the community composition in cacao plantations is the distance to the nearest forest and the fragmentation of the landscape on a larger scale. Small agroforests next to the forest allow birds to move into the plantations and back to the forest. The relatively high number of endemics and forest specialists occurring in plantations next to the forest even with planted shade trees (Clough et al. 2009) can be explained in this way. Some of these birds will never accept cacao plantations as breeding territory, like the Great Hornbill, *Rhyticeros cassidix* Temm., but use them for feeding. Others like the Black-naped Monach, *Artamus monachus* Bonaparte, an endemic forest specialist, use cacao agroforests for breeding (Clough et al. 2009). If such species can successfully benefit from plantations for breeding or if it is just a sink habitat just used by individuals that could not find a territory inside the forest remains unclear. Pangau-Adam et al. (2006) found higher predation rates on artificial nests in disturbed habitats but detailed investigations with real nests are still lacking.

A second explanation for the relatively high species number in agroforests next to the forest is that cacao plantations embedded in an unfragmented forest area allows large forest specialists such as *Spizaetus lanceolatus* Temm. & Schlegel, *Megapodius cumingii* Dillwyn or *Penelopides exarhatus* Temm. to persist in the area. These species need extensive forest areas because they have on average larger territories than smaller species and mostly use the plantations only for feeding or resting. In a more fragmented landscape with bigger plantations they would not use them.

Compared with primary forests, all types of plantations harbour a smaller number of endemics and forest specialists but a greater number of generalists. This shows the general difference in habitat quality between primary forests and agroforests. Especially ecological specialists like insectivores or species confined to one forest stratum, either to the ground or to the midstorey and canopy like *Meropogon forsteni* Bonaparte or *Coracina temminckii* Müller, need primary forest with its high habitat quality, characterised by well structured strata and a great variability of food resources. Such areas are of high conservation concern (Sodhi et al. 2005). About 15 percent of all species were never observed in plantations. For the protection of these mostly endemic forest specialists huge unfragmented primary forests are crucial (Castelletta et al. 2005). Additionally, 15-20 percent of bird species such as *Penelopides exarhatus* only rarely occur in cacao plantations next to the forest and can not survive without large expanses of (near-)primary forest.

10.2 The ecological role of birds in the cacao agroecosystem

The role of birds in the ecosystem is receiving ever-increasing interest on the part of basic and applied ecology alike. Indeed, the functional species richness and composition of birds in human-dominated landscapes is likely to play an important role for a number of key ecological services including predation, pollination, seed dispersal and scavenging (Sekercioglu 2006).

Insectivorous birds have been shown to be able to effectively reduce populations of herbivorous insects not only in tropical and temperate forests (Evelegh et al. 2001, Hooks et al. 2003, Loyn et al. 1983, Marquis and Whelan 1994, Van Bael et al. 2003), but also in agricultural systems (Mols and Visser 2002; Greenberg et al. 2000, Van Bael et al. 2008). Using exclusion designs, Kalka et al. (2008) and Van Bael et al. (2008) showed, in neotropical forests and agroforests respectively, that although part of the predation may be caused by bats, birds contribute to a large extent. A recent study demonstrates that the yield of Jamaican coffee farmers is increased substantially by bird predation on one of the main coffee pests (Kellermann et al. 2008). An interesting feature of the existing body of literature on the effect of birds in reducing herbivore populations is the correlation between the extent of that decrease and the number of insectivorous bird species (Van Bael et al. 2008). This suggests functional complementarity of bird species in providing this important ecosystem service. At least 23 species of insectivorous bird species forage, and a few have been observed to breed in cacao agroforests in Central Sulawesi (Clough et al. 2009). These species are mainly resident i.e. non-migratory species, in contrast to migratory insectivorous species which dominate herbivore removal effects in the neotropics (Van Bael et al. 2008, but see Perfecto et al. 2004). Using data from a large-scale bird survey conducted around the Lore Lindu National Park in cacao plots differing in management and distance to natural forest, we found that the number of insectivorous bird species that are likely to occur in a cacao plot increases both with the number of large shade trees and with proximity to the forest edge (Clough et al. 2009). How this affects herbivory in cacao is not yet known as data from ongoing bird exclusion experiments are not yet available. However, the negative correlation of cacao leaf herbivory with increasing canopy cover (Figure 9) suggests shade might cause reduction in herbivory because of the associated higher insectivorous bird species richness, which is also supported by data from neotropical coffee agroforests (Perfecto et al. 2004).

Birds also interact directly with plants by predating and dispersing seed, feeding on nectar and pollinating. Seed dispersion by frugivorous birds is essential for recolonisation of disturbed habitats by a diverse plant community. In the tropics, forest regeneration depends much more strongly on animal seed

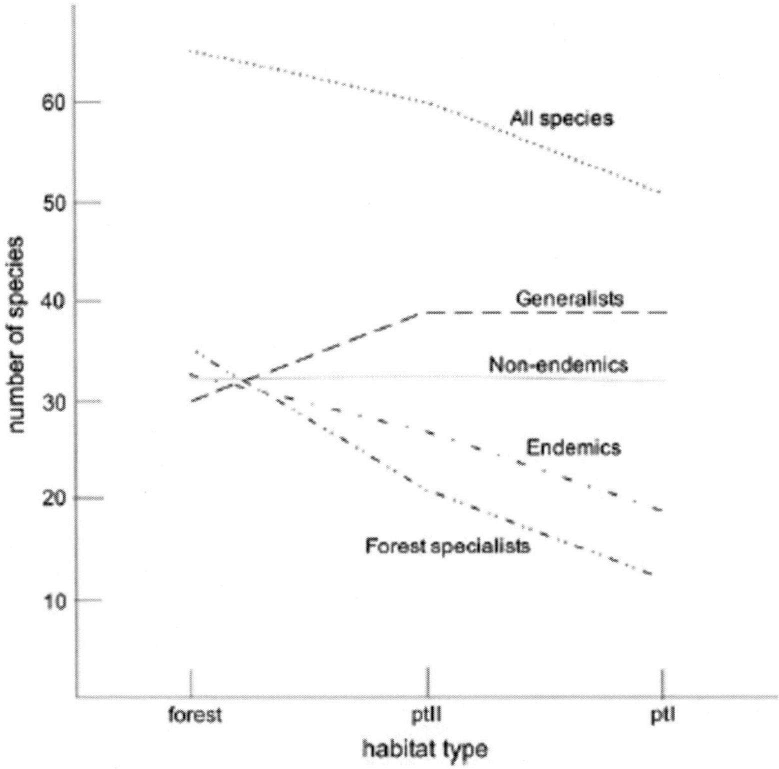

Fig. 9. Number of bird species in three habitat types (forest=near-primary forest, ptII=plantations with remnant forest trees, ptI=plantations without remnant trees) (Abrahamczyk et al. 2008).

dispersal than in temperate regions (e.g. Tabarelli and Peres 2002). As habitats get more disturbed, fewer frugivorous, potentially seed-dispersing bird species venture into them (Gomes et al. 2008) with potentially disastrous consequences predicted for forest tree regeneration (Sodhi et al. 2005). Our data show that the frugivorous bird community is less species rich at a distance from the forest edge and in cacao plantations with few or no shade trees. Data from Ecuador (Lozada et al. 2007) suggest tree cover in agroecosystems results in increased seed deposition by seed-dispersing birds, causing high seedling regeneration of species not locally present before. Avian seed dispersal may occasionally become a problem in agricultural systems such as cacao agroforests, as exemplified by flowerpeckers (*Dicaeum* spp.) spreading mistletoe species (Loranthaceae). In cacao plantations, mistletoe can cause considerable damage including tree death, although the most severe symptoms seem to be restricted to little-shaded and unshaded cacao trees (Yann Clough pers.obs.).

The fruits of these hemiparasitic plants are very attractive to flowerpeckers and depend on ingestion by birds to spread within and between trees. The movements of flowerpeckers may affect the spatial distribution of mistletoe, as reported for an Australian *Dicaeum* species by Ward and Paton (2007).

To summarise, even though many bird species are confined to natural forests, the bird community in cacao can benefit from shade tree cover, which may be instrumental in conserving more common, but functionally important species in human-dominated landscapes increasingly devoid of near-natural, forested habitat.

11 Amphibians and reptiles

The global amphibian assessment and a recent summary report of the ongoing global reptile assessment revealed amphibians and reptiles as the most threatened vertebrate taxa on the planet (IUCN 2008). While limited dispersal ability and a recently emerging disease (chytridiomycosis) further enhance extinction pressure on amphibians, habitat loss and climate change are the major threats for both groups with their distributions peaking in the tropics (Gibbons et al. 2000, Stuart et al. 2004, Wake and Vredenburg 2008; Sodhi et al. 2008). Effects of both threats are likely to be most severe for tropical ectotherms due to relatively constant environmental conditions in the geological past and, hence, high levels of local adaptation (Deutsch et al. 2008).

Despite the vast expansion of agricultural land in the tropics (Foley et al. 2005), surprisingly little is known about the impact of tropical land-use change on amphibians and reptiles (Gardner et al. 2007). Most studies were so far conducted in the neotropics and showed that, if canopy cover is high and natural habitat is sufficiently close, secondary habitats (such as coffee and cacao agroforest ecosystems) can sustain herpetological diversity (Pineda et al. 2005; Suazo-Ortuo et al. 2008). As open canopy alters the climatic regime in the plantation (Stratford and Robinson 2005), temperature is most likely limiting forest specialists in these habitats (Perfecto et al. 2007, Luja et al. 2008). Comparable studies from Southeast Asia are still scarce. Gillespie et al. (2005) studied herpetological diversity changes on offshore islands of Sulawesi, Indonesia and found decreasing species richness paralleled by increasing habitat disturbance. However, they did not identify the drivers for the observed patterns. In Central Sulawesi, Wanger et al. (2009) used a two step approach to understand the conservation value of cacao plantations. Their study of herpetological diversity patterns along a habitat gradient revealed that amphibian species richness decreases along with habitat complexity, leading only to slightly decreased richness in cacao agroforests compared to pristine habitat. In contrast, reptile species richness peaked in cacao agroforests but remained lower in all other habitats. While in amphibians Celebes toads were generally over-dominant, this over-dominance (*Eutropis grandis*) was found only in cacao agroforestry for reptiles (for examples of common and rare amphibian and reptile species see Figure 9). In a second step, a large-scale leaf-litter and log manipulation approach (Wanger et al. 2009; Figure 10) revealed that predominantly leaf litter thickness, branches and logs, the ratio between leaf litter and understory shrub cover, and annual average temperature in the plantations determined species occurrence. Experimental manipulation of leaf litter thickness changed herpetological diversity in the plantations but branch and log modifications did not yield detectable effects.

Overall, land-use change will negatively impact herpetological diversity in the tropics. But while in other tropical regions recommendations to mitigate these effects can be based on various studies (e.g. Ernst and Rödel 2005; Gardner et al. 2006; Suazo-Ortuo et al. 2008), results from Southeast Asia are too few to generalize. There, structurally complex agroforestry systems may sustain amphibian and reptile diversity similar to pristine habitats but benefit predominantly generalist species. In amphibians, beta diversity (i.e. species turnover) increases from pristine to strongly disturbed habitats and, therefore, urges conservation actions on a landscape scale. In contrast, reptile beta diversity is similar across all habitats but lowest in cacao agroforest, suggesting that habitat adjustments on the plot level may be sufficient (Wanger et al. 2009). The actual conservation value of cacao agroforest habitats depends, however, on a good knowledge of the driving biotic and abiotic factors. The dependence of diversity patterns on temperature in the plantations suggests that the interaction of land-use change and future climate change may have substantial impacts on amphibians and reptiles (Wanger et al. 2009).

12 Rats

The mammalian fauna of Sulawesi consist mainly of bats (Microchiroptera) and rats (Muridae). It is not particularly rich in species but shows a very high level of endemism. There are 127 mammal species, 79 of which are endemic. Endemism rises from 62 to 98 % if bats are excluded (Whitten *et al.* 2002), and most of these species are rats. In spite of the high level of endemism and the endangered natural habitats in Sulawesi, not much is known about the ecology of the murids and how they react to intensification of agroforestry and landscape fragmentation as existing studies focus mainly on taxonomic relationships (e.g. Musser 1969 a/b/c, Musser 1971, Musser 1991, Musser and Dagosto 1987).

There are but a few studies on small mammals and their ability to use agroforests as potential habitat. These studies were mainly conducted in South America (Estrada et al. 1993/1994) and India (Bali et al. 2007) and showed that mammals can use cacao agroforests as a potential habitat, but are less diverse in plantations than in forests and more frequently found in agroforest systems that are closer to the forest. Even in diverse agroforests, forest specialists are often missing because they need the native forest for a certain period of their life (Rice and Greenberg 2000). Studies on small mammals e.g. from

Fig. 10. Examples of rare and common amphibian and reptile species in the STORMA study region in Central Sulawesi. Above from left to right: *Callulops albotuberculatus* is a new and presumably rare microhylide frog, newly described from Mount Nokilalaki in Central Sulawesi (Iskandar, Brown, Wanger submit.). The Common Celebes Toad (*Ingerophrynus celebensis* Günther) is often encountered in a wide range of habitats throughout Sulawesi; the picture shows a young individual. Below from left to right: The Sulawesi Wolf Snake (*Lycodon stormi* Boettger) is rarely encountered throughout Sulawesi and little is still known about its biology. In contrast, the Painted Bronzeback (*Dendrelaphis pictus pictus* Gmelin) is a very agile snake, commonly encountered in plantations and human settlements (all photos by Thomas Wanger).

Borneo (Wells et al. 2007) or Venezuela (Ochoa 2000) showed that especially rare and specialised species are most likely to be affected by habitat change. This effect leads to a reduction of diversity even if losses in species richness are covered by an increasing number of individuals from species which are tolerant to disturbance (Cottingham et al. 2001, Ernest and Brown 2001).

Most native rats are reported to live in forests and are likely to show a certain degree of specialisation - such as most species of the genus *Rattus,* whose habitats range from tropical lowland to montane forest types and which tend to avoid human settlements (Nowak 1991). Consequently, murids can be expected to be highly influenced by deforestation through the loss of resources and habitats, the reduction of landscape connectivity and edge effects (Chiarello 2000, Laidlaw 2000, Lopes and Ferrari 2000). Rats are generally recognised as valuable indicators for forest fragmentation and habitat disturbance due to their relationship with forest cover and vegetative complexity (Harvey et al. 2006). On Sulawesi, this is all the more true given the importance of rats in the endemic mammalian fauna.

In two murid surveys conducted in 2006 and 2008 (Betz 2009, Weist 2008), we attempted to find out which species can use cacao agroforests as a potential habitat - farmers reported that rats feed on cacao pulp (see also Riley 2007), but the identity of the species was hitherto unknown. Secondly, our aim was to find out how murids were affected by ongoing land-use change, agroforest intensification and landscape context of plantations, i.e. proximity to near-natural forest habitat. We live-trapped murids in cacao agroforest systems differing in distance to natural forest and local management intensity, as well as in disturbed and native forest with contrasting vegetation structure. As a result of the total sample effort of 5760 trap-nights, 145 individuals belonging to five genera and 14 species were captured on 38 plots in Kulawi and Palolo valleys: 105 individuals were captured in cacao agroforest systems and 40 individuals were captured in forest systems. Among the 14 identified species were 12 endemic and two introduced species.

Species richness did not differ between the different habitat types but a change in species composition was clearly noticeable. The endemic species *Bunomys penitus* Miller & Hollister, *Bunomys.* sp., *Maxomys hellwaldii* Jentink and *Paruromys dominator* Thomas were only caught in (near-) primary forests. Interestingly *B. prolatus* was not caught in forest but in cacao agroforest. However this agroforest is close to forest with a high and very dense understory and this species was recorded to be common in the forest margins (Maryanto 2000, unpublished data). The endemic species *B. chrysocomus* Hoffmann, *R. marmosurus* Thomas, *R. hoffmanni* Matschie and *R. xanthurus* Gray are frequently caught in cacao agroforest systems and forest. The most frequently caught species was the endemic *Rattus marmosurus.* In cacao agroforests, abundance decreased with increasing distance to forest as the abundance in non-forest habitats is generally lower.

We found that *R. marmosurus, R. hoffmanni, R. xanthurus and B. chrysocomus* preferred to nest in fallen trees or subterranean nests hidden under

Fig. 11. Manipulation of cacao plantations to investigate the importance of leaf litter for the herpetofauna. Assistants clearing a plot of the entire leaf litter cover (top). The total amount of leaf litter removed from one plot (40 x 40 m) is enough to fill up two pick-ups like the one in the picture below (bottom; all photos by Thomas Wanger)

roots or stones, in places with higher understory vegetation density. As nests from *R. marmosurus* were found within and next to plantations, and home ranges were likely to be rather small, it seemed that *R. marmosurus* occurred as a resident in cacao plantations. While *R. marmosurus* usually nests in dead trees, its abundance could not be explained by the availability of dead wood. Occurrence of this species might be limited by the availability of dense understory growth which is found more frequently next to plantations which are at or near the forest edge. Our data indicate that the endemic species *R. xanthurus* is a generalist generalist that normally only occurs in primary forest. The present study showed that it was more commonly found in agroforest systems compared to forests. This species may have adapted to humans as it can be found in cacao agroforest systems close to the forest and to villages. This also implies the risk that *R. xanthurus* can be a potential disease carrier in the future for human beings as already known from *R. rattus* or *R. exulans* (e.g. Aplin et al. 2003).

In contrast to the endemic species mentioned above, the two introduced species *R. rattus* and *R. exulans* are commonly found across Southeast Asia. Both species are generalists and, in our study, were found mostly in cacao plantations. One individual of *R. rattus* was trapped in disturbed forest, suggesting that this species is able to spread out into forest habitats as found in studies conducted in Madagascar and Borneo (Ganzhorn 2003, Wells pers.comm.).

High occurrence of introduced species in cacao agroforest systems might have an additional negative effect on native species in Sulawesi because they can displace the native species. It is already known from other studies that the introduction of non-native species can cause unfavourable or unexpected consequences, including negative effects on natural forest and animal diversity (e.g. Amori and Clout 2003, but see Ganzhorn 2003).

Interestingly, the two non-native species are not (*R. rattus*) or even negatively (*R. exulans*) influenced by herb layer height while small mammal abundance is generally found to be positively influenced by herb layer height.

Comparing main species richness of the native species caught within cacao agroforests with that of forests (see also Maryanto and Yani 2001) it seems likely that almost all endemic species found in Lore Lindu remain in the forest although some species are frequently found in human-dominated landscapes around Lore Lindu NP. Only the species *B. chrysocomus*, *R. marmosurus* and *R. hoffmanni* seem to be able to adapt to cacao agroforest systems as a habitat, although we cannot, from our results, exclude the possibility that cacao acts as a population sink for these species. Most endemic small mammals from Sulawesi can be expected to be highly specialised to primary forest as a habitat and to be very sensitive to the changes in habitat characteristics that accompany the conversion of natural forest.

In conclusion, we suggest that cacao agroforests may be used as a buffer zone around protected areas for some endemic species but they show several limitations for conserving native small mammals as many endemic species remain restricted to the forest and the degree to which cacao agroforest systems

Table 1. Summary of effects of land-use change and intensification on plant and animal diversity based on data from Central Sulawesi, Indonesia (for details see text and references therein). Out of the 10 groups considered, 6 show similar or higher diversity in cacao plantations (CP) than natural forest (NF). In most cases diversity decreases from cacao agroforestry (CA) to cacao plantations (CP).

Species group	Natural forest (NF)	Secondary forest (SF)	Cacao agroforests (CA)	Cacao plantations (CP)
Trees	50-60 sp. per 0.25 ha	Similar diversity to NF, but composition differs	Intermediate diversity	Few species, many non-native
Rattan	Heavily exploited	Heavily exploited	Little or none	None
Lianas	6-12 sp. per 0.2 ha	Similar to NF	Similar to NF, but smaller, more herbaceous species	Species poor (0-5 sp. per 0.2 ha)
Herbs	Very high species richness (171-204 sp.), low density		High diversity (176 sp.), high density	Density high if no herbicide-use, many pantropic species, few shared with forest
Bryophytes	High local diversity (150 sp. on eight mature trees)		Loss of 70% of the NF species	Very species poor, removal by farmers
Ants	High diversity		Similar diversity as NF (163 sp. found), 75 % forest species retained, but endangered by invasive species	Few species lost, additional non-forest species, but endangered by invasive species
Bees	Few solitary sp. with low density, many, abundant social species		Increase of solitary species diversity and density, fewer social bees	Decrease of solitary species diversity compared to CA, fewer social bees
Dung beetles	17 species, 25% of species only at forest sites		Similar to NF, slightly different composition	
Fruit-feeding butterflies	Highest diversity, most endemic species	Diversity reduced by one third (older secondary forest)	Relatively high diversity (35 sp.)	Several species more abundant in less-shaded cacao
Birds	High species richness, especially endemics (altogether 224 sp. known in Sulawesi)		Increase in habitat generalists, fewer endemics and NF specialists	More granivore species, less insectivores, few endemics and NF specialists

may serve as a habitat depends on the proximity of the forest edge. Consequently, long- term conservation of endemic small mammals in Sulawesi is not possible without protected forest areas.

13 Conclusions

The agricultural expansion into natural forested areas is a major challenge for conservation (Sala et al. 2000). Needless to say, large expanses of near-primary forest are necessary to conserve numerous forest specialists. Notwithstanding this fact, agricultural landscapes offer the opportunity to combine conservation and economical benefits for local people. It has recently been highlighted (Bhagwat et al. 2008, Perfecto and Vandermeer 2008) that agroforestry has the potential to deliver both high biodiversity and economic welfare, but that this potential depends on the way agricultural management is conducted. This is most clearly reflected in the high variability in both relative species richness and similarity in species composition of agroforests compared to natural habitats serving as a reference (Bhagwat et al. 2008).

Our findings largely confirm that the management of cacao agroforests has a major impact on their suitability for diverse species assemblages across species groups. Most importantly, a dense and diverse shade cover is associated with high species richness in most groups. If we broaden the scale of investigation to include landscape context, we find that the proximity of natural habitat, in this case the forest edge, is another key predictor for species richness. This shows that not only shade management, but also the configuration of the landscape is crucial in determining species richness and composition of many taxonomic groups, of which several (e.g. insectivorous and seed-dispersing birds, parasitoids, pollinators) deliver ecosystem services to the agricultural sector. Our results suggest that ecosystem services delivered by shade tree cover and proximity of natural habitats outweigh potential disservices (Zhang et al. 2007) such as losses to herbivores.

One major impediment to yields in recent years, however, may benefit from enhanced shading: black pod disease *Phytophthora palmivora.* During years with heavy rainfall, as witnessed in 2007 and 2008, high levels of canopy cover due to a combination of insufficient pruning and high density of shade trees, together with insufficient sanitation of diseased material lead to unacceptably high levels of disease with incidences up to 90% in some cases (Clough unpublished). This is in line with results found in previous studies, which suggested trade-offs between conservation value and income mediated by shade (Steffan-Dewenter et al. 2007). If biodiversity-friendly cacao management is to be supported, farmers will need a combination of support in battling black pod disease, and premium prices to compensate for decreases in yield. Coffee certification in Central America (Mas and Dietsch 2003, 2004) has shown that this can be a viable path to improve farmer's livelihoods.

Acknowledgements
Helpful comments of anonymous reviewers and financial support of the German Research Foundation (DFG: SFB 552) are gratefully acknowledged.

References

Abrahamczyk S, Kessler M, Dadang DP, Waltert M, Tscharntke T (2008) The value of differently managed cacao plantations for forest bird conservation in Sulawesi, Indonesia. Bird Conserv Int 18: 348-362

Acebey A, Gradstein SR, Krömer T (2003) Species richness and habitat diversification of corticolous bryophytes in submontane rain forest and fallows of Bolivia. Journal of Tropical Ecology 18: 9-18

Achard F, Eva HD, Stibig HJ, Mayaux P, Gallego J, Richards T, Malingreau JP (2002) Determination of deforestation rates of the world's humid tropical forests. Science 297: 999-1002

Adi Prawoto A (1996) Infestation of cacao pod borer (*Conopomorpha cramerella* Snell.) in Central Sulawesi. Pelita Perkebunan 9: 79-84

Aizen MA, Garibaldi LA, Cunningham SA, Klein AM (2008) Long-term global trends in crop yield and production reveal no current pollination shortage but increasing pollinator dependency. Curr Biol 18: 1-4

Akiyama T, Nishio A (1996) Indonesia's cocoa boom: hands-off policy encourages smallholder dynamism. Bull Indones Econ Stud 33: 97-121

Altieri MA, Schmidt LL (1986) The dynamics of colonizing arthropod communities at the interface of abandoned, organic and commercial apple orchards and adjacent woodland habitats. Agric Ecosys Environ 16: 29-43

Amori G, Clout M (2003) Rodents on islands: a conservation challenge. In: Singelton GR, Hinds LA, Krebs CJ, Spratt DM (eds) Rats, mice and people: rodent biology and management, ACIAR Monograph No.96

Andren H (1994) Effects of habitat fragmentation on birds and mammals in landscapes with different proportion of suitable habitat: A review. Oikos 71: 355-366

Andresen E (2003) Effect of forest fragmentation on dung beetle communities and functional consequences for plant regeneration. Ecography. 26: 87-97

Antaranews (2008) Indonesia to produce one million tons of cocoa in 2013. 15th edn. Antara, Jakarta

Aplin KP, Chesser T, ten Have J (2003) Evolutionary biology of the genus *Rattus*: profile of an archetypal rodent pest, In: Singelton GR, Hinds LA, Krebs CJ, Spratt DM (eds) Rats, mice and people: rodent biology and management, ACIAR Monograph No.96

Arellano L, Favila ME, Huerta C (2005) Diversity of dung and carrion beetles in a disturbed Mexican tropical montane cloud forest and on shade coffee plantations. Biodivers Conserv 14: 601- 615

Ariyanti NS, Bos M, Kartawinata K, Tjitrosoedirdjo S, Guhardja E, Gradstein SR (2008) Bryophytes on tree trunks in natural forests, selectively logged forests and cacao agroforests in Central Sulawesi, Indonesia. Biol Conserv 141: 2516-2527

Bali A, Kumar A, Krishnaswamy J (2007) The mammalian communities in coffee plantations around a protected area in the western Ghats, India Biol Conserv 139: 93-102

Bawa, KS (1990) Plant-pollinator interactions in tropical rain forests. Annu Rev Ecol Syst 21: 299-422

Beer J, Muschler R, Kass D, Somarriba E (1998) Shade management in coffee and cacao plantations. Agric Ecosys Environ 38: 139-164

Betz L (2009) Murids along a land-use gradient in Central Sulwesi, Indonesia. Diploma thesis, University of Göttingen

Bhagwat SA, Willis KJ, Birks HJB, Whittaker RJ (2008) Agroforestry: a refuge for tropical biodiversity? Trend Ecol Evol 23: 261-267

Bigger M (1981) Observations on the insect fauna of shaded and unshaded Amelonado cocoa. Bull Entomol Res 71: 107-119

Bos MM, Steffan-Dwenter I, Tscharntke T (2007a) The contribution of cacao agroforests to the conservation of lower canopy ant and beetle diversity in Indonesia. Biodiver Conserv 16: 2429-2444

Bos MM, Höhn P, Shahabuddin S, Buchori D, Steffan-Dewenter I, Tscharntke T (2007b) Insect responses to forest conversion and agroforestry management In: Tscharntke T, Leuschner C, Guhardja E, Zeller M (eds) The stability of tropical rainforest margins: linking ecological, economic and social constraints of land-use and conservation, Springer, Berlin

Bos MM, Tylianakis JM, Steffan-Dewenter I, Tscharntke T (2008) The invasive Yellow Crazy Ant and the decline of forest ant diversity in Indonesian cacao agroforests. Biol Invas 10: 1399-1409

Brosi BJ, Armsworth PR, Daily GC (2008) Optimal design of agricultural landscapes for pollination services. Conserv Lett 1: 27-36

Brühl CA (2001) Leaf litter ant communities in tropical lowland rain forests in Sabah, Malaysia: effects of forest disturbance and fragmentation. Phd thesis. University of Würzburg

Brühl CA, Eltz T (2009) Fuelling the crisis: Species loss of ground-dwelling forest ants in oil palm plantations in Sabah , Malaysia (Borneo). Biodiversity & Conservation, online first DOI10.1007/s10531-009-9596-4.

Cannon CH, Summers M, Harting JR, Kessler PJA (2007) Developing conservation priorities based on forest type, condition, and threats in a poorly known ecoregion: Sulawesi, Indonesia. Biotropica 39: 747-759

Castelletta M, Thiollay JM, Sodhi NS (2005) The effects of extreme forest fragmentation on the bird community of Singapore. Biological Conservation 121: 135-155

Chazdon RL, Harvey C, Komar O, Griffith D, Ferguson B, Martinez-Ramos M, Morales H, Nigh R, Soto-Pinto L, van Breugel M, Philpott S (2008) Beyond reserves: A research agenda for conserving biodiversity in tropical human-modified landscapes. Biotropica (in press).

Chen KC, Tso IM (2004) Spider diversity on Orchid Island, Taiwan: A comparison between Habitats Receiving Different Degrees of Human Disturbance. Zool Stud 43: 598-611

Chiarello G (2000) Density and population size of mammals in remnants of Brazilian Atlantic forest. Conserv Biol 14: 1649-1657

Clough Y, Dwi Putra D, Pitopang R, Tscharntke T (2009) Local and landscape factors determine functional bird diversity in Indonesian cacao agroforestry. Biol Conserv

Coley PD, Aide TM (1991) Comparison of herbivory and plant defenses in temperate and tropical broad-leaved forests. In Price PW, Lewinsohn, TM, Fernandes GW, Benson WW (eds) Plant-animal interactions: Evolutionary ecology in tropical and temperate regions, Wiley & Sons, New York

Coley PD, Barone JA (1996) Herbivory and plant defenses in tropical forests. Annu Rev Ecol Syst 27: 305-335

Conway GR (1969) Ecological aspects of pest control in Malaysia. In: Farvar MT, Milton JP (eds) Careless Technology, Natural History Press, New York

Cottingham KL, Brown BL, Lennon JT (2001) Biodiversity may regulate the temporal variability of ecological systems. Ecol Lett 4: 72-85

Crone EE, Jones CG (1999) The dynamics of carbon-nutrient balance: effects of cottonwood acclimation to short-and long-term shade on beetle feeding preferences. J Chem Ecol 25: 635-656

Cronin JT (2003) Matrix heterogenity and host-parasitoid interactions in space. Ecology 84: 1506-1516

Daily GC, Soderqvist T, Aniyar S, Arrow K, Dasgupta P, Ehrlich PR, Folke C, Jansson A, Jansson BO, Kautsky N, Levin S, Lubchenco J, Maler KG, Simpson D, Starrett D, Tilman D, Walker B (2000) Ecology - the value of nature and the nature of value. Science 289: 395-396

Dammerman KW (1929) The agricultural zoology of the Malay Archipelago. De Bussy, Amsterdam

DeVries PJ (1988) Stratification of fruit-feeding nymphalid butterflies in a Costa Rican rainforest. J Res Lepidoptera 26: 98-108

Delabie JHC, Jahyny B, do Nascimento IC, Mariano CSF, Lacau S, Campiolo S, Philpott SM, Leponce M (2007) Contribution of cocoa plantations to the conservation of native ants (Insecta : Hymenoptera : Formicidae) with a special emphasis on the atlantic forest fauna of southern Bahia, Brazil. Biodivers Conserv 16: 2359-2384

Deutsch CA, Tewksbury JJ, Huey RB, Sheldon KS, Ghalambor CK, Haak DC, Martin PR (2008) Impacts of climate warming on terrestrial ectotherms across latitude. Proc Nat Acad Sci USA 105: 6668-6672

Direktorat Jenderal Perkebunan (2007) Agricultural statistics Direktorat Jenderal Perkebunan

Downum KR, Swain LA, Faleiro LJ (1991) Influence of light on plant allelochemicals - a synergistic defense in higher plants. Arch Insect Biochem Physiol 17: 201-211

Ernest SKM, Brown JH (2001) Delayed compensation for missing keystone species by colonization. Science 292: 101-104

Ernst R, Rödel MO (2005) Anthropogenically induced changes of predictability in tropical anuran assemblages. Ecology 86: 3111-3118

Estrada A, Coates-Estrada R, Meritt DJr., Montiel SS, Cuiel D (1993) Patterns of frugivore species richness and abundance in forest islands and in agricultural habitats at Los Tuxtlas, Mexico. Vegetatio 107: 245-257

Estrada A, Coates-Estrada R, Merritt DA (1994) Non-flying mammals and landscape changes in the tropical rain forest region of Los Tuxtlas, Mexico. Ecography 17: 229-241

Evans HC (1973) Invertebrate vectors of *Phytophthora palmivora* causing Black Pod Disease of Cocoa in Ghana. Ann Appl Biol 75: 331-345

Evelegh NC, Majer JD, Recher HF (2001) The effects of reducing bird predation on canopy arthropods of Marri *Eucalyptus calophylla* saplings on the Swan coastal plain, Western Australia. J R Soc West Aust 84: 12-21

Fagan WF, Hakim AL, Ariawan H, Yuliyantiningshi S (1998) Interactions between biological control efforts and insecticide applications in tropical rice agroecosystems: The potential role of intraguild predation. Biol Contr 13: 121-126

Faria D, Paciencia M, Dixo M, Laps R, Baumgarten J (2007) Ferns, frogs, lizards, birds and bats in forest fragments and shade cacao plantations in two contrastino landscapes in the Atlantic forest, Brazil. Biodivers Conserv 16: 2335-2357

Fermon H, Waltert M, Vane-Wright RI, Mühlenberg M (2005) Forest use and vertical stratification in fruit-feeding butterflies of Sulawesi, Indonesia: impacts for conservation. Biodivers Conserv 14: 333-350

Floren A, Deeleman-Reinhold C (2005) Diversity of arboreal spiders in primary and disturbed tropical forests. J Arachnol 33: 323-333

Floren A, Linsenmair KE (2005) The importance of primary tropical rain forest for species diversity, an investigation using arboreal ants as an example. Ecosystems 8: 559-567

Foley JA, Defries R, Asner GP, Barford C, Bonan G, Carpenter SR, Chapin FS, Coe MT, Daily GC, Gibbs HK, Helkowski JH, Holloway T, Howard EA, Kucharik CJ, Monfreda C, Patz JA, Prentice IC, Ramankutty N, Snyder PK (2005) Global consequences of land use. Science 309: 570-574

FWI/WRI/GFW (2002) The state of the forest: Indonesia. World Resources Institute, Washington

GAPKI, Gabungan Pengusaha Kelapa Sawit (2006) Palm oil statistics, Jakarta

Ganzhorn J (2003) Effects of introduced Rattus rattus on endemic small mammals in dry deciduous forest fragments of western Madagascar. Anim Conserv 6: 147-157

Gardner TA, Ribeiro-Junior MA, Barlow J, Cristina T, Avila-Pires S, Hoogmoed MS, Peres CA (2006) The value of primary, secondary, and plantation forests for a neotropical herpetofauna. Conserv Biol 21: 775-787

Gardner TA, Barlow J, Peres CA (2007) Paradox, presumption and pitfalls in conservation biology: The importance of habitat change for amphibians and reptiles. Biol Conserv 138: 166-179

Gardner TA, Barlow J, Araujo IS, Avila-Pires TC, Bonaldo AB, Costa JE, Esposito MC, Ferreira LV, Hawes J, Hernandez MIM, Hoogmoed MS, Leite RN, Lo-Man-Hung NF, Malcolm JR, Martins MB, Mestre LAM, Miranda-Santos R, Overal WL, Parry L, Peters SL, Ribeiro MA, da Silva MN F, Motta CDS, Peres CA (2008) The cost-effectiveness of biodiversity surveys in tropical forests. Ecol Lett 2: 139-150

Geist HJ, Lambin EF (2002) Proximate causes and underlying driving forces of tropical deforestation. Bioscience 52: 143-150

Gibbons JW, Scott DE, Ryan TJ, Buhlmann KA, Tuberville TD, Metts BS., Greene JL, Mills T, Leiden Y, Poppy S, Winne CT (2000) The global decline of reptiles, Deja Vu amphibians. Bioscience 50: 653-666

Gillespie G, Howard S, Lockie D, Scroggie M, Boeadi (2005) Herpetofaunal richness and community structure of offshore islands of Sulawesi, Indonesia. Biotropica 37: 279-290

Godfray HCJ (1994) Parasitoids: behavioral and evolutionary ecology. Princeton University Press, New Jersey

Gomes LGL, Oostra V, Nijman V, Cleef AM, Kappelle M (2008) Tolerance of frugivorous birds to habitat disturbance in a tropical cloud forest. Biol Conserv 141: 860-871

Gove AD (2007) Ant biodiversity and the predatory function (A response to Philpott and Armbrecht, 2006). Ecol Entomol 32: 435-436

Gradstein SR, Kessler M, Pitopang R (2007) Tree species diversity relative to human land uses in tropical rain forest margins in Central Sulawesi. In: Tscharntke T, Leuschner C, Zeller M, Guhardja E, Bidin A (eds) The stability of tropical rainforest margins, linking ecological, economic and social constraints of land use and conservation, Springer, Berlin

Greenberg R, Bichier P, Angon AC, Macvean C, Perez R, Cano E (2000) The impact of avian insectivory on arthropods and leaf damage in some guatemalan coffee plantations. Ecology 81: 1750-1755

Hartemink AE (2005) Nutrient stocks, nutrient cycling, and soil changes in Cocoa ecosystems: a Review. Adv Agron 86: 227-253

Harteveld M, Hertel D, Wiens M, Leuschner C (2007) Spatial and temporal variability of fine root abundance and growth in tropical moist forests and agroforestry systems (Sulawesi, Indonesia). Ecotropica 13: 111-120

Harvey CA, Gonzales JG, Somarriba E (2006) Dung beetle and terrestrial mammal diversity in forest, indigenous agroforestry systems and plantain monocultures in Talamanca, Costa Rica. Biodivers Conserv 15: 555-585

Hawkins BA (1994) Pattern and process in host-parasitoid interactions. Cambridge University Press, Cambridge

Hawkins BA, Cornell HV, Hochberg ME (1997) Predators, parasitoids, and pathogens as mortality agents in phytophagous insect populations. Ecology 78: 2145-2152

Henle K, Davies KF, Kleyer M, Margules C, Settele J (2004) Predictors of species sensitivity to fragmentation. Biodiv Cons 13: 207-251

Hertel D, Leuschner C, Harteveld M, Wiens M (2007) Fine root mass, distribution and regeneration in disturbed primary forests and secondary forests of the moist tropics. In: Tscharntke T, Leuschner Ch, Zeller M, Guhardja E, Bidin A (eds) The stability of tropical rainforest margins: linking ecological, economic and social constraints of land use and conservation. Springer, Berlin

Hill JK (1999) Butterfly spatial distribution and habitat requirements in a tropical forest: impacts of selective logging. J Appl Ecol 36: 564-572

Ho CT, Khoo KC (1997) Partners in biological control of cocoa pests: mutualism between *Dolichoderus thoracicus* (Hymenoptera: Formicidae) and *Cataenococcus hispidus* (Hemiptera: Pseudococcidae). Bull Entomol Res 87: 461-470

Hoehn P, Tscharntke T, Tylianakis JM, Steffan-Dewenter I (2008) Functional group diversity of bee pollinators increases crop yield. Proc R Soc London Ser B 275: 2283-2291

Hooks CR, Pandey RR, Johnson MW (2003) Impact of avian and arthropod predation on lepidopteran caterpillar densities and plant productivity in an ephemeral agroecosystem. Ecol Entomol 28: 522-532

Horgan FG (2005) Effects of deforestation on diversity, biomass and function of dung beetles on the eastern slopes of the Peruvian Andes. For Ecol Manag 216: 117-133

Hosang ML (2004) Interactions between natural enemies, herbivores and cacao in Palolo Valley, Central Sulawesi. Int Cacao Organ 10.01.2009.

IUCN (2008) Red List of threatened species. (http: //www.iucnredlist.org). accessed 14 December 2008

Jactel H, Brockerhoff EG (2007) Tree diversity reduces herbivory by forest insects. Ecol Lett 10: 835-848

Jocqué R, Alderweireldt M (2005) Lycosidae: the grassland spiders. Acta Zool Bulgarica S1: 125-130

Kaimowitz D, Sheil D (2007) Conserving what and for whom? Why conservation should help meet basic human needs in the tropics. Biotropica 39: 567-574

Kalka MB, Smith AR, Kalko EKV (2008) Bats limit arthropods and herbivory in a tropical forest. Science 320: 71

Kalshoven, L.G.E. (1950) De plagen van de cultuurgewassen in Indonesi. Van Hoeve, Den Haag

Kehlenbeck, K (2007) Rural homegardens in Central Sulawesi, Indonesia: an example for a sustainable agro-ecosystem? Phd thesis, Faculty for Agricultural Sciences, University of Göttingen

Kellermann JL, Johnson MD, Stercho AM, Hackett SC (2008) Ecological and economic services provided by birds on Jamaican blue mountain coffee farms. Conserv Biol 22: 1177-1185

Kessler M, Kessler PJA, Gradstein SR, Bach K, Schmull M, Pitopang R (2005) Tree diversity in primary forest and different land use systems in Central Sulawesi, Indonesia. Biodivers Conserv 14: 547-560

Kessler M, Abrahamczyk S, Bos M, Buchori D, Putra DD, Gradstein SR, Höhn P, Kluge J, Orend F, Pitopang R, Shahabuddin S, Schulze CH, Sporn SG, Steffan-Dewenter I, Tjitrosoedirdjo SS, Tscharntke T (2009, accepted) Alpha- versus beta-diversity of plants and animals along a tropical land-use gradient. Int J Ecol Appl

Khoo KC, Ho CT (1992) The influence of *Dolichoderus thoracicus* (Hymenoptera, Formicidae) on losses due to *Helopeltis theivora* (Heteroptera, Miridae), Black Pod Disease, and mammalian pests in cocoa in Malaysia. Bull Entomol Res 82: 485-491

Klein AM, Steffan-Dewenter I, Tscharntke T (2002a). Predator-prey ratios on cocoa along a land-use gradient in Indonesia. Biodiv Conserv 11: 683-693

Klein AM, Steffan-Dewenter I, Buchori D, Tscharntke T (2002b) Effects of land-use intensity in tropical agroforestry systems on coffee flower-visiting and trap-nesting bees and wasps. Conserv Biol 16: 1003-1014

Klein A-M, Steffan-Dewenter I, Tscharntke T (2003a) Fruit set of highland coffee increases with the diversity of pollinating bees. Proc R Soc London Ser B 270: 955-961

Klein A-M, Steffan-Dewenter I, Tscharntke T (2003b) Pollination of *Coffea canephora* in relation to local and regional agroforestry management. J Appl Ecol 40: 837-845

Klein A-M, Steffan-Dewenter I, Tscharntke T (2003c) Bee pollination and fruit set of *Coffea arabica* and *C canephora* (Rubiaceae) Am J Bot 90: 153-157

Klein A-M, Steffan-Dwenter I, Tscharntke T (2006) Rain forest promotes trophic interactions and diversity of trap-nesting Hymenoptera in adjacent agroforestry. J Anim Ecol 75: 315-323

Klein A-M, Vaissire BE, Cane JH, Steffan-Dewenter I, Cunningham SA, Kremen C, Tscharntke T (2007) Importance of pollinators in changing landscapes for world crops. Proc R Soc London Ser B 274: 303-313

Klein A-M, Cunningham SA, Bos M, Steffan-Dewenter I (2008) Advances in pollination ecology from tropical plantation crops. Ecology 89: 935-943

Klein AM, Mueller CM, Hoehn P, Kremen C (2009, in press) Understanding the role of species richness for pollination services. In: Bunker D, Hector A, Loreau M, Perrings C, Naeem S (eds) The consequences of changing biodiversity - solutions and scenarios, Oxford University Press, Oxford

Klein AM (in press) Nearby rainforest promotes coffee pollination by increasing spatiotemporal stability in bee species richness. For Ecol Manag.

Koh LP (2007) Impending disaster or sliver of hope for Southeast Asian forests? The devil may lie in the details. Biodiv Conserv 16: 3935-3938

Koh LP (2008) Birds defend oil palms from herbivorous insects. Ecological Applications 18: 821-825

Kremen C, Williams NM, Thorp RW (2002) Crop pollination from native bees at risk from agricultural intensification. PNAS 99: 16812-16816

Kruess A, Tscharntke T (1994) Habitat fragmentation, species loss, and biological control. Science 264: 1581-1584

Kruess A, Tscharntke T (2000) Species richness and parasitism in a fragmented landscape: experiments and field studies with insects on *Vicia sepium*. Oecologia 122: 129-137

Laidlaw RK (2000) Effects of habitat disturbance and protected areas on mammals of peninsular Malaysia. Conserv Biol 14: 1639-1648

Langer A, Hance T (2004) Enhancing parasitism of wheat aphids through apparent competition: a tool for biological control. Agric Ecosys Environ 102: 205-212

Larsen TH, Forsyth A (2005) Trap spacing and transect design for dung beetle biodiversity studies. Biotropica 37: 322-325

LaSalle J, Gauld ID (1993) Hymenoptera: Their diversity, and their impact on the diversity of other organisms. In: LaSalle J, Gauld ID (eds) Hymenoptera and Biodiversity, CAB International, London

Leuschner C, Wiens M, Harteveld M, Hertel D, Tjitrosemito S (2006) Patterns of fine root mass and distribution along a disturbance gradient in a tropical montane forest, Central Sulawesi (Indonesia). Plant Soil 283: 163-174

Lewinsohn TM, Roslin T (2008) Four ways towards tropical herbivore megadiversity. Ecol Lett 11: 398-416

Lo-Man-Hung NF, Gardner TA, Ribeiro MA Jr, Barlow J, Bonaldo AB (2008) The value of primary, secondary and plantation forests for Neotropical epigeic arachnids. J Arachnol 36: 394-401

Lopes MA, Ferrari SF (2000) Effects of human colonization on the abundance and diversity of mammals in eastern Brazilian Amazonia. Conserv Biol 14: 1658-1665

Loyn RH, Runnalls RG, Forward GY, Tyers J (1983) Territorial bell miners and other birds affecting populations of insect prey. Science 221: 1411-1413

Lozada T, De Koning GHJ, Marche R, Klein AM, Tscharntke T (2007) Tree recovery and seed dispersal by birds: comparing forest, agroforestry and abandoned agroforestry in coastal ecuador. Perspect Plant Ecol Evol Syst 8: 131-140

Luja VH, Herrando-Perez S, Gonzalez-Solis D, Luiselli L (2008) Secondary rain forests are not heavens for reptile species in tropical Mexico. Biotropica 40: 747-757

MacArthur RH, Wilson EO (1967) The Theory of Island Biogeography. Princeton University Press, New Jersey

Maiorana VC (1981) Herbivory in sun and shade. Biol J Linn Soc 15: 151-156

Maloney D, Drummond FA, Alford R (2003) Spider Predation in Agroecosystems: Can spiders effectively control pest populations? MAFES Techn Bull 190: 1070-1524

Mansour F, Ross JW, Edwards GB, Whitcomb WH, Richman DB (1982) Spiders of Florida citrus groves. Fla Entomol 65: 514-522

Mansour F, Richman DB, Whitcomb WH (1983) Spider management in agroecosystems: habitat manipulation. Environ Manage 7 : 43-49

Marc P, Canard A (1997) Maintaining spider biodiversity in agroecosystems as a tool in pest control. Agric Ecosyst Environ 62: 229-235

Marc P, Canard A, Ysnel F (1999) Spiders (Araneae) useful for pest limitation and bioindication. Agr Ecosyst Environ 74: 229-273

Marchiori CH, Lussari MA, Rosa DC, Penteado-Dias (2007) Parasitoid Hymenoptera collected during the diurnal and nocturnal periods in Itumbiara, Gois. Braz J Biol 67: 581-582

Marino PC, Landis DA (2000) Parasitoid community structure: implications for biological control in agricultural landscapes.In: Ekbom B (ed.) Interchanges of insects between agricultural and surrounding habitats. Kluwer Academic Publishers, Dordrecht, The Netherlands

Marino PC, Landis DA, Hawkins BA (2006) Conserving parasitoid assemblages of North American pest Lepidoptera: Does biological control by native parasitoids depend on landscape complexity?. Biol Control 37: 173-185

Marquis RJ, Whelan CJ (1994) Insectivorous birds increase growth of white oak through consumption of leaf-chewing insects. Ecology 75: 2007-2014

Mas AH, Dietsch TV (2003) An index of management intensity for coffee agroecosystems to evaluate butterfly species richness. Ecol Appl 13: 1491-1501

Mas AH, Dietsch TV (2004) Linking shade coffee certification to biodiversity conservation: butterflies and birds in Chiapas, Mexico. Ecol Appl 14: 642-654

Mason WRM, Huber JT (1993) Order Hymenoptera. In: Gouled H, Huber JT (eds) Hymenoptera of the world: an identification guide to families, Minister of Supply and Services, Ontario

Matson PA, Parton WJ, Power AG, Swift MJ (1997) Agricultural intensification and ecosystem properties. Science 277: 504-509

Matteson PC (2000) Insect pest management in tropical Asian irrigated rice. Ann Rev Entomol 45: 549-574

Mattson WJ (1980) Herbivory in relation to plant nitrogen-content. Ann Rev Ecol Syst 11: 119-161

McGregor AJ, Moxon JE (1985) Potential for biological - control of tent building species of ants associated with *Phytophthora palmivora* pod rot of Cocoa in Papua-New-Guinea. Ann Appl Biol 107: 271-277

Mendoza CAO, Rios AM, Cano EB, Cortes JLN (2005) Dung beetle community (Coleoptera: Scarabaeidae: Scarabaeinae) in a tropical landscape at the Lachua Region, Guatemala. Biodivers Conserv 14: 801-822

Mitchell PL (2004) Heteroptera as vectors of plant pathogens. Neotrop Entomol 33: 519-545

Mogea J (2002) Preliminary study on the palm flora of the Lore Lindu National Park, Central Sulawesi, Indonesia. Biotropia 18: 1-20

Mole S, Ross JAM, Waterman PG (1988) Light-induced variation in phenolic levels in foliage of rain-forest plants - 1. chemical changes. J Chem Ecol 14: 1-21

Mols CM, Visser ME (2002) Great tits can reduce caterpillar damage in apple orchards. J Appl Ecol 39: 888-899

Murphy F, Murphy J (2000) An introduction to the spiders of south east Asia. Malaysian Nature Society. Selangor Press Sdn Bhd, Kuala Lumpur

Musser GG (1969a) Notes on the taxonomic status of *Rattus aspinatus* Tate and Archbold and *Mus callitrichus* Jentik. Results of the Archbold Expeditions No.89. Am Mus Novit 2365: 1-9

Musser GG (1969b) A new genus and species of murid rodents from Celebes, with a discussion of its relationships. Results of the Archbold Expeditions No.91. Am Mus Novit 2384: 1-44

Musser GG (1969c) Taxonomic notes on *Rattus dollmanni* and *Rattus hellwaldii* (Rodentia, Muridae) of Celebes. Results of the Archbold Expeditions No.92. Am Mus Novit 2386: 1-24

Musser GG (1971) Taxonomic association of *Mus feberi jentik* and *Rattus xanthurus* (Gray), a species only known from Celebes. Zoologische Mededelingen Leiden 45(9): 107-118

Musser GG (1991) Sulawesian rodents: description of new species of *Bunomys* and *Maxomys* (Muridae, Murinae) Am Mus Novit 3001: 1-41

Musser GG, Dagosto M (1987) The identity of *Tarsius pumilus*, a pigmy speciesendemic to montane mossy forest of Central Sulawesi. Am Mus Novit 2867: 1-53

Myers N, Mittermeier RA, Mittermeier CG, Da Fonseca GAB, Kent J (2000) Biodiversity hotspots for conservation priorities. Nature 403: 853-858

Naumann ID (1991) Order Hymenoptera in insect of australia. Melbourne University Press, Melbourne

Navarrete D, Halffter G (2008) Dung beetle (Coleoptera: Scarabaeidae: Scarabaeinae) diversity in continuous forest, forest fragments and cattle pastures in a landscape of Chiapas, Mexico: the effects of anthropogenic changes. Biodivers Conserv 17: 2869-2898

Nichols E, Larsen TH, Spector S, Davis A, Vulinec K ,Escobar F (2007) Global dung beetle response to tropical forest modification and fragmentation: a quantitative literature review and meta-analysis. Biol Conserv137: 1-19

Nowak MN (1991) Walker`s mammals of the world. The Johns Hopkins University Press, London

Noyes JS, Hayat M (1984) A review of the genera of Indo-Pacific Encyrtidae (Hymenoptera: Chalcidoidea). Bull Br Mus Nat His (Entomol) 48: 131-395

Noyes JS (1989) A study of five methods of sampling Hymenoptera (Insecta) in a tropical rainforest, with special reference to the parasitica. J Nat His 23: 285-298

Ochoa J (2000): Effects of logging on small-mammal diversity in the lowland forests of the Venezuelan Guyana region. Biotropica 32: 146-164

Okuma C, Wongsiri T (1973) Second report on the spider fauna of the paddy fields in Thailand. Mushi 47: 1-17

Olschewski R, Tscharntke T, Bentez PC, Schwarze S, Klein A-M (2006) Economic evaluation of pollination services comparing coffee landscapes in Ecuador and Indonesia. Ecol Soc. http: //www.ecologyandsociety.org/vol11 /iss1/art7/ Accessed 15.01.2009

Olschewski R, Tscharntke T, Bentez PC, Schwarze S, Klein A-M (2007) Economic evaluation of ecosystem services as a basis for stabilizing rainforest margins? The example of pollination services and pest management in coffee landscapes. In: Tscharntke T, Leuschner C, Zeller M, Guhardja E, Bidin A (eds) The stability of tropical rainforest margins, linking ecological, economic and social constraints of land use and conservation, Springer, Berlin

Omaliko CPE (1984) Dung decomposition and its effects on the soil component of a tropical grassland ecosystem. Trop Ecol 25: 214-220

Owens IPF, Bennett PM (2000) Ecological basis of extinction risk in birds: habitat loss versus human persecution and introduced predators. Proc Nat Acad Sci USA 97: 12144-12148

Pangau-Adam MZ (2003) Avian nest predation in forest margin areas in Lore Lindu National Park, Central Sulawesi, Phd-thesis, University of Göttingen.

Pérez-de La Cruz M, Sanchez-Soto S, Ortiz-Garcia CF, Zapata-Mata R, Cruz-Perez A (2007) Diversidad de insectos capturados por araas tejedoras (Arachnida: Araneae) en el agroecosistema cacao en Tabasco, México. Neotrop Entomol 36: 90-101

Perfecto I, Vandermeer J (2008) Biodiversity conservation in tropical agroecosystems - a new conservation paradigm. Ann New York Acad Sci 1134: 173-200

Perfecto I, Mas A, Dietsch T, Vandermeer J (2003) Conservation of biodiversity in coffee agroecosystems: a tri-taxa comparison in southern Mexico. Biodivers Conserv 12: 1239-1252

Perfecto I, Vandermeer JH, Bautista GL, Nunez GI, Greenberg R, Bichier P, Langridge S (2004) Greater predation in shaded coffee farms: the role of resident neotropical birds. Ecology 85: 2677-2681

Perfecto I, Armbrecht I, Philpott SM, Soto-Pinto L, Dietsch TV (2007) Shaded coffee and the stability of rainforest margins in northern Latin America. In: Gerold G, Fremerey M, Guhardja E (eds) Land use, nature conservation and the stability of rainforest margins in South Asia, Springer, Berlin

Pfeiffer M, Ho CT, Teh CL (2008) Exploring arboreal ant community composition and co-occurrence patterns in plantations of oil palm Elaeis guineensis in Borneo and Peninsular Malaysia. Ecography 31: 21-32.

Philpott SM, Armbrecht I (2006) Biodiversity in tropical agroforests and the ecological role of ants and ant diversity in predatory function. Ecol Entomol 31: 369-377

Pineda E, Moreno C, Escobar F, Halffter G (2005) Frog, bat and dung beetle diversity in the cloud forest and coffee agroecosystems of Veracruz, Mexico. Conserv Biol 19: 400-410

Pitopang R, Keler PJA, Gradstein SR, Guhardja E, Tjitrosoedirdjo SS, Wiriadinata H (2004) Tree composition in secondary forest of Lore Lindu National Park, Central Sulawesi, Indonesia. In: Gerold G, Fremerey M,

Guhardja E (eds) Land use, nature conservation and the stability of rainforest margins in South Asia, Springer, Berlin

Pitopang R, Tjitrosoedirdjo SS, Setiadi D (2008) Structure and composition of understory plant assemblages of six land use types in the Lore Lindu National Park, Central Sulawesi, Indonesia. Bangladesh J Bot 15: 1-12

Poulsen AD, Pendry CA (1995) Inventories of ground herbs at three altitudes on Bukit Belalong, Brunei, Borneo. Biodivers Conserv 4: 745-757

Putz FE (1984) The natural history of lianas on Barro Colorado Island, Panama. Ecology 65: 1713-1724

Quicke DLJ (1997) Parasitic Wasps. Chapman and Hall, London

Reitsma R, Parrish JD, McLarney W (2001) The role of cacao plantations in maintaining forest avian diversity in southeastern Costa Rica. Agrofor Syst 53: 185-193

Rodrigues ASL, Brooks TM (2007) Shortcuts for biodiversity conservation planning: the effectiveness of surrogates. Ann Rev Ecol Evol Syst 38: 713-737

Rogo L, Odulaja A (2001) Butterfly populations in two forest fragments at the Kenya coast. Afr J Ecol 39: 266-275

Root RB (1973) Organization of a plant-arthropod association in simple and diverse habitats: The fauna of collards (*Brassica oleracea*). Ecol Monogr 43: 95-124

Rice RA, Greenberg R (2000) Cacao cultivation and the conservation of biological diversity. Ambio 29: 167-173

Ricketts TH, Daily GC, Ehrlich PR, Michener CD (2004) Economic value of tropical forest to coffee production. Proc Nat Acad Sci USA 101: 12579-12582

Ricketts TH, Regetz J, Steffan-Dewenter I, Cunningham SA, Kremen C, Bogdanski A, Gemmill-Herren B, Greenleaf SS, Klein AM, Mayfield MM, Morandin LA, Ochieng A, Viana BF (2008) Landscape effects on crop pollination services: are there general patterns?. Ecol Lett 11: 499-515

Riihimaki J, Kaitaniemi P, Koricheva J, Vehvilainen H (2005) Testing the enemies hypothesis in forest stands: the important role of tree species composition. Oecologia 142: 90-97

Riley EP (2007) The human-macaque interface: conservation implications of current and future overlap and conflict in Lore Lindu National Park, Sulawesi, Indonesia. American Anthropologist 109: 473-484

Ruf F, Schroth G (2004) Chocolate forests and monocultures: a historical review of Cocoa growing and its conflicting role in tropical deforestation and forest conservation. In: Schroth G, Da Fonseca GAB, Harvey CA, Gascon C, Lasconcelos HL, Izac AN (eds) Agroforestry and Biodiveristy Conservation in Tropical Landscapes, Island Press, Washington

Russel-Smith A, Stork NE (1994) Abundance and diversity of spiders from the canopy of tropical rainforests with particular reference to Sulawesi, Indonesia. J Trop Ecol 10 : 545-558

Saaksjarvi IL, Ruokolainen K, Tuomisto H, Haataja S, Fine PVA, Cardenas G, Mesones I, Vargas V (2006) Comparing composition and diversity of parasitoid wasps and plants in an Amazonian rain-forest mosaic. J Trop Ecol 23: 167-176

Sahari B (2004) Effect of isolation on temporal dynamic of insect community structure and parasitic hymenoptera diversity in cacao agroforestry systems at the margin of Lore Lindu National Park, Central Sulawesi. PhD-thesis, Bogor Agricultural University

Sala OE, Chapin FS, Armesto JJ, Berlow E, Bloomfield J, Dirzo R, Huber-Sanwald E, Huenneke LF, Jackson RB, Kinzig A, Leemans R, Lodge DM, Mooney HA, Oesterheld M, Poff NL, Sykes MT, Walker BH, Walker M, Wall DH (2000) Biodiversity - global biodiversity scenarios for the year 2100. Science 287: 1770-1774

Samways ML (1994) Insect conservation biology. Chapman & Hall, London

Schoonhoven LM, Jermy T, van Loon JJA (1998) Insect-plant biology from physiology to evolution, 15th edn. Chapmann & Hall, London

Schroth G, Harvey C ,Vincent G (2004) Complex agroforests: their structure, diversity, and potential role in landscape conservation. In: Schroth G, Fonseca G, Harvey C, Gascon C, Vasconcelos H, Izac A. (eds) Agroforestry and conservation of biodiversity in tropical landscapes, Island Press, Washington

Schulze CH, Steffan-Dewenter I, Tscharntke T (2004a) Effects of land use on butterfly communities at the rain forest margin: a case study from Central Sulawesi. In: Gerold G, Fremerey M, Guhardja E (eds) Land use, nature conservation and the stability of rainforest margins in Southeast Asia, Springer, Berlin

Schulze CH, Waltert M, Kessler PJA, Pitopang R, Shahabuddin, Veddeler D, Steffan-Dewenter I, Mühlenberg M, Gradstein SR, Tscharntke T (2004b): Biodiversity indicator groups of tropical land-use systems: comparing plants, birds, and insects. Ecol Appl 14: 1321-1333

See YA, Khoo KC (1996) Influence of *Dolichoderus thoracicus* (Hymenoptera: Formicidae) on cocoa pod damage by *Conopomorpha cramerella* (Lepidoptera: Gracillariidae) in Malaysia. Bull Entomol Res 86: 467-474

Sekercioglu CH (2006) Increasing awareness of avian ecological function. Trend Ecol Evol 21: 464-471

Settle WH, Ariawan H, Astuti ET, Cahyana W, Hakim AL, Hindayana D, Lestari AS, Pajarningsih, Sartanto (1996) Managing tropical rice pests through conservation of generalist natural enemies and alternative. Prey Ecol 77: 1975-1988

Shahabuddin S, Schulze CH, Tscharntke T (2005) Changes of dung beetle communities from rainforests towards agroforestry systems and annual cultures. Biodivers Conserv 14: 863-877

Shahabuddin S (2007) Effect of land use on dung beetles (Coleoptera: Scarabaeidae) diversity and dung decomposition in Central Sulawesi, Indonesia. PhD-thesis, Bogor Agricultural University

Shahabuddin S, Tscharntke T, Schulze CH (2008) Biodiversity and body size of dung beetles colonizing introduced vs endemic dung along a tropical land-use gradient. J Trop Ecol (Accepted)

Shahabuddin S, Hidayat P, Manuwoto S, Tscharntke S, Schulze CH (In Prep) Ecosystem function of coprophagous beetles: effect of land-use, body size, species diversity and biomass

Shaw MR, Hochberg ME (2001) The neglect of parasitic Hymenoptera in insect conservation strategies: The British fauna as a prime example. J Insect Conserv 5: 253-263

Siebert SF (2000) Survival and growth of rattan intercropped with coffee and cacao in the agroforests of Indonesia. Agrofor Syst 50: 95-102

Siebert SF (2002) From shade- to sun-grown perennial crops in Sulawesi, Indonesia: implications for biodiversity conservation and soil fertility. Biodivers Conserv 11: 1889-1902

Siebert SF (2002) Demographic effects of collecting rattan cane and their implications for sustainable harvesting. Conserv Biol 18: 424-431

Siebert SF (2005) The abundance and distribution of rattan over an elevational gradient in Sulawesi, Indonesia. For Ecol Manag 210: 143-158

Simanjuntak H (2001) Musuh alami hama dan penyakit tanaman Kakao. Proyek Pengendalian Hama Terpadu Perkebunan Rakyat. Departemen Pertanian, Jakarta

Slade EM, Mann DJ, Villanueva JF, Lewis OT (2007) Experimental evidence for the effects of dung beetle functional group richness and composition on ecosystem function in a tropical forest. J Anim Ecol 76: 1094-1104

Smith RF (1970) The vegetation structure of Puerto Rico rain forest before and after short-term gamma irradiation. In: Odum H (ed) A tropical rain forest, U.S. Atomic Energy Commission, Oak Ridge

Snyder WE, Ives AR (2003) Interaction between specialist and generalist natural enemies: parasitoids, predators, and pea aphid biocontrol. Ecology 84: 91-107

Sodhi NS, Koh LP, Brook BW, Ng PKL (2004) Southeast Asian biodiversity: an impending disaster. Trends Ecol Evol 19: 654-660

Sodhi N, Pin Koh L, Prawiradilaga D, Tinulele I, Dwi Putra D, Han Tong Tan T (2005) Land use and conservation value for forest birds in Central Sulawesi. Biol Conserv 122: 547-558

Sodhi NS, Brooks TM, Koh LP, Acciaioli G, Erb M, Tan AKJ, Curran LM, Brosius P, Lee TM, Patlis JM, Gumal M, Lee RJ (2006) Biodiversity and human livelihood crises in the Malay Archipelago. Conserv Biol 20: 1811-1813

Sodhi NS, Bickford D, Diesmos AC, Lee TM, Koh LP, Brook BW, Sekercioglu CH, Bradshaw CJA (2008) Measuring the meltdown: drivers of global amphibian extinction and decline. PLoS ONE 3: e1636

Soerjani M, Korstermans AJGH, Tjitrosoepomo G (1987) Weeds of rice in Indonesia. Balai Pustaka, Jakarta

Sperber CF, Nakayama K, Valverde MJ, Neves FS (2004) Tree species richness and density affect parasitoid diversity in cacao agroforestry. Basic Appl Ecol 5: 241-251

Sporn SG (2008) Epiphytic bryophytes in natural forests and cacao agroforests of Central Sulawesi, Indonesia. Phd thesis, University of Göttingen

Sporn SG, Bos M, Gradstein SR (2007) Is productivity of cacao impeded by epiphytes? An experimental approach. Agric Ecosys Environ 122: 490-493

Stattersfield A, Crosby N, Long A, Wege D (1998) Endemic bird areas of the world. Priority areas for biodiversity conservation. Birdlife International, Cambridge, UK

Steffan-Dewenter I, Tscharntke T (1999) Effects of habitat isolation on polinator communities and seed set. Oecologia 121: 432-440

Steffan-Dewenter I, Kessler M, Barkmann J, Bos M, Buchori D, Erasmi S, Faust H, Gerold G, Glenk K, Gradstein R, Guhardja E, Harteveld M, Hertel D, Höhn P, Kappas M, Köhler S, Leuschner C, Maertens M, Marggraf R, Migge-Kleian S, Mogea J, Pitopang R, Schaefer M, Schwarze S, Sporn S, Steingrebe A, Tijtrosoedirdjo S, Tijtrosoemito S, Twele A, Weber R, Woltmann L, Zeller M, Tscharntke T (2007) Tradeoffs between income, biodiversity, and ecosystem functioning during tropical rainforest conversion and agroforestry intensification. Proc Nat Acad Sci USA 104: 4973-4978

Stratford JA, Robinson DW (2005) Gulliver travels to the fragmented tropics: geographic variation in mechanisms of avian extinction. Front Ecol Environ 3: 91-98

Strong D (1974) Rapid asymptotic species accumulation in phytophagous insect communities: the pests of cacao. Science 185: 1064-1066

Stuart SN, Chanson JS, Cox NA, Young BE, Rodrigues ASL, Fischman DL, Waller RW (2004) Status and trends of amphibian declines and extinctions worldwide. Science 306: 1783-1786

Suaz-Ortuno I, Alvarado-Diaz J, Martinez-Ramos M (2008) Effects of conversion of dry tropical forest to agricultural mosaic on herpetofaunal assemblages. Conserv Biol 22: 362-374

Tahvanainen JO, Root RB (1972) The influence of vegetational diversity on the population ecology of a specialized herbivore, *Phyllotreta cruciferae* (Coleoptera: Chrysomelidae). Oecologia 10: 321-346

Taylor B, Griffin MJ (1981) The role and relative importance of different ant species in the dissemination of Black Pod Disease of cocoa. In: Gregory PH, Maddison AC (eds) Epidemiology of *Phytophthora* on Cocoa in Nigeria. Commonwealth Mycological Institute, Kew, Surrey

Thies C, Steffan-Dewenter I, Tscharntke T (2003) Effects of landscape context on herbivory and parasitism at different spatial scales. Oikos 101: 18-25

Tjitrosoedirdjo SS, Mahfudz S, Tjitrosemito S, Chozin MA (2004) Study of weed dynamic in some cropping systems in forest margins of Lore Lindu National Park. Jurnal Gulma Tropika 2: 21-26

Tooker JF, Hanks LM (2000a) Flowering plant hosts of adult hymenopteran parasitoids of Central Illinois. Ann Entomol Soc Am 93: 580-588

Tooker JF, Hanks LM (2000b) Influence of plant community structure on natural enemies of Pine Needle Scale (Homoptera: Diaspididae) in urban landscapes. Environ Entomol 29: 1305-1311

Tsai ZI, Hunag PS, Tso IM (2006) Habitat management by aboriginals promotes high spieder diversity on Asian tropical island. Ecography 29: 84-94

Tscharntke T, Klein AM, Kruess A, Steffan-Dewenter I, Thies C (2005) Landscape perspectives on agricultural intensification and biodiversity - ecosystem service management. Ecol Lett 8: 857-874

Tscharntke T, Bommarco R, Clough Y, Crist TO, Kleijn D, Rand TA, Tylianakis JM, van Nouhuys S, Vidal S (2007) Conservation biological control and enemy diversity on a landscape scale. Biol Control 43: 294-309

Tuomisto H, Ruokolainen K, Aguilar M, Sarmiento A (2003) Floristic patterns along a 43-km long transect in an Amazonian rain forest. J Ecology 91: 743-756.

Tylianakis JM, Rand TA, Kahmen A, Klein A-M, Buchmann N, Perner J, Tscharntke T (2008) Resource heterogeneity moderates the biodiversity-function relationship in real world ecosystems. PLoS Biol 6: 947-956

Van Bael SA, Brawn JD, Robinson SK (2003) Birds defend trees from herbivores in a neotropical forest canopy. Proc Nat Acad Sci USA 100: 8304-8307

Van Bael SA, Bichier P, Ochoa I, Greenberg R (2007) Bird diversity in cacao farms and fragments of western Panama. Biodivers Conserv 16: 2245-2256

Van Bael SA, Bichier P, Greenberg R (2007) Bird predation on insects reduces damage to the foliage of cocoa trees (*Theobroma cacao*) in western Panama. J Trop Ecol 23: 715-719

Van Bael SA, Philpott SM, Greenberg R, Bichier P, Barbier NA, Mooney KA, Gruner DS (2008) Birds as predators in tropical agroforestry systems. Ecology 89: 928-934

Vane-Wright RI, de Jong R (2003) The butterflies of Sulawesi: an annoted checklist for a critical island fauna. Zool Verh Leiden 343: 3-267.

Van Mele P, Cuc NTT (2001) Farmers' perceptions and practices in use of *Dolichoderus thoracicus* (Smith) (Hymenoptera: Formicidae) for biological control of pests of *Sapodilla*. Biol Control 20: 23-29

Veddeler D, Schulze CH, Steffan-Dewenter I, Buchori D, Tscharntke T (2005) The contribution of tropical secondary forest fragments to the conservation of fruit-feeding butterflies: effects of isolation and age. Biodivers Conserv 14: 3577-3592

Wake DB, Vredenburg VT (2008) Are we in the midst of the sixth mass extinction? A view from the world of amphibians. Proc Nat Acad Sci USA 105: 11466-11473

Wallner WE (1987) Factors affecting insect population dynamics differences between outbreak and non-outbreak species. Ann Rev Entomol 32: 317-340

Waltert M, Mardiastuti A, Muhlenberg M (2004) Effects of land use on bird species richness in Sulawesi, Indonesia. Conserv Biol 18: 1339-1346

Wanger TC, Saro A, Iskandar D, Brook B, Sodhi NS, Clough Y, Tscharntke T (2009) Conservation value of cacao agroforestry for amphibians and reptiles in Southeast Asia: combining correlative models with follow-up field experiments. Journal of Applied Ecology 46: 823-832

Wanger TC, Iskandar D, Motzke I, Brook BW, Sodhi NS, Clough Y, Tscharntke T (submitted) Land-use changes affect community composition and traits of tropical amphibians and reptiles in Sulawesi (Indonesia).

Ward MJ, Paton DC (2007) Predicting mistletoe seed shadow and patterns of seed rain from movements of the mistletoebird, *Dicaeum hirundinaceum*. Austral Ecol 32: 13-121

Weist M (2008) Diversity of small mammals in different types of cacao agroforest systems in Central Sulawesi, Indonesia. Diploma thesis, University of Göttingen

Wells K, Kalko EKV, Lakim MB, Pfeiffer M (2007) Effects of rain forest logging on species richness and assemblage composition of small mammals in Southeast Asia. J Biogeogr 34: 1087-1099

Whitcomb WH, Exline H, Hite M (1963) Comparison of spider populations of ground stratum in Arkansas pasture and adjacent cultivated field. Arakansas Acad Sci Proc 17: 32-37

Whitfield JB (1998) Phylogeny and evolution of host-parasitoid interactions in Hymenoptera. Ann Rev Entomol 43: 129-151

Whitten AJ, Muslimin M, Henderson GS (2002) The Ecology of Sulawesi, 15th edn. Periplus, Jakarta

Wielgoss AC, Tscharntke T, Buchori D, Fiala B, Clough Y (submitted) Drivers of ant diversity losses in Indonesian cacao agroforestry: shade reduction, enhanced temperature and the dominance of an invasive dolichoderine ant species.

Wood BJ (2002) Pest control in Malaysia's perennial crops: A half century perspective tracking the pathway to integrated pest management. Integr Pest Manag Rev 7: 173-190.

Zhang W, Ricketts TH, Kremen C, Carney K, Swinton SM (2007) Ecosystem services and dis-services to agriculture. Ecol Econ 64: 253-260

The potential of land-use systems for maintaining tropical forest butterfly diversity

Christian H. Schulze*, Stefan Schneeweihs, and Konrad Fiedler

Department of Population Ecology, Faculty of Life Sciences, University of Vienna, Rennweg 14, A-1030 Vienna, Austria

*corresponding author: C. H. Schulze, email: christian.schulze@univie.ac.at

Summary

Globally, tropical rainforest species are under increasing threat due to ongoing deforestation. We summarize published studies on effects of forest disturbance and conversion on tropical butterfly assemblages with a focus on forest species. Additionally, we use existing faunal monographs to extract information on the habitat affiliations of tropical forest butterflies. Based on these data, we quantify the importance of human-modified habitats for the conservation of tropical butterflies inhabiting forests across all major tropical regions. Although studies on effects of land use and forest conversion are biased by several methodological problems, the emerging pattern clearly indicates that (old-grown) secondary forests represent an important habitat type for maintaining a large proportion of forest butterflies in all major tropical regions. In contrast, agroforestry systems are only capable of protecting a small proportion of mostly widespread forest species tolerant against strong disturbance. Extinction risk of forest butterflies depends on the utilization of certain plant growth types, while host plant specialization alone does not have significant effects. Specifically, butterflies using trees as larval host plants are more likely to be absent in logged forest, whereas species that tend to be more abundant in logged forest are those with larvae feeding on lianas and grasses. Greater resource requirements may be one reason for a higher extinction risk of larger butterflies after forest fragmentation although not all studies found a relationship between extinction risk and body size. Butterfly species with narrow geographic ranges are particularly prone to local and global extinction. While widespread species dominate species assemblages of highly modified land-use systems and agroforests, endemic butterflies are more commonly found in old-growth forests. Our analyses covering all four major tropical regions underline the severe effects of forest modification. Land-use systems are

T. Tscharntke et al. (eds.), *Tropical Rainforests and Agroforests under Global Change*,
Environmental Science and Engineering, 73–96, DOI 10.1007/978-3-642-00493-3_3,

only of minor importance for protecting forest butterfly species. Furthermore, they may only represent ecological sinks for a substantial proportion of the few forest species that are reported to occur in this habitat type. The future of tropical forest butterflies in all four major tropical regions hence depends on the strict protection of remaining large blocks of natural forests.

Keywords: forest butterflies, forest conversion, secondary forests, agroforestry systems, tree plantations, butterfly conservation, tropical biodiversity, range size

1 Introduction

Globally, rainforest species are under increasing threat due to ongoing deforestation, which is progressing at an unprecedented scale (Sodhi & Smith 2007). Between 2000 and 2005 alone, a total of 27.2 million hectares of humid tropical forests were cleared (Hansen et al. 2008). Deforestation and forest fragmentation have been identified as the main drivers of biodiversity loss in the tropics. However, predictions of imminent extinctions have to consider the potential of a certain fraction of rainforest species to utilize secondary habitats such as secondary forests and agroforestry systems as refuge. For example, a study on the conservation value of tropical primary, secondary, and plantation forests for 15 taxonomic groups in northeastern Brazilian Amazonia (Barlow et al. 2007a) showed that areas of native regeneration and exotic tree plantations can provide complementary conservation services, but also provided clear evidence for the irreplaceable value of primary forests to protect tropical biodiversity. Similar conclusions were drawn from comparable studies, including a larger number of taxonomic groups, from Sulawesi (Schulze et al. 2004b, Kessler et al. 2009) and Cameroon (Lawton et al. 1998, Stork et al. 2003). However, only for few groups of organisms, sufficient data are available that allow for more general and robust conclusions on the importance of land-use systems for maintaining tropical biodiversity. The species-rich arthropods pose a particular challenge in that respect, yet butterflies provide a notable exception. During the last two decades studies on effects of forest conversion on butterfly diversity and species composition have been conducted in all major tropical regions (e.g. Barlow et al. 2007b, Dunn 2004, Fermon et al. 2005, Koh 2007, Lawton et al. 1998, Lien and Yuan 2003, Ramos 2000, Sundufu and Dumbuya 2008).

This chapter intends to summarize published studies on effects of forest disturbance and subsequent land-use change on tropical butterfly assemblages with a focus on forest species. In addition to the literature review, we use faunal monographs to extract information on the habitat affiliations of tropical forest butterflies. Based on these data, we quantify the importance of human-modified habitats on a global scale across all major tropical regions. Particularly, we aim to evaluate the conservation value of anthropogenically modified forests (selectively logged forests and regeneration forest after complete forest clearance) and agroforestry systems for maintaining forest butterfly diversity. From a conservation perspective, forest butterfly species deserve the highest attention (Koh 2007). Therefore, we did not include young fallows, pastures and other openland habitats in our review and analyses. These strongly disturbed habitats are of minor importance to preserve the diversity of tropical forest butterflies (e.g. Lawton et al. 1998, Schulze et al. 2004b).

2 The conservation value of modified tropical forests

Tropical secondary forests often harbour extremely diverse butterfly faunas including a large proportion of true forest species. For example, the number of forest species recorded in logged forest in Borneo decreased by less than 10% compared to primary forest (Koh & Wilcove 2008). Particularly in fruit-feeding nymphalid butterflies some studies have even failed to find any significant differences of species richness between natural and adjacent secondary forest (e.g. Borneo: Beck & Schulze 2000). However, there are marked differences in the assemblage composition between the two forest types (Hamer et al. 2003, Cleary et al. in press).

The response to habitat disturbance and modification can differ significantly between (Lawton et al. 1998; Schulze et al. 2004b) and within taxonomic groups (Barlow et al. 2007a), such as butterflies. Of 20 studies on the impact of forest disturbance on butterflies reviewed by Koh (2007), seven reported higher species richness/diversity in near-natural forest than in disturbed habitats, nine the opposite trend, further three reported no effect of forest disturbance and one a strong effect of seasonality on impacts of logging. These apparently contrasting responses of butterfly assemblages to forest modification may not represent true differences but may often result from methodological problems related to the sampling design (Hill and Hamer 2004, Koh 2007). Hamer & Hill (2000) showed that disturbance had opposite effects on species richness at large and small spatial scales. The probability of observing a positive effect of disturbance on diversity increased as the sampling scale decreased. Butterfly data from logged and unlogged forest in Maluku Province (Indonesia) indicated that recorded species richness increases at a significantly faster rate with increasing spatial scale in undisturbed forest (Hamer & Hill 2002). This was most likely the result of higher habitat heterogeneity in undisturbed forests compared to anthropogenically modified forest types. This was confirmed by a subsequent study demonstrating that changes in butterfly assemblages were associated with changes in vegetation structure following selective logging. The logging activities resulted in much lower habitat heterogeneity with less dense shade and fewer open gaps in logged forest. A decline of vegetation complexity most likely is related to a decline of plant diversity, which may not only reduce the structural diversity of niches but particularly the richness of potential larval food plants. This again highlights the need to sample at sufficiently large spatial scales to account for impacts of disturbance on heterogeneity in forest environments (Hamer et al. 2003) and emphasizes that the understanding of species responses to natural variation in environmental conditions within undisturbed forest is crucial to interpreting responses of species to anthropogenic habitat modification.

The apparent strength of disturbance effects on butterfly species richness can also depend on the survey techniques and, consequently, the sampled fraction of the butterfly communities. For example, major, but conflicting, differences in species accumulation rates were found under different disturbance conditions between forest butterflies sampled by transects and caught by fruit baits in Trinidad (Wood & Gillman 1998). Significantly higher species richness was found in disturbed forest when considering transect count data but significantly lower richness when the analysis was based on fruit trap data. This most likely reflects the fidelity of much of the fruit-feeding guild to closed canopy forest, while disturbed forests were found to lack that distinct portion of forest butterflies (Wood & Gillman 1998).

Another methodological problem may arise from the prominent vertical stratification of forest butterfly assemblages (DeVries 1988, DeVries et al. 1997, 1999, Fermon et al. 2005, Schulze et al. 2001). This vertical stratification is shaped by a variety of factors such as the availability of larval food plants, light-level preferences of adults, predator avoidance strategies, microclimate (e.g. temperature and humidity) and the micro-distribution of adult resources (e.g. fruits, nectar) (DeVries 1988, DeVries 1994, Schulze et al. 2001). Several butterfly species that occur within the closed forest in the canopy might treat a sunny forest edge as a similar interface (DeVries 1988). Forest disturbance can alter patterns of vertical stratification in butterfly assemblages, as reported from Sulawesi (Indonesia) (Fermon et al. 2005). As a consequence a higher proportion of canopy species can be recorded (e.g. by bait trapping) in lower forest strata or even close to the ground in disturbed forests or at forest edges. When the sampling design does not account for this potential bias, results may contradict each other, actually only reflecting a better inventory of the canopy fauna in a certain habitat or at a certain study sites (e.g. Horner-Devine et al. 2003).

The likelihood of survival of forest butterflies after forest disturbance has been shown to depend on life-history traits such as larval host plant use. For example, butterflies using trees as larval host plants were more likely to be absent in logged forest in Borneo. By contrast, species that tended to be more abundant in logged forest were those with larvae feeding on lianas and grasses (Cleary et al. in press). The availability of such plant growth types used as larval resource is strongly related to forest structure, which significantly changes after forest disturbance, e.g. through selective logging (Cleary et al. in press). Therefore, the dependence of butterflies on these plant growth types appears to be a better predictor for their survival, than the number of used host plant species. For example, Schulze et al. (2004a) did not find a significant difference between the response of monophagous vs. polyphagous butterflies to forest disturbance and modification.

Based on their studies in Trinidad, Wood & Gillman (1998) concluded that the optimal strategy for safeguarding tropical butterfly species richness under natural forest management regimes would be to maintain a mosaic of habitats that include areas of undisturbed primary forest and a network of other forest patches varying in management regime and disturbance level. Other studies have also emphasized that certain types of disturbed forest, such as selectively logged forests, can make an important contribution to the conservation of tropical butterfly diversity, when they are managed in a way that maintains a sufficiently high environmental heterogeneity (Hamer et al. 2003).

3 The conservation value of agroforestry systems

Particularly traditionally managed, often complex agroforests appear to have a high potential to act as refuge for biodiversity (Perfecto et al. 1996, Moguel and Toledo 1999), even though they are no substitutes for natural habitat on whose proximity they often depend for attaining high levels of wild biodiversity. A study on ants, birds and fruit-feeding butterflies in coffee agroforestry systems in Mexico emphasized the importance of management intensity for maintaining biodiversity (Perfecto et al. 2003). Butterflies appeared to be particularly sensitive to forest conversion. The decrease of butterfly species richness was closely related to the decrease of shade cover (Perfecto et al. 2003). The decrease of shade cover also proved to be a good predictor for changes of species richness in tropical butterfly assemblages in other studies (Schulze et al. 2004a). Therefore, the ongoing intensification of agroforestry, which usually results in a significant decrease of shade cover (e.g. Steffan-Dewenter et al. 2007), is most likely globally devaluating the conservation value of this land-use system for forest butterflies as well as for other forest-dependent arthropods.

It was demonstrated for large-scale *Eucalyptus* plantations that the richness of the native understorey vegetation had the strongest independent effect on the richness, abundance and composition of fruit-feeding butterflies, as well as a subset of species that had been recorded in nearby primary forests (Barlow et al. 2008). However, the overall pattern was strongly influenced by the most abundant subfamily (Satyrinae), while the abundance of any other subfamily, or non-Satyrinae species was not related to vegetation richness. This highlights the importance of accounting for possible phylogenetic dependency when examining butterfly-environment relationships. Similar relationships between understorey plant diversity and butterfly species richness were also documented for butterflies studied along a land-use gradient in Sulawesi (Schulze et al. 2004a) and moths sampled across a forest disturbance gradient in Borneo (Schulze 2000, Beck et al. 2002).

4 The importance of habitat fragmentation

Both the conversion of natural habitats and the subsequent fragmentation are recognized as the major drivers for the increasing rate of species extinctions in recent decades (Groombridge 1992, Henle et al. 1996, Turner 1996). However, not all species are affected equally, but some are at greater risk of extinction in fragmented landscapes than others. A better understanding of these differential sensitivities of species to forest modifications has wide implications for ecological theory and for the setting of conservation priorities (Henle et al. 2004).

As also documented for other animal groups, tropical forest butterflies exhibit a variety of population-level responses to forest fragmentation, ranging from population increase to extinction (Shahabuddin and Ponte 2005, but compare Singer and Ehrlich 1991). Shahabuddin and Terborgh (1999) studied the effects of isolation and fragmentation on fruit-feeding butterflies of forested islands in a water reservoir in eastern Venezuela. Whereas some islands showed reduced abundance and species diversity in comparison to unfragmented sites, others did not. However, isolation status affected both butterfly abundance and diversity. Species composition varied significantly between continuous forest, islands close to the coastline of the reservoir, and islands strongly isolated. Additionally, strong interspecific differences were observed in species' responses to fragmentation. Also a study on fruit-feeding butterflies in South-eastern Brazil found a strong effect of forest fragmentation on species richness, although species richness was surprisingly similar within the studied large forest block and the forest fragments of fruit-feeding butterflies in South-eastern Brazil (Uehara-Prado et al. 2007).

Small forest butterfly species appear to be less sensitive to forest fragmentation than larger ones. For example, small-sized satyrines were relatively resistant against fragmentation in Venezuela, while other medium-sized and large species were more vulnerable (Shahabuddin and Terborgh 1999). Similarly, Benedick et al. (2006) reported that larger species are more adversely affected by rain forest fragmentation in Borneo. The association between large body size and high extinction risk may be due to greater resource requirements, lower population densities, or a stronger tendency to disperse away from habitat fragments in larger species (Benedick et al. 2006, Shahabuddin and Ponte 2005). However, not all studies found a relationship between extinction risk and body size (e.g. Koh et al. 2004).

Furthermore, Shahabuddin and Terborghs (1999) data indicated that tropical forest butterfly species may exist as mainland-island metapopulations, in which small habitat fragments require regular recolonization from source populations in large islands and mainland habitat. However, the forested islands in a water reservoir were all small (0.11.15 ha) and results may not be transferable to habitat islands embedded in a landscape matrix consisting of terrestrial habitats. Various biological attributes are underlying differences in extinction vulnerability among butterfly species in habitat fragments. For example, a study of butterflies in forest fragments in Venezuela showed that larger species are more vulnerable to extinction from forest fragments than smaller ones. Surprisingly, rarer species proved to be not more vulnerable to extinction, emphasizing that rarity *per se* may not be always an important correlate of vulnerability to extinction, at least amongst fruit-feeding butterflies. Contrary to expectation, faster-flying species were more and not less vulnerable to extinction from small habitat fragments (Shahabuddin and Ponte 2005). In a long term, forest fragmentation likely increases extinction risk through reduced genetic diversity as documented for a fruit-feeding understorey satyrine butterfly studied in forest remnants in East Malaysia (Benedick et al. 2007).

5 Secondary forests and land-use systems: an ecological sink for forest species?

One major problem in estimating the importance of secondary forests, agroforests and other land-use systems is that most studies do not provide data adequate to evaluate if forest species are capable to maintain self-sustaining populations in human-dominated habitats. This is due to that most studies only refer to records of adult butterflies but not, for example, to larval stages indicating the presence of suitable larval habitats (food plants). In other words, it is often impossible to predict to which extent such habitats represent ecological sinks for forest species rather than source areas supporting a long-term survival within a landscape without remaining forest. If disturbed habitats only represent ecological sinks, then the importance of disturbed forests and agroforestry patches may be much lower than indicated by the majority of studies on the impact of forest modification and land use on the richness of tropical forest butterfly assemblages in human-dominated habitats.

Only few studies have quantified effects of the distance to closed forest or forest remnants on forest butterfly species richness in secondary forests (Veddeler et al. 2005) and agroforests (Horner-Devine et al. 2003, C. H. Schulze et al. unpublished data) and thus far indicate that at least some human-dominated habitats apparently represent ecological sinks for a substantial proportion of forest butterflies. In fruit-feeding nymphalid butterflies in Central Sulawesi richness in secondary forest patches increased significantly with vegetation succession, but was not related to isolation from closed mature forest (Veddeler et al. 2005). However, distances of sampled forest fragments to the nearest closed forest were small (all $<$1700 m). This is not surprising when considering that many fruit-feeding nymphalid species are strong on the wing and, consequently, should have high dispersal abilities. For the fruit-feeding butterfly *Hamadryas februa* even distances of up to 2.7 km between forest fragments and the potential source area do not appear to be an important barrier for movements (Shahabuddin et al. 2000). Often butterfly densities may not be constrained by colonization capabilities but rather by lack of appropriate habitat quality (e.g. host plants) and concomitant high rates of emigration from islands (Shahabuddin et al. 2000). Therefore, the study by Veddeler et al. (2005) is not adequate to evaluate if secondary forests may be able to maintain self-sustaining populations of a substantial proportion of forest species because for a large fraction of forest species most of the surveyed forest fragments should still have been within the range of frequent dispersal movements.

In contrast, species richness of forest butterflies in agroforestry systems has been found to suffer significantly through isolation from forest remnants. To test for effects of distance from forest remnants on butterfly diversity, Horner-Devine et al. (2003) surveyed the butterfly fauna of coffee plantations located near (<2.5 km), but either contiguous with small forest fragments or lacking adjacent forest, and far (>6 km) from the large forest remnant. Both types of coffee plantations near the large forest remnant differed from those far from it in species composition but not in species richness. Area of forest cover within a radius of 50100 m of the sampling site was significantly correlated to species richness of frugivorous butterflies during the dry season but was not correlated to richness of frugivorous butterflies in the wet season or of non-frugivorous butterflies in either season. This indicates that for a certain fraction of forest butterflies, their successful survival in secondary habitats may differ between seasons and, consequently, depends on regular colonization from adjacent forest. Forest patches embedded in a matrix of agricultural land may provide either microclimate or food resources that are particularly important for certain butterflies during the dry season (Horner-Devine et al. 2003). Nearby forest can even increase species richness of forest butterflies in oil palm plantations (Koh 2008), which are usually characterized by a very depauperate butterfly fauna (Fitzherbert et al. 2008, Koh and Wilcove 2008). This emphasizes that small, isolated forest fragments may help retain butterfly diversity in the tropical countryside and increase the conservation value of agricultural landscapes. Relatively large tracts of forest remain important, however, because they maintain rare and endemic species (Horner-Devine et al. 2003; but see Shahabuddin and Ponte 2005). In contrast, Perfecto et al. (2003) did not find a significant correlation between species richness of fruit-feeding butterflies, mainly represented by forest species, in coffee agroforestry systems and distance to forest in Southern Mexico. However, the fauna in the studied agroforestry systems was extremely depauperate and the remaining small species pool may have comprised only species that are tolerant against disturbance and able to colonize a wide variety of land-use systems.

6 Range-restricted species

Butterfly species with narrow geographic ranges are less able to make use of human-modified environments than are more widely distributed species (Neotropics: Horner-Devine et al. 2003, Thomas 1991, but see Wood & Gillman 1998; Oriental region: Barua 2007, Cleary et al. in press, Fermon et al.

2001, 2005, Ghazoul 2002, Hamer et al. 1997, Hamer and Hill 2000, Hamer et al. 2003, Hill et al. 2005, Lien & Yuan 2003, Schulze et al. 2004a, Spitzer et al. 1993). Whereas widespread species dominate species assemblages of highly modified land-use systems and agroforests (e.g. Horner et al. 2003, Schulze et al. 2004a), endemic butterflies are more commonly found in later successional forests (Leps and Spitzer 1990, Spitzer et al. 1993). Endemics might be specialists that respond less flexibly to habitat modifications and, therefore, might be at greater risk through disturbance (Koh et al. 2004). Besides maintaining lower species richness, this bias can additionally limit the conservation value of non-pristine habitats. Species with restricted geographic ranges are of special conservation concern due to a higher global extinction risk caused by the local loss of suitable habitats.

Table 1. Faunal monographs used in this study for extracting information on distribution and habitat affiliations of tropical rain forest butterflies, and number of butterfly species included in the analysis.

Tropical region	Species number	Subregion/country	Reference
Australian region	160	Australia	Braby (2000a, b)
		Papua New Guinea	Parsons (1999)
		Solomon Islands	Tennent (2002)
Oriental region	263	Hong Kong	Bascombe et al. (1999)
		Malayan Peninsula	Eliot (1991), Otsuka (2001)
		Borneo	Otsuka (2001)
		Ceylon	d'Abrera (1998)
Afrotropical region	653	West Africa	Larsen (2005)
		Kenya	Larsen (1991)
		Tanzania	Congdon & Collins (1998), Kielland (1990)
Neotropical region	709	Costa Rica	DeVries (1987, 1997)
		Venezuela	Neild (1996, 2008)

In general, endemic forest species tend to be restricted to closed forest canopies, whereas non-endemic forest butterflies tend to be forest gap specialists (Spitzer et al. 1997, Hill et al. 2001). The butterfly fauna of gaps is partly represented by canopy species present in gaps. Gap species have higher dispersal rates (as measured by recapture rates) and there is evidence that gap butterflies have relatively larger and broader thoraxes, indicating a flight morphology adapted for stronger flight abilities (Hill et al. 2001). Therefore, habitat modification decreasing canopy cover resulting in a higher density of artificial gaps is likely to increase the number of widespread species but most likely will cause a decline in understorey species with often restricted distributions. This hypothesis was corroborated by a subsequent study, which reported that areas of dense shade, which were more common in unlogged forest, supported more species of Satyrinae and Morphinae with restricted geographical distributions (Hamer et al. 2003).

7 The global pattern

All studies summarized thus far pertain to small spatial scales: individual forest sites, nature reserves, or landscape sections. It emerges from these studies that conversion of natural tropical forest into secondary forest or agroforestry areas alters species composition of butterfly communities at the cost of habitat specialists and range-restricted species, whereas local species diversity may even increase as long as natural forests remain in close vicinity (for similar patterns in nocturnal moths see Fiedler et al. 2007). Although modern faunal monographs for various tropical regions are available, analyses of the human impact on larger geographic scales such as entire biogeographical regions are scarce. We therefore extracted information on habitat affiliations of butterflies from the literature covering all major tropical regions, viz. the Australian, the Oriental, the Afrotropical and the Neotropical region (Tab. 1). Based on the selected faunas and the large number of species included we assume that our sample is representative to yield robust patterns although our literature survey does not cover the total area of the four tropical regions (and, hence, does not include the entire diversity of forest butterflies). We excluded all species only occurring in dry forest and above 1,000 m a.s.l. from the analysis.

Table 2. Habitat categories used to classify habitat affiliations of forest butterflies.

Habitat type	Definition
Forest	Primary forest, natural forest, old-grown forest, undisturbed forest, perhumid forest, rainforest
Secondary Forest	Secondary forest, selectively logged forest, disturbed forest, forest edge, roadcut, large clearing, large light-gap, riparian edge
Agroforest	Agroforest, tree plantation, palm plantation
Scrubland	Dense vegetation predominantly as results of anthropogenic disturbance, bushland
Openland	Open, non-shaded habitats, with predominately low herbaceous vegetation, including annual cultures (e.g. paddy fields), young fallows and pastures

From the selected faunal monographs (see Tab. 1), we extracted information for all forest species on their habitat affiliation. It was noted if they only occur in old-grown natural forest or, additionally, can be found in secondary forests, agroforests, scrubland and openland (Tab. 2). The size of species distribution ranges relative to the biogeographical realm inhabited was assigned to one of five range scores (Kudrna 1989; see Tab. 3). Similar range size scores proved to be useful in other studies analyzing effects of disturbance on butterflies with different distributions (e.g. Charrette et al. 2006). We excluded two butterfly families (Lycaenidae and Hesperiidae) because: (1) these families are not yet covered in any comprehensive modern treatment of Neotropical butterfly faunas; (2) due to their small body size, often elusive behaviour, high species number, and difficulties in correctly identifying them in the field, butterflies of these two families are usually severely under-sampled in tropical field surveys. In total, we collected data on habitat affiliation and relative range size for 1,781 forest butterfly species (representing more than 17 % of the global richness of the analysed families).

The mean proportional representation of forest butterfly species (including Papilionidae, Pieridae, Nymphalidae and Riodinidae) per biogeographical region decreases dramatically from natural forest towards secondary forest and even more strongly towards agroforestry systems (Fig. 1). The latter maintain a forest species richness which is similar to the one found in scrubland and openland habitats. While in secondary forests still about half of the forest species can be found, less than 20 % of the total forest butterfly richness can be observed in all other habitat types. This relative loss of forest species from natural forest to secondary forest and agroforests, scrubland and openland is similarly strong when quantified separately for Papilionidae, Pieridae, Nymphalidae and Riodinidae (Fig. 1b-e), although the loss of forest species in the latter family appears to be even more severe than in all other families.

A Generalized Linear Model (GLM) testing for effects of region, habitat and family on range size of forest butterflies allowing for pairwise interactions between all three variables revealed highly significant effects. All three variables and their pairwise interactions except of the interaction family x habitat were related to the mean size of species ranges. Habitat had the strongest effect on range size scores, followed by the variables region and family (Tab. 4). Range size scores of forest butterfly species increased from natural forest to secondary forests. An even stronger increase could be seen towards highly modified habitats such as agroforestry systems, scrubland and openland. The pattern was similar in Papilionidae, Pieridae and Nymphalidae (Fig. 2a-c). Only in Riodinidae no clear response of range size to habitat modification was found (Fig. 2d). This may be due to the particularly small number of forest species from this family that also occur in disturbed forests and other modified habitats (Fig. 1e).

Range size scores for forest butterfly species increased with increasing forest modification and land-use intensity in all four tropical regions (Fig. 3), although the most distinct pattern was found for the Australian and Oriental region (Fig. 3a-b). As indicated by the 95 % confidence intervals of the least squares means of range size scores, Australian region forest species that are able to persist in non-forest habitats had a significantly larger distribution range than forest butterflies that occur only in both forest types (Fig. 3a). In the Oriental region, the least squares means of range size scores for forest but-

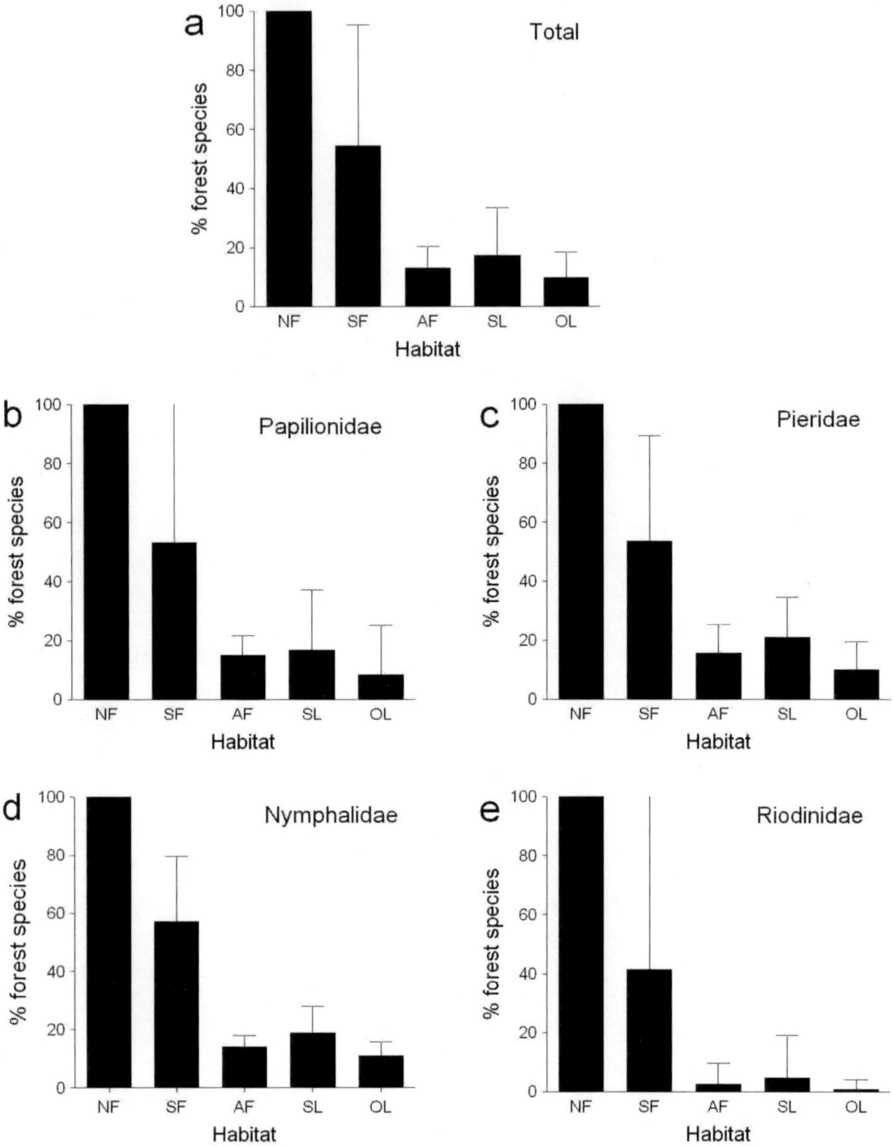

Fig. 1. Mean proportion of forest species (+ standard deviation) that are able to persist in different habitat types (NF natural forest, SF secondary forest, AF agroforest, SL shrubland, OL openland) quantified for all forest species (a) and separately for forest species of the butterfly families Papilionidae (b), Pieridae (c), Nymphalidae (d) and Riodinidae (e).

Table 3. Scores for range sizes of forest butterflies occurring in the four tropical regions.

Region	Range size score	Distribution
Australian	1	Endemic to parts of Northern Australia, New Guinea, Sulawesi or restricted to archipelagos
	2	Entire New Guinea, Northern Australia or Sulawesi, respectively, including satellite islands
	3	Two larger islands, or one larger island plus Australia or one of the larger archipelagos (e.g. Solomon islands)
	4	Australian region
	5	Australian and other tropical region(s)
Oriental	1	One of the Sundaland islands and/or its satellite islands, Sri Lanka, or Peninsular Malaysia
	2	Entire Sundaland, Sri Lanka plus India, Peninsular Malaysia plus Thailand etc.
	3	From Sri Lanka and India to Peninsular Malaysia, Neomalaya
	4	Oriental region
	5	Oriental and other tropical region(s)
Afrotropical	1	Endemic to forest blocks west or east of Dahomey gap, west Kenia plus Uganda, or coastal forests of Kenia plus Tanzania
	2	Restricted to West Africa, Central Africa, or East Africa, respectively
	3	Occurring in two of the regions mentioned above
	4	Entire tropical Africa
	5	Africa and other biogeographical region
Neotropical	1	Endemic to parts of Central America, Northwestern South America, Northeastern South America, or Amazonia
	2	Mexico plus Central America, Venezuela to Guianas, Northern South America, entire Amazonia
	3	Occurring in two of the regions mentioned above
	4	Entire Neotropical region
	5	Neotropical and other biogeographical region

terflies were still significantly higher in agroforestry systems, scrubland and openland than natural forest, whereas forest species in secondary forests were characterized by intermediate range size scores (Fig. 3b).

Focusing on the importance of secondary forests and agroforests, we then used only three habitat categories and tested for effects of region, family and habitat affiliation on range size scores of forest butterflies using (1) species restricted to natural forest; (2) species occurring in natural forest and secondary forests; (3) species occurring in both forest types plus extending to agroforests. Again, GLMs indicated that all three independent predictor variables signifi-

Table 4. Results of GLM testing for effects of the variables region (4 categories: Australian, Oriental, Afrotropical, Neotropical), habitat (5 categories: natural forest, secondary forest, agroforest, scrubland, openland), family (4 families: Papilionidae, Pieridae, Nymphalidae, Riodinidae) and all two-way interactions of these variables on range size scores of forest butterflies.

Variables included	df	MS	*F*	*p*
Constant	1	5006.69	4993.397	< 0.001
Region	3	13.56	13.525	< 0.001
Family	2	7.69	7.665	< 0.001
Habitat	4	32.34	32.252	< 0.001
Region x Family	6	3.98	3.972	< 0.001
Region x Habitat	12	5.13	5.112	< 0.001
Family x Habitat	8	0.53	0.525	0.838
Error	2031	1.00		

Table 5. Results of GLM testing for effects of the variables region (4 categories: Australian, Oriental, Afrotropical, Neotropical), habitat affiliation (3 categories: only natural forest, natural forest plus secondary forest, both forest types plus agroforests) and family (4 families: Papilionidae, Pieridae, Nymphalidae, Riodinidae) and all two-way interactions of these variables on range size scores of forest butterflies.

Variables included	df	MQ	*F*	*p*
Constant	1	2051.85	2275.60	< 0.001
Region	3	7.93	8.80	< 0.001
Family	3	2.93	3.25	0.021
Habitat affiliation	2	50.33	55.82	< 0.001
Region x Family	9	1.49	1.65	0.095
Region x Habitat affiliation	6	10.14	11.25	< 0.001
Family x Habitat affiliation	6	1.44	1.60	0.144
Error	1718	0.90		

cantly affect range size scores. Furthermore, the interaction Region x Habitat proved to affect range size scores (Tab. 5). Particularly in the Australian and Oriental region a prominent shift of range sizes from species restricted to natural forests towards species also colonizing secondary forests and agroforests was visible, which was less conspicuous in Afrotropical and Neotropical forest butterflies (Fig. 4).

Our results emphasize that when viewed on larger spatial scales, the potential loss of forest butterfly species as a result of forest modification and conversion is even more dramatic than indicated by local case studies on effects of forest disturbance and modification on butterfly species richness. Due to the restriction of a large number of endemic butterflies to undisturbed forest habitats, tropical deforestation most likely affects endemic species more

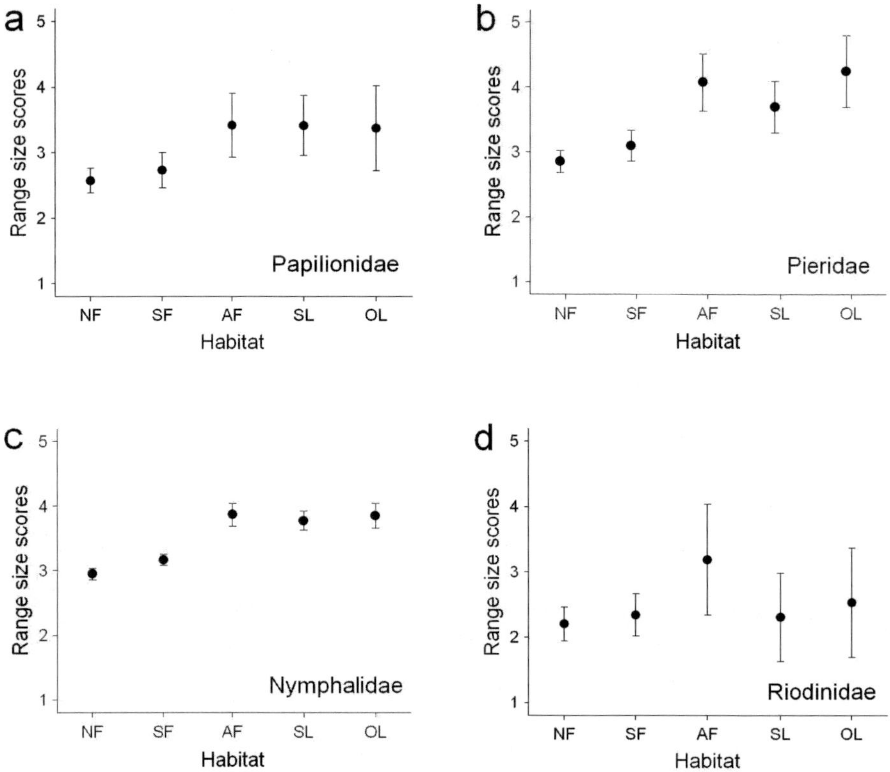

Fig. 2. Least squares means of range size scores (± 95% confidence intervals) for forest butterfly species in different habitat types (NF natural forest, SF secondary forest, AF agroforest, SL scrubland, OL openland) in the families Papilionidae (a), Pieridae (b), Nymphalidae (c) and Riodinidae (d).

heavily, and consequently butterfly faunas of different regions will become increasingly similar as local endemics are exterminated (Thomas 1991).

8 The future of tropical forest butterflies

Our results show severe and similar species loss of forest butterflies after conversion of forests to agroforest, both on small scale and across all major tropical regions. Some land-use systems, such as oil palm plantations represent a particularly serious threat to biodiversity, as demonstrated for Southeast Asian birds and butterflies (Koh & Wilcove 2008). For both groups, the conversion of either primary or secondary (logged) forests to oil palm would result in significant biodiversity losses, whereas conversion of pre-existing cropland (rubber) to oil palm causes fewer losses (Koh & Wilcove 2008). To safeguard

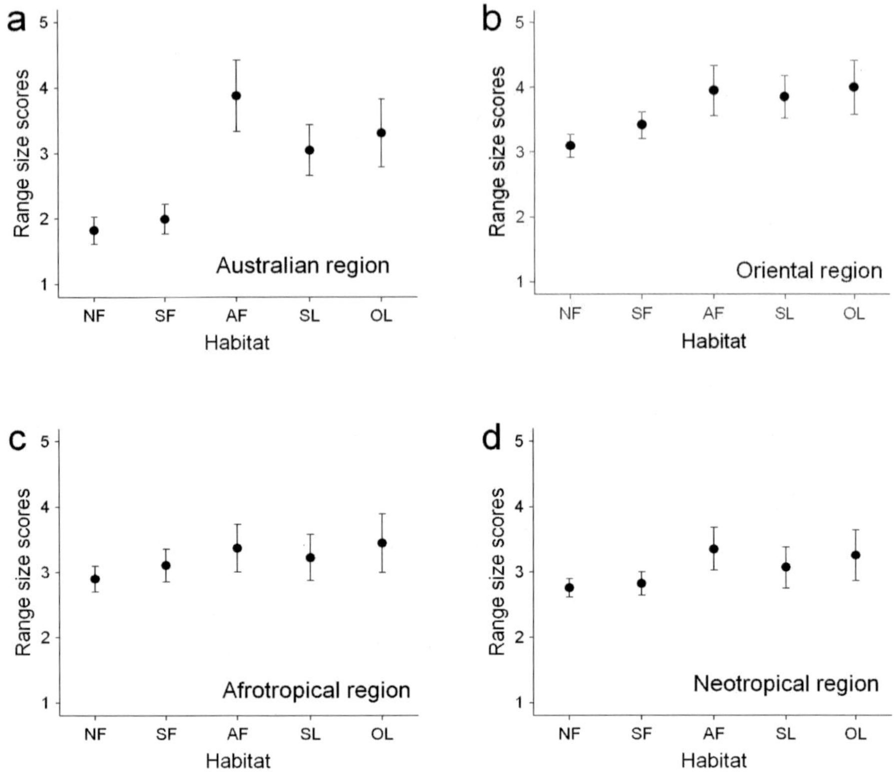

Fig. 3. Least squares means of range size scores (± 95% confidence intervals) for forest butterfly species in different habitat types (NF natural forest, SF secondary forest, AF agroforest, SL scrubland, OL openland) in the Australian (a), Oriental (b), Afrotropical (c) and Neotropical region (d).

biodiversity, more fine-scale and spatially explicit data on land-use change need to be collected and analyzed to determine the extent and nature of any further conversion of forests to land-use systems (Koh & Wilcove 2008). Additionally to primary forest, secondary forests should be protected against conversion to land-use systems by restricting future expansions of agroforestry and agriculture to pre-existing cropland or highly degraded habitats. Clearly, agroforestry systems hold much less promise for maintaining high tropical butterfly diversity as compared to other groups of organisms for which agroforestry has been promoted in recent years (e.g. Perfecto et al. 2003, Rice and Greenberg 2000, Schroth et al. 2004). We assume that larval hostplant relationships are probably the major causal factor responsible for the poor representation of tropical forest butterflies in agroforests: despite some structural complexity that mimics natural forest, agroforests apparently lack resources

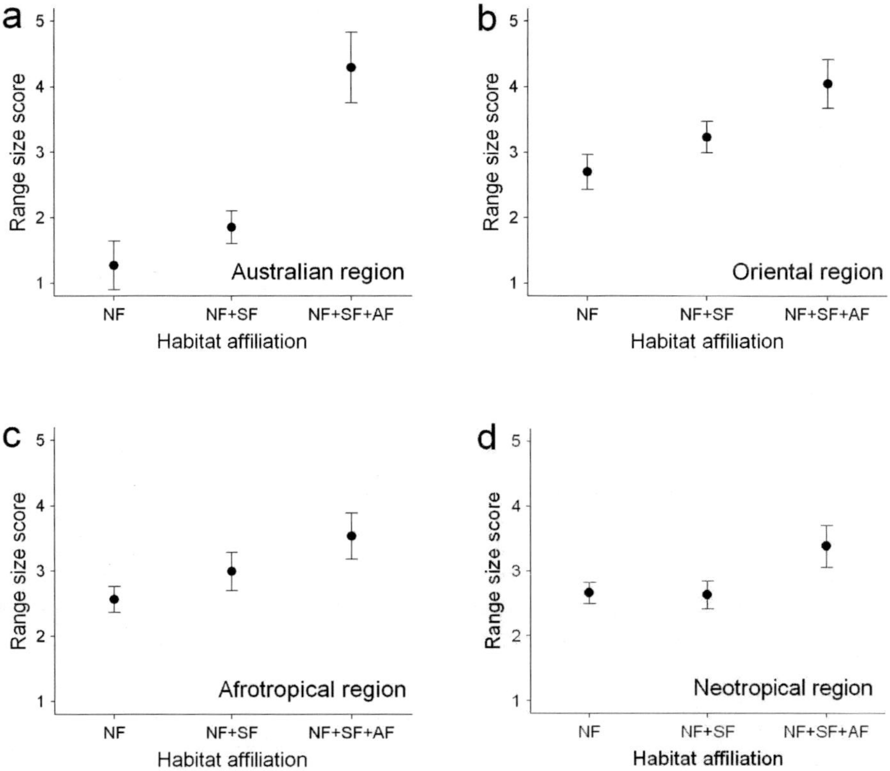

Fig. 4. Least squares means of range size scores (± 95% confidence intervals) for forest butterfly species restricted to natural forest (NF), occurring in NF and secondary forest (NF+SF), and occurring additionally in agroforests (NF+SF+AF), in the Australian (a), Oriental (b), Afrotropical (c) and Neotropical region (d).

that would be important to maintain populations and communities of tropical forest butterflies at a near-intact level.

Range-restricted butterflies do not only appear to be particularly sensitive against disturbance caused by human activities, but, as reported for Bornean forest butterflies, they may be more strongly affected by climatic events such as the El Niño Southern Oscillation (ENSO; e.g. Charrette et al. 2006, Cleary and Grill 2004, Cleary and Mooers 2006). Cleary and Genner (2004) studied effects of the droughts associated with the 199798 ENSO event, which hit large parts of South-East Asia und resulted in large-scale fires affecting millions of hectares of rain forest. The forest fires dramatically altered structure and composition of butterfly communities and resulted in a major decline in species richness (Cleary and Genner 2004, 2006, Cleary et al. 2004). After the fires, butterfly assemblages in all landscapes were dominated by large-winged generalist species, while assemblages needed 1-2 years until smaller-sized and more specialist species recovered. Despite intensive sampling, species endemic to Borneo that were present before fires were still absent in 2000 (Cleary and Genner 2004). The observation that the recovery of the forest butterfly fauna was related to the geographical distance between sampling sites indicates that it depended upon colonization from nearby habitats (Cleary and Genner 2004). The increasing fragmentation and isolation of remaining forest patches will further prevent a fast colonization of forest islands from habitat patches less affected by catastrophic events such as forest fires in the future. The fact that anthropogenic forest conversion and habitat fragmentation both interact as determinants of the fraction of initial butterfly diversity that can be expected to occur at any given site exacerbates the conservation perspective. For true forest species, the only relevant option at present seems to be the preservation of as large blocks as near-natural forest as possible.

A recently published study by Larsen (2008) on forest butterflies in Africa west of the Dahomey Gap reported that apparently no butterfly species has yet gone regionally extinct in West Africa. Yet, he also emphasized that the long-term survival of fauna and flora is wholly dependent on the continued existence of sufficient forest in reasonable condition. The remaining forest area of high quality is very small and now contained almost exclusively within protected areas. Its continued conservation is vital for protecting the last refuges for forest biodiversity in West Africa. The situation described by Larsen (2008) is of global relevance, and hence the future of tropical biodiversity depends on the implementation of conservation measures translating these experiences into sustainable protection of the worlds remaining rainforests. This is also emphasized by our analyses of the proportional representation of forest butterflies in agroforestry systems. Land-use systems are of only minor importance for protecting forest species in all major tropical regions. Furthermore, they may only represent ecological sinks for a substantial proportion of the few forest species, which are reported to occur in this habitat type. At least for sensitive environmental indicators such as tropical forest butterflies, all these surrogate habitats now appear as far less promising than has been though even a few years ago.

References

Barlow J, Gardner TA, Araujo IS, Ávila-Pires TC, Bonaldo AB, Costa JE, Esposito MC, Ferreira LV, Hawes J, Hernandez MIM, Hoogmoed MS, Leite RN, Lo-Man-Hung NF, Malcolm JR, Martins MB, Mestre LAM, Miranda-Santos R, Nunes-Gutjahr AL, Overal WL, Parry L, Peters SL, Ribeiro-Junior MA, da Silva MNF, da Silva Motta C, Peres CA (2007a) Quantifying the biodiversity value of tropical primary, secondary and plantation forests. Proc Natl Acad Sci USA 104: 18555-18560

Barlow J, Overal WL, Araujo IS, Gardner TA, Peres CA (2007b) The value of primary, secondary and plantation forests for fruit-feeding butterflies in the Brazilian Amazon. J Appl Ecol 44:1001-1012

Barlow J, Araujo IS, Overal WL, Gardner TA, da Silva Mendes F, Lake IR, Peres CA (2008) Diversity and composition of fruit-feeding butterflies in tropical Eucalyptus plantations. Biodiv Conserv 17: 1089-1104

Barua KK (2007) Diversity and habitat selection of Papilionidae in a protected forest reserve in Assam, Northeast India. Doctoral thesis, University of Göttingen, Göttingen, Germany

Bascombe MJ, Johnston G, Bascombe FS (1999) The butterflies of Hong Kong, Academic Press, London, San Diego

Beck J, Schulze CH (2000) Diversity of fruit-feeding butterflies (Nymphalidae) along a gradient of tropical rainforest succession in Borneo with some remarks on the problem of pseudoreplicates. Trans lepid Soc Japan 51:89-98

Beck J, Schulze CH, Linsenmair KE, Fiedler K (2002) From forest to farmland: diversity of geometer moths along two habitat gradients on Borneo. J Trop Ecol 18:33-51

Benedick S, Hill JK, Mustaffa N, Chey VK, Maryati M, Searle SB, Schilthuizen M, Hamer KC (2006) Impacts of rain forest fragmentation on butterflies in northern Borneo: species richness, turnover and the value of small fragments. J Appl Ecol 43:967-977

Benedick S, White TA, Searle JB, Hamer KC, Mustaffa N, Chey VK, Mohamed M, Schilthuizen M, Hill JK (2007) Impacts of habitat fragmentation on genetic diversity in a tropical forest butterfly on Borneo. J Trop Ecol 23:623-634

Braby MF (2000a) The Butterflies of Australia: Their Identification, Biology and Distribution, Volume one, CSIRO Publishing, Collingwood, Australia

Braby MF (2000b) The Butterflies of Australia: Their Identification, Biology and Distribution, Volume two, CSIRO Publishing, Collingwood, Australia

Charrette NA, Cleary DFR, Mooers A (2006) Range-restricted, specialist Bornean butterflies are less likely to recover from ENSO-induced disturbance. Ecology 87:2330-2337

Cleary DFR, Genner MJ (2004) Changes in rain forest butterfly diversity following major ENSO-induced fires in Borneo. Global Ecol Biogeogr 13:129-140

Cleary DFR, Grill A (2004) Butterfly response to severe ENSO-induced forest fires in Borneo. Ecol Entomol 29:666-676

Cleary DFR, Genner MJ (2006) Diversity patterns of Bornean butterfly assemblages. Biodiv Conserv 15:517-538

Cleary DFR, Mooers A (2006) Burning and logging differentially affect endemic vic. widely distributed butterfly species. Divers Distr 12:409-416

Cleary DFR, Mooers A, Eichhorn KAO, van Tol J, de Jong R, Menken SBJ (2004) Diversity and community composition of butterflies and odonates in an ENSO-induced fire affected habitat mosaic: a case study from East Kalimantan, Indonesia. Oikos 105:426-446

Cleary DFR, Genner MJ, Koh LP, Boyle TJB, Setyawati T, de Jong R, Menken SBJ (in press) Butterfly species and traits associated with selectively logged forest in Borneo. Basic Appl Ecol, doi:10.1016/j.baae. 2008.03.004

Congdon C, Collins S (1998) Kiellands butterflies of Tanzania, Supplement, African Butterfly Research Institute, Nairobi, Kenya

d'Abrera B (1998) The butterflies of Ceylon, Hill House Publishers, Melbourne, London

DeVries PJ (1987) The butterflies of Costa Rica and their natural history: Papilionidae, Pieridae, Nymphalidae, Princeton University Press, Princeton, USA

DeVries PJ (1988) Stratification of fruit-feeding nymphalid butterflies in a Costa Rican rainforest. J Res Lepid 26:98-108

DeVries PJ (1994) Patterns of butterfly diversity and promising topics in natural history and ecology. Pp. 187-194 in McDade LE (ed) La Selva: ecology and natural history of a Neotropical rain forest. University of Chicago Press, Chicago

DeVries PJ (1997) The butterflies of Costa Rica and their natural history, Volume 2: Riodinidae, Princeton University Press, Princeton, USA

DeVries PJ, Walla TR, Greeney HF (1999) Species diversity in spatial and temporal dimensions of fruit-feeding butterflies from two Ecuadorian rainforests. Biol J Linn Soc 68:333-353

Dunn RR (2004) Managing the tropical landscape: a comparison of the effects of logging and forest conversion to agriculture on ants, birds, and Lepidoptera. Forest Ecol Manag 191:215-224

Eliot JN (ed) (1992) The butterflies of the Malayan Peninsula by A. Steven Corbet and H.M. Pendlebury, Malayan Nature Society, Kuala Lumpur

Fermon H, Schulze CH, Waltert M, Mühlenberg M (2001) The butterfly community of the Noyau Central, Lama Forest (Republic of Benin), with notes on its ecological composition and geographic distribution. African Entomol 9:177-185

Fermon H, Waltert M, Vane-Wright RI, Mühlenberg M (2005) Forest use and vertical stratification in fruit-feeding butterflies of Sulawesi, Indonesia: impacts for conservation. Biodiv Conserv 14:333-350

Fiedler K, Hilt N, Brehm G, Schulze CH (2007) Moths at tropical forest margins how mega-diverse insect assemblages respond to forest disturbance and recovery. Pp 39-60 in: Tscharntke T, Leuschner C, Zeller M, Guhardja E, Bidin A (eds), The stability of tropical rainforest margins: Linking ecological, economic and social constraints of land use and conservation. Springer, Berlin, Heidelberg

Fitzherbert EB, Struebig MJ, Morel A, Danielsen F, Brühl CA, Donald PF, Phalan B (2008) How will oil palm expansion affect biodiversity? Trends Ecol. Evol. 23:538-545

Ghazoul J (2002) Impact of logging on the richness and diversity of forest butterflies in a tropical dry forest in Thailand. Biodiv Conserv 11:521-541

Groombridge B. 1992. Global Biodiversity. Chapman & Hall, London.

Hamer KC, Hill JK (2000) Scale-dependent effects of habitat disturbance on species richness in tropical forests. Conserv Biol 14:1435-1440

Hamer KC, Hill JK, Lace LA, Langan AM (1997) Ecological and biogeographical effects of forest disturbance on tropical butterflies of Sumba, Indonesia. J Biogeogr 24:67-75

Hamer KC, Hill JK, Benedick S, Mustaffa N, Sherratt TN, Maryati M, Chey VK (2003) Ecology of butterflies in natural and selectively logged forests of northern Borneo: the importance of habitat heterogeneity. J Appl Ecol 40:150-162

Hansen MC, Stehman SV, Potapov PV, Loveland TR, Townshend JRG, DeFries RS, Pittman KW, Arunarwati B, Stolle F, Steininger MK, Carroll M, DiMiceli C (2008) Humid tropical forest clearing from 2000 to 2005 quantified by using multitemporal and multiresolution remotely sensed data. Proc Natl Acad Sci USA 105:9439-9444

Henle K, Poschlod P, Margules C, Settele J (1996) Species survival in relation to habitat quality, size, and isolation: summary, conclusions and future directions. Pp. 373381. In Settele J, Margules CR, Poschlod P, Henle K (eds), Species survival in fragmented landscapes. Kluwer, Dordrecht

Henle K, Davies KF, Kleyer M, Margules C, Settele J (2004) Predictors of species sensitivity to fragmentation. Biodiv Conserv 13:207-251

Hill JK, Hamer KC (2004) Determining impacts of habitat modification on diversity of tropical forest fauna: the importance of spatial scale. J Appl Ecol 41:744-754

Hill JK, Hamer KC, Tangah J, Dawood M (2001) Ecology of tropical butterflies in rainforest gaps. Oecologia 128:294-302

Hill JK, Hamer KC, Lace LA, Banham WMT (1995) Effects of selective logging on tropical forest butterflies on Buru, Indonesia. J Appl Ecol 32:754-760

Horner-Devine MC, Daily GC, Ehrlich, PR (2003) Countryside biogeography of tropical butterflies. Conserv Biol 17:168-177

Kessler K, Abrahamczyk S, Bos M, Putra DD, Gradstein SR, Höhn P, Orend F, Pitopang R, Saleh S, Schulze CH, Sporn SG, Steffan-Dewenter I,

Tscharntke T (2009) Alpha and beta diversity of plants and animals along a tropical land-use gradient. Ecological Applications, in press
Kielland J (1990) The butterflies of Tanzania, Hill House, Melbourne, London
Koh LP (2007) Impacts of land use change on south-east Asian forest butterflies: a review. J Appl Ecol 44:703-713
Koh LP (2008) Can oil palm plantations be made more hospitable for forest butterflies and birds? J Appl Ecol 45:1002-1009
Koh LP, Wilcove DS (2008) Is oil palm agriculture really destroying tropical biodiversity?. Conserv Lett 1: 60-64
Koh LP, Sodhi NS, Brook BW (2004) Ecological correlates of extinction proneness in tropical butterflies. Conserv Biol 18:1571-1578
Kudrna O (ed.) (1986) Butterflies of Europe. Vol. 8: Aspects of the conservation of butterflies in Europe. Aula-Verlag, Wiesbaden
Larsen TB (1991) The butterflies of Kenya and their natural history, Oxford University Press, Oxford, England
Larsen TB (2005) The butterflies of West Africa, Apollo Books, Stenstrup, Denmark
Larsen TB (2008) Forest butterflies in West Africa have resisted extinction so far (Lepidoptera: Papilionoidea and Hesperioidea). Biodiv Conserv 17:2833-2847
Lawton JH, Bignell DE, Bolton B, Bloemers GF, Eggleton P, Hammond PM, Hodda, M, Holt RD, Larsen TB, Mawdsley NA, Stork NE, Srivastava DS, Watt AD (1998) Biodiversity inventories, indicator taxa and effects of habitat modification in tropical forest. Nature 391:72-76
Leps J, Spitzer K (1990) Ecological determinants of butterfly communities (Lepidoptera, Papilionoidea) in the Tam Dao mountains, Vietnam. Acta Entomol Bohem 87:182-194.
Lewis OT, Wilson RJ, Harper MC (1998) Endemic butterflies on Grande Comore: habitat preferences and conservation priorities. Biol Conserv 85:113-121
Lien VV, Yuan D (2003) The differences of butterfly (Lepidoptera, Papilionoidea) communities in habitats with various degrees of disturbance and altitudes in tropical forests of Vietnam. Biodiv Conserv 12:1099-1111
Moguel P, Toledo VM (1999) Biodiversity conservation in traditional coffee systems in Mexico. Conserv Biol 12:1-11
Neild AFE (1996) The butterflies of Venezuela, Part 1, Meridian Publications, Greenwich, London
Neild AFE (2008) The butterflies of Venezuela, Part 2, Meridian Publications, Greenwich, London
Otsuka K (2001) A field guide to the butterflies of Borneo and South East Asia, Hornbill Books, Kota Kinabalu, Malaysia
Parsons M (1999) The butterflies of Papua New Guinea: Their systematics and biology, Academic Press, London, San Diego
Perfecto I, Rice RA, Greenberg R, Van der Voort ME (1996) Shade coffee: a disappearing refuge for biodiversity. BioScience 46:598-608

Perfecto I, Mas A, Dietsch T, Vandermeer J (2003) Conservation of biodiversity in coffee agroecosystems: a tri-taxa comparison in southern Mexico. Biodiv Conserv 12:1239-1252

Ramos FA (2000) Nymphalid butterfly communities in an Amazonian forest fragment. J Res Lepid 35: 29-41

Rice RA, Greenberg R (2000) Cacao cultivation and the conservation of biological diversity. Ambio 29:167-173

Schulze CH (2000) Auswirkungen anthropogener Störungen auf die Diversität von Herbivoren Analyse von Nachtfalterzönosen entlang von Habitatgradienten in Ost-Malaysia. Doctoral thesis, University of Bayreuth (Germany)

Schulze CH, Linsenmair KE, Fiedler K (2001) Understorey versus canopy: patterns of vertical stratification and diversity among Lepidoptera in a Bornean rain forest. Plant Ecol 153:133-152

Schulze CH, Steffan-Dewenter I, Tscharntke T (2004a): Effects of land use on butterfly communities at the rain forest margin: a case study from Central Sulawesi. Pp. 281-297. In: Gerold G, Fremerey M, Guhardja E (eds.), Land use, nature conservation and the stability of rainforest margins in Southeast Asia. Springer, Berlin

Schulze CH, Waltert M, Kessler PJA, Pitopang R, Shahabuddin, Veddeler D, Steffan-Dewenter I, Mühlenberg M, Gradstein SR & Tscharntke T (2004b) Biodiversity indicator taxa of tropical land-use systems: comparing plants, birds and insects. Ecol Appl 14:1321-1333

Schroth G, Da Fonseca GAB, Harvey CA, Gascon C, Vasconcelos HL, Izac AMN (eds.) (2004) Agroforestry and biodiversity conservation in tropical landscapes. Island Press, Washington

Shahabuddin G, Ponte CA (2005) Frugivorous butterfly species in tropical forest fragments: correlates of vulnerability to extinction. Biodiv Conserv 14:1137-1152

Shahabuddin G, Terborgh JW (1999) Frugivorous butterflies in Venezuelan forest fragments: abundance, diversity and the effects of isolation. J Trop Ecol 15: 703-722

Shahabuddin G, Herzner GA, Aponte R C, del C. Gomez M (2000) Persistence of a frugivorous butterfly species in Venezuelan forest fragments: the role of movement and habitat quality. Biodiv Conserv 9:1623-1641

Singer MC, Ehrlich PR (1991) Host specialization of satyrine butterflies, and their responses to habitat fragmentation in Trinidad. J Res Lepid 30:248-256

Sodhi NS, Smith KG (2007) Conservation of tropical birds: mission possible? J Ornithol 148:305-309

Spitzer K, Novotny V, Tonner M, Leps J (1993) Habitat preferences, distribution and seasonality of the butterflies (Lepidoptera, Papilionoidae) in a montane tropical rain forest, Vietnam. J Biogeogr 20:109-121

Spitzer K, Jaros J, Havelka J, Leps J (1997) Effect of small-scale disturbance on butterfly communities of an Indochinese montane rainforest. Biol Conserv 80:9-15

Stork NE, Srivastava DS, Watt AD, Larsen TB (2003) Butterfly diversity and silvicultural practice in lowland rainforests of Cameroon. Biodiv Conserv 12:387-410

Sundufu AJ, Dumbuya R (2008) Habitat preferences of butterflies in the Bumbuna forest, Northern Sierra Leone. J Insect Sci 8.64, 17 pp.

Tennent J (2002) The butterflies of the Solomon Islands: Systematics and biogeography, Storm Entomological Publications, Norfolk, England

Thomas CD (1991) Habitat use and geographic ranges of butterflies from the wet lowlands of Costa Rica. Biol Conserv 55:269-281

Turner IM (1996) Species loss in fragments of tropical rain forest: a review of the evidence. J Appl Ecol 33:200-209

Uehara-Prado M, Brown Jr KS, Freitas AVL (2007) Species richness, composition and abundance of fruit-feeding butterflies in the Brazilian Atlantic Forest: comparison between a fragmented and a continuous landscape. Global Ecol Biogeogr 16:43-54

Veddeler D, Schulze CH, Steffan-Dewenter I, Tscharntke T (2005) The contribution of tropical secondary forest fragments to the conservation of fruit-feeding butterflies: Effects of isolation and age. Biodiv Conserv 14:3577-3592

Wiemers M, Fiedler K (2008) Butterfly diversity of the Piedras Blancas National Park and its vicinity a preliminary assessment (Lepidoptera: Papilionoidea & Hesperioidea). Pp. 277-294. In: Weissenhofer A, Huber W, Mayer V, Pamperl S, Weber A & Aubrecht G (eds.), Natural and Cultural History of the Golfo Dulce Region, Costa Rica. Stapfia 88. Biologiezentrum des Oberösterreichischen Landesmuseums, Linz

Wood B, Gillman MP (1998) The effects of disturbance on forest butterflies using two methods of sampling in Trinidad. Biodiv Conserv 7:597-616

Insect pollinator communities under changing land-use in tropical landscapes: implications for agricultural management in Indonesia

Bandung Sahari[1], Akhmad Rizali[1,2,3], and Damayanti Buchori[1,2*]

[1] Peka Indonesia Foundation (Indonesian Nature Conservation Foundation)-Wildlife Trust Alliance. Jl. Uranus Blok H No 1 Perum IPB Sindang Barang 2, Bogor, West Java-Indonesia
[2] Department of Plant Protection, Bogor Agricultural University, Kampus IPB Dramaga, Bogor-West Java-Indonesia
[3] Agroecology, University of Göttingen, Waldweg 26, 37073 Göttingen, Germany

*corresponding author: D. Buchori, email: dami@indo.net.id

Summary

The destruction of rainforests can affect many ecosystem services, e.g pollination services for wild and crop plants in landscapes with high proportion of natural habitats. We discuss evidence of the impact of landuse change toward insect pollinator communities and pollination based in the tropics, with an emphasis on case-studies from Indonesia. Some studies showed that species richness of flower visiting bees, pollination, and fruit set were found to be negatively correlated with increasing rainforest isolation and land-use intensity. However, others demonstrated an opposite pattern. Species composition changes significantly between habitat types, which may be relevant in the context of environmental changes as species composition can be an important variable to ensure pollination services and fruit set.

Keywords: pollination, tropical landscape, Indonesia, land use change, agroecosystem, habitat fragmentation

1 Introduction

The tropical landscape, with its vast forests, has traditionally been associated with sources of diversity. This is true for many groups of insects, such as ants (Brühl 2001), moths (Beck and Schulze 2000; Beck et al. 2002), butterflies

T. Tscharntke et al. (eds.), *Tropical Rainforests and Agroforests under Global Change*, Environmental Science and Engineering, 97–114, DOI 10.1007/978-3-642-00493-3_4,

(Schulze et al. 2004), and some functional groups of bees (Roubik 2001). Biodiversity acts as genetic bank that is essential for agriculture, medicine, and also plays a critical role in providing services for human population such as food, genetic resources, and nutrient cycling (Kremen 2005; Chemini and Rizzoli 2003; Baumgärtner 2007). Alterations of that biodiversity may negatively affect ecosystem and landscape processes (Chapin et al. 1998; Turner et al. 2007).

Today, tropical landscapes are suffering from the increasing rate of land-use change and habitat disturbance (Klein et al. 2002a; Cairns et al. 2005). The vastness of tropical forests and other natural ecosystems are now replaced by agricultural and urban areas isolated from big patches of tropical rainforests (Klein et al. 2002a). The transformation of forest to mixed agricultural matrix can affect biodiversity, species composition, and ecosystem services (Ricketts et al. 2004; 2008).

2 Pollination: Its Significance and Current State

Pollination is an ecosystem service closely related to food quantity and quality (Losey and Vaughan 2006; Klein et al. 2007; Kremen et al. 2007; Gallai et al. 2008; Ashworth et al. 2009). Since 35% of the global plant-based food supply is derived of animal-mediated pollination crops (Klein et al. 2007), pollinators are an important functional group of taxa that should be conserved and their economic significance should be appreciated much more. Furthermore, the decline in pollinator diversity may lead to a "pollinator crisis" meaning that the availability of pollinators is low and can not ensure high pollination services of all crop and wild plant species. This may impact in a worst-case food productivity and security scenario (Steffan-Dewenter et al. 2005).

A study by Kremen et al. (2002) demonstrated how the decreasing number of species and abundance of pollinators may not be sufficient to adequately pollinate watermelons in California. Their study showed that watermelon pollination was only temporally secured when farms were surrounded by more than 30% of semi-natural habitat ensuring high bee diversity (Kremen et al. 2002, 2004). A study by Olschewski et al. (2006) demonstrated that rainforest isolation can affect coffee yield and net revenues of coffee growers in Indonesia. Gallai et al. (2008) showed that pollinator loss may result in the decreasing capacity of agricultural crops to provide diverse sources of food at the world scale.

Today in many areas of the world, farmers import honey bees to pollinate crops to fulfill pollination requirements for cash crops instead of relying on wild insect pollinators (Kremen et al. 2002; Cunningham et al. 2002; Ruz 2002; Klein et al. 2008). This applied especially to non-native crops, such as almond depending heavily on managed honey bees in California (Klein et al. 2008). The importance of imported pollinators was also supported by Raju and Rao (2002) who identified that pod and seed yields of *Acacia* may be enhanced by introducing manageable bees together with their nesting requirements. Overall, drastic landscape changes seem to result in a shift of pollination services by wild bees being replaced by "managed bees". In fact, Thompson (2001) pointed out that there is a constant chronic critical pollination deficit that has been increasing to a higher level compared to the past. This "pollination crisis" continues today, and is spreading to other parts of the world. The situation demands careful management of pollinators, since the increase in scarcity may lead to pollinator crisis that will affect food production and supply, and in fact, effect the productivity of plants overall (Delaplane and Mayer 2000; Cunningham et al. 2002; Olschewski et al. 2006)

The importance of pollination for agricultural productivity and the way in which habitat modification affects the diversity of pollinators and its services have been shown by many studies in several regions (reviewed by Ricketts et al. 2008; Gallai et al. 2008; Ashworth et al. 2009). However for many tropical countries such as Indonesia the information is largely lacking. Even though Indonesia is an agricultural country, where many crops, fruits and vegetables are grown locally for both local and international markets (BPS 2009), studies on pollination and the importance of pollinators for crop production are few and scattered. The main studies on pollination in Indonesia have focused on crop species such as coffee, mustard and oil palm. Klein et al. (2003a,b,c) studied lowland and highland coffee pollination in Sulawesi. Pardede (1990) studied the biology and pollination behaviour of *Elaeidobius kamerunicus* Faust (Coleoptera: Curculionidae) on oil palm in North Sumatra, while Atmowidi et al. (2007) and Sahari et al, (unpublished) studied mustard pollination in West Java. There are several other studies that have not been well documented on durian (*Durio zibethinus*), mango (*Mangifera indica*), avocado (*Persea Americana*) and watermelon (*Citrullus lanatus*)(Kasno et al. 2002; Kasno et al. 2004). We argue that it is important for local scientists to start collecting information on the pollination and breeding system of many more crop species and crop varieties and to understand the threats pollinators are facing in the changing environments and how this is related to crop production. In parallel, local scientists need to bring their knowledge to the local stakeholders. Since Indonesia harbors amazing various unique habitat types and climates (tropical island country with mountainous agricultural landscapes), and suffers complicated environmental problems (high extent of landuse conversion, pesticide application, and habitat fragmentation), responses of pollinators to habitat disturbances may be locally specific. A detailed understanding of pollinator communities and pollination and how this is related to food security and productivity is critical for developing appropriate conservation strategies in this country. After reviewing the significance of pollinator communities and pollination services, we present and discuss several case studies conducted in Indonesia.

3 Pollination, pollinator insect communities, and land-use change

Pollination is the mutual interacting processes between flowering plants and flower-visiting animals transferring compatible pollen (Gituru at al 2002; Driscoll 2003; Cauich et al. 2004; Faheem et al. 2004; Agbagwa et al. 2007; Klein et al. 2007; Mitchell et al. 2009) Flowers attract insect pollinators with pollen (Yi-Bo et al. 1998; Delaplane and Mayer 2000; Cauich et al. 2004), nectar or oil (Delaplane and Mayer 2000; Gardener and Gillman 2002), resin (for nest construction) (Roubik 1993), and other floral traits (odor, color and morphology) (Riffel et al. 2008; Schiestl and Schlüter 2009). Pollination occurs when pollen grains are distributed by the insects harvesting pollen (Driscoll 2003; Karron et al. 2009; Schlumpberger et al. 2009). However, not all flower-visiting animals are effective pollinators, some of them are pollen and nectar robbers and contribute less to pollination services (Irwin et al. 2001). Insects, and especially bees, are known to be the most important pollinators (Roubik 1993; Losey and Vaughan 2006; Klein et al. 2007) due to their high abundance and effectiveness in pollen transfer. Many wild bee species contribute to pollination of crops known for sunflower, cotton, lucerne (El-sarrag et al. 1993), coffee (*Coffea* spp., Klein, Steffan-Dewenter & Tscharntke 2003b), watermelon (*Citrullus lanatus* (Thunb. Kremen et al. 2002), sweet pepper (Cruz et al. 2005), mustard (*Brassica rapa*, Atmowidi et al. 2007), broad bean (*Vicia faba* L.var. *major*) (Aouar-sadli et al. 2008), tomato (*Solanum lycopersicum* L., Cauich et al. 2004; Greenleaf and Kremen 2006), and pumpkin (Hoehn et al. 2008). These wild bee visits provide a supplement to honeybee pollination and insurance against honeybee declines (Winfree et al. 2007).

The role of insect pollinator communities for pollination is presented by many studies (e.g. El-sarrag et al. 1993; Ghzawi et al. 2003; Cruz at al. 2005;Atmowidi et al. 2007; Klein et al. 2007; Aouar-sadli et al. 2008). Several studies have tested the relationship between fruit set (%) and pollinators by observing caged and non-caged flowers. Some recent studies were conducted on lowland and highland coffee (Klein et al. 2003c), okra (Ghzawi et al. 2003), sweet orange (*Citrus sinensis* L., Malerbo-Souza et al. 2004), mustard (Atmowidi et al. 2007), and broad bean (Aouar-sadli et al. 2008). For example, in Mexico, around 85% (145 of 171) of fruit or seed consumed species depend to some degree on pollinators for successful fruit/seed production (Ashworth et al. 2009). Agricultural production in USA is highly dependent on insect pollination that is provided by both managed bee (Apis mellifera) and native insects (Cunningham et al. 2002; Losey and Vaughan 2006). In the US, native pollinators – mainly bees – may be responsible for almost \$3.07 billion of fruits and vegetables (Losey and Vaughan 2006), however some agricultural crops may not be affected by pollinators as they are either wind-pollinated, self pollinated, or reproduce vegetatively (Cunningham et al. 2002; Ghazoul 2005a). Greenleaf and Kremen (2006) through their work in North California, identified that bee pollination increases the number of tomatoes produced both in terms of quantity and volume (size), compared with hand and self pollination. A similar pattern was also documented for sweet orange in Brazil (Malerbo-Souza et al. 2004). The role of pollinators for the productivity of several agricultural crops was also reviewed by Slaa et al. (2006), who showed that the presence of pollinators increased the number of seed production of six agricultural crops (broccoli, rape, cauliflower, endive, chicory, leek).

An insect pollinator community can be assessed in terms of richness, composition, and changes of both abundance and species composition over time (succession). Each of these components (richness, composition, and succession) is unique and the ramifications of their being should be viewed in a detailed way, especially in relation to pollination. Several studies have shown positive relationship between pollinator species richness and fruit set (Steffan-Dewenter and Tscharntke 1999; Klein et al. 2003b; Cruz et al. 2005; Hoehn et al. 2008), suggesting that pollinators are complementary in their behaviour, spatial, and temporal activity patterns or other important traits. Other studies have shown that ecological substitution may happen in the field, as shown by Kremen et al. (2002) who identified that the functional role of the most effective bee species could be replaced by another bee species, and by Dick (2001), who documented that African honey bees replaced native insect pollinators in isolated patches.

Species composition is another factor that is important in pollination. Since species composition can vary over space and time, the (numerical) dominance of certain species is hypothetically another factor that will affect pollination, particularly since ecological services can be upheld by a few dominant species. Under the existence of dominant species, their presence will be the main factor affecting the ecosystem services, as shown in (Dick 2001). Others have shown that species identity can be more responsible in pollination services than species richness and visitation rate. Klein et al. (2003a,b) identified that rare solitary bees are more effective in pollinating on a per-visit basis than social bees.

At a global level, there are two major threats for pollinator diversity: (1) destruction and fragmentation of natural and semi-natural habitats (Mustajarvi et al. 2001; Steffan-Dewenter and Westphal 2008), and (2) landuse intensification in agricultural landscapes (Kremen et al. 2002; Steffan-Dewenter and Westphal 2008). Both habitat fragmentation and land-use intensification can affect pollinator communities at different spatial scales and interact with each other (Mustajarvi et al. 2001; Steffan-Dewenter and Westphal 2008).

In terrestrial ecosystems, habitat isolation is commonly related to distance effects of modified habitat toward the remaining nearest natural habitat. As such, habitat isolation may effect species diversity on a different scale, depending on the ecology of the different species. Effect of habitat isolation on species richness of insect pollinators have been documented by many studies (reviewed by Ricketts et al. 2008). There is a common pattern where increasing isolation resulted in the decrease of insect pollinator richness and visitation rate (Ricketts et al. 2008). Ricketts et al. (2008) reviewed 19 cases, 16 of which indicate a decline of pollinator richness with increasing isolation from natural habitat both in the tropic and temperate biomes. This review also provided a strong evidence of a decline in visitation rate by non-managed pollinators with increasing isolation from natural habitat (negative response from 20 cases of 22 observed cases).

Responses of different pollinator taxa to habitat types or isolation to natural habitats may be contrasting. Klein et al. (2002a) found that only honeybees and stingless bees were negatively affected by increasing land-use intensity and that solitary bee species profit from increasing management intensity of coffee and cacao agroforests. A further study conducted by Greenleaf and Kremen (2006) documented the abundance of two important pollinators of tomato, *Bombus vosnesenskii* and *Anthophora urbana* that responded differently to habitat isolation. *A. urbana* did not show negative response to habitat isolation (Greenleaf and Kremen 2006). Ricketts et al. (2008) showed that 10 studies which sampled social and solitary bees separately found some evidence that visitation rates of social bees declined more steeply than those of solitary bees. The overall decay rate is more negative for social than solitary bees in all 10 studies. Those findings indicated that a pool of species showed a collective response toward habitat isolation, however since different species may response differently toward habitat modification as shown by Greenleaf and Kremen (2006), taxon-based observation should be taken into account in measuring habitat isolation effect.

Table 1. Production of *B. rapa* crops under opened and caged condition in the forest margin of Salak Mountain, West Java (100 plants were observed per site every hour for 15 menits between 07.30 am and 2.30 pm).

Parameters	Opened*)	Caged*)	Increased yield (%)
Plant height (cm)	113.9 ± 16.3a	116.3 ± 13.5a	
Number of racemes per plant	16.6 ± 15.1a	10.0 ± 9.6b	65.6
Number of pods per plant	14.9 ± 6.4a	5.4 ± 3.3b	178.8
Number of seeds per pod	12.8 ± 3.2a	6.5 ± 2.4b	98.2
Number of seeds per plant	3319.7 ± 3123.9a	321.5 ± 308.4b	932.5
Seed weight per plant (gram)	6.4 ± 6.8a	0.6 ± 0.6b	931.9

Note: Data courtesy of Atmowidi et al. (2007)
*) Numbers followed by the same letter in the same rows are not significantly different, (t test, $p<0.05$)

Townsend and Levey (2005) showed that corridors connecting between fragment patches can facilitate pollen transfer. Corridors increase the movement of insect pollinators into patches of habitat and thereby increase pollen transfer for two species of plants, one pollinated by butterflies (*Lantana camara*) and the other by bees and wasps (*Rudbeckia hirta*).

4 Land-use changes and pollination from Indonesia

4.1 Mustard pollination in highland agricultural landscapes around Halimun-Salak Mountain National Park

Two different pollination studies have been carried out in this area. A study conducted by Atmowidi et al. (2007) in January-May, 2006 investigated the relationship between pollinator visits and reproductive success of mustard. In this study, observation was conducted on 100 mustard plants for the whole flowering period under open field condition and caged plants. Each flowers were observed individually, and the flower-visiting pollinators were observed every 15 minutes during the time period between 07.30 am and 02.30 pm The study showed that there is a strong relationship between pollinator diversity and reproductive success. Number of pods, seeds per pod, and seed weight per plant was higher in plants freely accessible to pollinators , compared to caged plants. The increase can be up to three times of the original production (see Table 1).

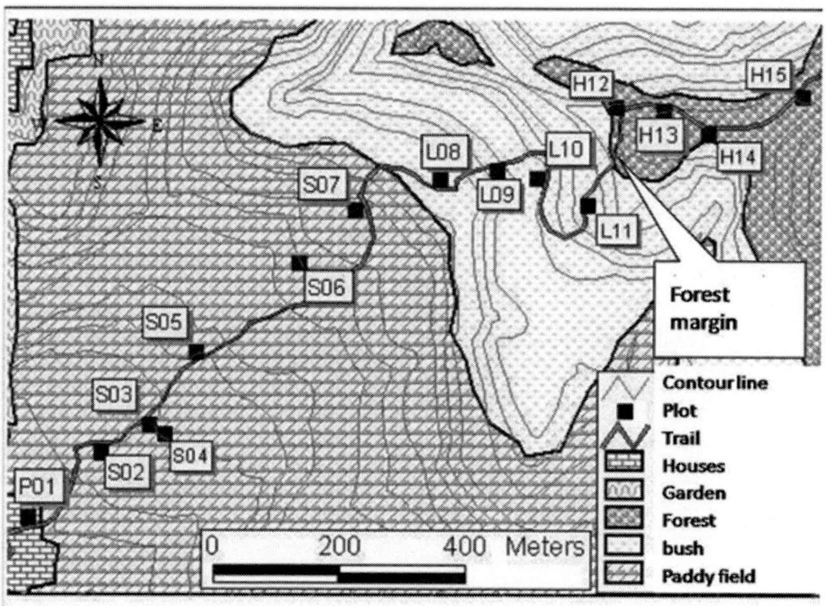

Fig. 1. Study sites of the area showing observation plots that are situated in different distances from the nearest forest.

Study on the effects of habitat isolation on pollinator diversity was conducted in the Halimun Salak National Park from December 2006 toSeptember

2007 (Sahari et al. unpublished). To study the effects of habitat isolation on species richness and composition of flower visiting-bee, approximately 15 plots were established. These plots are situated in different distances from the forest (between 0 m and 1078 m, which expresses three different habitats: forest, near forest (86m-714m), and far away from forest (822m-1078m, see Fig 1). About 20 mustard crops were set up in each plot. For each plot, during flowering season (about 25-30 days), flower visiting-bee were collected every 30 minutes between 06.00 am and 14.00 am.

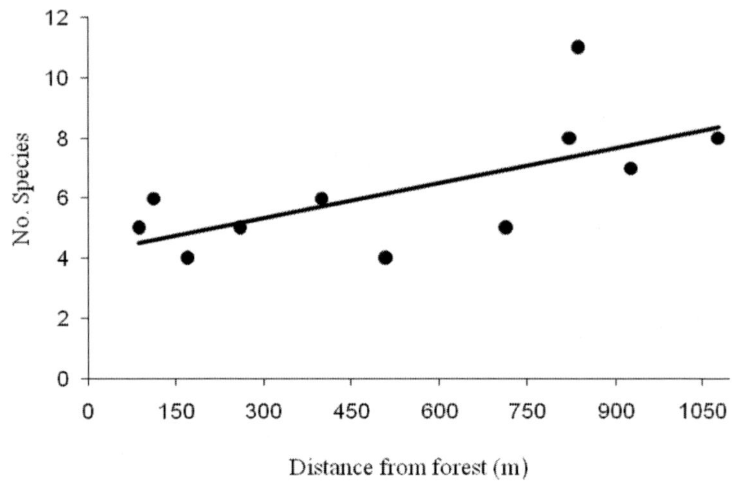

Fig. 2. Relationship between the number of flower-visiting bee species and forest distance ($y = 0.0039x + 4.1988$; $r^2 = 0.43$; $N = 11$; $P = 0.003$) in the forest margin of Halimun Salak Mountain, Indonesia.

The result of the study (Sahari et al. unpublished) indicated that increasing isolation of patches was not followed by decreasing numbers of flower-visiting bee species, instead the species number increased with increasing isolation to the nearest forest (see Fig.2). We also found a pronounced change of species composition between habitat types. Species composition was significantly different between forest, habitat near forest, and habitat far away from the forest. This can be seen in a two-dimensional scaling plot based on Srensen indices for measuring similarity of species composition between 15 sites of *B. rapa* plantations (see Fig. 3). The community similarity between sites of the same habitat type was higher than between sites of different habitat types. A similar result was also shown by Kremen et al. (2002), who revealed that species composition of wild bees changed with agricultural intensification and isolation between years.

Our study identified that there was a change of species composition and species rank across time (Sahari et al. unpublished, see also Hoehn et al..

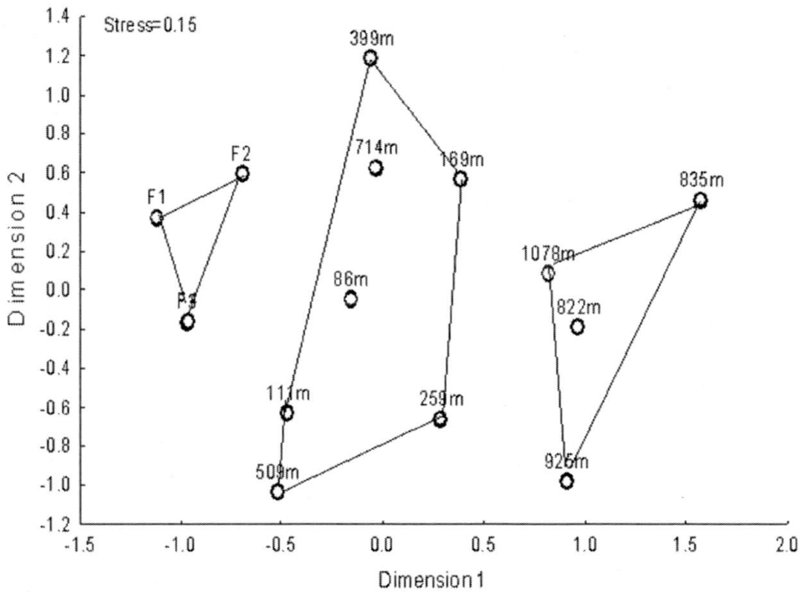

Fig. 3. Two-dimensional scaling plot based on Serensen indices for measuring similarity of species composition between 15 sites of *Brassica rapa* plantations. Connecting lines indicate three defined groups of agroforestry systems: (1) forest (F1-F4) (2) near forest (86m-714m) and (3) far away from forest margin (822m-1078m) in the forest margin of Halimun Salak Mountain, West Java. Insect pollinators were sampled from 20 plants per each site. For quantifying the similarity of different samples, we pooled the samples from the 20 plants per site. Srensen indices were used to quantify the similarity of species composition between sampled sites. Based on Srensen values we performed non-linear multidimensional scaling (MDS) which led to a two-dimensional projection of distances between the sites.

The community structure of insect pollinators' of *B. rapa* changed from the early to late flowering periods (see Fig 4). The temporal dynamic implies that composition of pollinators may be a successional phenomenon, i.e. there are changes of species composition across time that may reflect the phenology of plants and/or the changes that are happening at the landscape scale. This changes can result in functional substitution or complementation of species within existing pollinator communities. Hence, the functional roles of pollinators may stay constant across time.

Our results revealed a change of proportion (rank of dominance) of species composing the community of flower- visiting bees among forest habitat, forest margin, patches of near forests, and patches isolated from forest (see Fig 5). The changes could affect the rank order of the functional importance that species and the number of species require to provide sufficient pollination, as has been found by Kremen et al. (2002).

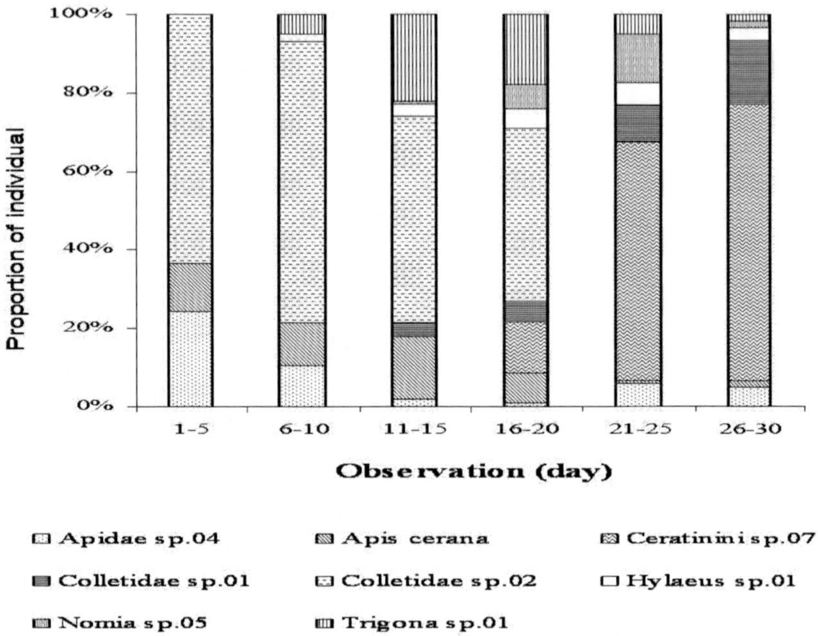

Fig. 4. Species succesion of insect pollinator visiting *B. rapa* crops, across observation days in the forest margin of Halimun-Salak Mountain.

Our survey identified that abundance of *Apis cerana* (social bee) decreased with increasing distance from the forest, while in contrast that of Ceratini (tribe) sp. increased with increasing distance from the forest (see Fig.6). This was also supported by Ramirez (2005) who identified that pollination overlap increased significantly from forest to disturbed areas. The result of this study indicated that generalist pollinators dominated the habitat and not specialised pollinators.

4.2 Pumpkin pollination

In an observational study conducted in Central Sulawesi, Indonesia, Hoehn et al (2008) found that seed set per fruit per pumpkin patch was positively correlated with bee species richness. Increasing number of seeds per fruit increases the fruit size (Hoehn et al 2008). The authors of this study showed that different flower-visiting bee species may compose different functional guilds that provide different pollination services depending on temporal and spatial niche (i.e. time of the day and position of the flower). In this case, the change of species composition is expected to dramatically change the shape of response.

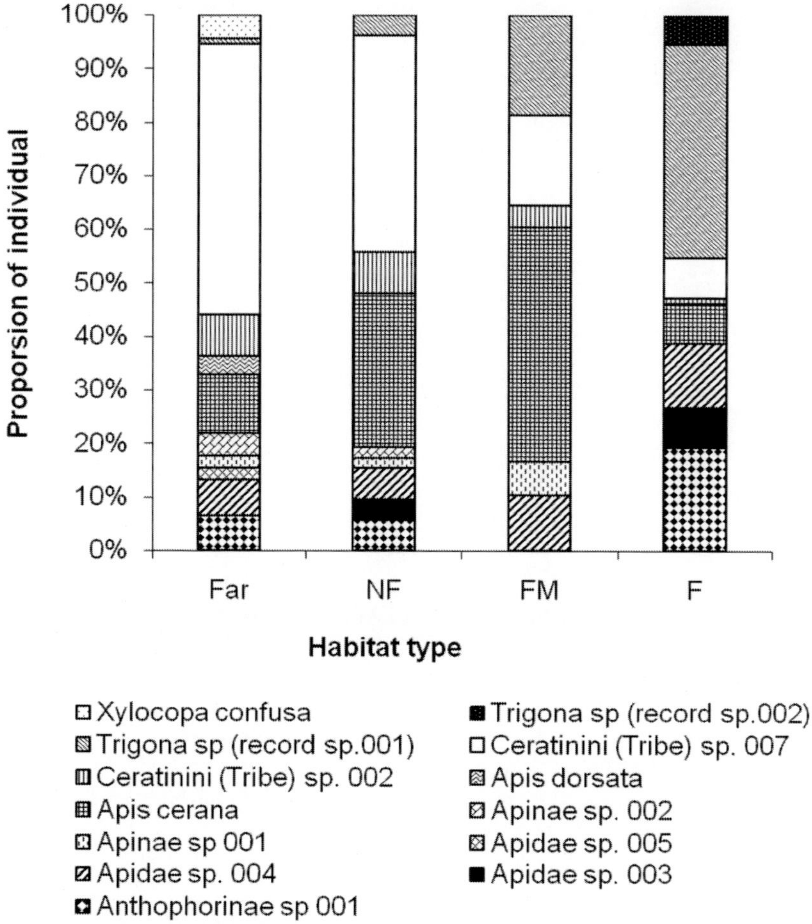

Fig. 5. Community structure of flower-visiting bee community visiting *B. rapa* crops in four different habitat types, (1) far away from forest (Far), (2) Near to forest (NF), (3) Forest margin (FM), and (4) Forest (F) of the forest margin of Halimun Salak area.

4.3 Coffee pollination

Klein et al. (2003c) identified the breeding and pollination system of lowland and highland coffee in Sulawesi, Indonesia and showed that fruit set of open pollinated coffee flowers increased with increasing number of bee species (Klein et al. 2003a,b). Solitary bees are more efficient pollinators on a per-visit basis than social honey and stingless bees on highland and lowland coffee (Klein et al. 2003a,b). Overall species richness, temporal and spatial variability in overall bee species richness, and similarity of coffee-flower visiting bee communities were negatively affected by rainforest isolation (Klein et al. 2003a,b,

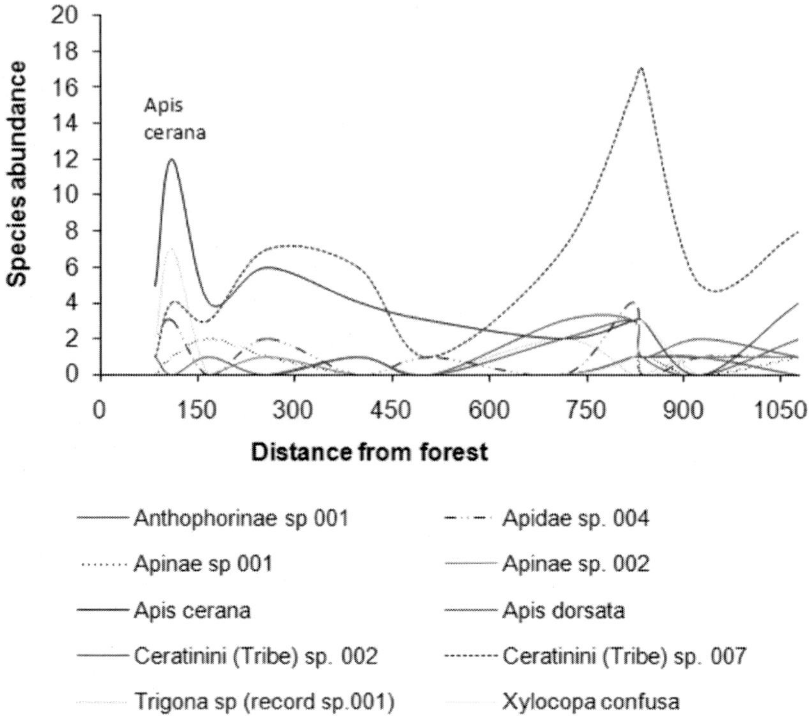

Fig. 6. Species abundance across sites that are situated in different distance from the forest of Halimun Salak Mountain, West Java.

Klein et al. 2009). However, in an adjacent research region in Sulawesi, Klein et al. (2002a) found that species richness of solitary bees and solitary cavity-nesting bees and wasps increased with increasing land-use intensity of coffee and cacao agroforests. The studies therefore indicate that rainforest isolation and land-use intensity are not necessarily characterized by lower species richness.

5 Conclusions

In summary, the effect of landscape change and rainforest isolation on crop pollination services shows the following: (1) Species richness and community composition are strongly affected by landscape changes; (2) These changes can affect the ecological functions played by different insect pollinator assemblages. If the change is related to the loss of pollinators, then it follows that the loss of species will result in the loss of ecosystem services; (3) temporal dynamics of pollinator communities may reflect a complementary services strategy played by different species composing the community; (4) Landscape changes and a changes of species composition may not significantly affect crop production since there can be ecological substitution mechanisms working in the plant-pollinator system.

Based on the studies on the effect of changing landscape on pollination and insect pollinator species richness, there are evidences on how habitat modification affects the services provided by insect pollinators, and this in turn may ultimately lead to pollination decline. Declining populations of native insect pollinators can create pollination deficit that would have a tremendous effect on decreasing "free service" for agricultural production (Thomson 2001; Sinu and Shivanna 2007). The decrease of free pollination services may cause an agricultural crisis for crops that are pollinator dependent, pollinator limited and pollinator specific (Ghazoul 2005) but see Aizen et al. 2008. More research is necessary, especially in countries such as Indonesia, to explain possible relationships and mechanisms between insect pollinators, crop production, and food supply in order to acquire appropriate approach for the conservation of insect pollinator communities and, ultimately, the pollination function.

For Indonesia, land-use change may result in considerable decreases in pollination services unless pollinator communities are conserved. Since agricultural productivity is mostly produced by small-scale farmers, conservation of pollinator in agricultural landscapes should be farmer-based. Eco-agriculture practices by enhancing habitat complexity that restores natural ecological processes should be established to maintain high extent of agrobiodiversity i.e pollinator diversity. Community-based pollinator conservation by developing farmer group that promotes eco-agriculture practices is a promising model for tropical developing countries to secure sustainable food supply by enhancing functional pollinator diversity. Creation of corridors between habitat patches may help in pollination.

Acknowledgements
We are grateful to Yann Clough and Alexandra Klein for their constructive comments and discussions on previous version of the manuscript, to Teja Tscharntke for his constant support in pursuing this publication. We also would like to thank Tri Atmowidi for sharing some part of his work and to Noer Hasanah, Nina Soeharina, Yus Rahman, Rizki Maulana for their help during the fieldwork. This work was supported by the Whitley Award, UK.

References

Agbagwa KO, Indukwu BC, Mensah SI (2007) Floral Biology, Breeding System, and Pollination Ecology of Cucurbita moschata (Duch. ex Lam) Duch. ex Poir. Varieties (Cucurbitaceae) from Parts of the Niger Delta, Nigeria. Turk J Bot. 31: 451-458

Aizen MA, Garibald LA, Cunningham SA, Klein AM (2008) Long term global trends in crop yield and production reveal no current pollination shortage but increasing pollinator dependency. Current Biology 18: 1572-1575

Aouar-sadli M, Louadi K, Doumandji SE (2008) Pollination of the broad bean (*Vicia faba* L.var. *major*) (Fabaceae) by wild bees and honey bees (Hymenoptera: Apoidea) and its impact on the seed production in the Tizi-Ouzou area (Algeria). African J. Agric Research. 3 (4): 266-272

Ashworth L, Quesada M, Casas A, Aguilar R, Oyama K (2009) Pollinator-dependent food production in Mexico. Biol. Conserv., doi:10.1016/j.biocon.2009.01.016

Atmowidi T, Buchori D, Manuwoto S et al (2007) Diversity of pollinator insects in relation to seed set of mustard (Brassica rapa L: Cruciferae). Hayati J Biosci 14: 155-161

Baumgärtner S (2007) The insurance value of biodiversity in the provision of ecosystem services. Nat Resource Modelling. 20 (1): 87-127

Beck J, Schulze CH (2000) Diversity of fruit feeding butterflies (Nymphalidae) along a gradient of tropical rainforest succession in Borneo with some remarks on the problem of pseudoreplicates. Trans Lepid Soc Japan 51: 89-98

Beck J, Schulze CH, Linsenmair KE et al (2002) From forest to farmland: diversity of geometer moths along two habitat gradients on Borneo. J Trop Ecol 8: 33-51

Bos MM, Veddeler D, Bogdanski AK et al (2007) Caveats to quantifying ecosystem services: fruit abortion blurs benefits from crop pollination. Ecol Appl 17: 1841-1849

Brühl CA (2001) Leaf litter ant communities in tropical lowland rain forests in Sabah, Malaysia: effects of forest disturbance and fragmentation. [PhD thesis]. University of Würzburg, Germany

Biro Pusat Statistik (2009) Produksi Tanaman Pangan 2008.

Cairns CE, Villanueva-Gutiérrez R, Koptur S, Bray DB (2005) Bee Populations, Forest Disturbance, and Africanization in Mexico. BIOTROPICA 37 (4): 686692

Cauich O, Javier J, Quezada-Eua JJG, Macias-Macias, JO, Reyes-Oregel V, Medina-Peralta S, Parra-Table V (2004) Behavior and Pollination Efficiency of *Nannotrigona perilampoides* (Hymenoptera: Meliponini) on Greenhouse Tomatoes (*Lycopersicon esculentum*) in Subtropical Mexico. J. Econ Entomol. 97 (2): 475-481

Chapin FS, Sala OE, Burke IC et al (1998) Ecosystem consequences of changing biodiversity. Biosci 48: 45-52

Chemini C, Rizzoli A (2003) Landuse change and biodiversity conservation in the alp. J Mt Ecol 7: 1-7

Cruz DO, Freitas BM, Antônio da Silva L et al. (2005) Pollination efficiency of the stingless bee *Melipona subnitida* on greenhouse sweet pepper. Pesq Agropec Bras 40: 1197-1201

Cunningham SA, FitzGibbon F, Heard TA (2002) The future of pollinators for Australian agriculture. Aust J Agric Res 53: 893-900

Delaplane KS, Mayer DF (2000) Crop Pollination by Bees. CABI Publishing, New York

Driscoll C (2003) Pollination ecology of *Tetratheca juncea* (Tremandraceae): finding the pollinators. Cunninghamia 8 (1): 133140

Dick CW (2001) Genetic rescue of remnant tropical trees by an alien pollinator. Proc R Soc Lond B 268: 2391-2396

El-sarrag MSA, Ahmed HM, Siddig MA (1993) Insect pollinators of certain crops in the Sudan and the effect of pollination on seed yield and quality. J King Saud Univ Agric Sci 5(2): 253-263

Faheem M, Aslam M, Razaq M (2004) Pollination ecology with species reference to insects. J Res (Sci) 15: 395-409

Gardener MC, Gillman MP (2002) The taste of nectar a neglected area of pollination ecology. OIKOS 98 (3): 552-557

Ghazoul J (2005a) Buzziness as usual? Questioning the global pollination crisis. TRENDs in Ecology and Evolution. 20 (7):367-373

Ghazoul J (2005b) Response to Steffan-Dewenter et al: Questioning the global pollination crisis. TRENDs in Ecology and Evolution 20 (12): 652-653

Ghzawi AA, Zaittoun ST, Makadmeh I, Ragman AM, Tawaha (2003) The impact of wild bees on the pollination of eight okra genotypes under semi-arid Mediterranian Conditions. Int J Agri Biol 5 (4): 408-410

Gituru WR, Wang Q, Wang Y, Guo Ybot (2002) Pollination ecoloy, breeding system, and coservation of Caldesia Grandis (Alismataceae), an endangered marsh plant in China Bull Acad Sin. 43: 231-240

Goulson D, Lye GC, Darvill B (2008) Decline and Conservation of Bumble Bees. Annu Rev Entomol. 53: 191208

Greenleaf SS, Kremen C (2006) Wild bee species increase tomato production and respond differently to surrounding land use in Northern California. Biol Conserv 133: 81 87

Hoehn P, Tscharntke T, Tylianakis JM, Steffan-Dewenter I. (2008) Functional group diversity of bee pollinators increases crop yield. Proc R Soc B 275: 22832291

Huston MA (2005) The three phases of landuse change: implication for biodiversity. Ecol Appl 15: 1864-1878

Irwin RE, Brody AK, Waser NM (2001) The impact of floral larceny on individuals, populations, and communities. Oecologia 129: 161168

Karron JD, Holmquist KG, Flanagan RJ, Mitchell RJ (2009) Pollinator visitation patterns strongly influence among-flower variation in selfing rate. Annals of Botany. 103: 13791383

Kasno, Hidayat P, Widayanti S, Suwardi (2004) Exploration on insect pollinators in improving the fruit production of mango and avocado. [Report]. SEAMEO BIOTROP, Bogor

Kasno, Hidayat P, Widayanti S (2002) Exploration of insect pollinators for highly exportable fruit crop:insect pollinators of durian and watermelon flowers. [Report]. SEAMEO BIOTROP, Bogor

Klein AM, Lira LF, Cardoso RS, Ferreira PCG et al (2003) Long-term population isolation in the endangered tropical tree species *Caesalpinia echinata* Lam. revealed by chloroplast microsatellites. Mol Ecol 12: 3219-3225

Klein AM, Müller C, Hoehn P, Kremen C (2009) Understanding the role of species richness for crop pollination services. In: Naeem S, Bunker DE, Hector A, Loreau M, Perrings C (Eds, 2009): Biodiversity, ecosystem functioning, and human wellbeing – an ecological and economic perspective. Oxford University Press, Oxford, pp 195-208

Klein AM, Olschewski R, Kremen C (2008) The Ecosystem Service Controversy: Is There Sufficient Evidence for a "Pollination Paradox"? GAIA 17:1216

Klein A-M, Steffan-Dewenter I, Buchori D et al (2002a) Effects of land-use intensity in tropical agroforestry systems on coffee flower-visiting and trap-nesting bees and wasp. Conserv Biol 16: 1003-1014

Klein A-M, Steffan-Dewenter I, Tscharntke T (2002b) Predator-prey ratios on cocoa along a land-use gradient in Indonesia. Biodiv Conserv 11: 683-693

Klein A-M, Steffan-Dewenter I, Tscharntke T (2003a) Pollination of *Coffea canephora* in relation to local and regional agroforestry management. J Appl Ecol 40: 837-845

Klein A-M, Steffan-Dewenter I, Tscharntke T (2003b). Fruit set of highland coffee increases with the diversity of pollinating bees. Proc R Soc Lond. B. 270: 955961

Klein A-M, Steffan-Dewenter I, Tscharntke T (2003c) Bee pollination and fruit set of *Coffea arabica* and *C. canephora* (Rubiaceae). Am J Bot 90: 153-157

Kremen C, Williams NM, Thorp RW (2002) Crop pollination from native bees at risk from agricultural intensification. PNAS 99: 16812-16816

Kremen C (2005) Managing ecosystem services: what do we need to know about their ecology? Ecol Letters 8: 468-479

Kremen C, Ostfeld RS (2005) A call to ecologists: measuring, analyzing,and managing ecosystem services. Front Ecol Environ 3 (10): 540548

Losey J E, Vaughan M (2006) The economic value of ecological services by insects. Biosci 56: 311-323

Loreau M, Naeem S, Inchausti P et al (2001) Biodiversity and Ecosystem Functioning: Current Knowledge and Future Challenges. Sci compass 294: 804-808

Malerbo-Souza DT ,Nogueira-Couto RH, Couto LA (2004) Honey bee attractants and polliation in sweet orange, Citrus sinensis (L.) Osbeck, var. Pera-rio J. Venom. Anim Toxins incl Trop Dis 10 (2): 144-153

Mitchell RJ, Irwin RE, Flanagan RJ, Karron JD (2009) Ecology and evolution of plantpollinator interactions. Annals of Botany. 103: 13551363

Memmott J, Waser NM, Price MV (2004) Tolerance of pollination networks to species Extinctions. Proc R Soc Lond B. 271: 26052611

Moguel P, Toledo VM (1999) Biodiversity conservations in traditional coffee systems of Mexico. Conserv Biol 13: 11-22

Mustajarvi K, Siikamaki P, Rytkonen S, Lammi A (2001) Consequences of plant population size and density for plantpollinator interactions and plant performance. J. Ecol. 89: 8087

Olschewski R, Tscharntke T, Benitez PC et al (2006) Economic Evaluation of Pollination Services Comparing Coffee Landscapes in Ecuador and Indonesia. Ecol Soci 11: 7

Pardede DB (1990) Bioecology of *Elaeidobius kamerunicus* Faust and its pollination success in oil palm plantation, North Sumatra. [PhD dissertation, in Indonesian]. Bogor Agricultural University. Bogor

Perfecto I, Mas A, Dietsch T et al (2003) Conservation of biodiversity in coffee agroforestry systems: a tri-taxa comparison in southern Mexico. Biodiv Conserv 12: 1239-1252

Perfecto I, Vandermeer J, Hanson P et al (1997) Arthropod biodiversity loss and the transformation of tropical agroecosystem. Biodiv Conserv 6: 935-945

Ramirez N (2005) Temporal overlap of flowering species with the same pollinating agent class: the importance of habitats and life forms. Int J Botany 1: 27-33

Raju AJS, Rao SP (2002) Pollination ecology and fruiting behaviour in *Acacia sinuata* (Lour.). Merr. (Mimosaceae), a valuable non-timber forest plant species. Curr Sci 82: 1466-1471

Roubik DW (1993) Direct costs of forest reproduction, bee-cycling and the efficiency of pollination modes. J. Biosci 18 (4): 537-552

Roubic DW (2001) Ups and downs in pollinator populations: When is there a decline. Conservation Ecology 5 (1): 2 (http:www.consecol.org/vol5/iss1/art2/)

Rice RA, R Greenberg (2000) Cacao cultivation and the conservation of biological diversity. Ambio 29: 167-173

Riffell JA, Alarcón R, Abrell L (2008) Floral trait associations in hawkmoth-specialized and mixed pollination systems. Communicative & Integrative Biology 1 (1): 6-8

Rizali A, Buchori D, Triwidodo H (2002) Insect diversity at the forest margin-rice field interface: indicator for a healthy ecosystem (in Indonesian). Hayati J Biosci 9: 41-48

Ruz L (2002) Bee Pollinators Introduced to Chile: a Review . IN: Kevan P, Imperatriz Fonseca VL (Eds) - Pollinating Bees - The Conservation Link Between Agriculture and Nature - Ministry of Environment / Brasília, p.155-167

Sinu PA, Shivanna KR (2007) Pollination biology of large cardamom (*Amomum subulatum*). Curr Sci 93: 548-552

Schiestl FP, Schlüter PM (2009) Floral Isolation, specialized pollination, and pollinator behavior in Orchids. Annu Rev Entomol. 54: 42546

Schlumpberger BO, Cocucci AA, More M, Se'rsic AN, Raguso RA (2009) Extreme variation in floral characters and its consequences for pollinator attraction among populations of an Andean cactus. Annals of Botany. 103: 14891500

Schulze CH, Steffan-Dewenter I, Tscharntke T (2004) Effects of land use on butterfly communities at the rain forest margin: a case study from Central Sulawesi. In: Gerold G, Fremerey M, Guhardja E (Eds) Land Use, Nature Conservation and the Stability of Rainforest Margins in Southeast Asia. Springer, Berlin

Slaa EJS, Sanchez Chaves LA, Malgodi-Braga KS, Holfstede FE (2006) Stingless bees in applied pollination: practice and perspectives. Apidologie 37 : 293315

Steffan-Dewenter I, Potts S, Packer L (2005) Pollinator diversity and crop pollination services are at risk. Trends Ecol. Evol. 20:

Steffan-Dewenter I, Westphal C (2008) The interplay of pollinator diversity, pollination services and landscape change. J Appl Ecol 45: 737741

Thomson JD (2001) Using pollination deficits to infer pollinator declines: Can theory guide us? Conserv Ecol 5: 6

Turner WR, Brandon K, Brooks TM, Costanza R, daFonseca GAB, Portela R (2007) Global conservation of biodiversity and ecosystem services. BioScience. 57 (10): 868-873

Yi-Bo L, Yan-Long P, Kai Yu P et al (1998) A study on pollination biology of Paeonia Suffruticosa Subsp. Spontanea (Paeoniacea). Acta Phytotaxonomia Sinica 36: 134-144

Structure and management of cocoa agroforestry systems in Central Sulawesi across an intensification gradient

Jana Juhrbandt[1*], Thomas Duwe[2], Jan Barkmann[1], Gerhard Gerold[2], and Rainer Marggraf[1]

[1] Environmental and Resource Economics, Department of Agricultural Economics and Rural Development, Georg-August-University Göttingen, Platz der Göttinger Sieben 5, D-37073 Göttingen, Germany

[2] Department of Landscape Ecology, Institute of Geography, Georg-August-University Göttingen, Goldschmidtstr. 5, 37077 Göttingen, Germany

*corresponding author: J.Juhrbandt, email: jjuhrba@gwdg.de

Summary

Central Sulawesi is a major cocoa producing region in Indonesia. Nevertheless, very little is known about the basic socio-economic and pedological properties of cocoa agroforestry systems in the region. In the vicinity of Lore Lindu National Park (LLNP), 144 cocoa plots covering an intensification gradient were selected for an intensive 1-year cocoa management study including a subset of 48 plots for extended soil analyses.

Local cocoa plots are mostly established by converting natural forest lands, and they are increasingly intensified by removal of their natural shade tree cover. Soil nutrient status is mostly sufficient but total P availability and stagnant soil water conditions limit yields. Phytosanitary and soil amelioration management are often suboptimal and may need to be improved. Marketing of cocoa beans takes place mostly via small traders from the same village. Farm gate prices account for around 70% of world prices. No price incentives exist for enhancing bean quality by better processing.

Cocoa bean yield varies strongly by season. A structural intensification index integrating data on canopy closure, cocoa tree density, number of native forest trees and number of intercrops, was positively correlated with yields. Labour input also increases yield. Labour input was not correlated with the structural intensification index but expenses for material inputs and hired labour as well as biophysical parameters as rainfall and phosphor content

T. Tscharntke et al. (eds.), *Tropical Rainforests and Agroforests under Global Change*, Environmental Science and Engineering, 115–140, DOI 10.1007/978-3-642-00493-3_5,

were. The strong economic incentive for farmers to intensify cocoa agroforests threatens local biodiversity and ecosystem functioning.

Keywords: Cocoa agroforests, plot structure, shade management, yields, yield determinants, intensification, soil analysis

1 Introduction

Although already introduced to Java at the beginning of the 18th century and to Northern Sulawesi in the late 18th century, Indonesian cocoa did not play a major role at world markets until the 1990ies (Durand 1995, Pomp and Burger 1995). Between 1980 and 1994, Indonesia's cocoa production increased at an average rate of 26 percent p.a. Presently, Indonesia is the third largest producer of cocoa after Ivory Coast and Ghana with over 480.000 metric tons (MT) produced 2007/2008 (13% of global production; ICCO 2008). Most recent production increases are in bulk cocoa produced by cocoa hybrid lines (FAO 2003). Smallholders from Central Sulawesi, Southeast Sulawesi and South Sulawesi provinces produce nearly 75 percent of the national cocoa bean output (Akiyama and Nishio 1996, COPAL 2008) providing the main source of income for over 400.000 farming households (Panlibuton and Meyer 2004).

Indonesias cocoa expansion was favoured by the availability of suitable land providing "forest rents" (high initial soil fertility and low levels of pests and abundance of pollinators), low production cost, a relatively good transport infrastructure, favourable macroeconomic policies, and the entrepreneurship of smallholders (Akiyama and Nishio 1996, Ruf et al. 1995). Low taxation and efficiently working local cocoa marketing channels result in relatively high producer ("farm gate") prices.

In the past three decades, cocoa area increased from zero (1979) to ˜18.000 ha around and inside Lore Lindu National Park (LLNP) in Central Sulawesi (Maertens 2003), which is part of the Wallacea biodiversity "hotspot" (Myers et al. 2000). Agricultural expansion of perennial cropping systems - mainly cocoa - often after illegal slashing of natural forest were identified as the main drivers of regional forest conversion (Erasmi et al. 2004, Koch et al. 2008). Although traditionally wet rice is grown around LLNP, cocoa agroforestry already provides substantially more income (Schwarze 2004). Intensification of cocoa agroforestry systems appears as a financially favoured strategy, as yields can nearly be doubled when decreasing canopy cover from medium (50-65%) to zero-shade conditions (Steffan-Dewenter et al. 2007, Schneider et al. 2006). Consequently, a shift takes place from multilayer agroforestry systems with diverse shade canopies to rather simple structured cocoa plantations with only one or two planted shade tree species (Siebert 2002, Steffan-Dewenter et al. 2007).

Although data availability on long term cocoa agroforestry performance under different management intensities is scarce, declining cocoa yields in sun-grown systems are known, for example, from Ghana. Pest and disease pressures are usually mounting a few years after the introduction of widespread, intensified cocoa cropping (Ahenkorah et al. 1974, Ahenkorah et al. 1987). In the project region, intensification may also put soil fertility and sustainable production at risk (Belsky and Siebert 2003). With regard to soil fertility, local cocoa agroforestry appears less demanding than annual upland crops (Dechert et al. 2004). First reports of partly dramatic yield losses have appeared in recent years from Sulawesi, however (Reuters 2009).

During the last decade, agroforestry systems have repeatedly been evaluated with respect to conservation aspects. Shaded coffee agroforestry systems with natural forest trees can maintain some of the original rainforest biodiversity (Perfecto et al. 1996, Moguel and Toledo 1999). Recent results from the LLNP area indicate that this is partly also true for shaded cocoa agroforestry systems (Steffan-Dewenter et al. 2007). Although substantial loss in specialized forest species takes place during the initial conversion of rainforests to agroforests, high to intermediate shade cocoa plots may still serve as important habitat for the native flora and fauna. A further shift to intensive, low shade plots is expected to result in high additional biodiversity losses (Steffan-Dewenter et al. 2007).

Previous studies of agriculture in the LLNP region focused on the intermediate section of the intensification gradient. In order to analyse the implications of the ongoing intensification process as well as pro-biodiversity policy options requiring high shade cocoa agroforestry systems, data across the entire intensification gradient are necessary, though. Thus, we designed and conducted a detailed, one-year cocoa management study that documents plot establishment and structure, soil nutrient status, pest and disease pressure, as well as plot management, yields, processing and marketing. In this contribution, we focus on descriptive data and bivariate correlations between plot and management variables including cocoa yield. This is the first step in a comprehensive set of analyses on ecology-economy trade-offs in cocoa cropping around LLNP. With the initial "forest rents" (Ruf 1995) steadily declining, such knowledge is indispensable for a systematic search for sustainable land use options that improve rural incomes without unnecessarily jeopardizing biological diversity.

2 Methods

2.1 Study area and sampling

The study was conducted in 12 villages in the vicinity of LLNP in Central Sulawesi, Indonesia. The selected villages are part of a 13 village random sample (Zeller et al. 2002). The villages are located in four valleys covering altitudes from 75 to 1275 m a.s.l.: Palu valley (Maranata, Pandere and Sidondo II), Palolo Valley (Berdikari, Bulili and Sintuwu), Napu valley (Watumaeta, Wuasa, Wanga and Rompo) and Kulawi valley (Bolapapu and Lempelero). This region provides near to optimal agro-climatic conditions for cocoa farming, which includes an annual precipitation of 1500-2000 mm, a dry season of not more than 3 months, and temperatures with 30-32°C mean maximum and 18-21°C mean minimum. Soil depth should not be less than 1.5 m and soil pH should be between 6.0 and 7.5 (Wood 1985a). In mountainous regions of LLNP, annual average precipitation reaches 2500 mm (Berlage 1949 in Leemhuis 2005). In the sampled villages, rainfall varies between 1215 mm (Sigimpu, 640 m a.s.l.) and 1900 mm (Talabosa, 1090 m a.s.l.). Temperatures range from 21°C (Wuasa, 1133 m a.s.l.) to 27.4°C (Pandere, 93.3 m a.s.l.) (Daily meteorological data from 2002-2006, STORMA-B1, H. Kreilein.)

In each of the 12 villages, a sample of one cocoa plot of each of 12 cocoa producing households was selected, resulting in a total sample size of 144 plots. The cocoa agroforestry plots were not randomly selected but systematically chosen to represent the entire intensification gradient of high to low canopy closure (CC) values. Canopy closure is the proportion of the sky hemisphere obscured by vegetation when viewed from a single point (Jennings et al. 1999). Plot selection was accomplished in two waves in 2006 guided by German researchers with prior experience in the project region, and supported by local staff. Site selection was conducted based on farmer assessments of plot canopy closure and on-site verifications by hemispherical convex densiometer measurements (Model-C, Robert E. Lemmon). Per village, three plots were identified for each of 4 *a priori* defined shading categories: (near) natural forest cover (>85% CC; category "1"), dense shade cover (>65% CC; "2"); medium shade cover (>35% CC; "3"); low to zero shade (0-35% CC; "4"). For all plots, structural and management data were sampled (2.2, 2.3), and soil analysis was conducted for a subset of 48 plots (2.4).

2.2 Agroforest structure

Plots were characterised in terms of plot history and structure including cocoa tree density, intercrops and shade trees. Plots were geo-referenced and photographed, and their layout sketched. Shade tree cover, i.e. CC, was monitored three times from 2006 to 2008. We measured CC as the average of 8-16 randomly selected points per plot using a hemispherical convex densiometer.

Canopy closure itself can already be viewed as a proxy for intensification in cocoa agroforestry (Juhrbandt and Barkmann 2008). However, intercrops such as banana or coconut also contribute to CC, and a dense upper canopy may even consist of trees if a single planted shade tree species without conservation value (e.g., *Glyricidia sp.*). Specifically for analyses in a biodiversity conservation context, CC is a very rough an indicator. For a compact albeit more comprehensive inclusion of structural plot parameters, we turned to a Management Intensity Index (MI) suggested by Mas and Dietsch (2003). Adapting their concept, our MI includes the planting density of cocoa trees as well as the total number of native forest tree species and intercrop species per plot besides CC. Each of the four components of the index was normalised, and values added. Resulting MI scores range from 0 to 4 with 4 indicating the most intensive system.

2.3 Agroforest management

Farmers were contracted to prepare weekly records on yields and several yield determining factors from January to December 2007. In each village, one particularly collaborative farmer was employed to support the preparation of the records. Every month, local university graduates collected and checked the management record sheets. Surveyed parameters include capital and labour used for: plot management activities (including phytosanitary measures), cocoa pod and bean processing, for changes in plot structure, intercropping, fertilizer input, pesticide input, fungicide and herbicide input. Finally, yield of fresh pods and proceeds from dry bean marketing were recorded.

Adoption of agricultural innovations as well as farmer perceptions on soil fertility, and on the impact of pests, diseases, dryness and tree age on cocoa production were surveyed additionally. Particularly, farmers estimates on yield losses due to Cocoa Pod Borer attacks (CPB) *(Conopomorpha cramerella)* and Black Pod Disease (BPD) *(Phytophthora palmivora* L.*)* were gathered for the beginning of infestation and in 2007.

2.4 Soil analyses

One plot per shading category in each village was selected for soil analyses, resulting in a subset of 48 cocoa plots. Accessible and homogeneous plots were preferentially selected. In order to locate the sampling plot, 6 to 15 Pürckhauer profiles were analyzed for each plot. Based on this on-site analysis, a representative 20m x 20m sampling plot was chosen. Within the sampling plot a 1m x 1m x 1m soil profile was excavated. Soils were classified into two water condition categories:
0= Dry to fresh sites: Groundwater level 2-3m, soil profile shows no stagnant moisture.
1= Moist and groundwater sites: Groundwater level 1m or less, close to rivers, or flooded after heavy rainfall; soil profile showed strong stagnant moisture or gleyic conditions.

Within the sampling plot, three 5m x 5m subplots were defined surrounding the soil profile. Mixed samples were taken at three depths (0-10cm, 10-30cm, 30-50cm) by five Pürckhauer profiles per subplot. These depths cover the main distribution of roots and soil nutrient stocks (De Geus 1973, Hartemink 2005).

Measured soil parameters which are essential to judge soil nutrient status in the tropics include: the total amount of Carbon (C_t), Nitrogen (N_t) and Phosphorus (P_t), the amount of available Phosphorus (P_{av}) (cf. Bray and Kurtz 1945), exchangeable Calcium (Ca_{ex}), Potassium (K_{ex}), Magnesium (Mg_{ex}) and Aluminium (Al_{ex}), and the effective Cation Exchange Capacity (CEC_{eff}). Lanfer (2003) provides a simple classification scheme in terms of general soil nutrient status (Tab.1). The classification scheme is based on a synthesis of several dedicated studies. Nutrient concentrations were converted into kg ha^{-1} (sampled thickness [m] x bulk density [kg m^{-3}] x nutrient concentration [kg kg^{-1}] x area [m ha^{-1}]).

Table 1. Classification for different soil parameters, derived from different studies (see below).

parameter level	C_t	N_t	P_{av}	Ca_{ex}	K_{ex}	Mg_{ex}	Al_{ex}	CEC_{eff}
low	<1.5	<0.10	<3	<0.4	<0.15	<0.2	<0.3	<4
medium	1.5-4.5	0.10-0.15	3-7	0.4-4	0.15-0.3	0.2-0.8	0.3-1.0	4-8
high	>4.5	>0.15	>7	>4	>0.3	>0.8	>1.0	>8
source	(1)	(2)	(1), (4)	(1)	(1)	(1)	(3)	(1)

(1) Cochrane & Sanchez (1982), (2) Guamn (1999), (3) Iniap, (in Lanfer 2003). (4) Bray 1945. P_t is not included in this classification system.
C_t and N_t in [%], P_{av} in [ppm] and Ca_{ex}, K_{ex} , Mg_{ex}, Al_{ex} and CEC_{eff} in [cmolkg^{-1}].

2.5 Data analyses

Labour, inputs and outputs were aggregated at a monthly and yearly level for further analyses. All parameters were normalized on per ha basis, except the number of native forest tree species and the number of intercrops. As most species are not homogeneously distributed, species richness is not increasing continuously with area, so that an up- or downscaling of species richness with area would lead to biased results. Cocoa yields are calculated as kilograms sun-dried cocoa beans per hectare sold to small traders, middlemen or collection centres.

Gross margins (US$ per ha) are calculated as the differences between revenues (sale of cocoa beans and intercrops) and variable input costs. Variable costs include expenses for pesticides and fertilizers, transport costs, paid labour, seeds and other material. Returns to labour are calculated as US$ gross margin per hour of total working time.

Pearson correlation analysis and regression analysis was used to identify linear relationships between canopy closure and several CC-dependant variables as well as between yields and several yield determining factors. For linear regression analysis, OLS was used as estimator and influential observations were excluded using Cooks distance measures (>4/n are problematic cases). One-way ANOVA with Tukey post-hoc tests were applied to determine group differences in marketing analysis. All statistical analyses were carried out with SPSS 16.0 and Stata 9.2.

3 Results

3.1 Structure and management of cocoa plots

Plot establishment

35.4% of the farmers reported that they established their cocoa plots by converting natural forest land. Cocoa agroforestry usually follows a few seasons of dry land agricultural crops. 22.9% reported to have converted other perennial crop systems (coconut or coffee), and 25% reported that they converted land with annual crop. 28.5% were purchased as established cocoa plots between 1970 and 2005. Between 1995 and 2005, the average price was 582.8 US$/ha. Even after adjustment for inflation, plot prices significantly increased since 1995 ($P < 0.001$, inflation adjusted according to International Monetary Fund, World Economic Outlook Database, April 2009).

Plot structure

The entire CC gradient was covered although no zero shade plots were found in the Palu valley where cocoa plots are often grown under coconut trees. CC ranged from 1.6% to 98.6% in 2008 (average CC 42.4% in 2008). Cocoa plot size ranged between 0.4 and 3.3 ha (0.63 ha on average) with 75% of the plots smaller than 1 ha. With substantial variability, mean planting density was 854 (STD 346.2) cocoa trees per ha. Planting density was highest in Palolo. Cocoa tree age varied between 3 and 27 years. In Palolo, cocoa trees were slightly older on average reflecting a longer cocoa cropping history.

We found high variability of intercrops and shade trees. Native forest trees were present on 66% of the plots with up to 9 different species per plot. A high share of forest trees was found in Kulawi valley plots. A total of 80 different native forest tree species were identified by farmers. 91.2 % of the plots were intercropped with 1 to 5 intercrops (mean 1.8). In total, 20 different intercrops were found. Intercrops were predominantly bananas and perennials such as fruit trees, coconut or coffee. Vanilla or vegetables were also frequently grown. Highest diversity of intercrops was found in the Palu valley.

Pest and disease pressure

Cocoa Pod Borer (CPB) *(Conopomorpha cramerella)* and Black Pod Disease (BPD) *(Phytophthora palmivora* L.*)* spread rapidly: Farmers reported that BPD and CPB arrived in their villages around the year 2000 or later. In 2007 BPD and CPB occurred on 100% and on 99% of all plots respectively. Farmers estimated yield losses of 24.3% on average due to CPB (median 20%, maximum 70%), and 20.5% due to BPD. BPD and CPB induced yield losses are correlated ($r = 0.45$, $P<0.001$). Plots farther away from the forest edge showed lower CPB yield losses ($r = -0.215$, $P=0.01$). Yield losses due to BPD decreased with higher altitude ($r = -0.32$, $P<0.001$).

Recommended cultural control techniques to combat CPB and BPD include 4 major steps: frequent harvest (at 14 days interval), cocoa pruning, sanitation of pod husks (burning or burying) and fertilization (ACDI/VOCA 2005). None of the household applied all 4 steps, and only 8.4% reported employing 3 steps. Whereas 37.8% of farmers practiced sanitation of pod husks, not a single household harvested at a 14-day-interval. In contrast 51.7% of the farmers used pesticides. Farmers who had suffered high yield losses already when CPB first occurred at their plots, reported spraying pesticides more frequently later ($r=0.28$, $P=0.013$).

Table 2. Stocks and available nutrients from 48 cocoa plots within the first 30 cm of the topsoil.

	C_t	N_t	P_t	P_{av}	Ca_{ex}	K_{ex}	Mg_{ex}	Al_{ex}	CEC_{eff}
Mean	52.068	5.024	2.193	12	6.889	664	1.148	301	341
SD	15.831	1.752	1.312	10	2.904	292	773	333	137
Max	98.914	8.612	7.279	41	13.132	1.515	3.357	1.463	648
Median	49.902	4.992	1.856	10	6.168	636	765	202	298
Min	25.257	1.756	458	1	1.578	247	407	42	130

About half of the households were able to reduce yield losses occurring by CPB (51%) and BPD (53.8%) attacks since begin of infestation. More frequent pruning of cocoa trees helped reducing BPD ($r = -0.177$, $P=0.041$), higher fertilizing and pesticide application frequency reduced CPB yield losses ($r = -0.2$, $P=0.018$ and $r = -0.176$, $P=0.038$ respectively), whereas an increase in CC between 2007 and 2008 led to increasing CPB yield losses ($r=0.279$, $P=0.001$).

Soil characterization

Cocoa plots are located on lower slopes, alluvial fans and at the border of the alluvial basins, resulting in geologically young topsoil. According to the

WRB/FAO (2007) soil classification, the following soil types were found: Cambisols, Gleysols, Phaeozemes, Stagnosols and Fluvisols. Cambisol was found on 30 of 48 plantations in different specifications (gleyic, eutric, stagnic, fluvic, endoskeletic). A catena of soil types was identified from the slope to the basin: Cambisol followed by stagnic/ gleyic Cambisol and in the basins Gleysol, and fluvic/ gleyic Cambisol. A comprehensive characterisation of the stocks and available nutrients from the 48 investigated plots is provided in Tab. 2.

Table 3. Nutrient status distribution of cocoa plots (n) according to Lanfer (2003); numbers vary due to data gaps (missing analysis).

	C_t	N_t	P_{av}	Ca_{ex}	K_{ex}	Mg_{ex}	Al_{ex}	CEC_{eff}
Low	38	18	34	1	0	1	12	1
Medium	6	19	6	3	12	0	12	29
High	0	11	4	36	28	39	16	10

All units are kg $ha^{-1}0.3m^{-1}$ except CEC_{eff} [kmol ha^{-1} $0.3m^{-1}$]; av. = available, ex. = exchangeable. Many plots are classified as low nutrient plots mainly for C and Pav. For N_t and exchangeable Al, plots are evenly distributed across all three categories. For the remaining nutrients and CEC_{eff}, most plots attain a medium to high nutrient status (Tab. 3).

Only 43% of surveyed households ever fertilized their cocoa plots, and in 2007 even only 27.3% did so. 30.1% of farmers stated that soil fertility was already reduced compared to the time when they started to manage the plot.

Cocoa yields

2007 yields showed a broad range (7-1613 kg/ha) with an average of 476.9 kg/ha and pronounced seasonality (Fig.1).

Canopy closure, presence of stagnant soil water conditions, number of native forest tree species and number of intercrops were correlated in this order with decreasing yields (Tab. 4). Increasing yields were correlated with cocoa tree density and labour input. Among themselves, plot structural parameters were frequently correlated. Aggregated in a Management Index (MI) as a proxy for plot structure intensification, they revealed substantial explanation power for cocoa yield variation (Fig.2).

Influences of soil parameters surveyed on a subset of plots (n=48) on cocoa yields are not very strong. In regression analysis, only total soil phosphor content is a yield determinant (Tab 5). The model improves when a dummy for stagnant soil water conditions is included, which has a negative influence on yields.

Table 4. Correlation of plot structure parameters and cocoa yields.

	cocoa yield [kg ha^{-1} yr^{-1}]	planting density of cocoa trees [ha^{-1}]	no. forest tree species per plot	no. inter-crops per plot	canopy cover [%]	material input expenses [IDR ha^{-1}]	total work time [h ha^{-1} yr^{-1}]
cocoa yield [kg ha^{-1} yr^{-1}]	1						
planting density [cocoa trees ha^{-1}]	0.337**	1					
no. forest tree species per plot	-0.205*	n.s.	1				
no. intercrops per plot	-0.197*	-0.226**	n.s.	1			
canopy cover [%]	-0.396**	-0.169*	0.182*	0.279**	1		
material input expenses [IDR ha^{-1}]	n.s.	n.s.	n.s.	n.s.	-0.141	1	
total work time [h ha^{-1} yr^{-1}]	0.248**	0.191*	n.s.	n.s.	n.s.	n.s.	1

Pearson correlations; n.s.: not significant; displayed correlation coefficients significant at $p<=0.1$; *: $p<=0.05$; **: $p<=0.01$.

Table 5. Regression analysis, dependant variable: cocoa yield. N=43.

Cocoa dry bean yield [kg ha^{-1}] vs.	R^2	p	Coeff. total soil P	Coeff. stagnant water
Total soil P [kg ha^{-1}]	0.21	0.002	0.456 (p=0.002)	
Total soil P [kg ha^{-1}] stagnant water (0/1)	0.27	0.002	0.453 (p=0.002)	-0.25 (p=0.07)

Processing and marketing

Of four main post-harvest activities, cracking of the cocoa husk followed by extraction of the beans require, by far, most labour per ha (Fig.3).

71.8% of all cocoa beans were sold within the same village, mostly to small traders or middlemen. Only 14.4% of sales were done at a cocoa collection centre in the provincial capital Palu.

Producer prices at farm gate rose quickly in early 2007, and peaked in July (Fig.4) closely following world market price (fob, ICCO monthly averages; ICCO 2008). Farmers of the project region received on average 70.2% (minimum: 62.4%, maximum 77.4%) of the world market price. The linear correlation between local and world market prices is very high (r=0.834, P=0.01).

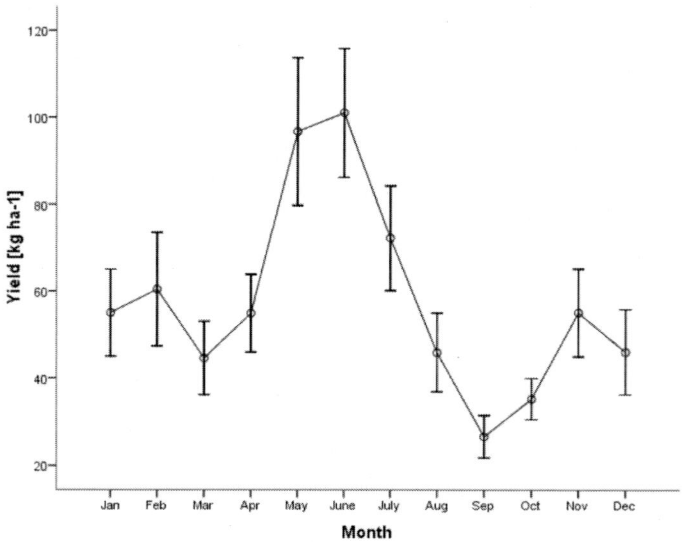

Fig. 1. Average monthly yields in 2007 (n=143, error bars show standard deviations).

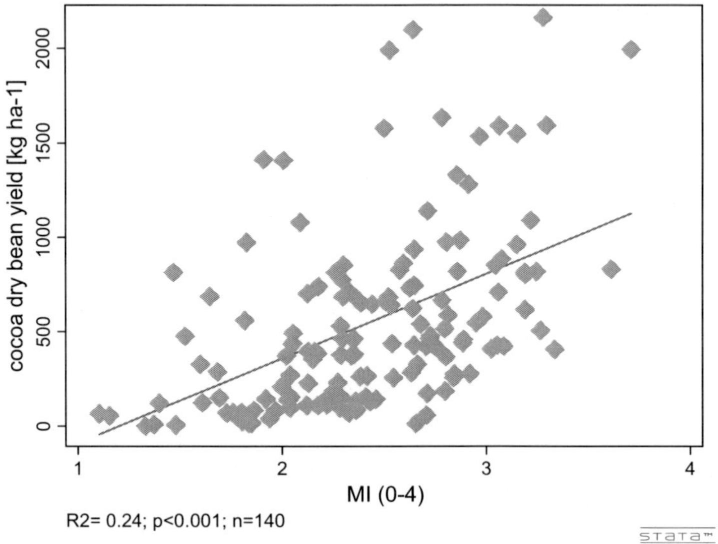

Fig. 2. Cocoa dry bean yield 2007 in relation to a Management Index MI (0-4) composed of plot structural parameters.

One-way-ANOVA with Tukey post-hoc tests showed the following significant cocoa price differences ($p<0.01$): Prices were higher when cocoa beans

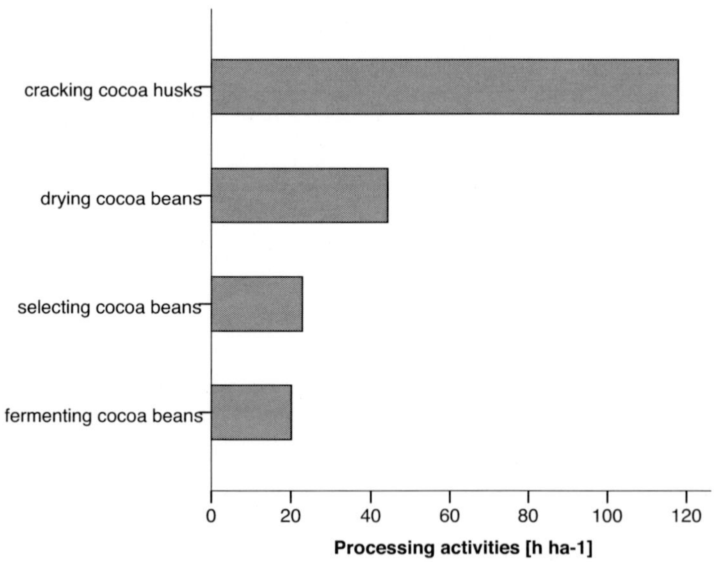

Fig. 3. Labour requirement for cocoa bean processing (means of n=144 plots).

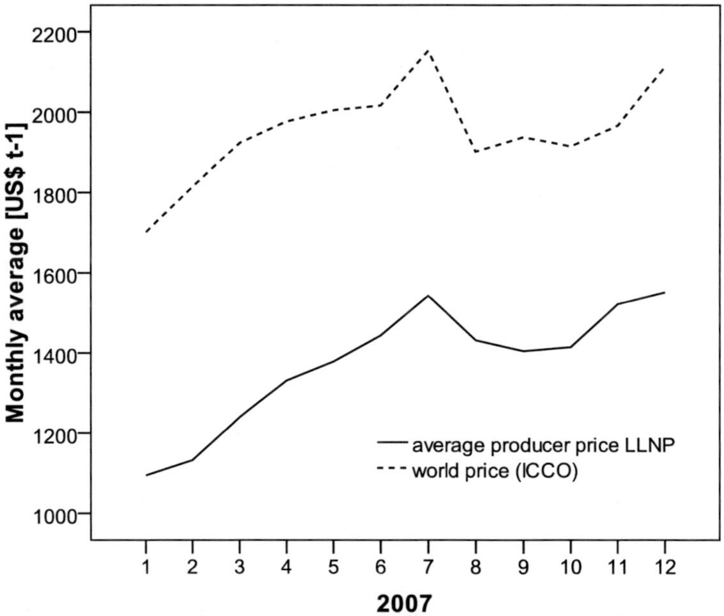

Fig. 4. Cocoa producer prices in the LLNP region and world market prices in 2007 (monthly averages in US$/t dry beans; world prices according to ICCO 2008).

were sold directly Palu compared to sales in the neighbour village (+11.1%) or at the home village (+16.2%). Higher prices were also achieved by selling directly to a big merchant compared to selling to a middleman (+13.7%) or to a small trader (+15.5%). Prices gained in Palolo valley were significantly higher than in the other valleys (+6.4%).

3.2 Shade management and Intensification

Shade Management

From 2006 to 2008, a reduction in CC was measured on 72% of all cocoa plots. For example, 65% of the households eliminated shade trees on their cocoa plots in 2007 alone. Plots with initially high CC in 2006 tended to show the highest decreases. In contrast, nearly shade-free plots in 2006 tended to increased in CC (Fig.5). Between 2007 and 2008 alone, canopy closure decreased from 64.3% to a mere 42.3% on average.

Canopy closure is more closely related to the total number of intercrops on a plot (r=0.279; P=0.001) than to the number of native forest tree species (r=0.182; P=0.029). Planted shade trees which have no food usage are often times leguminous, N_2 fixing trees. 92.7% of all plots include planted leguminous trees, mostly *Glyricidia sp.* or *Erythrina sp.*

Structural intensification vs. labour and material inputs

The average annual labour input for cocoa plots was 86.2 person-days per hectare (minimum: 9.4; maximum 339.1). Weeding belongs to the most labour demanding activities accounting for up to 72.2% of the total working time (mean 19.5%). Weeding is neither correlated with MI nor with the use of herbicides. Furthermore, we found no correlation between labour for any of the activities and material inputs. In concert with increasing yields, relatively more time is used for harvesting along an increasing MI gradient (r=0.39, P<0.01).

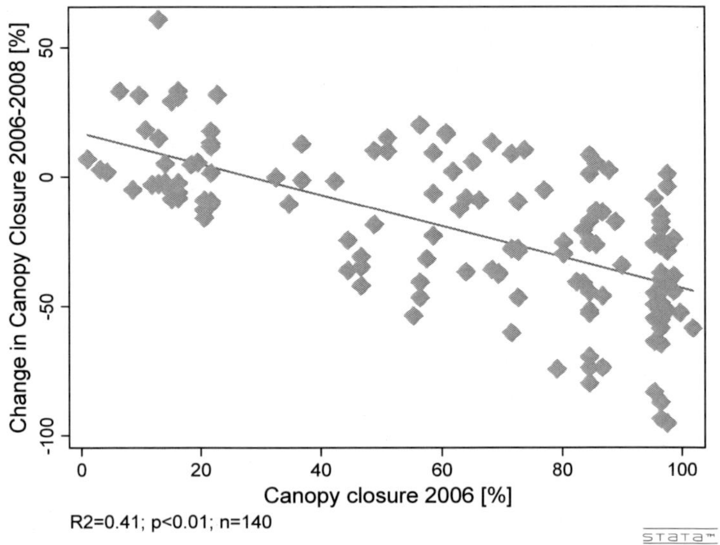

Fig. 5. Change in Canopy closure from 2006-2008 in relation to initial CC 2006 in % (R^2=0.412, P=0.05).

Structural intensification according to MI goes along with higher inputs of hired labour (r=0.18, p=0.04) and expenses for material inputs such as pesticides and fertilizer (r=0.14, p=0.09). In contrast, no relationship was found between MI and total working time.

Structural intensification vs. pest pressure

Initial yield losses when infestation with CPB or BPD began are negatively correlated with 2008 CC (CPB: r=-0.24, P=0.004; BPD: r=-0.17, P=0.04). *Current* yield losses, in contrast, are positively correlated with higher CC for CPB (r= 0.19, P=0.02). Regression analysis reveals some evidence for decreasing yield loss due to pest pressure along the intensification gradient (BPD: R^2=0.08, p=0.001, n=128; CPB: R^2=0.05, p=0.01, n=136).

Intensification vs. soil fertility and rainfall

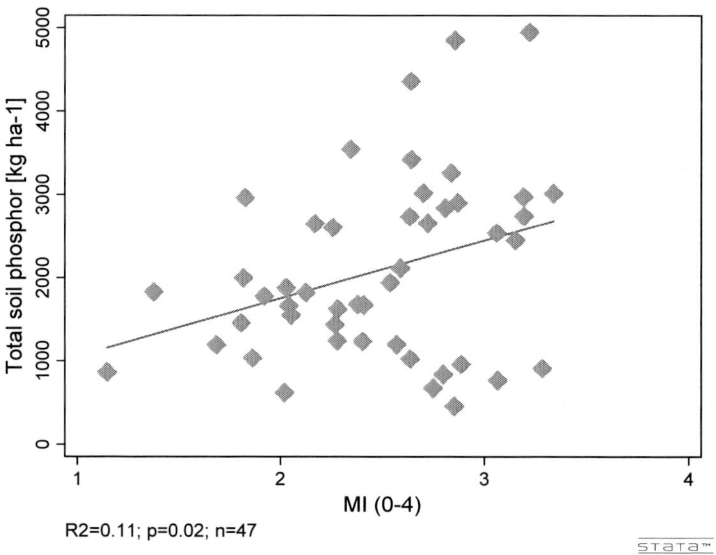

Fig. 6. Total soil phosphor content in relation to MI.

Soil fertility in terms of total phosphor (P_t) content is higher on more intensively managed plots (Fig.6). For available phosphor (P_{av}), the relationship is even weaker (R^2=0.07, p=0.08, n=45). In contrast, total soil potassium (K_t) content is decreasing (R^2= 0.15, p=0.01, n=46).

Management intensity increases significantly with average yearly rainfall (Fig 7).

Intensification vs. gross margins and returns to labour

The share of intercrops in plot revenues is very small (4.6% on average), only 12% of households have a share higher than 10%. Gross margins are closely related to yields (r=0.916, p<0.001), and therefore display a similar relationship to the MI (Fig.2). Also when related to labour input, returns to labour are increasing along the intensification gradient (Fig. 8).

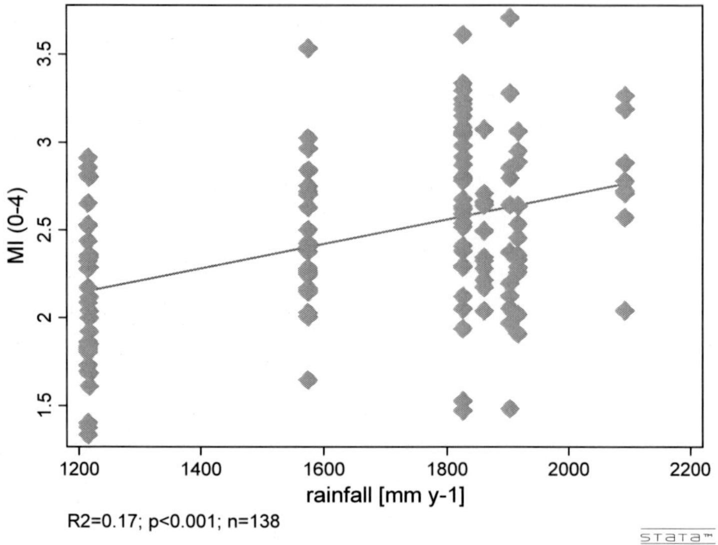

Fig. 7. MI in relation to rainfall.

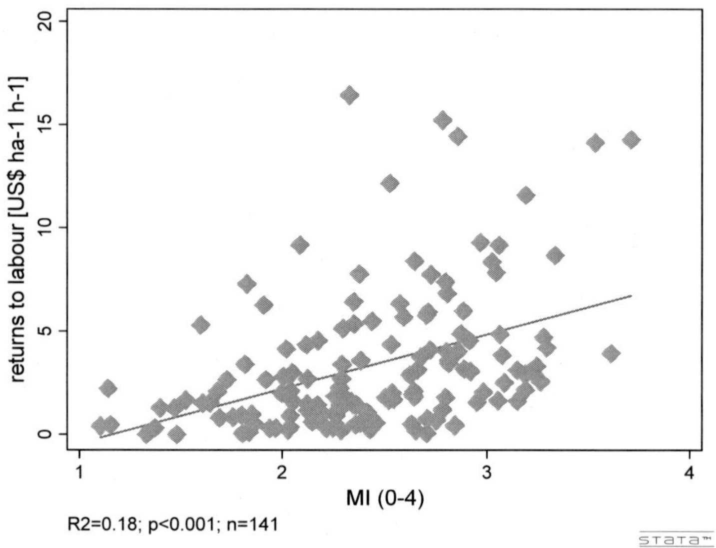

Fig. 8. Returns to labour along a structural intensification (MI) gradient.

4 Discussion

With this contribution, we give a comprehensive overview on the structure and management of cocoa agroforests across virtually the entire gradient of canopy cover and management intensities found around Lore Lindu National Park in Central Sulawesi. In the following, we discuss the influence of tree age, pest pressure, management and soil fertility on cocoa yields, followed by an analysis of the special role the intensification process plays in the LLNP region and its economic and ecological impacts.

Plots surveyed in this study roughly fall within the typical range of plot sizes on Sulawesi, where 95% of the production is grown on plots of 0.5 to 1.5 ha (Taher 1996, Panlibuton and Meyer 2004). Globally, 70% of cocoa is produced by small farms (Donald 2004).

In the LLNP region, directly age-related yield declines need not be a main issue for years to come.

Cocoa trees are relatively young and mostly in an early producing phase. In Malaysia, highest cocoa yields were found for trees between 15 and 20 years, but the profitable life span may reach 50 years (Montgomery 1981 in Wessel 1985). High pest and disease pressure (see below), can reduce the economically viable lifespan of cocoa trees, however (Wessel 1985, Lass 1985b).

Corroborating recent data by Neilson (2007), we found substantial CPB-related yield declines. Together with losses due to BPD, yield losses in Central Sulawesi are comparable to global average estimates of pest and disease induced yield losses of ˜30% (Padwick 1956 in Duguma et al. 2001). In West Africa, for example, yield losses vary between 10-80% being highest in Cameroon (50-80%; Bakala and Kone 1998 in Duguma et al. 2001).

In spite of the pests and disease pressure and farmer trainings on integrated pest management by the Sustainable Cocoa Enterprise Solutions for Smallholders (SUCCESS) Alliance in 2005 (ACDI/VOCA 2005), our results indicate a low level of adoption of integrated pest management practices. Farmers rely strongly on the effect of pesticides while important but labour-intensive activities such as sanitary pruning, frequent harvests and, above all, the removal of affected pods are only rarely practiced (Taher 1996, Lass 1985a).

The soils of the investigated cocoa plots have high stocks of soluble minerals (Ca, Mg and K; cf. Wood 1980). This is not uncommon in the humid tropics if high geologic activity results in young soils. The average nitrogen contents are moderate compared to literature values (Hartemink 2005, Wood 1980). The lack of any influence on yield indicates that these nutrients are usually not limiting. Only for wet, stagnant soil water conditions and available P, influences were found.

Available phosphorus (56 kg ha^{-1} on average) was low compared to other cocoa sites (cf. Hartemink 2005), and lower than recommended for successful cocoa growing (Wood 1980). The results of the classification confirm the impression about the nutrient status, but also reveal that humus content is low.

Improved organic matter management would also benefit N and P contents. Phosphorus is a key factor for sustainable agriculture in the Tropics and may limit production (IRRI 1990, Appiah 2004, Ojeniyi et al. 1982). P deficits can also reduce N uptake in cocoa (Lockard and Asomaning 1964, Smith 1992) and the fixation of N_2, for example by leguminous shade trees (Mappaona and Kitou 1995). We find, in fact, that high P availability positively influences cocoa yields in the investigated sample of cocoa plots. Most available P is usually stored in an organically bound fraction, i.e. in litter and/or humus (White and Ayoub 1983), and is lost easily by water or wind erosion (Brams 1973). An application of P fertiliser could be particularly useful in combination with improved humus and soil cover management.

The resulting average annual yield in our sample (476.9 kg/ha) falls within the low end of the yield spectrum reported for the entire island of Sulawesi (400 to 800 kg/ha; Panlibuton and Meyer 2004), and are also far below the reported Indonesian average in 2007 (801 kg/ha; FAOSTAT 2009). Ten years earlier spectacular yields of 2500 kg/ha on average on alluvial soils without intercropping or shade trees and around 1500 kg/ha in the uplands were common on Sulawesi (Ruf et al. 1995). The interpretation of our average yield value should be interpreted with caution, however, as we did not draw a random sample but purposefully oversampled low- and high-shading plots. In a representative study in the project area, farmers reported average annual yields of 531 kg/ha (data from 204 cocoa producing households with productive cocoa plots/ minimum age of 4 years; van Edig and Schwarze 2007 unpubl.) Nevertheless, our average yield is close to the average yield in West Africa (495 kg/ha in 2007), and higher than average yields from South America (393 kg/ha; FAOSTAT 2009).

Yields in our sample depend on several structural parameters of the plots. Low CC, as well as a reduced number of forest trees and intercrops, and a higher cocoa tree planting density significantly lead to higher yields. If combined to a structural management intensity index (MI), the influence is strong. In addition to structural parameters, total working time dedicated to cocoa management influences yields positively. Total labour input is not related to intensification in our study. Moreover, average labour input in 2007 (86.2 person-days/ha) accounts for only about one third of the labour demand for cocoa as estimated by the Indonesian Ministry of Agriculture (235 person-days/ha). Thus, labour inputs appear extremely low. As compared to wet rice farming, cocoa farming is seen as a labour saving option of land-use by many farmers (personal observation, Juhrbandt 2007). Extremely low labour input was also reported for (early) mixed cocoa systems in Malaysia (Lass 1985c).

The strong positive impact of labour indicates that substantial income effects may be realised without additional intensification in terms of structural modifications or higher inputs of plant protection agents. More labour investment could also raise the quality of the produced beans, for example, by a better fermentation process (Panlibuton and Meyer 2004). Fermenting is a crucial step in cocoa bean processing, influencing bean quality substantially.

Currently, there exists only a single market for almost all levels of bean quality, with little price differentiation. Thus smallholder farmers have no incentives to invest more labour into an improved quality.

Cocoa farmers of the project region receive a high share of the world market price as also reported by Panlibuton and Meyer (2004). This high share is related to the relatively competitive market situation of the Indonesian cocoa sector and is much higher than for cocoa produced in other countries or for other commodities produced in Indonesia (Akiyama and Nishio 1996). In West Africa, for example, the farm gate price is usually only about 50% to 63% of world market prices (Panlibuton and Meyer 2004).

A clear trend towards low shade tree covers can be recognized across the project region for virtually all but the already least shaded plots.

The effects of light and plant nutrition on cocoa yields are interrelated: Cocoa responds well to increased light if nutrients and pest pressure are not limiting (Almeida and Valle 2007, Wessel 1985, Ahenkorah et al. 1974). Long-term experimental shade and fertilizer trials in Ghana Ahenkorah et al. 1974, Ahenkorah et al. 1987) had indicated that sun-grown cocoa plantations may not be able to sustain high yields over long periods. The deterioration of unshaded cocoa trees was more rapid and severe than under no-shade conditions. Faster nutrient depletion was suggested as an explanation (Ahenkorah et al. 1974). Confirming the results from Dechert et al. (2004), our data do not provide evidence for particular nutritional stress. The cultivation of cocoa in full sun may still be unsustainable e.g., because of higher weather risks (Belsky and Siebert 2003), however, in our 1-year-study we did not find any immediate disadvantages of low- or no-shading cocoa farming.

A radical reduction of shade canopies in intensive cocoa agroforestry can affect biological diversity and ecological functioning (Siebert 2002, Schroth and Harvey 2007, Franzen and Borgerhoff Mulder 2007). While there is strong evidence that the current structural intensification in the project region has negative impacts on biological diversity of forest species (Steffan-Dewenter et al. 2007) the functional consequences of intensification are much more difficult to quantify. Regarding impacts on hydrology, Kleinhans (2003) found that compared to primary or older secondary forest, cocoa plots have only a small negative impact on dry season hydrology, for example, while annual cultures have a much more strongly negative effect. Yet, different cocoa systems were not considered. Regarding the impacts on the ecological function of natural pest control, our data do not aloud to draw clear conclusions on the role of structural intensification, although there is some evidence of reduced yield loss towards the more intensive systems. In contrast, Bos et al. (2007) found that fruit losses due to pathogenic infections and insect attacks increase with the homogenization of the agroforests, supporting the hypothesis that agricultural homogenization increases the risk of pest outbreaks.

Yields, gross margins and returns to labour turn out to increase sharply with intensification in terms of canopy thinning and plot structure simplification, indicating a strong economic incentive to intensify cocoa production by the removal of shade trees. However, shade trees in agroforestry systems often provide secondary products such as fruits and timber contributing to household income and nutrition (Rice and Greenberg 2000, Gockowski et al. 2004, Franzen and Borgerhoff Mulder 2007). Belsky and Siebert (2003) report that an increasing share of local residents in Central Sulawesi values crop diversification. Yet, the share of intercrops sales in cocoa plot revenues was found to be usually very small in the LLNP region.

5 Outlook: Current status of Cocoa agroforests in Central Sulawesi and road ahead

Compared to other cocoa producing regions in the world, Central Sulawesi is still a relatively young production frontier but our results show that pest and disease pressure have started to reduce the forest rents obtainable a few years ago, and that selected nutrient deficiencies can be observed (cf. Ruf 1995). In face of these growing agro-ecological challenges in cocoa cropping, the search for "sustainable" land use options becomes particularly important. Complementing the natural habitat of protected areas by biologically rich buffer zones, high to medium shade agroforestry could play an important role for integrated biodiversity conservation strategies in the Tropics (Schroth et al. 2004; Schroth and Harvey 2007; Steffan-Dewenter et al. 2007). However, results from the cocoa management study show that - in line with the high financial attractiveness of structural intensification - a clear trend towards simplified, low-shade cocoa agroforestry is prevalent in the project area. Without suitably gauged offers of financial incentives, for example via a price premium for certified "biodiversity-friendly" produced cocoa (Barkmann et al. 2007) or direct compensation payments for environmental services ("PES"; Schneider et al. 2006, Mas and Dietsch 2003, Dahlquist 2007), this trend is likely to continue.

From an agronomic point of view, our results show a potential for substantial improvement, especially with regard to pest and disease management, P fertilisation and humus management as well as cocoa bean processing. The high returns to additional labour inputs in our sample indicate that a higher labour input in the respective activities is particularly called for. This may result in a further income and welfare-improvement of smallholders without necessitating a further simplification of the cocoa agroforests or higher inputs of chemical plant protection agents. Unfortunately, the farmer field schools of SUCCESS Alliance have been discontinued in 2005, and the establishment of PES or certification programs needs to overcome high implementation costs.

From a research perspective, further integrated in-depth analyses of the pattern found in this study will contribute to a better understanding of the drivers and impacts of the "intensification syndrome" that we observe in the LLNP region. But also for some fundamental issues in environmental economics and conservation science, the study introduced here is likely to yield essential data. Particularly, we intend to quantify the private net benefits of ecosystem conversion along the entire intensification gradient from forests to monocultures. Complementing data from the project region on major biodiversity and functional ecosystem responses of intensification will allow us to compare these financial benefits to much of the environmental benefits lost (cf. Balmford et al. 2002).

References

ACDI/VOCA (2005) Indonesia Global Development Alliance. Project profile, ACDI/VOCA, SUCCESS Alliance, Mars inc., USAID: 1

Ahenkorah Y, Akrofi GS, Adri AK (1974) The end of the first cocoa shade and manurial experiment at the Cocoa Research Institute of Ghana. Journal of Horticultural Science 49: 43-51

Ahenkorah Y, Halm BJ, Appiah MR, Akrofi GS, Yirenkyi JEK (1987) Twenty years' results from a shade and fertilizer trial on Amazon cocoa (*Theobroma cocoa*) in Ghana. Experimental Agriculture 23: 31-39

Akiyama T, Nishio A (1996) Indonesia's Cocoa Boom: Hands-off policy encourages smallholder dynamism, SSRN

Almeida AF, de Valle RR (2007) Ecophysiology of the cocoa tree. Brazilian Journal of Plant Physiology 19: 425-448

Appiah MR (2004) Evaluation of fertilizer application on some peasant cocoa farms in Ghana. Ghana Journal of Agricultural Science 33 (2): 183-190

Balmford A, Bruner A, Cooper P, Costanza R, Farber S, Green RE, Jenkins M, Jefferiss P, Jessamy V, Madden J, Munro K, Myers N, Naeem S, Paavola J, Rayment M, Rosendo S, Roughgarden J, Trumper K, Turner RK (2002) Economic reasons for conserving wild nature. Science 297 (5583): 950-953

Barkmann J, Schneider E, Schwarze S (2007) Sweet as chocolate: Stabilisation of ecosystem services by production of cocoa in high-shade agroforestry systems in Central Sulawesi (Indonesia). Book of Abstracts, Tropentag 2007, Witzenhausen, Germany, 9-11 October 2007, p 322

Belsky JM, Siebert SF (2003) Cultivating cocoa: Implications of sun-grown cocoa on local food security and environmental sustainability. Agriculture and Human Values 20 (3): 277-285

Bos MM, Steffan-Dewenter I, Tscharntke T (2007) Shade tree management affects fruit abortion, insect pests and pathogens of cacao. Agriculture, Ecosystems & Environment 120 (2/4): 201-205

Brams E (1973) Soil organic matter and phosphorus relationships under tropical forests. Plant and Soil 39 (2): 465-468

Bray RH, Kurtz LT (1945) Determination of total, organic and available forms of Phosphorus in soils. Soil Science 59: 39-45

COPAL (2008) Cocoa Info. A Weekly Newsletter of Cocoa Producers' Alliance, Issue No. 264 1st 4th January 2008

Dahlquist R, Whelan M, Winowiecki L, Polidoro B, Candela S, Harvey C, Wulfhorst J, McDaniel P, Bosque-Pérez N (2007) Incorporating livelihoods in biodiversity conservation: A case study of cocoa agroforestry systems in Talamanca, Costa Rica. Biodiversity and Conservation 16 (8): 2311-2333

Dechert G, Veldkamp E, Anas I (2004) Is soil degradation unrelated to deforestation? Examining soil parameters of land use systems in upland Central Sulawesi, Indonesia. Plant and Soil 265: 197-209

Donald PF (2004) Biodiversity impacts of some agricultural commodity production systems. Conservation Biology 18 (1): 17-38

Duguma B, Gockowski J, Bakala J (2001) Smallholder cocoa (Theobroma cocoa Linn.) cultivation in agroforestry systems of West and Central Africa: challenges and opportunities. Agroforestry Systems 51 (3): 177-188

Durand F (1995) Farmer strategies and agricultural development: The choice of cocoa in Eastern Indonesia. In: Ruf, F, Siswoputanto PS (eds.) Cocoa Cycles. The economics of cocoa supply. Woodhead Publishing. Cambridge: 315-338

Erasmi S, Twele A, Ardiansyah M, Malik A, Kappas M (2004) Mapping deforestation and land cover conversion at the rainforest margin in central Sulawesi, Indonesia. EARSeL eProceedings 3 (3): 388-397

FAO (2003) FAO Commodities and trade technical paper, Medium-term prospects for agricultural commodities, Projections to the year 2010, Food and Agriculture Organization of the United Nations, Rome

FAOSTAT (2009) http://faostat.fao.org/site/567/default.aspx#ancor

Franzen M, Borgerhoff Mulder M (2007) Ecological, economic and social perspectives on cocoa production worldwide. Biodiversity and Conservation 16(13): 3835-3849

Gockowski J, Weise S, Sonwa DJ, Tchtat M, Ngobo M (2004) Conservation because it pays: shaded cocoa agroforests in West Africa. IITA-HFC. Yaoundé, Cameroon

Hartemink AE (2005) Nutrient stocks, nutrient cycling, and soil changes in cocoa ecosystems: a review. Advances in Agronomy 86: 227-253

ICCO (2008) Quarterly Bulletin of Cocoa Statistics, Vol. XXXIV, No. 3, Cocoa year 2007/08, published August 2008-12-10

IRRI (1990) Phosphorus requirements for sustainable agriculture in Asia and Oceania Proceedings of a Symposium 6-10 March 1989

Jennings SB, Brown ND, Sheil D (1999) Assessing forest canopies and understorey illumination: canopy closure, canopy cover and other measures. Forestry 72 (1): 59-74

Juhrbandt J, Barkmann J (2008) Yield determinants in cocoa agroforestry systems in Central Sulawesi: Is shade tree cover a good predictor for intensification? In: Grosse M, Lorenz W, Tarigan S, Malik A (eds.) Tropical Rainforests and Agroforests under Global Change. Proceedings International Symposium, October 5-9, 2008, Kuta, Bali, Indonesia. Universitätsverlag Göttingen 2008

Kleinhans A (2003) Einfluss der Waldkonversion auf den Wasserhaushalt eines tropischen Regenwaldeinzugsgebietes in Zentral Sulawesi (Indonesien). Dissertation, Universität Göttingen

Koch S, Faust H, Barkmann J (2008) Differences in power structures controlling access to natural resources at the village level in Central Sulawesi (Indonesia). Austrian Journal of South-East Asian Studies 1 (2): 59-81

Lanfer N (2003) Landschaftsökologische Untersuchungen zur Standortbewertung und Nachhaltigkeit von Agrarökosystemen im Tieflandsregenwald Ecuadors. Band 9. EcoRegio, Göttingen

Lass RA (1985a) Maintenance and improvement of mature cocoa farms. In: Wood GAR, Lass RA (eds.) 1985: Cocoa. 4. ed., Longman, Harlow, UK

Lass RA (1985b) Diseases. In: Wood GAR, Lass RA (eds.) 1985: Cocoa. 4. ed., Longman, Harlow, UK

Lass RA (1985c) Labour usage. In: Wood GAR, Lass RA (eds.) 1985: Cocoa. 4. ed., Longman, Harlow, UK

Leemhuis C (2005) The impact of El Niño Southern Oscillation Events on water resource availability in Central Sulawesi, Indonesia - A hydrological modelling approach. Geographisches Institut. Göttingen, Georg-August-Universität zu Göttingen. Dissertation: 172

Lockard RG, Asomaning EJA (1964) Mineral nutrition of cocoa (Theobroma Cocoa L.) I. Deficiency symptoms and nutrient levels in plants grown in sand culture. Plant and Soil 21 (2): 142-152

Maertens M (2003) Economic modeling of agricultural land-use patterns in forest frontier areas: Theory, empirical assessment and policy implications for Central Sulawesi, Indonesia. Fakultät für Agrarwissenschaften. Göttingen, Georg-August-Universität

Mappaona SY, Kitou M (1995) Difference in Phosphorus response among tropical green manure legumes grown under limed and unlimed soil conditions. Soil Science and Plant Nutrition 41 (1): 9-19

Mas AH, Dietsch TV (2003) An index of management intensity for coffee agroecosystems to evaluate butterfly species richness. Ecological Applications 13 (5): 1491-1501

Moguel P, Toledo VM (1999) Biodiversity conservation in traditional coffee systems of Mexico. Conservation Biology 13: 11-21

Myers N, Mittermeier RA, Mittermeier CG, da Fonseca GAB, Kent J (2000) Biodiversity hotspots for conservation priorities. Nature 403 (6772): 853-858

Neilson J (2007) Global markets, farmers and the state: Sustaining profits in the Indonesian cocoa sector. Bulletin of Indonesian Economic Studies 43 (2): 227 - 250

Ojeniyi S, Egbe N, Omotoso T (1982) Effects of nitrogen and phosphorus fertilizers on unshaded Amazon cocoa in Nigeria. Fertilizer Research 3: 13-16

Panlibuton H, Meyer M (2004) Value chain assessment: Indonesia cocoa. Accelerated microenterprise advancement project (AMAP) microREPORT #2 (June). Prepared by Action for Enterprise and ACDI/VOCA for US-AID, Washington, DC

Perfecto I, Rice RA, Greenberg R, Van der Voort ME (1996) Shade coffee: A disappearing refuge for biodiversity. Bioscience 46 (8): 598-608

Pomp M, Burger K (1995) Innovation and Imitation: Adoption of cocoa by Indonesian smallholders. World Development 23: 423-431

Reuters (2009) World's biggest cocoa growers face aging trees. FlexNews, Food Industry News 09/03/2009. Reuters

Rice RA, Greenberg R (2000) Cocoa cultivation and the conservation of biological diversity. AMBIO: A Journal of the Human Environment 29 (3): 167-173

Ruf F (1995) From "forest rent" to "tree capital": Basic "laws" of cocoa supply. In: Ruf F, Siswoputanto PS (eds.) Cocoa Cycles. The economics of cocoa supply. Woodhead Publishing. Cambridge: 1-54

Ruf F, Yoddang J, Ardhy W (1995) The "spectacular" efficiency of cocoa smallholders in Sulawesi: Why? Until when? In: Ruf F, Siswoputanto PS (eds.) Cocoa Cycles. The economics of cocoa supply. Woodhead Publishing. Cambridge: 339-375.

Schneider EM, Barkmann J, Schwarze S (2007) Sweet as Chocolate: Stabilisation of ecosystem services by production of cocoa in high-shade agroforestry systems in Central Sulawesi (Indonesia), Tropentag 2007 Proceeding: 323

Schroth G, Harvey CA, Vincent G (2004) Complex agroforests: their structure, diversity, and potential role in landscape conservation. In: Schroth G, da Fonseca GAB , Harvey CA, Gascon C, Vasconcelos HL, Izac AMN (eds.) Agroforestry and biodiversity conservation in tropical landscapes, Island Press: 227-260

Schroth G, Harvey C (2007) Biodiversity conservation in cocoa production landscapes: an overview. Biodiversity and Conservation 16 (8): 2237-2244

Schwarze S (2004) Determinants of income generating activities of rural households: a quantitative study in the vicinity of the Lore Lindu National Park in Central Sulawesi/ Indonesia. Doctoral thesis, Institute of Rural Development, Georg-August Universität Göttingen

Siebert SF (2002) From shade- to sun-grown perennial crops in Sulawesi, Indonesia: implications for biodiversity conservation and soil fertility. Biodiversity and Conservation 11 (11): 1889-1902

Smith VH (1992) Effects of nitrogen: phosphorus supply ratios on nitrogen fixation in agricultural and pastoral ecosystems. Biogeochemistry 18: 19-35

Steffan-Dewenter I, Kessler M, Barkmann J, Bos MM, Buchori D, Erasmi S, Faust H, Gerold G, Glenk K, Gradstein SR, Guhardja E, Harteveld M, Hertel D, Hohn P, Kappas M, Kohler S, Leuschner C, Maertens M, Marggraf R, Migge-Kleian S, Mogea J, Pitopang R, Schaefer M, Schwarze S, Sporn SG, Steingrebe A, Tjitrosoedirdjo SS, Tjitrosoemito S, Twele A, Weber R, Woltmann L, Zeller M, Tscharntke T (2007) Tradeoffs between income, biodiversity, and ecosystem functioning during tropical rainforest conversion and agroforestry intensification. PNAS 104 (12): 4973-4978

Taher S (1996) Factors influencing smallholder cocoa production A management analysis of behavioural decision-making processes of technology adoption and application. PhD Thesis, Wageningen

Wessel M (1985) Shade and nutrition. In: Wood GAR, Lass RA (eds.) 1985: Cocoa. 4. ed., Longman, Harlow, UK

White R, Ayoub A (1983) Decomposition of plant residues of variable C/P ratio and the effect on soil phosphate availability. Plant and Soil 74 (2): 163-173

Wood GAR, Urquhart DH (eds., 1980). Cocoa. 3. ed., London

Wood GAR (1985a) Environment. In: Wood GAR, Lass RA (1985, eds.) Cocoa. 4. ed., Longman, Harlow, UK

Zeller M, Schwarze S, van Rheenen T (2002) Statistical sampling frame and methods used for the selection of villages and households in the scope of the research programme on Stability of Rainforest Margins in Indonesia (STORMA). STORMA Discussion Paper Series No 1. Bogor, Indonesia: Universities of Göttingen and Kassel, Germany and the Institut Pertanian Bogor and Universitas Tadulako, Indonesia

Land tenure rights, village institutions, and rainforest conversion in Central Sulawesi (Indonesia)

Jan Barkmann[1*], Günter Burkard[†,2], Heiko Faust[3], Michael Fremerey[2], Sebastian Koch[1,3**], and Agus Lanini[4]

[1] Environmental & Resource Economics, Department of Agricultural Economics and Rural Development, Georg-August-University Göttingen, Platz der Göttinger Sieben 5, 37073 Göttingen, Germany
[2] Rural Sociology, Faculty of Organic Agricultural Sciences, University of Kassel, Steinstr. 19, 37213 Witzenhausen, Germany
[3] Department of Cultural and Social Geography, Institute of Geography, Georg-August-University Göttingen, Goldschmidtstr. 5, 37077 Göttingen, Germany
[4] Faculty of Law, Universitas Tadulako, Kampus Bumi Tondo, Palu 94118, Indonesia

*corresponding author: Jan Barkmann, email: jbarkma@gwdg.de
**current address: Didactics of Biology, Albrecht-von-Haller Institute for Plant Sciences, Georg-August-University Göttingen, Waldweg 26, 37073 Göttingen (Germany)

Summary

Small-scale agriculture continues to account for a significant share of global forest conversion, especially, in remote forest frontier areas such as the mountainous forests of Central Sulawesi. Although much information on the proximate driving forces of deforestation can be obtained from analyses of the household level, the analysis of supra-household level institutional phenomena is indispensable for an adequate understanding of forest conversion. This is particularly true for the interaction of local institutions governing land access and tenure rights. In this contribution, we synthesize results from several related studies conducted around Lore Lindu National Park (LLNP) that investigate this relation. Contrary to the theoretical argument that tenure security fosters resource conservation, we find that the high tenure security of formal land titles increasingly available in the region attracts migrants who are aggressive buyers of land for petty capitalist cocoa production. At the end

T. Tscharntke et al. (eds.), *Tropical Rainforests and Agroforests under Global Change*, Environmental Science and Engineering, 141–160, DOI 10.1007/978-3-642-00493-3_6,

of the process, a substantial share of the autochthonous population finds itself either landless or is forced to cut marginal forest land, often inside LLNP. Restricting land ownership to traditional forms of community land rights avoids the formation of a class of newly landless locals but it comes at costs in terms of social discrimination and lost agricultural income.

Keywords: customary community, land access, tenure security, common property resources, land title, cocoa production

1 Introduction

Small-scale agriculture continues to account for a large share of global forest conversion, especially, in remote forest frontier areas (de Sherbinin et al. 2008; de Sherbinin et al. 2007; Chomitz & Griffiths 1996) such as the mountain forests of Central Sulawesi. In its analyses of the driving forces of deforestation in Central Sulawesi, socio-economic STORMA research has analyzed household level as well as supra-household level phenomena. Studies using household data investigated, e.g., the influence of agricultural intensification, household demographic factors, ethnicity and migration on land use change (Maertens et al. 2006; Weber et al. 2007), analyzed income generation from forest products (Schwarze et al. 2007) and from the cash crop cocoa (Schippers and Faust 2009), or highlighted the multiple dimensions of poverty (Schwarze et al. 2005; van Edig et al. 2007). There is also a large body of work on the supra-household level, namely on issues such as settlement and colonization history (Weber and Faust 2006), regional political ecology (Adiwibowo 2008), the interaction of migration and cacao cropping (Sitorus 2004), or the institutional management of resource use conflicts (Thamrin 2007; Mappatoba 2008; Koch et al. 2008). Most recently, a dedicated volume with eleven articles on the social organization of forest management was published based on data surveyed from 2000 to 2006 by several socio-economic STORMA projects (Burkard & Fremerey 2008[1]).

In this contribution, we aim at advancing the integration of some supra-household level results with key findings from household level data extending analyses such as by Abdulkadir-Sunito & Sitorus (2007) or Weber et al. (2007). In particular, we focus on two central institutional and community aspects, namely tenure/land use rights and the corresponding informal village institutions. We start with a brief background section on institutions for the governance of open access and common pool resources (section 2), introduce

[1] Günter Burkard (†12.10.2008) was closely involved in the joint integrative analysis of the causes of deforestation in the LLNP area among the socio-economic STORMA projects. Because we draw heavily on results from his studies and on ideas developed in discussion with him, we include him as a co-author although he had no opportunity to rectify any mistakes this manuscript may contain.

the study area (section 3.1), and outline the methods used (section 3.2). Next, we give an extended account of the legal framework for land tenure systems in the project area and in Indonesia (section 4). In section 5, the interaction of informal, customary village institutions on formal land tenure and land access is explained for three contrasting villages. The quantitative differences in relative deforestation rates between these villages are described in section 6. The findings of the preceding sections are integrated and discussed in the final section 7.

2 Governing open access and common pool natural resources

Institutions are defined as any form of constraint that shapes human interaction (North 1990:4). Institutions play a decisive role in the sustainable (or non-sustainable) utilization of many natural resources (Acheson 2006; Stern et al. 2002; Ostrom 1990). One institutional arrangement affecting tropical deforestation is an ill-defined property rights structure which rewards individual resource exploitation while damaging overall resource availability (cf. Hardin 1968, Mishan 1969). Such divergences between individual rationality and group rationality as observed in the "commons dilemma" easily occur in so-called open access and common pool resources (Berkes & Folke 1998:6). Common pool resources share characteristics with private goods and with open access goods. Open access goods are defined by rivalry in consumption and non-excludability (Gibson et al. 2000:6). Rivalry in consumption (also called subtractability) means that a unit of the resource "consumed" by one user cannot be consumed by another. Non-excludability means that it is very difficult or prohibitively costly to exclude a potential user of the resource from actual consumption.

Several studies concluded that neither privatization nor government control *necessarily* lead to sustainable resource use of common property resources because the establishment of effective control remains a challenge if users are structurally difficult to exclude (Dietz et al. 2003; Ostrom et al. 1999). In contrast, there is growing evidence that certain types of joint local management institutions with clearly defined rules can be successful in averting the "tragedy of the commons". Ostrom (1990: 90) points out that the successful long-term management of common pool resources by local communities is historically characterized by certain design principles, among them (i) resource extraction rates in tune with long-term sustainable yields, (ii) resource extraction monitoring, (iii) graduated sanctions in case of violations of local resource use regulations, and (iv) minimal recognition of rights to local resource governance.

3 Study area and methods

3.1 Study area

Located in the humid Tropics in Central Sulawesi, Lore Lindu National Park (LLNP) covers ~2,290 km^2 from ~200-2610 m a.s.l. (Figure 1). The LLNP area is characterized by rift valleys and a steep rainfall gradient from 500 to 2500 mm per year. First established as an UNESCO Man & Biosphere Reserve in 1978, it was declared a national park in 1982. Its permanent borders were not fixed until the end of the 1990s. Approximately 136,000 citizens, mainly agricultural smallholders, live in 119 villages within the study area of 7220 km^2 around LLNP (Erasmi et al. 2004; Maertens 2003:22ff). Central Sulawesi is one of the poorest provinces in Indonesia, and the LLNP area is considered a distant, disadvantaged place with fragile ecosystems and poverty stricken villages (Li 1999: 34; Li 2007).

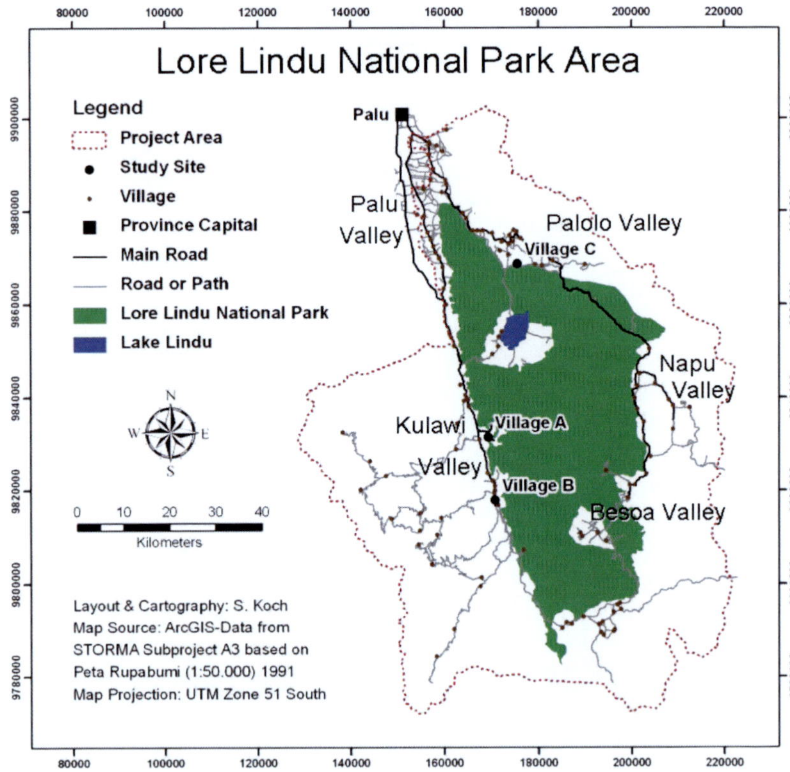

Fig. 1. Lore Lindu National Park area with the location of the main valleys and the three main research villages (see sections 4 to 6; changed after Koch et al. 2008)

3.2 Methods outline

The research took place in several villages located in four of the five valleys bordering LLNP: Palolo Valley, Napu Valley, Besoa Valley and Kulawi Valley. Most villages belong to a set of 12 villages that were selected for intensive socio-economic research by stratified random sampling (Zeller et al. 2002). Three contrasting villages (Toro, Nopu, Lempelero) used for quantitative and qualitative in depth studies were identified based on the household survey as well as on additional research on cultural landscape change (Weber 2006; Faust et al. 2003). In the following, we refer to these villages (not necessarily in the same order) as villages A, B, and C. The villages are located at the northern and western edge of LLNP.

In villages A, B, and C, a household census covering virtually all households was carried out in 2004 (n=898). Inter alia, the census instrument included questions on ethnicity, on current household land holdings, and details of land transactions (mode of acquisition of the land, land received from whom). For each of the villages, we calculated which fraction of all plots that was cut from the forest ("cleared"), remained in the hands of the same ethnic group, or changed hands between the local population, Bugis migrants from southern Sulawesi, and non-Bugis migrants (NBM) (see Fig. 2).

Furthermore, we conducted a qualitative in depth study on the influence of village elites and local institutions on natural resource access relying on semi-structured key informant interviews (n=3*10; Koch et al. 2008). Cluster analyses of the household data performed by Schippers & Faust (2009) helped to select households for qualitative interviews.

In village A and in nine additional villages, a variety of qualitative social science methods was applied to investigate land tenure regimes in Central Sulawesi (observation, qualitative interviews, focus group discussions) by Lanini. Additionally, we draw essential information from studies by Burkard (2002, 2008a/b) conducted in three further villages of the 12 villages set located at the northern and eastern edge of LLNP (Rompo, Watumaeta, Sintuwo).

4 Land rights and their development in the project area

Property rights can include distinct use rights, rights to exclude others, management rights, and rights to sell the resource (Aggarwal 2006). In many traditional societies, property rights over natural resources are exercised by privileged individuals but are fundamentally held by the local community (Bromley 1992: 4; Gibbs & Bromley 1989: 31). The Indonesian *Basic Agrarian Law* (BAL; UU5; 24 Sept. 1960) was established on principles of nineteen different systems of customary "*adat*" law in Indonesia (Vollenhoven in Soekanto & Taneko 2002; Ter Haar in Hadikusuma 2003). BAL Article 5 mandates that customary law prevails as long as was it does not collide with national interests, specific stipulations of BAL, or with religious rules. The provisions of Article 5 and Article 2 grant a special type of property right to land held in customary land tenure systems (*hak ulayat*). Specifically, this encompasses a right of allocation, i.e., a customary authority (*lambaga adat*) can allocate natural resources to members of its community. Because of a mythically interpreted relation of traditional communities and their ancestral lands, migrants are principally not allowed to own or utilize such lands unless explicit permission is granted by the customary authority.

Under the traditional form of land tenure, individuals can obtain individual land rights if a plot is continuously cultivated, particularly, if the plot was also opened by the individual. Land left unused reverts back to the allocation right of the community. In accordance with customary law, the land individually owned can be object of various kinds of land permanent or temporary transactions including opening a plot, selling, purchasing, bequeathing or leasing.

However, the BAL grants customary land rights only subject to restrictive conditions (Burkard 2008*b*:143ff). In line with similar regulations in the Foresty Law (UU41/1999, Article 67), traditional land rights can only be exercised by *customary law communities* (*masyarakat hukum adat*). This legal term stresses a certain level of institutional formality of the customary community as judged against bureaucratic standards. Article 67 [1] recognizes the existence of a *customary law community* if it fulfills several criteria. Among them, the community must

- continue to exist as a legal community based on customary law;
- have an institution acting as a customary authority;
- have a clearly defined jurisdiction;
- continue to harvest forest products on their ancestral lands for subsistence needs.

In contrast to this restrictive legal provisions, political discourse in Indonesia is dominated by the term *masyarakat adat*, translated as indigenous people or indigenous community. This is also the self-description of many communities in the project area. It stresses mainly the "indigenousness" of the respective community. It does not suffice to support a formal claim to customary allocation rights, however (Burkard 2008*b*:142ff). Although the National Agency for Land Affairs (BPN: *Badan Pertanahan Nasional*), which is charged with managing land tenure issues, has been successfully lobbied to account for the concerns of traditional communities, it lacks recognized procedures to support the claims of traditional communities against national park priorities (Burkard 2008*b*:145f). Thus, customary allocation rights remain effectively restricted to customary law communities. Consistent with a strict interpretation of national law, the governor of Central Sulawesi decreed in 1991 that no customary land (*tanah adat*) exists in the province, and all respective forest land would be under a form of state land right.

Not all land in the study area is part of LLNP. The official forest classification distinguishes four types of forest differing in protection status (Burkard 2002):

- national Park (LLNP, *Taman Nasional Lore Lindu*),
- protection forest (*hutan lindung*),
- limited production forest (*hutan produksi terbatas*), and
- community forest (*hutan kemasyarakatan*).

Only the community forest may be converted to agricultural land. In the protection forest, properly licensed collection of rattan is the only legal use. In judging the intensity of resource use conflicts involving LLNP, one needs to consider topography. Some villages directly border upland LLNP, and do not have any other forest at their disposal as all lowland forests have long been converted to agricultural lands. Other villages, although in the same valley, are located at the opposite edge of the valley not adjacent to LLNP. Their adjacent upland forests are placed in less restrictive forest categories (cf. Tab. 2 in Burkard 2002).

In the eastern part of the project area (Palolo, Napu and Besoa Valleys) swidden agriculture was partly practiced until the 1990s (Burkard 2002:7ff). Villagers formed groups of cultivators (*horobo*) around a kinship-based core group. Plots were either worked individually or in teams with strictly reciprocal labor sharing. The cultivator group had control over a sector of land consisting of several, roughly concentric spheres around the settlement with consecutively longer rotation periods (Burkard 2008*a*). Land in the innermost circle was used for annuals (*bonde*) and/or in a short fallow system (*holua*). These types of land were embedded in young secondary forest (< 7 yrs, *lopo lehe*), followed by older secondary forest (< 30 yrs, *lopo matua*), and finally "primary" forest from which only trees and non-timer forest products were selectively harvested (*pandulu*).

Nowadays, land scarcity is pervasive, and the formerly concentric patterns have dissolved with much of the former near primary forest under short rotation and even permanent agriculture – partly even inside LLNP (cf. Burkard 2008*a*:24).

A family could access land traditionally via three main avenues, resulting in different land categories:

- *mapandulu*: primary forest opened with consent of the respective cultivator group. Because of the mystical aspects associated with primary forests including (small) "holy places", the opening of primary forest needed additional permission from the village leader;
- *rape bolo:* land borrowed. The pioneer cultivator grants access to his land to other households on a long-term basis. Also the term *tanah niinda* is used for land under such an arrangement;
- *sosora:* inherited land, usually divided by the eldest son among siblings. Only wet rice plots or plots planted with perennials could be inherited

or claimed as *sosora* land. In the face of widespread land scarcity, many siblings have started to work on part of their parents' land before formal inheritance. In addition to individually inherited land, older secondary forest may fall into the *budel* category of inherited land that belongs to the family of the pioneer cultivator without being inherited to a special household heir. Such plots are the source of much conflict and tenure insecurity.

From village A at the western border of LLNP, a set of different local terms are reported to denote the various categories of land subject to their shifting cultivation history and property rights status (Shohibuddin 2008). Here, the revitalisation of customary law resulted in the definition of "traditional" land and forest categories that match the requirements of the Indonesian official national park zonation. Particularly, short- and long-term fallows including secondary forests around the village could be equated with the official "Traditional Utilisation Zone" (Shohibuddin 2008). Also in village A, the pioneer cultivator gains permanent ownership rights of the plot cleared.

In the research region, five main forms of agrarian tenancy arrangements are currently found (Burkard 2008*a*:39ff). Corresponding to the *rape bolo* land (borrowed land, see above), we find long-term lend out/borrow in arrangements *(pinjam garap)*. As written contracts were not used in the past, long-term borrowing has resulted in numerous conflicts when the heirs of a borrowed plot denied its borrowed status, and/or even registered the plot to obtain a formal land title. Under the formerly prevailing open access situation, strict monitoring of one's own land lend out was no essential task. However, the highly productive, permanent paddy rice fields have always been rented out for only one or two seasons. Until recently, the *pinjam garap* lend out/borrow in arrangement was regarded as a "rural welfare mechanism" because the tenant would be given the land for free (Burkard 2008*a*:40). The tenant would be responsible for all inputs, and could fully enjoy the harvest. While such arrangements were extended beyond kinship groups in the past, it is now restricted to kin in more and more villages.

Among the Kaili ethnics living at the northern edge of LLNP, a similar tenancy arrangement with a welfare function is called *tana niinda*. Based on mutual trust when established, a borrower in need is required to return the land once the situation of his household has improved. Particularly, when the children of the initial borrower claim the plot as their heritage, conflict in the Kaili community arises.

The second land arrangement is *share tenancy* (*bagi hasil* = "distribution of the harvest"). Along with fixed rent tenancy (see below), this is a recent innovation imported from densely populated Java (Burkard 2008*a*:42). Depending on details, the landlord receives a fixed amount of the harvest, usually between 40 and 50%.

Fixed-rate tenancy (*bapajak*) is usually a very short term arrangement (one year), in which the tenant pays the landlord a fixed amount of money upfront.

While being regarded as incompatible with local traditions in some villages, the first arrangements of this kind can be found in the project region. The forth arrangement, future harvest participation (*bapetak*), is highly insecure as a landless worker invests his/her labor in a wet rice plot in order to receive a (low) share of the future harvest. Fixed-rate tenancy is often practiced among kin groups, and provides at least some limited job security.

The fifth arrangement is land-share tenancy (*bagi tanah*). In land-share tenancy, the tenant prepares a larger plot with a perennial plantation according to exact written specifications on input sharing, location etc. Most importantly, the contract contains a description of that share of the land that the tenant can keep after the initial plot is divided in two parts.

While these five arrangements differ strongly in terms of tenure security, most provide access to land not only for the autochthonous population but also for migrants. For example, "free" land borrowing (*pinjam garap*) is at times still extended to migrants nowadays; in these cases it does regularly include hidden benefits for the landlord, however (Burkard 2008*a*:41). The most important arrangement involves assigning a secondary forest plot to the migrant tenant. This plot has to be cleared by the migrant, and must be reverted to the landlord in cleared condition. Furthermore, tenancy arrangements can be a way for migrants to overcome widespread norms mandating infra-ethnic land transfers (Sitorus 2004). Land-sharing (*bagi tanah*) is reported exclusively from migrant tenants (Burkard 2008*a*:43). This arrangement ideally suits the needs of low-cash migrants planning to plant cacao trees and requiring a certain level of tenure security.

For the involved migrants from Bugis and Toraja ethnicity, entering into tenancy arrangements is not a mere survival strategy but the first step of a land accumulation strategy. Although often starting from tenancy arrangements years ago, many Bugis migrants have accumulated sufficient funds to buy land from cash-stripped locals (Sitorus 2004). Thus, purchase transactions have become an increasingly important means to secure land access. Land purchases by Bugis migrants started in the second part of 1970s mainly in the Palolo Valley with "native" Kaili or Kulawi sellers (Sitorus 2004). In the 1980s land accumulation by purchase accelerated, and started to spread to the more remote valleys in the LLNP area.

Because of their limited social capital in terms of kinship or ethnic relations to local village elites (Koch et al. 2008), migrants from non-autochthonous ethnics rely extensively on official land titles to prove ownership. Land holders can apply to the regional BPN office for official land titles (Burkard 2008*a*:26). The certificates are essential to use land as collateral for bank loans. For proving land ownership, also "records of land transaction" issued by the district head can be used. They are more secure than tax receipts that only indicate that the plot is registered at the village office, and that tax was paid. Finally, the village headman issues written ownership letters or informal letters of land utilisation (*surat keterangan pengolaehan tanah*) that only serve to prevent tenure disputes within the village (Abdulkadir-Sunito & Sitorus 2007).

5 The role of informal institutions and village elites in governing forest land access

We turn now to an in-depth description of the interplay of informal village institutions of three contrasting villages. For each of the three villages, a brief description is given first.

5.1 The "traditional" village A

Village A represents a relatively static village with low immigration, and a high share of autochthonous ethnics. The village is one of the oldest settlements in the LLNP area and has a strong emphasis on traditions. In addition to a Council of Traditional Leaders (*lembaga adat*), an indigenous women's organization is active here since the 1990s (OPANT – *Organisasi Perempuan Adat Ngata Toro*). Land beyond the narrow valley bottom is scarce because access to land and to all other forest resources was officially suspended in 1982 with the declaration of the national park (Fremerey 2002; Burkard 2002).

As the result of negotiations with the national park administration, village A was granted far-reaching self-governance rights to about 23 km^2 of LLNP forest land in 2001. In this "community" forest, village authorities monitor and regulate forest resource utilization. The "community" forest is divided into six zones. In one zone, for example, cultivation is strongly prohibited because it serves as a habitat for scarce flora and fauna, and as a water source. In a traditional utilization zone, shifting cultivation with up to 25 years of fallow can be practiced. While the LLNP administration recognizes village A's self-governance of its claimed traditional lands, the community is not recognized by BPN (or the provincial government) as a formal customary law community. Burkard (2008b:146ff) argues that the revitalisation of traditional *adat* institutions was foremost a strategic move of the village elite to gain the status of a customary law community by molding its institutions (and its history) as to fit the official requirements set out in the national Forestry Law (cf. Shohibuddin 2008).

All interviewed individuals highlighted the strong influence of local institutions mainly exerted by the Council of Traditional Leaders. A clan of pioneer cultivator families dominates almost all positions in the village leadership. Besides dominating the *lembaga adat*, its members also occupy the positions of the village headman (*kepala desa*) and of the constitutional village council (*Badan Perwakilan Desa*/BPD). To be a member of these governance bodies, candidates must be indigenous: " [...] the members of village government should be the indigenous of Kulawi" (teacher; indigenous, village A). Although the village headman and the traditional leaders are elected, positions are often passed from one family member to another.

These local institutions are effectively in control of access to local natural resources. If a household has too little land to cover its basic needs, the household head will appear at a Council of Traditional Leaders (*lembaga adat*) meeting, explain his/her cause, and hope for the appropriation of a forest plot. The land assigned to households in need is located inside LLNP. In addition to access to land and forest plots, the *lembaga adat* grants permissions for the extraction of timber and non-timber forest products including rattan. Regularly, punishments are imposed by the formal village leadership if villagers violate resource use regulations.

The same procedures apply principally to migrants intending to settle in the village. Poor, recent migrants are discriminated against, however. This is obvious in cases of smaller land appropriations, or more restricted access to other forest resources: "[...] some people are being pressured not to take forest products while others were allowed to do. So sometimes I intend to ask where the justice is!" (migrant, village A).

Immigration by members of other ethnic groups is strictly discouraged by very restrictive regulations on land purchases. The village government hinders villagers from obtaining private land titles. Differing from other villages in the research area, land transactions must be reported in advance to the village headman. Together with the Council of Traditional Leaders, the village headman will decide whether any proposed land transaction will be allowed or not. Furthermore, the size of land that may be sold is restricted by the Council of Traditional Leaders. Finally, it is not allowed to resell purchased plots except to the previous owner.

5.2 The "transitional" village B

Village B is less static than village A. It was relatively recently established at the western edge of LLNP. The population of village B doubled within the last ten years displaying high demographic dynamics with many migrants, particularly, southern Sulawesi Bugis. Village B represents a transitional village type, in which immigration has started to impact village life but is not dominating it. Its community forest is located outside LLNP.

Traditional institutions and power structures still appear to be in place. The Council of Traditional Leaders guards local customs. It grants access to the community forest to autochthonous households. Also village regulations concerning natural resource use and designed by village headman and Council of Traditional Leaders exist. The regulations are not documented in writing, though, and are not fully implemented. There is a lack of monitoring and enforcement mechanisms. Fines and punishments are rarely imposed.

Similar to village A, migrants are excluded from influential positions in the village leadership. "[For] migrants it is impossible to be involved in traditional leadership" (village official, local, village B). Migrants are not as strictly discriminated against as in village A, however. For example, land transactions are not restricted in village B. Although migrants are not allowed to convert community forest into new agricultural plots, Bugis migrants – as well as better-off autochthonous households – acquire land via purchase from poorer, local households. Groups of autochthonous, partly land-stripped households

(15-20 persons) then collectively prepare new agricultural land via conversion of uphill (primary or secondary) forest.

5.3 The "post-transitional" village C

Village C is a young village in the Palolo Valley. Local migrants from surrounding villages were the most important group of pioneer cultivators before and during the 1970s. They cleared existing lowland forest for the establishment of wet rice fields. At the onset of the cacao boom in Central Sulawesi the 1980s (*revolusi cokelat*; Sitorus 2004), Palolo was a favorite destination for regional migrants, mostly Bugis well versed in cacao cultivation. Today, village C's agriculture consists predominantly of intensive cacao plantations.

Traditional power relationships are replaced by economic power structures dominated mostly by Bugis migrants. The village headman and the BPD form the official village government, and even a Council of Traditional Leaders exists. However, neither the formal nor the customary institutions are as powerful as in villages A or B. The low importance of these institutions is reflected by the fact that BPD members are not known by many inhabitants of village C, be they migrants or locals.

Autochthonous as well as Bugis interviewees agreed that a widespread laissez-faire attitude on natural resource use prevails. Every household is regarded as responsible for itself. "There are no regulations about the use of the forest and its products in this village" (village headman, village C). Rules and regulations with regard to the national park exist only at the regional and national level; they are neither monitored nor is any sanctioning established in the village. There are some attempts to dissuade smallholders from farming steep slopes inside LLNP. In absence of other forest resources and "free" land outside LLNP, respective regulations are disregarded more often than honored.

Virtually without institutional restrictions, Bugis migrants – as well as some better-off local households – acquire land via purchase from poorer, local households. The land-stripped local households, in turn, acquire new land by illegally clearing primary or old secondary forest inside LLNP. These new plots are of inferior land use quality, and of a highly precarious tenure status. In contrast to village A or B, newly converted plots are reported to the village headman only after establishment because permissions from the village leadership are not needed. In order to minimize land conflicts within the community, the village headman usually issues a letter that documents the establishment of a certain plot for a small fee. "It is common here that everybody goes to the forest without permission" (local, village C). Furthermore, while our respondents report that fines and punishments were imposed in the past, this is not the case today.

6 Quantitative impact on rainforest conversion

For an analysis of the impact of land tenure rights and village institutions on rainforest conversion, we turn to the data of the village census. In its move to stylize itself as a traditional law community, the village institutions of village A prevent the population from obtaining official land titles. Very little land is appropriated to migrants, and land sales are approved of only under restrictive circumstances. Only LLNP land is available for agricultural expansion. This combination of factors has not completely eliminated forest clearing but has virtually closed the circulation of land within the local population of the village: Only locals themselves own land bought from locals (Fig. 2, "A" bars). Self reported rates of relative forest clearing are roughly half as high here compared to villages B and C, and not a single plot was reported to be sold to Bugis migrants.

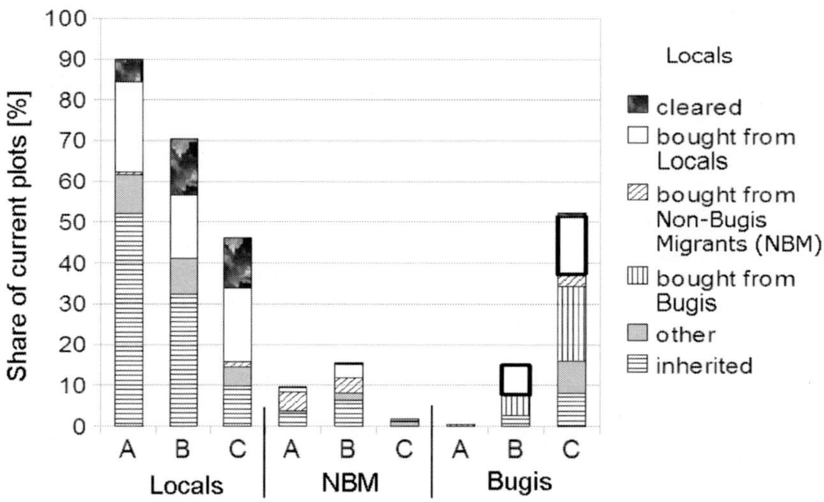

Fig. 2. Acquisition of currently owned plots and relative rainforest conversion ("cleared") in three contrasting villages A, C, and C; the two largest sets of inter-ethnic land transfer - from Locals to Bugis in villages B and C - are highlighted (data from STORMA A1 village census 2004).

Village B has community forests outside LLNP at its disposal. Thus, absolute land scarcity is lower than in the other villages. Similar to village A, only autochthonous households are allowed to clear community forest, though. However, relative rainforest conversion rates are much higher. We trace this back to a more liberal handling of land titeling and land sales to Bugis (see highlighted segment) and non-Bugis migrants: new forest conversion roughly offsets these sales. In village C, the dominance of Bugis buyers of land is even more pronounced, particularly for land bought from locals (see highlighted segment). Although LLNP forest land is converted, the relative conversion rate is virtually as high as in village B. Here, Bugis migrants selectively buy land outside LLNP, for which official land titles are issued. Added up, the Bugis migrants own nearly as much land as the local population. If locals collectively had not replaced land sold to Bugis by converted forests, the Bugis would already be the biggest landowners. In terms of the number of plots, the Bugis have already passed the local population (Fig. 2).

7 Discussion and conclusion

In principle, the Indonesian official land laws allow for the self-governance of traditional communities and their ancestral lands (*tanah adat*). Bureaucratic hurdles to be recognized as a customary law community (*masyarakat hukum adat*) are high, however, and the provincial government of Central Sulawesi denies the existence of *tanah adat* in the province.

Customary forms of land tenure in the project region were shaped by the requirements of swidden agriculture at an open forest frontier, i.e., without pervasive land scarcity. This means that only certain (weak) community prerogatives existed regarding the opening of primary forest. A request for forest conversion was rarely denied, and individuals, households, and families of the pioneer cultivators acquired certain individualized property rights (usage, lending, inheritance). Thus, the traditional land rights regime was characterized by a complex mix of open access, common property, and individual property aspects. Given the fact that the forest frontier was not closed, that a valuable cropping system could be established, and that the marked property of a household was generally respected, the development of a detailed traditional governance regime for common property resources cannot be expected to be found (cf. Ostrom 1990) – and could not be ascertained empirically to exist in the past (Burkard 2002).

Also under the current situation of growing land scarcity across much of the project region, a general strengthening of common property resource institutions cannot be observed. Remaining forest resources at the village level – be they within or outside of LLNP – are still under some form of community prerogative (village A and B, Rompo), but individual ownership can be established. There are also examples where forests reverted back to a *de facto* open access situation (village C), or where individual rent seeking led to generous land transactions in favor of migrants (Watumaeta). Furthermore, increasingly more "economic" forms of tenancy and tenure proliferate. Particularly, for Bugis migrants, purchase transactions have become a prime means to access land. Certain land tenancy arrangements (*bagi tanah, bapajak*) were even "imported" from other parts of Indonesia to suit the needs of migrants without much cash on hand.

In combination with an efficient production technology, the proliferation of secure private land rights enabled petty capitalist surplus production of the cash crop cocoa by Bugis migrants and some local households. Further facilitated by a lower culturally mandated status consumption for weddings and funerals, the Bugis migrant community established itself as an economically leading group in many villages of the research area (Abdulkadir-Sunito & Sitorus 2007). Within three decades, a substantial share of the land base in village C, and adjacent places such as Berdikari and Sintuwu was transferred to Bugis. A corresponding transition of autochthonous households "from landlords to landless" can be observed. Although the factual prohibition of land transfers to economically successful Bugis reduces economic efficiency and *average* wealth creation in village A and in Rompo[2], it also restricts the evolution of a class of landless or agriculturally marginalized autochthonous farmers (see also Li 1999).

Does the differentiation between common pool resource (LLNP land in village A, community forest of village B) and *de facto* open access resource (LLNP land in village C) help to resolve the documented differences in deforestation rates (Campbell et al. 2001; see section 6)? While relative deforestation rates are very high in village B, they are much lower in village A and roughly as high in village C. Differences regarding monitoring and applied sanctions in case of land use violations could play a role (cf. Ostrom 1990). In the traditional village A (likewise in Rompo), the Council of Traditional Leaders (*lembaga adat*) draws up land use regulations, and is actively involved in their enforcement. In contrast, the *lembaga adat* hardly plays any role in village C (likewise in Sintuwu), and is not involved in (virtually nonexistent) monitoring. However, the *lembaga adat* in village B successfully restricts access to its community forest to local inhabitants. Thus, our results suggest that the main proximate cause for the observed differences is mainly a strong village leadership in village A that uses *restrictions in private property of land* (no official land titling) to prevent a complex interaction of land purchases by migrants and forest conversion by locals.

Accordingly, an exception to the growing dominance of private land rights or "economic" tenancy arrangements in the project area is village A. Here, the (re-?) creation of strong customary institutions is an element in the struggle over claimed land inside LLNP. Although not legally recognized, this land is claimed as traditional land implying that no full private property can be established, and no land titles can be issued.

If the relatively low deforestation rates in village A can be sustained under internal population growth and under pressure by a population cut off from rural credit by lacking collateral remains to be seen. A long term retarding effect is likely to arise from the acceptance of village A's traditional

[2] The village A leadership justifies current restrictions by the need to alleviate long-term, intergenerational land scarcity as observed in neighboring villages where locals have sold out their land holdings to migrants (Burkard 2008*b*:160f).

lands by the LLNP administration: It includes much agricultural and forest land mapped as regenerating swidden cultivation land (traditional use zone within LLNP; Shohibuddin 2008). Because all of this land is recognized as ancestral land already opened many years ago, none of the families settling later in the village can claim the land. As a consequence, the families of the first settlers have additional support for the claim that they have opened the land in former times – and now own it within the traditional land ownership categories. This development further strengthens land access of the local elite disadvantaging less fortunate households. While these consequences appear socially problematic, they are likely to retard deforestation.

These considerations strengthen the suggestion that general judgments on the socio-economic desirability of the investigated land tenure systems are difficult to draw from the presented results. In nearly classical form, the advantages and disadvantages of a paternalistic, near-feudal system of land tenure have to be compared to the advantages and disadvantages of a land tenure system facilitating capitalist social dynamics.

In sum, our contribution highlights the importance of the analysis of supra-household level phenomena for an understanding of forest conversion at the Central Sulawesi rainforest margin. This is particularly true for the interaction of local institutions governing land access and tenure rights. With respect to the theoretical argument that tenure security fosters resource conservation, we observe a contrary effect of the introduction of formal land titles at the community level in the project region: It is the high tenure security of formal land titles that attracts migrants who are successful buyers of land for petty capitalist cocoa production. The autochthonous population may find itself landless at the end of the process – or in an "uphill battle" for marginal land inside LLNP (Barkmann et al. *in prep*).

Acknowledgements
We thank DFG-funded collaborative research center SFB 552 – STORMA "Stability of Rainforest Margins in Indonesia" as well as Bundesministerium für wirtschaftliche Zusammenarbeit und Entwicklung (BMZ) and Otto Vahlbruch-Stiftung for funding, and our local interviewees and assistants for their much appreciated cooperation.

References

Acheson J (2006) Institutional Failure in Resource Management. Annual Review of Anthropology 35: 117-134

Adiwibowo S (2008) Dongi-Dongi: Culmination of a Multi-dimensional Ecological Crisis: A Political Ecology Perspective. In: Burkard G, Fremerey M (Eds.) A Matter of Mutual Survival – Social Organisation of Forest Management in Central Sulawesi, Indonesia. Southeast Asian Modernities. LIT Verlag, Berlin, p. 307-356

Abdulkadir-Sunito M, Sitorus F (2007) From Ecological to Political Buffer Zone: Ethnic Politics and Forest Encroachment in Upland Central Sulawesi. In: Tscharntke T, Leuschner C, Zeller M, Guhardja E, Bidin A (Eds.) The stability of tropical rainforest margins, linking ecological, economic and social constraints of land use and conservation. Springer Verlag, Berlin, p.167-180

Aggarwal RM (2006) Globalization, local ecosystems, and the rural poor. World Development 34 (8): 1405-1418

Barkmann J, Faust H, Koch S, Schippers B, Schwarze S (in preparation): "Uphill battles": Capital driven social stratification in the Lore Lindu Biosphere Reserve, Central Sulawesi/Indonesia.

Berkes F, Folke C (1998) Linking social and ecological systems for resilience and sustainability. In: Berkes F, C Folke (Eds.) Linking practices and social mechanisms for building resilience. Cambridge University Press, Cambridge, p. 1-25

Bromley D (1992) Common Property as an Institution. In: Bromley D (Ed.) Making the Commons Work - Theory, Practice, and Policy. ICS Press, San Francisco, p. 3-16

Burkard G (2002) Natural Resource Management in Central Sulawesi: Past Experience and Future Prospects. STORMA Discussion Paper Series No. 8, Göttingen and Bogor

Burkard G (2008a) "Stability" or "Sustainability"? Changing Conditions for Socio-economic Security and Processes of Deforestation in a Rainforest Margin. In: Burkard G, Fremerey M (Eds). A Matter of Mutual Survival – Social Organisation of Forest Management in Central Sulawesi, Indonesia. Southeast Asian Modernities, LIT Verlag, Berlin, p. 15-58

Burkard G (2008b) Customary Law Communities, Property Relations, and the Management of Common-Pool-Resources: Evidence from Toro Village, Central Sulawesi, Indonesia. In: Burkard G, Fremerey M (Eds). A Matter of Mutual Survival – Social Organisation of Forest Management in Central Sulawesi, Indonesia. Southeast Asian Modernities, LIT Verlag, Berlin, p. 133-173

Burkard G, Fremerey M (Eds.) (2008) A Matter of Mutual Survival – Social Organisation of Forest Management in Central Sulawesi, Indonesia. Southeast Asian Modernities, LIT Verlag, Berlin

Campbell B, Mandondo A, Nemarundwe N, Sithole B, De Jong W, Luckert M, Matose F (2001) Challenges to proponents of common property resource systems: despairing voices from the social forests of Zimbabwe. World Development 29 (4): 589-600

Chomitz K, Griffiths C (1996) Deforestation, shifting cultivation, and tree crops in Indonesia: nationwide patterns of smallholder agriculture at the forest frontier. Research Project on Social and Environmental Consequences of Growth-Oriented Policies. Poverty, Economics, Growth (PEG) Working Document no. 4, The Worldbank, Washington D.C.

de Sherbinin A, Carr D, Cassels S, Jiang L (2007) Population and Environment. Annual Review of Environment and Resources 32: 345-373

de Sherbinin A, VanWey LK, McSweeney K, Aggarwal R, Barbieri A, Henry S, Hunter LM, Twine W, Walker R (2008) Rural household demographics, livelihoods and the environment. Global Environmental Change 18 (1): 38-53

Dietz T, Ostrom E, Stern PC (2003) The Struggle to Govern the Commons. Science 302 (5652): 1907-1912

Erasmi S, Twele A, Ardiansyah MA, Kappas M (2004) Mapping Deforestation and Land Cover Conversion at the Rainforest Margin in Central Sulawesi, Indonesia. EARSeL eProceedings 3 (3): 388-397

Faust H, Maertens M, Weber R, Nuryartono N, van Rheenen T, Birner R (2003) Does Migration lead to Destabilization of Forest Margins? STORMA Discussion Paper Series No. 11, Göttingen and Bogor

Fremerey M (2002) Local Communities as Learning Organizations - The case of the village of Toro, Central Sulawesi, Indonesia. STORMA Discussion Paper Series No. 6, Göttingen and Bogor

Gibbs C, Bromley D (1989) Institutional Arrangements for Management of Rural Resources. In: Berkes F (Ed.) Common Property Resources - Ecology and Community Based Sustainable Development. Belhaven Press, London, p. 22-32

Gibson C, McKean M, Ostrom E (2000) Explaining Deforestation: The Role of Local Institutions. In: Gibson C, McKean M, Ostrom E (Eds.) People and Forests - Communities, Institutions, and Governance. MIT Press, Cambridge, 1-26

Hardin G (1968) The Tragedy of the Commons. Science 162 (3859): 1243-1248

Hadikusuma H (2003) Pengantar Ilmu Hukum Adat Indonesia [Introduction to the Science of Adat Law of Indonesia], Mandar Madju, Bandung

Koch S, Faust H, Barkmann J (2008) Differences in Power Structures Regarding Access to Natural Resources at the Village Level in Central Sulawesi (Indonesia). Austrian Journal of South-East Asian Studies 1 (2): 59-81

Li TM (Ed.) (1999) Transforming the Indonesian Uplands – Marginality, Power and Production. Amsterdam, OPA, Amsterdam

Li TM (2007) The Will to Improve – Governmentality, Development, and the Practice of Politics. Duke University Press, Durham (SC)

Maertens M (2003) Economic Modeling of Agricultural Land-Use Patterns in Forest Frontier Areas - Theory, Empirical Assessment and Policy Implications for Central Sulawesi, Indonesia. Ph.D. Thesis, Georg-August-Universität Göttingen, Germany

Maertens M, Zeller M, Birner R (2006) Sustainable agricultural intensification in forest frontier areas. Agricultural Economics 34 (2): 197-206

Mappatoba M (2008) Co-Management as a strategy to balance community and Conservation interests in the LLNP. Contributed paper presented at the international symposium "Tropical Rainforests and Agroforests under Global Change", Bali, Indonesia, October 5 - 9, 2008

Mishan E (1969) The Relationship between Joint Products, Collective Goods, and External Effects. The Journal of Political Economy 77 (3): 329-348

North D (1990) Institutions, Institutional Change and Economic Performance. Cambridge University Press, Cambridge

Ostrom E (1990) Governing the Commons. Cambridge University Press, Cambridge

Ostrom E, Burger J, Field CB, Norgaard RB, Policansky D (1999) Revisiting the Commons: Local Lessons, Global Challenges. Science 284 (5412): 278-282

Schippers B, Faust H (2009) Migrants as cash crop-"pioneers"? – Socio-cultural change and land use in Central Sulawesi. STORMA Discussion Paper Series No. 29, Göttingen and Bogor

Schwarze S, Schippers B, Faust H, Wardhono A, Weber R, Zeller M (2005) The Status of Toro Village in the Lore Lindu Region: Is it Really Exceptional? A Comparative Quantitative Study of Socio-Economic Indicators. STORMA Discussion Paper Series No. 14, Göttingen and Bogor

Schwarze S, Schippers B, Weber R, Faust H, Wardhono A, Zeller M, Kreisel W (2007) Forest Products and Household Incomes: Evidence from Rural Households Living in the Rainforest Margins of Central Sulawesi. In: Tscharntke T, Leuschner C, Zeller M, Guhardja E, Bidin A (Eds) The stability of tropical rainforest margins: Linking ecological, economic and social constraints of land use and conservation, Springer Verlag Berlin, p. 209-224

Shohibuddin M (2008) Discoursive Strategies and Local power in the Politics of Natural Resource Management: The Case of Toro Community. In: Burkard G, Fremerey M, (Eds). A Matter of Mutual Survival – Social Organisation of Forest Management in Central Sulawesi, Indonesia. Southeast Asian Modernities, LIT Verlag, Berlin, p. 91-132

Sitorus F (2004) "Revolusi Cokelat": Social Formation, Agrarian Structure, and Forest Margins in Upland Sulawesi, Indonesia. In: Gerold G, Fremerey M, Guhardja E (Eds.) Land Use, Nature Conservation and the Stability of Rainforest Margins in Southeast Asia. Springer Verlag, Berlin, p. 105-118

Soekanto S, Taneko S (2002) Hukum Adat Indonesia [The Adat Law of Indonesia]. Raja Grafindo Persada, Jakarta

Stern PC, Dietz T, Dolsak N, Ostrom E, Stonich S (2002) Knowledge and Questions After 15 Years of Research. In: Ostrom E, Dietz T, Dolsak N, Stern PC, Stonich S, Weber E (Eds.) The Drama of the Commons. National Academy Press, Washington, D.C., p. 445-489

Thamrin TS (2007) The management of Conflicts over Natural Resources - A case study from the Lore-Lindu National Park, Central Sulawesi, Indonesia. STORMA Discussion Paper Series, No. 22, Göttingen and Bogor

Weber R (2006) Kulturlandschaftswandel in Zentralsulawesi. Universitätsverlag Göttingen, Göttingen

Weber R, Faust H (2006) Kulturelle Aspekte der Landnutzung in Indonesien. Geographica Helvetica 61 (4): 237-245

Weber R, Faust H, Schippers B, Mamar S, Soetarto E, Kreisel W (2007) Migration and ethnicity as cultural impact factors on land use change in the rainforest margins of Central Sulawesi, Indonesia. In: Tscharntke T, Leuschner C, Zeller M, Guhardja E, Bidin A (Eds) The stability of tropical rainforest margins: Linking ecological, economic and social constraints of land use and conservation. Springer Verlag Berlin, p. 417-436

Van Edig X, Schwarze S, Zeller M (2007) Indicator based poverty assessment for rural Central Sulawesi. Quarterly Journal of International Agriculture 46: 145-158

Zeller M, Schwarze S, van Rheenen T (2002) Statistical Sampling Frame and Methods Used for the Selection of Villages and Households in the Scope of the Research Programme on Stability of Rainforest Margins in Indonesia (STORMA). STORMA Discussion Paper Series No. 1, Göttingen and Bogor

Rural income dynamics in post-crisis Indonesia: evidence from Central Sulawesi

Jan Priebe[1*], Robert Rudolf[1], Julian Weisbrod[1], Stephan Klasen[1], Iman Sugema[2], and Nunung Nuryartono[2]

[1] University of Göttingen, Faculty of Economics, Platz der Göttinger Sieben 3, 37073 Göttingen, Germany
[2] Institut Pertanian Bogor, Indonesia, 16710 Bogor, International Center for Applied Finance and Economics, Kampus IPB Baranang Siang, Gedung Utama Lt. II, Jalan Raya Paiajaran

*corresponding author: J. Priebe, email: jpriebe@uni-goettingen.de

1 Summary

Indonesia is an emerging economy characterized by increasing rural-urban income disparities and high poverty rates, particularly in rural areas. Despite a large part of the Indonesian population residing in rural areas, no studies currently exist that thoroughly analyze the factors determining rural income dynamics for the years after the Asian Financial Crisis of 1997/98. Utilizing a new panel data set collected in Central Sulawesi, this study aims to shed more light on rural livelihoods and therefore investigates the socio-economic factors driving the income process of Indonesian farmers at rainforest margins. Our results indicate that a sharp increase in rural incomes took place in the post-crisis period. While traditional agriculture still constitutes the backbone of household incomes, the ability to alleviate poverty and to enjoy income growth has been strongly associated with a household's ability to diversify into the non-agricultural sector of the economy, to focus on higher value-added agricultural activities and its capability to invest into new production techniques. Comparing our results to the national SUSENAS household data for Central Sulawesi we find that our findings enhance the understanding of rural income generation processes in Indonesia.

Keywords: poverty, income dynamics, diversification, Indonesia

T. Tscharntke et al. (eds.), *Tropical Rainforests and Agroforests under Global Change*,
Environmental Science and Engineering, 161–176, DOI 10.1007/978-3-642-00493-3_7,

2 Introduction

After the severe financial crisis in Indonesia in 1998, average income levels saw a recovery to, and subsequently above pre-crisis levels. Nevertheless, poverty rates in Indonesia are still above pre-crisis levels while rural-urban disparities have been constantly increasing over time (World Development Report, 2008). These developments in turn imply two things: Firstly, potentially high political and social strain, and secondly, a high incidence of poverty in rural areas. Moreover, the World Development Report 2008 gives a succinct summary of the heterogeneity of rural income generating activities and its potential determinants, constraints and drivers. On the one hand, it clearly acknowledges the fact that higher agricultural productivity is crucial to raise income in "agriculturally-based" countries and for the poorest of rural households (see also Mellor, 1976; Ravallion and Datt, 1996, 1998a; Timmer, 2002; Huppi and Ravallion, 1991; Suryahadi and Sumarto, 2003; Fan, Zhang and Zhang, 2004; Fan, Gulati, and Thorat, 2007; Timmer, 1997, 2004; Thirtle et al., 2003; Majid, 2004). On the other hand, it makes it equally clear, as confirmed by related literature (Ravallion and Datt, 1996, 1998b, 2002; Lanjouw and Lanjouw, 2001), that for countries undergoing a rapid structural transformation rural non-agricultural activities account for roughly half of rural income and can be most conducive towards income growth and poverty reduction. These general research findings also conform with results obtained by numerous other Southeast Asian studies, e.g. Estudillo, Sawada and Otsuka (2006) for the Philippines, Cherdchuchai and Otsuka (2006) for Thailand, Nargis and Hossain (2006) for Bangladesh as well as with well-known studies from Lanjouw and Lanjouw (2001) and Ersado (2005).

In the case of Indonesia, few studies have analyzed the link between employment sectors, individual and household characteristics and their impact on (rural) household income levels. Conducting a panel analysis drawing on data from the Indonesian Family Life Surveys (IFLS) for 1993 and 2000, McCulloch, Weisbrod and Timmer (2007) find that while agriculture remains crucial for income growth, in particular of the poorest households, a gradual diversification of economic activities, characterized by an ever greater reliance on non-agricultural sources, is taking place. Moreover, Dewi et al. (2005) investigate the determinants of rural non-agricultural income on the village level in East Kalimantan and find that better infrastructure, the closeness to transmigration[1] sites and deforestation (1992-1996) positively correlate with non-agricultural income in this region characterized by high logging activity. In a further step, they find that overall village welfare rises with higher economic diversity (especially through higher non-agricultural income), agro-

[1] Transmigration refers to governmentally induced migration. Through several programs aiming at decentralization and the development of remote islands, mainly Javanese were encouraged to settle on other islands from the early 19th century on by providing them land and housing.

suitability of land, land use intensity, forest cover in the initial period (1992) and village population size.

However, all the above literature only refers to pre-crisis or within-crisis Indonesia and therefore does not contribute to the question on how rural households have fared in the *post-crisis* period.

We aim to contribute to the existing literature on rural Indonesia in several ways. Our study draws on a highly detailed rural household panel data set (STORMA) for Central Sulawesi (CS) comprising three waves from 2001 to 2006. Benefiting from the unprecedented wealth of detail in STORMA, we are able to (1) analyze the determinants of rural income generating activities more thoroughly and (2) analyze their dynamics. Furthermore, in order to identify whether our results might hold additional lessons for Indonesian rural income processes, we (3) generalize our findings by comparing our results with national household data from the 2002 and 2005 round of SUSENAS, the national household survey.

Our paper is structured as follows: Section 2 introduces our data sources and estimation strategy. Section 3.1 provides a comparison of some key socio-economic descriptive statistics obtained from STORMA and SUSENAS for the post-crisis period. In section 3.2 we estimate and analyze cross-sectional income level regressions based on STORMA 2001, 2004 and 2006. Regressions will comprise a core set of explanatory variables that are found in STORMA and SUSENAS. In addition, the regressions are extended in a further step to incorporate additional variables that are *not* included in SUSENAS and that can enhance our understanding of the determinants of rural income generating activities. Section 4 summarizes and concludes.

3 Data and Methodology

3.1 Data

Previous analyses of the rural economy in Indonesia has been limited by insufficient data availability. In this study we address the above mentioned shortcomings, utilizing a three wave rural panel household survey (2001, 2004, 2006) conducted as part of the STORMA project in rural Central Sulawesi. The study area encounters about 110 villages in four sub-districts (Kecamatan) bordering the Lore Lindu National Park in Central Sulawesi (CS). Household surveys were conducted in a total of 12 villages in this research area. A multi-stage sampling design was used based on the proximity of the villages to the Lore Lindu Park, population density and ethnic composition.[2] Originally, 294 households in 12 villages were interviewed. Due to some technical problems in the 2004 round, only 258 households could be contacted. In the 2006 round

[2] A more detailed description about the selection procedure of villages and households in the study area can be found in Zeller, Schwarze, and van Rheenen (2002).

still 271 of the original 294 households could be re-interviewed. In total, the sample comprises 256 households who were interviewed in each of the three waves.

For reasons of comparability in the considered time horizon, we will analyze the rural economy vis-à-vis the 2002 and 2005 national SUSENAS household data sets. SUSENAS covers about 200,000 Indonesian households per round and is representative at the district (Kabupaten) level. Furthermore, the survey is conducted annually, with a complete income and expenditure module every three years (here 2002 and 2005), which covers about 60,000 households. Yet the prevailing panel data structure within SUSENAS is kept for a maximum of two years, so that the panel structure never covers two consecutive income/expenditure modules. This constitutes one of SUSENAS' main disadvantages.

3.2 Methodology

The main purpose of this analysis is to isolate the determinants of rural income and its changes in an emerging rural economy. Thus, we follow the common practice of using the logarithm of per-capita income as our dependent variable in a standard income regression framework that empirically estimates the determinants of this income. In order to achieve comparability between STORMA and SUSENAS our income variable contains only those income components that are available in both data sets. In particular, we exclude imputations of rents from housing and remittances from the construction of our income variables.

Furthermore, we decided to use only those households from SUSENAS that are located within the 1st to the 7th deciles of the household per-capita income distribution. We drop the richest 30 percent of SUSENAS to facilitate the comparability with STORMA. This procedure is motivated by the fact that SUSENAS comprises semi-urban areas that are classified as rural and that are much richer than typical rural Indonesian households.

With respect to the regressions on the determinants of income levels we apply a standard model concerning the determinants of log per-capita income that stems from the labor-supply literature. We follow this approach and model the log per-capita income level as a function of individual characteristics of the household head and general household characteristics. Further on, the household's main employment sector enters the regression in the form of sector binary variables which take the value one if a household earns more than 50 percent of its total household income in that sector and zero otherwise. We follow the method proposed by Barrett et al. (2001) who classify income sources according to sectors (agricultural and non-agricultural) and functions (wage and self-employed).

The level equations are estimated by Ordinary Least Squares (OLS). The estimated equation is depicted in equation (1), where Y stands for household per-capita income, X for a set of household characteristics for which information is available in SUSENAS *and* STORMA, and Ψ for a set of variables that are *exclusively* available in STORMA.

$$log(Y_i) = \alpha + X_i\beta + \Psi_i\lambda + u_i \quad (1)$$

While OLS gives consistent and efficient results in the absence of endogeneity issues, unobserved heterogeneity or omitted variable bias, however if one of the problems occurs this might cause significant bias in the estimated coefficients.[3] Nevertheless, OLS still provides us with important general insights into the nature of the rural income generating processes.

In order to address the problems raised above regarding the reliability of our OLS estimates, we introduce additional variables into our STORMA income equation that are not covered by SUSENAS and that from a theoretical point of view should have a substantial effect on a households' capability to generate income, e.g. proxies for credit constraints and asset and land ownership.

4 Results

4.1 Descriptive Results

We start our empirical part with the descriptive analysis of the key variables, found in STORMA and SUSENAS. Table 1 depicts mean values of important household characteristics and its respective standard deviations for the three STORMA surveys and the two SUSENAS surveys where total household income was available. Two main findings can be directly derived from the table. Firstly, keeping in mind that only deciles 1 to 7 from CS-SUSENAS are considered here rural households in the STORMA region started on average economically below the provincial rural SUSENAS average. Secondly, STORMA households have shown a remarkable catching-up process since then and enjoy substantially higher per-capita income levels in 2006.

Concentrating on the comparability between the three STORMA waves and the two restricted CS-SUSENAS waves, we see quite similar mean estimates for most demographic and economic variables. While household size and the number of men seem to be still a little higher in the STORMA research area, age, sex and education structure of household heads are highly comparable. In both data sets, a similar trend can be observed towards rising levels in the years of schooling.

[3] This paper explicitly treats the mitigation of the problem of omitted variable bias. The other two methodological concerns, endogeneity and unobserved heterogeneity, are tackled in further work by the authors. The related panel regressions can be obtained on request from the authors. They mainly confirm the cross-section results of this paper.

Table 1. Comparison of Demographic and Income Means (STORMA vs. CS-SUSENAS)

	STORMA '01	STORMA '04	STORMA '06	CS-SUSENAS '02*	CS-SUSENAS '05*
Household Size	5.42	5.19	4.56	4.49	4.53
	(2.00)	(1.96)	(1.93)	(1.58)	(1.67)
Age of HH Head	43.8	46.5	48.1	40.3	42.4
	(14.0)	(14.1)	(13.6)	(12.6)	(12.4)
Sex of HH Head	0.95	0.93	0.91	0.96	0.90
	(0.21)	(0.26)	(0.29)	(0.21)	(0.30)
Dependency Ratio	0.70	0.75	0.74	0.77	0.80
	(0.58)	(0.60)	(0.70)	(0.63)	(0.65)
Number of Men	1.85	1.86	1.37	1.39	1.37
	(1.03)	(1.10)	(0.87)	(0.79)	(0.83)
Years of Schooling of HH Head	6.77	6.79	6.78	5.97	6.82
	(3.36)	(3.37)	(3.35)	(3.82)	(3.12)
Max. Years of Schooling of a HH Member	8.68	8.67	8.43	7.84	8.50
	(2.87)	(2.89)	(2.87)	(3.22)	(2.91)
Total Per-Capita Income	95076	93187	119586	96197	89877
	(106003)	(131061)	(123391)	(29569)	(35093)
Agricultural Self-employed Income, per capita	60266	52751	68005	60651	54617
	(68679)	(77544)	(81073)	(42831)	(43628)
Non-Agricultural Self-employed Income, p.c.	10906	11062	19678	13307	13722
	(64371)	(40068)	(68299)	(30015)	(30663)
Agricultural Wage Income, p.c.	8319	4820	8200	12397	
	(17016)	(11164)	(18353)	(27955)	215389**
Non-Agricultural Wage Income, p.c.	15583	23652	22659	9842	(36768)
	(46465)	(102055)	(63891)	(28030)	
N	256	256	256	523	530

All monetary values are real values and in Indonesian Rupiahs. Incomes are monthly. Standard deviations in parentheses. Shares of the four income sources are with respect to total household income, *not* per capita. * SUSENAS means cover the deciles 1 to 7 of the original income distribution for rural Central Sulawesi. Deciles 8 to 10 were dropped due to comparability reasons. ** In the SUSENAS 2005 questionnaire no distinction is made between agricultural and non-agricultural wage employment.

How did the income situation evolve over time? In the CS-SUSENAS samples for 2002 and 2005 we observe mean values which are equivalent to the 2001 and 2004 STORMA data respectively. The STORMA region then proved itself to be very dynamic in the subsequent period when it witnessed a remarkable 28 percent real per-capita income increase within two years (2004-06).

The initial income decline between 2001 and 2004was partly due to the fact that the STORMA project regionwas hit exceptionally severe by the coffee crisis in the 1990s and as a consequence suffered agricultural income stagnation or even decline.[4] During the period 2002-05 (SUSENAS) and 2001-2004 (STORMA), real agricultural incomes decreased by 10 percent in the SUSENAS sample and by 17 percent in the STORMA sample. Hence, farmers had to adopt several coping strategies to deal with the situation, which eventually led to the growth spurt between 2004 and 2006. Two main strategies can be observed: First, from 2001 onwards a strong diversification of income generating activities into the non-agricultural sector occurred. Both non-agricultural sub-sectors (self-employed and wage employed) rose significantly by 2004, by 12 and 57 percent respectively. Second, households in the STORMA region decided to shift their cash crop production from coffee to the more productive cocoa responding to changes in world market prices.

Table 2. Agricultural Self-employed Diversification - Mean Incomes

Sector	STORMA '01	STORMA '04	STORMA '06
Livestock	6190	3350	5026
Gathering	10527	4249	2931
Cropping	44752	46549	60048
Annual crops	21859	18588	26146
Perennial Crops	22892	27961	33901
Cocoa	13278	24280	28307
Coffee	11433	1752	2861
N	256	256	256

All values are monthly in per-capita terms and real IDR with base year 2001.

The extent of agricultural intensification versus diversification is considered in Table 2. STORMA households increased their income from cropping by 34 percent between 2001 and 2006. This development was facilitated through a mixture of relative price changes and technology transfer. Hence, more and more farmers shifted from coffee to cocoa which promised higher returns per

[4] Agricultural returns decline was mainly caused by two factors: Firstly, while world coffee prices persisted on very low levels between 2001 and 2004, world cocoa prices rose gradually during the same period. This induced a decline of traditional coffee returns and a change in planting decisions of farmers. Secondly, cocoa returns could only compensate for the whole loss with a time lag due to the minimum four to five years until they bear first pods.

hectare and more attractive market prices. The decision to switch from coffee to cocoa turned out to be very beneficial for many farmers in the STORMA region during the observed period. Table 2 shows that in 2001 both cocoa and coffee played almost equal roles in terms of incomes obtained from perennial crops, while by 2006 coffee production had disappeared almost completely in contrast to the sharply rising cocoa production in the region.

Although agricultural incomes remain the backbone of the rural economy as we have seen, it is also the case that rural non-agricultural activities are growing in importance. The increasing engagement of households in the non-agricultural sector that took place in the STORMA region between 2001 and 2006 represented a coping strategy on the one hand and an example of structural change and economic recovery on the other. The income source transition matrix in Table 3 further underlines that the number of households in our sample that generated more than half of their incomes from non-agricultural activities rose from 41 in 2001 to 54 in 2006. At the same time, the number of households who generated more than half of their incomes in agriculture decreased from 207 in 2001 to 187 in 2006.[5] The transition matrix furthermore allows to track certain groups of households over time and to see the consequences of their sectoral activities. From the matrix we see that a household who was primarily engaged in the agricultural sector in 2001 and then shifted the main occupation into the non-agricultural sector was able to increase its per-capita income on average to a much larger extent than if it would have stayed agriculturally dependent. For example, from those 178 households mainly active in self-employed agriculture in 2001 (average per-capita income in 2001: 87,580 Indonesian Rupiah (IDR)), 133 households remained in that sector by 2006 (av. per-capita income in 2006: 105,969 IDR), while 14 households changed to agricultural wage employment (65,992 IDR in 2006) and 12 households changed to each of the sectors non-agricultural self-employment (198,022 IDR in 2006) and non-agricultural wage employment (180,752 IDR in 2006). Despite the limited number of observation in our sample it seems that rural households that participate in non-agricultural activities are better off.

4.2 Multivariate Analysis - Cross-Sectional Income Regressions

We start our multivariate analysis based on a rather simple kind of income equation, thus neglecting issues of omitted variable bias and unobserved heterogeneity. This enables us to show the potential bias that arises when important omitted independent variables are correlated with covariates in the model and when individuals or groups have different intercepts. In other words, we

[5] Households that derive most of their income from agricultural sources are those that were classified into the agricultural wage and agricultural self-employed category. Households from the mixed category were considered as well of most the income came from agricultural sources.

Table 3. Income Sector Transition Matrix: Average real per-capita income of households engaged mainly in the respective sector in 2006 by 2001 sector

				STORMA 2006									
				Agric. Self-employed		Non-agric. Self-employed		Agric. wage		Non-agric. wage		Mixed	
		Starting inc.	# of obs.	Income	# of obs.	Income	# of obs.	Income	# of obs.	Income	# of obs.	Income	# of obs.
STORMA	**2001**	Agricultural Self-employed	87580	178	105969	133	198022	12	65992	14	180752	12	87955	7
		Non-agric. Self-employed	178477	15	153616	6	243471	4		0	142350	5		0
		Agricultural wage	52744	29	76603	17	423417	1	52292	4	81924	4	72583	3
		Non-agric. wage	129681	26	72397	8	224819	2	35164	1	239930	13	87608	2
		Mixed	163616	6	120554	4	195338	1		0		0	72661	1
				254	103448	168	220927	20	61486	19	186104	34	83178	13

All monetary values are real, per-capita values and in 2001 Indonesian Rupiahs. Incomes are monthly. The category Mixed' refers to those households for which none of the four previous categories exceeds 50 percent of total household income.

estimate the bias that would arise when estimating rural income determinants with SUSENAS. The estimated model is the one specified under equation (1) in section 2.2 except that the set of variables represented by Ψ will not be included. Variables that are included in X are the best variables available in SUSENAS to explain the income of rural households in Indonesia.[6] We use the data of those variables from our STORMA data sets and estimate the determinants of the log of per-capita income of our households with OLS. The results for each of our survey waves are shown in Table 4.

Table 4. Basic Level Regressions.

	LN REALPERCAPITAINCOME		
	STORMA '01	STORMA '04	STORMA '06
Sex	0.096	0.102	-0.180
Age	0.024	- 0.006	0.070
Age^2	- 0.000	0.000	-0.001
Maxeducation	0.452	0.837***	-0.019
HH Size	- 0.155***	- 0.224***	-0.142**
Dependencyratio	- 0.125	0.087	-0.287**
Numberofmen	0.108	0.273***	0.161*
Agriselfemployed	0.523**	0.786***	0.310*
Nonagriwage	1.006***	1.413***	1.026***
Nonagriselfemployed	0.860**	1.263***	1.109***
Mixed	1.280***	1.169***	-0.036
_cons	10.337***	10.141***	10.233***
N	256	256	256
Adj. R-squared	0.10	0.20	0.16

Deciles 8 to 10 were dropped due to comparability reasons.
Significance levels: ***/**/* denote 0.01, 0.05 and 0.1; respectively (robust t-statistics used).
Variables: Age refers to the square of the age variable and is included to control for non-linear effects on the dependent variable. Sex refers to a dummy variable taking the value 1 if the head of the household is male while _cons relates to the constant in the regression equation. Maxeducation refers to the maximum years of schooling of a household member.

OLS estimates for STORMA show that the effects of most of the included covariates are very similar. Key determinants of the income generating process are a subset of the household characteristics, in particular household size, the number of men in a household and the education variable, all of which are statistically significant and take signs as expected from economic

[6] The selection of variables is based on economic theory. Therefore, variables that represent income production factors, e.g. human and physical capital were included in addition to control variables for the demographic structure of the household.

theory. Furthermore, the belonging of a household to a specific economic sector plays an important role. Households that are predominantly engaged in the non-agricultural sector seem to do much better than households deriving most of their incomes from the agricultural sector as indicated by the positive coefficients that increase in size from agricultural self-employed over non-agricultural wage to non-agricultural self-employed.[7] For all covariates mentioned above signs and rankings do not switch over survey rounds indicating a robust relationship of the underlying factors that determine rural income processes.

Extending our previous regressions we include a set of covariates previously described with Ψ that allows us to examine the effect of the prevalent omitted variable bias in the baseline regressions presented above. Especially, such an exercise will allow us to get a clearer picture of the size and direction of the effect on the coefficients estimated in table 4. Variables included in our case in Ψ were chosen based on theoretical economic grounds covering different domains. As a proxy for market proximity and available infrastructure we include the household's distance to the next road. Furthermore, we incorporate the value of a household's physical assets as an indicator for the ability to invest and produce at the optimal production frontier and to diversify into other economic sectors. In addition we decided to include covariates that a priori should have a strong effect on the revenue from agricultural production and that capture the household's ability to diversify its agricultural portfolio into cash crops (*area of cocoa, area of land).*

The results from the full estimation of equation 1 are shown in table 5. By looking at the values of the *adjusted* R^2 one can already see that the inclusion of the four additional variables has improved the explanatory power of our OLS estimation substantially. Most of this improvement is clearly attributable to the inclusion of *area of cocoa* and *value of assets* as additional covariates which show economically and statistically significant positive coefficients. Moreover, the infrastructure variable shows the expected sign in all of the three waves although only being significant in the 2006 round.

With respect to the effect of the introduction of our Ψ set of variables on the estimated β coefficients, we observe very mixed effects. While the coefficient of the education variable tends to become smaller or negative, it turns out to be non-significant in two of the three rounds suggesting that in the more detailed model of the rural economy the level of education plays a smaller role than in the less detailed specification. Yet interestingly, it is significant in 2004, when non-agricultural wage incomes had risen significantly. Similar to education, the *agricultural self-employed* variable is no longer significant in the 2001 and 2006 estimations either. As it could be expected variables included in Ψ, like the possession of assets, are positively correlated with a higher level of education and the means to diversify into profitable non-agricultural

[7] All coefficients on the agricultural sector have to be interpreted in reference to the agricultural wage sector which is the left-out category.

Table 5. STORMA - Extended Level Regressions.

	LN REALPERCAPITAINCOME		
	STORMA '01	STORMA '04	STORMA '06
Sex	0.087	0.095	-0.303
Age	0.016	0.002	0.064
Age^2	0.000	0.000	-0.001
Maxeducation	0.021	0.069**	0.066
HH Size	- 0.107***	- 0.190***	-0.160
Dependencyratio	0.200	0.104	0.180
Numberofmen	0.069	0.110	0.176**
Agriselfemployed	0.244	0.557**	-0.179
Nonagriwage	0.417	1.195***	0.627**
Nonagriselfemployed	0.780***	1.241***	0.556**
Mixed	1.074***	1.080***	-0.535
Area Owned	0.000	0.001	0.000
Area Cocoa	0.002**	0.003***	0.002
ln real Value of Assets	-0.009	-0.021	-0.001***
Distance to road	0.022*	0.029	0.138***
_cons	10.769***	9.678***	9.536***
N	256	256	256
Adj. R-squared	0.13	0.34	0.26

Significance levels: ***/**/* denote 0.01, 0.05 and 0.1 respectively (robust t-statistics used).
We controlled for spatial differences using kecamatan (sub-district) dummies.

activities. This thus also implies that the effect of education on incomes works to a considerable extent via the accumulation of assets. Similarly, the *area of land* and *area of cocoa* are correlated with the *agricultural self-employed* dummy. Controlling for these factors therefore diminishes the impact of those variables on the household income generating process.

Furthermore, the effect of estimating the full specification of equation (1) on our demographic household characteristics is noteworthy. While almost no change can be observed in the coefficients on household size, the number of men in a household or the age of the household head, the effect of the dependency rate on household per-capita income becomes smaller due to the fact that households with a higher dependency rate are on average less endowed with physical or agricultural assets.

Reflecting on the results from the cross-sectional regressions we can safely state that the inclusion of important covariates in STORMA therefore does not only allow us to estimate equation (1) in its unrestricted form and to remove or lessen the bias in our estimated β coefficients, it further facilitates the interpretation and understanding of the rural income generating process for other parts of rural Indonesia.

5 Conclusions

Indonesia's rural economy had not been studied to a great extent in the past, neither before the Asian Financial Crisis nor afterwards. This is a great neglect since most of Indonesia's poverty is being found in rural areas. We present new estimates from the detailed STORMA data set from Central Sulawesi and thus shed more light on the rural economy. In particular we aimed to contribute to the debate whether it is agricultural productivity growth or the diversification into rural non-agricultural activities that is most conducive to rural income generation in Indonesia's transforming rural economy.

Our study period was marked by a general upswing of the Indonesian economy after the crisis. We estimated cross-section determinants of rural per-capita incomes for the years 2001, 2004 and 2006. Our findings suggest a significant positive effect on the level of income of being employed in the non-agricultural sector compared to the agricultural sector. Moreover, we showed explicitly that the cultivation of cocoa had a significant positive impact on incomes in our study area. Furthermore we found that while years of schooling lose in importance when new variables enter the model, it indirectly affects incomes via capital endowments which are positively correlated with per-capita incomes.[8]

In terms of the debate described above we find that both factors were at work. On the one hand, agriculture still remains the major income source for most of the rural population. Agricultural productivity growth in the form of shifts to high value-added agriculture increased incomes. Less productive activities like gathering and coffee production seem to phase out. On the other hand, non-agricultural are gaining in importance nowadays. The number of rural households engaged in non-agricultural activities is increasing and these households presented both higher average incomes and higher income growth than agricultural households. It seems that further diversification into the non-agricultural sector can therefore be highly beneficial to further reduce rural poverty rates.

We showed that STORMA data is superior to SUSENAS data for the analysis of rural income generation processes since it offers a large set of additional variables designed exclusively for the understanding of the rural sector. Thus, we are able to control explicitly for omitted variable bias. STORMA results can hold additional lessons for rural Indonesia that SUSENAS was not able to provide up to now. STORMA's panel data structure offers further methodological advantages that were not treated in this paper.

[900?]Are there any policy implications that can be deduced from our findings? The evidence shows the importance of the rural non-agricultural sector in providing income and reducing poverty. Measures could be taken

[8] Here the question of causality arises. Yet, we show in further work that causality runs mainly from *value of assets* to *per-capita income*. Related Panel-IV-Regressions can be obtained from the authors on request.

to support a faster transition of the rural economy towards more agricultural downstream industries, better connections to semi-urban and urban areas and towards more employment in rural services sectors. Likewise it remains crucial to further raise productivity in the agricultural sector since this is the sector which nourishes the poorest of the poor. Access to capital still constitutes one of the main binding factors, thus a better supply of rural households with financial capital and agricultural production technology can be a means to ensure higher efficiency in production. Lastly, investment into human capital should focus more on the quality of education than on the quantity whose direct effect on incomes is relatively small. Education might also focus more on market-specific knowledge transmission.

Acknowledgements
We would like to thank two referees and participants at the International Symposium "Tropical Rainforests and Agroforests under Global Change" in Bali for helpful comments and discussion. Funding from the German Research Foundation as part of CRC 552 (STORMA) is gratefully acknowledged.

References

Barrett CB, Bezuneh M, Aboud A (2001) Income diversification, poverty traps and policy shocks in Cte d'Ivoire and Kenya. Food Policy 26 (4): 367-384

Cherdchuchai S, Otsuka K (2006)Rural income dynamics and poverty reduction in Thai villages from 1987 to 2004. Agricultural Economics 35 (s3): 409-423

Dewi S, Belcher B, Puntodewo A (2005) Village economic opportunity, forest dependence, and rural livelihoods in East Kalimantan, Indonesia. World Development 33 (9): 1419-1434

Ellis F (2003) Rural Livelihoods and Diversity in Developing Countries. Oxford University Press

Ersado L (2005) Income diversification before and after economic shocks: Evidence from urban and rural Zimbabwe. Development Southern Africa 22 (1): 27-45

Estudillo JP, Sawada Y, Otsuka K (2006) The green revolution, development of labor markets and poverty reduction in the rural Philippines, 1985-2004. Agricultural Economics 35 (s3): 399-407

Fan S, Zhang L, Zhang X (2004) Reforms, investment and poverty in rural China. Economic Development and Cultural Change 52 (2): 395-421

Fan S, Gulati A, Thorat S (2007) Investments, subsidies and pro-poor growth in rural India. IFPRI Discussion Paper

Huppi M, Ravallion M (1991) The sectoral structure of poverty during an adjustment period: Evidence for Indonesia in the mid-1980s. World Development19 (12): 1653-1678

Lanjouw JO, Lanjouw P (2001) The rural non-farm sector: Issues and evidence from developing countries. Agricultural Economics 26 (1): 1-23

Majid N (2004) Reaching Millennium Goals: How well does agricultural productivity growth reduce poverty? ILO Employment Strategy Papers

Mellor JW (1976) The new economics of growth: A strategy for India and the developing world.

McCulloch N, Weisbrod J, Timmer CP (2007) Pathways out of Poverty During an Economic Crisis: An Empirical Assessment of Rural Indonesia. World Bank Policy Research Working Paper 4173

Nargis N, Hossain M (2006) Income dynamics and pathways out of rural poverty in Bangladesh, 1988-2004. Agricultural Economics 35 (s3): 425-435

Ravallion M, Datt G (1996) How important to India's poor is the sectoral composition of economic growth? World Bank Economic Review10 (1): 1-25

Ravallion M, Datt G (1998a) Farm productivity and rural poverty in India. Journal of Development Studies 34 (4): 62-85

Ravallion M, Datt G (1998b) Why have some Indian states done better than others at reducing rural poverty? Economica 65 (257): 17-38

Ravallion M, Datt G (2002) Why has economic growth been more pro-poor in some states of India than in others? Journal of Development Economics 68 (2): 381-400

Suryahadi A, Sumarto S (2003) Poverty and vulnerability in Indonesia before and after the economic crisis. Asian Economic Journal 17 (1): 45-64

Suryahadi A, Suryadarma D, Sumarto S, Molyneaux J (2006a) Agricultural Demand Linkages and Growth Multiplier in Rural Indonesia. SMERU Working Paper

Suryahadi A, Suryadarma D, Sumarto S (2006b) Economic Growth and Poverty Reduction in Indonesia: The Effects of Location and Sectoral Components. SMERU Working Paper

Thirtle C, Lin L, Piesse J (2003) The impact of research-led agricultural productivity growth on poverty reduction in Africa, Asia and Latin America. World Development 31 (12): 1959-1975

Timmer CP (1997) Farmers and markets: The political economy of new paradigms. American Journal of Agricultural Economics 79 (2): 621-627

Timmer CP (2002) Agriculture and economic development, in: Handbook of Agricultural Econnomics 2 (1): 1487-1546

Timmer CP (2004) The road to pro-poor growth: The Indonesian experience in regional perspective. Bulletin of Indonesian Economic Studies 40 (2): 177-207

Woolard I, Klasen S (2004) Determinants of Income Mobility and Household Poverty Dynamics in South Africa. IZA DP No. 1030

World Bank (2008) World Development Report: Agriculture for Development. IBRD, The World Bank. Washington, D.C.

Zeller M, Schwarze S, van Rheenen T (2002) Statistical Sampling Frame and Methods Used for the Selection of Villages and Households in the Scope of the Research Program on Stability of Rainforest Margins in Indonesia. STORMA Discussion Paper Series 1, STORMA, University of Göttingen.

Gender division of labor in agroforestry activities within households: a case of Wonogiri - Central Java - Indonesia

Herien Puspitawati[1]* and Ma'mun Sarma[2]

[1] Department of Family and Consumer Sciences, Faculty of Human Ecology, Bogor Agricultural University, Jalan Lingkar Akademik Kampus IPB, Darmaga, Bogor 16680, Jawa Barat, Indonesia
[2] Department of Management, Faculty of Economics and Management, Bogor Agricultural University, Indonesia

*corresponding author: H. Puspitawati, email: herien_puspitawati@email.com

Summary

This chapter provides a brief description on the gender division of labor in agroforestry activities within households. The introduction describes the underlying background of gender roles in agroforestry activities. In general it was found that there is an imbalance of gender partnership in labor division applied at cultivation, processing and marketing of cashew nuts between men and women. The unequal gender role has also existed at the accessibility and control levels towards collector traders and wholesale buyers; however, an equal gender role has been found at the accessibility and control towards farming activities, and the role of women is even more dominant in processing than that of men. It was also found that the unequal gender partnership at the community level is due to the socio-cultural constraints. Finally, research and policy recommendations describe the focus of research related to gender in agroforestry, the capacity building for regional and national government, and also community and gender empowerment. The community and gender empowerment includes the agreement among communities (both men and women) in the village to reformulate its regional planning, the increase of women's potential skills and knowledge, and the strengthening of farmer group/institutions.

*Keywords:*Gender roles, gender empowerment, farming activities, agroforestry

T. Tscharntke et al. (eds.), *Tropical Rainforests and Agroforests under Global Change*, Environmental Science and Engineering, 177–189, DOI 10.1007/978-3-642-00493-3_8,

1 Introduction

1.1 Underlying Background

State promotion of agroforestry production is widespread in Southeast Asia. It frequently aims at economic development and sometimes environmental protection, particularly in rural areas. State promotion is often on a large scale and can fail to consider or monitor the various localized impacts of promoting a particular agroforestry activity. In many cases, the emphasis is on production rather than maximizing benefits to poor farm households, which involves marketing issues such as improving bargaining power, value adding and product development.

Java is the most populous island in Indonesia, covering just six percent of its land but housing almost 60 percent of the country's 215 million or so people. Moreover, as the centre of industry and the main food production area in Indonesia, Java plays a leading role in the national economy. The government has launched a number of initiatives to ameliorate environmental degradation in Java and other parts of the country. One of them, which has been running since the early 1980s, is reforestation of upland areas through agroforestry. Through this program, the government provides materials and other support to assist upland farmers to plant tree crops such as coffee and the cashew nut. Combined with a growing market, this policy has lead to extensive participation of community households in the agroforestry production.

Farmers in Java are generally small landholders using a low level of production technology and limited access to market information. They face a range of constraints including: tedious, complex and labor intensive processing; inadequate grading and packing, and a lack of market information throughout the marketing chain. In addition, they generally bargain individually with middlemen, which give them less bargaining power. Despite this, cashew production contributes significantly to a processor's family income, and is well established.

Wonogiri district is located in Central Java province at the latitude of 7°32' to 8°15', with the longitude of 110°41' to 111°18' and the altitude is about 400m above sea level. The district has the size area of 182 237 Ha (BPS Provinsi Jawa Tengah 2000). In the year of 2000, the district's population was 966 414 people (BPS Pusat 2001). This means that the population density was 530 persons per km^2. Most of the area of Wonogiri district is upland with very low soil fertility. Severe shortages of fresh water for drinking and agriculture at the dry season have been another problem that hinders economic development of this district. The severity of economic problems in this district has forced a significant part of its population to migrate to other areas of Indonesia, such as Jakarta and Lampung, for jobs. This has been the underlying factor behind the fact that its population density affects the socio-economic problems.

The program's environmental goal is to control soil erosion in river catchments. It is also expected to have a positive effect on incomes, particularly through improved land productivity, water conditions, and soil fertility. Thus, the agroforestry program is also considered to be a poverty-alleviation program. Based on this reasoning, the study examines the gender division of labor within households in agroforestry activities.

1.2 The Objectives of the Study

The objectives of the study are:

1. To examine the gender roles in production, processing, and marketing of cashew nuts.
2. To propose strategies to improve market bargaining power of women so as to improve their contribution to their family income.

2 Literature studies

Gender Mainstreaming was defined by the United Nations Economic and Social Council in 1997 as a strategy for making women's as well as men's concerns and experiences an integral dimension of their policies and programs. The Government of Indonesia's increased efforts to Gender Mainstream into its national development process is an affirmation of the commitment to gender equality and the rights of both women and men (KPP-UNDP 2007).

As a strategy, Gender Mainstreaming aims at promoting a fairer distribution of resources, opportunities and benefits of development processes and population programs. As Indonesia is now implementing a decentralized approach in the country's development, the need for mainstreaming gender concerns in local government's programs and policies has become more important (KPP-BKKBN-UNFPA 2004).

The meaning of gender is different from the meaning of the type of sex. However, both gender and the type of sex involve both men and women. The type of sex in general is applied to identify the difference between men and women from biological anatomy that it is natural. Meanwhile, gender discusses the differences of role distributions and function between men and women that are decided by the community/culture since they were born and that it is not natural. At this point, gender does not discuss women only, but also men in relation to cooperation/partnership and the role distribution between men and women to achieve the same goal. Therefore, gender discusses the problems of both men and women in society's life. Thus, gender is the community perspectives toward roles, functions, and responsibilities among men and women are a result of social and cultural constructions that can be changed along with the dynamics of time and community aspirations (KPP-BKKBN-UNFPA 2004).

The application of gender roles in the households' life is shown by the power of men and women, with women typically being the less powerful partners. The differences in power between men and women are directly related to differences in their income and other resources. The work of women at home is unremunerated, and therefore not regarded as "real" work, and often constrains women's opportunities to earn outside income. Thus, the public world of work is a men's domain, but the private world of home is a women's domain (Renzetti & Curran 1995).

Gender roles in agricultural work are influenced by the gender division of labor. In some countries women are virtually full-time farmers, while in other countries women do little work in fields. In some places, women are active farmers and work side-by-side with the men in the fields and in other places women are work separately from men in the fields (Osteergaard 1992).

3 Research methodology

3.1 Location of the study

The study was conducted at Rejosari village, located in the important cashew producing sub-district of Jatisrono Wonogiri District of Central Java, Indonesia during January-March 2006 (as part of the SEANAFE Project on Markets for Agroforestry Tree Products) (SEANAFE. 2007a; SEANAFE. 2007b; SEANAFE-IPB. 2008).

3.2 Types of Information Required and Their Sources

In this study, both primary and secondary information have been used to achieve the study's goals. The required primary information was obtained from a variety of respondents. The respondents included farmers, middlemen, processors, exporters, relevant government agencies, financial institutions, farming cooperatives and NGO's. The respondents were interviewed in-depth, either individually or collectively through FGD's (Focus Group Discussions). Meanwhile, the required secondary information will be obtained from various agencies (private and public agencies) and relevant publications.

3.3 Methods of Analysis

The information collected was analyzed by using a variety of analytical tools. The tools of analysis include descriptive analysis, by using gender analysis approach.

Gender analysis is a process for the analysis of data and information systematically about men and women to identify and indicate the status, function, role and the responsibility of men and women and its affected factors (KPP-BKKBN-UNFPA 2004).

4 Findings

4.1 The Meaning of Gender and Family

The fact that Wonogiri upland is upstream of the Solo River has been another important reason for the government to make this upland area a target for the agroforestry program. The Solo River is an essential source of fresh water for industry and people not only in Central Java, but also in East Java province. The important role of this river for the population and economic development of these provinces has, in fact, led the government of Indonesia to place the management of this river under a special management authority, called **Perum Jasa Tirta 2.**

The agroforestry program implemented in the Wonogiri upland areas has been quite successful, especially in promoting the cultivation of cashew nut tree crops on private farmland. This can be discerned from the data which highlight some aspects of cashew nut tree crops in Wonogiri district. From the data it can be seen that the area of cashew nut tree farms has increased quite significantly from 5 643 ha in 1999 to 7 738 in 2004. At the same period of time total production also has increased quite significantly, from 5 304.42 tons in 1999 to 10 833.20 tons in 2004. Similarly, average farm productivity has also increased significantly from 940 Kg/ha in 1999 to 1 400 Kg/ha in 2004.

In Wonogiri district, farmers produce cashew nuts not just for their own household consumption, but also for the market. For these small farmers the cultivation of cashew nuts tree crops on their small marginal land is for the purpose of obtaining cash income. Farmers are involved in the market for cashew nuts through selling of almost all of their production into their local market. The market for cashew nuts in this district is, in fact, growing over the last few years. This market phenomenon is reflected in the growing of cashew nuts of Wonogiri district to be shipped into overseas markets. The total of export of cashew nuts from Wonogiri district increased from 5 304 420 Kgs in 1998 to 10 833 200 Kgs in 2004. Similarly, the value of exports also increased from USD 6 129 221 in 1998 to USD 17 027 652 in 2004 (Bappeda Wonogiri 2005).

Most of these upland farmers are small-holders, with the average farm size of less than 0.50 ha. In fact, in 2004 the number of farmers that were involved in cashew nuts tree crop farming in Wonogiri district was 23 422 households. This meant that the average holding was 0.33 ha per farming household (Bappeda Wonogiri 2005).

Agriculture is one the most important sectors in Wonogiri. The contribution of the sector on regional gross domestic product (RGDP) is around 54 per cent in the last five years, followed by trade (8.7 per cent), industry (5.6 per cent). Amongst sub-sectors in the agricultural sector, the food crop is the most important, contributing to 43.7 per cent, followed by estate crops (5.2 per cent), cattle (1.9 per cent) and fishing (0.7 per cent). As mentioned before, the cashew nut is one of the most important crops in the estate crop sub-sector.

Wonogiri has also faced persistent problems of a high level of poverty. Based on the National Socio-Economic Survey conducted in 2004, the number of the people living below the poverty line is 272 795 people, around 24.4 per cent of total population. This poverty incidence is much higher than the national level of 16 per cent. Some other indicators of the high level of poverty in the region are the high level of illiteracy (5.3 per cent), child-mortality rate (14.6 of 1 000), and a high rate of undernourished people (4.8 per cent). With this situation, the government of Wonogiri has placed its poverty alleviation program as one of the central programs. Moreover, since agriculture is the most dominant sector, then the government has promoted agricultural development as a leading sector in combating poverty.

With a total population of 1.12 million people and a family size around 4.4 people per household, Wonogiri still faces serious problems of unemployment, although population growth has been actually very low, between 0.44 0.62 per cent per annum. The official unemployment level is 2.1 per cent, but the real unemployment level must be much higher. This is because the education level of the man-power from Wonogiri is low. The education level of the man-power has is low because almost 40 per cent of the population has only completed elementary school (BPS Wonogiri 2005).

Rejosari village located at Jatisrono sub-district, Wonogiri District, Central Java Province is an agricultural area surrounded by people's forests and hilly areas. In 2004, the population in the village was 1,075 head of families or 4,627 people consisting of 2298 (49.67 per cent) women and 2329 (50.33 per cent) men. Almost 60 per cent of the population is engaged in agriculture as farmers or farm workers.

The community of the Rejosari village is an agricultural community dominated by rural traditional life. Although in general the society's norm is patriarchal, in daily life the community has been practicing close cooperation between men (husband) and women (wife) at the family level since a long time ago.

It is acknowledged by the women's group (majority at the age of 30-40 years old) that the role of women in the traditional family is parenting children, helping in all domestic jobs and assisting agricultural jobs in the paddy field. Women born in 1950-1960 admit that they marry their husband from the same village/district and get married at around 17 years old. The women's constraint in general is the limitation of education and mobility. This situation prevents women from becoming independent in many ways psychologically as well as economically. This is proven by the fact that women from this village feel afraid of going alone to the outer cities. This means the accessibility of women to the economic activities; financial credit and training opportunities are limited. However, most women are aware that there is a big change right now shown by the increase of women's roles in all aspects in their village. Some of the young generations of women have started to go to the outer cities/regions for selling food (meat balls, rice and soup) and traditional herbs.

The division of roles in their daily family activities has been implemented respectively both in economic and domestic activities. The community of Rejosari village has been getting used to implement the **gender distribution of roles** starting from the family level and continuing to the community level, even though it is still at the traditional stage. The value of the division of gender roles has been practiced in their daily family management activities starting from collaboration between husband and wife in their domestic activities to the economics and community social activities. The values of gender equality and justice have also been implemented in formal education. Boys and girls are not discriminated against for entry into school from Primary to Junior High Schools. The community realizes that the value of boys and girls are the same, even though the community acknowledges that boys can be leaders for the family as is the root custom of Javanese culture and the majority of Moslem societies who have been practicing a patriarchal system.

The villagers consider the importance of family values in every aspect of human life. They realize that a human will not have a quality of life without family. If somebody has a problem, he/she will ask for help to his/her family. Somebody who is successful in the city, at the end he/she will return to his/her place of origin and finally to his/her large family. Thus, the values of "my family is my world" and "back to my family" are basic values of family life in Rejosari village.

The community of Jatisrono sub-district in general, and the community of Rejosari village in particular, have been popularly seen as tough and hardworking people, people who cooperate readily, love peace and maintain a sustainable environment. The existence of good work ethics of the community has been provided by the good organizational structure of village government, good village administration, active village cooperative, good facilities of infrastructure, bridges, and the village office. Most of the community infrastructures were built by local enterprise using the community's own finance. The people in the village, both men and women, young and old, were reported on participating in the construction of the village road in 1997. This activity is viewed as the community's close cooperation both in social and economic activities, with involving men and women. These activities make Rejosari Village one of the role models of community empowerment that involves gender equality and equity.

4.2 The Gender Division of Labor in Agroforestry Activities

The results of the survey showed that in general the partnership has existed in good condition in the activity of the cashew nut business, even though it is not balanced yet and perfect (Figure 1 & Figure 2). There is an unequal gender role in the accessibility and control as a collector trader and wholesale trader, whereas the roles of men are dominant over women as collectors, wholesale traders, access to credit and information, and marketing. However, the roles of women in the processing of cashew nuts and the use of machinery for processing are relatively higher than that of men.

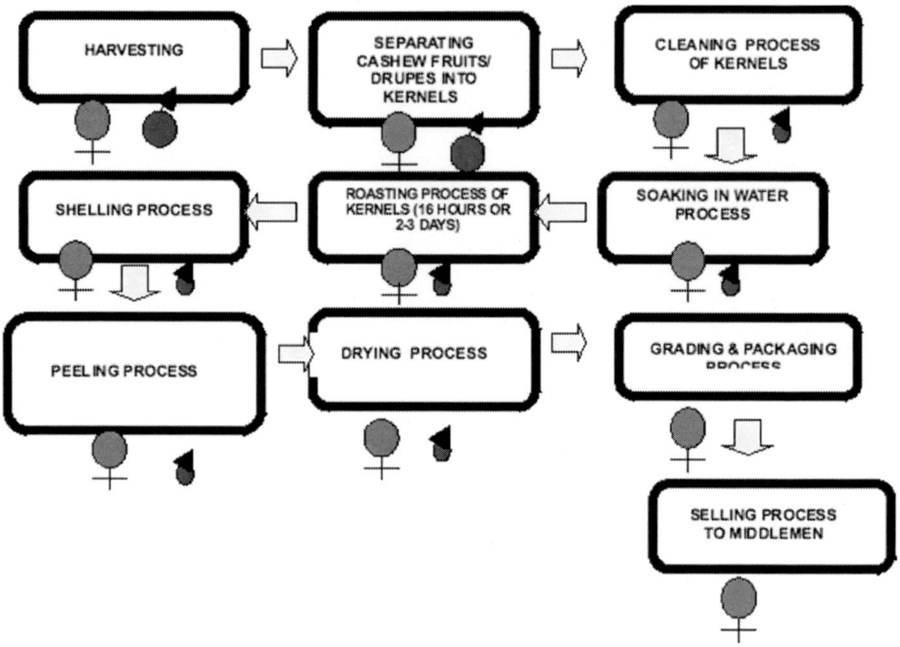

Fig. 1. Gender roles in the processing of Cashew nuts

In general, it was also found a good partnership existed between men and women in the accessibility of control towards resources and processing technology. The role of women is limited in the accessibility of information on pricing and training. Even the access of women to credit for the production and marketing does not yet exist in Rejosari village, Wonogiri District. On the other hand, the role of women is greatest in the overall manufacturing process.

The role of gender in the household activity is a good practice and almost equal between men and women. The role of gender in the activities of financial economics of the cashew nut business is presented in Table 1.

Based on the survey, it has been recognized that in general the role of women is more dominant than that of men in financial economic activity and the cashew nut business. On the financial activity of the cashew nut business, both men and women participate actively whether it is alone or altogether to budget for family expenses, to plan family finances, to manage family finances

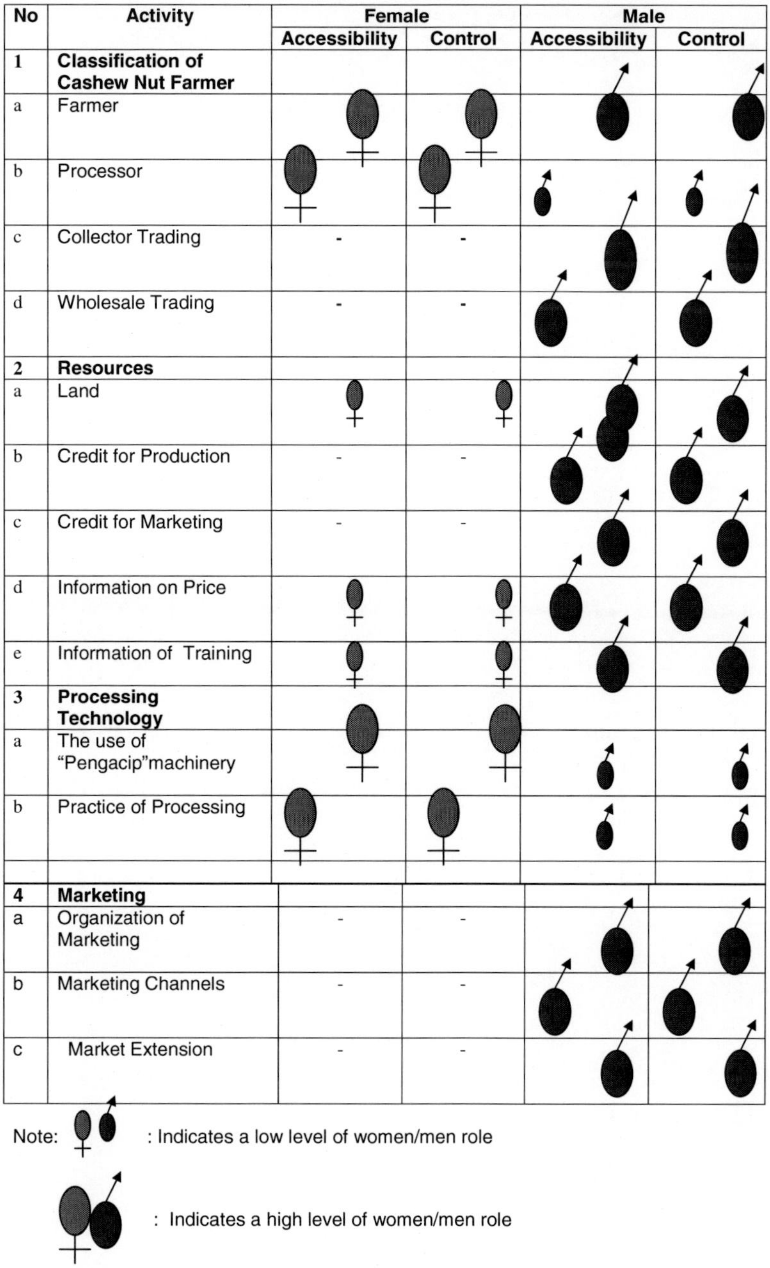

Fig. 2. Gender analysis of the Cashew nut business in Rejosari village, Wonogiri district.

Table 1. Distribution of Respondents Based on the Role of Gender in the Activities of Family Economics and Cashew Nut Business (n=28).

No	Category	Wife		Husband and Wife		Husband		Total	
		n	%	n	%	n	%	n	%
	FINANCE OF CASHEW NUT BUSINESS AND FAMILY								
1.	Decision of spending of family money	21	**75.0**	7	25.0	0	0.0	28	100
2.	Management of family money	18	**64.3**	9	32.1	1	3.6	28	100
3.	Planning of family finances	12	42.9	15	**53.6**	1	3.6	28	100
4.	Management of cashew nut money	17	**60.7**	10	35.7	1	3.6	28	100
5.	Planning of finance cashew nut business	12	**42.9**	12	**42.9**	4	14.2	28	100
6.	Borrowing/credit money for business	5	23.8	11	39.3	5	23.8	21	100
7.	Borrowing money for family needs	10	40.0	11	**44.0**	4	16.0	25	100
8.	Finding alternative solutions for financial problems	4	14.3	21	**75.0**	3	10.7	28	100
	ACTVITIES OF CASHEW NUT AND ECONOMIC BUSINESS								
9.	Activity of maintenance of cashew nut tree	6	28.6	4	19.0	11	**52.4**	21	100
10.	Activity of fertilizing cashew nut tree	7	33.3	2	9.5	12	**57.1**	21	100
11.	Activity of harvesting cashew nut tree	8	**38.1**	7	33.3	6	28.6	21	100
12.	Activity of drying cashew nut	8	34.8	13	**56.5**	2	8.7	23	100
13.	Activity of processing cashew nut	9	39.1	13	**56.5**	1	4.3	23	100
14.	Activity of selling cashew nut	23	**85.2**	1	3.7	3	11.1	27	100
15.	Receiving of payment from selling cashew nut	25	**89.3**	0	0.0	3	10.7	28	100
16.	Responsibility of public work/economics	2	7.1	16	**57.1**	10	35.7	28	100

for the cashew nut business, to borrow money for family needs, and to apply for a credit for their business. For the activity of the cashew nut business, both women and men participate actively whether it is alone or together in the activities of maintenance of cashew nut trees, fertilizing, harvesting, drying and processing. Women are dominant in selling the cashew nut and receiving the money, whereas men are responsible for public/economic activities. Below is a presentation of the proposal of marketing strategy of the cashew nut business (Table 2).

Table 2. Distribution of Respondents Based on the Proposal of Marketing Strategy for the Cashew Nut.

No	Strategy	Wife		Husband and Wife		Husband		Total	
		n	%	n	%	n	%	n	%
	FAMILY GENERATING INCOME								
1.	Raising finance by pawning of goods to plant cashew nut	2	33.3	3	**50.0**	1	16.7	6	100
2.	Searching for credit to plant cashew nut	2	28.6	4	**57.1**	1	14.3	7	100
3.	Increase working hours to plant cashew nut	3	14.3	13	**61.9**	5	23.8	21	100
4.	Asking assistance from children to plant cashew nut	6	42.9	7	**50.0**	1	7.1	14	100
5.	Selling assets to plant cashew nut	0	0.0	1	**100**	0	0.0	1	100
	CUTTING BACK EXPENSES								
6.	Reducing other costs to plant cashew nut	2	20.0	7	**70.0**	1	10.0	10	100
7.	Reduce the cost of transportation by walking/riding bicycle or join with friends	1	25.0	2	**50.0**	1	25.0	4	100
8.	Withdraw savings to plant cashew nut	1	20.0	4	**80.0**	0	0.0	5	100
	LOAN OR DEBT								
9.	Borrowing/Owing from family/neighbor to plant cashew nut	1	14.3	6	**85.7**	0	0.0	7	100
	SEARCHING INFORMATION								
10.	Searching for information for planting and processing cashew nut	0	0.0	4	30.8	9	**69.2**	13	100

The result of the survey consistently shows that the role of women is more dominant than that of men in the finance and economic activity in the cashew nut business and the cashew nut marketing strategy. It also recognized that both men and women are participating actively whether it is alone or together for implementing income generation strategies such as pawning assets to cultivate cashew nut, seeking of credit for cashew nut business, increase working hours to cultivate cashew nut trees, asking children to cultivate cashew nut trees, and selling assets to cultivate cashew nut trees. It was found that both men and women are participating actively whether it is alone or together for implementing cutting back strategies such as the reduction of cost of cultivation of cashew nut trees, reduce the transportation costs by cycling, walking or obtaining lift, and withdrawing savings to cultivate cashew nut trees. The role of information gathering and the forming of strategies to cultivate cashew nut trees is dominated by men.

5 Conclusios and recommendations

In conclusion, it is found that there was an imbalance of gender partnership in the division of labor applied at cultivation, processing and marketing of cashew nuts between men and women. This unequal gender role also existed in the accessibility and control with collector traders in the collection and wholesaling processes. However, an equal gender role already existed in accessibility and control of farming activities, while in the processing the role of women was the more dominant. It is also found that the unequal gender partnership at the community level was caused by socio-cultural constraints.

Based on the above findings the following recommendations are made. Firstly, agreements among communities in the villages should be used to reformulate regional planning of agroforestry so that it involves both women and men. Secondly, female empowerment strategies should be focused on the potential of their processing skills with expansion into higher processing skills *e.g.* the processing of broken nuts into nut-sandwich (chips) and cashew nut chocolate. Thirdly, farmer groups should be empowered by increasing the quality of women's skills and knowledge, especially in the area of marketing and processing of local commodities. Finally, it is proposed that there is a need for capacity building by regional government aimed at strengthening gender empowerment in agroforestry programs.

References

Bappeda Wonogiri (2005) Rencana Pembangunan Jangka Menengah Kabupaten Wonogori, Bappeda Wonogiri, Wonogiri

BPS Provinsi Jawa Tengah (2000) Jawa Tengah dalam Angka: Tahun 1999. Semarang: BPS Provinsi Jawa Tengah

BPS Pusat (2001) Penduduk Indonesia: Hasil Sensus Penduduk 2000. Jakarta: BPS Pusat

BPS Pusat (2004) Statistik Indonesia 2003. Jakarta: BPS Pusat

BPS Pusat (2005) Data dan Informasi Kemiskinan Tahun 2004: Kabupaten. Jakarta: BPS Pusat

BPS Wonogiri (2005) Wonogiri in Figures 2004. BPS Kabupaten Wonogiri, Wonogiri

Hutagaol M, Parulian, Adiwibowo S (2002) Degradasi Lingkungan dan Ketahanan Pangan Nasional: Investigasi Singkat Mengenai Peranan Kebijakan Pembangunan Nasional. In: Krisnamurthy Bayu et al. (eds.) Tekanan Penduduk, Degradasi Lingkungan dan Ketahanan Pangan. Bogor: Pusat Studi Pembangunan,pp 106-131

KPP-BKKBN-UNFPA (2004) Bunga Rampai: Panduan dan Bahan Pembelajaran Pelatihan Pengarusutamaan Gender dalam Pembangunan Nasional. Kementerian Pemberdayaan Perempuan Republik Indonesia Cooperation with BKKBN and UNFPA

KPP-UNDP (2007) Modul: Pengarusutamaan Gender dalam Pembangunan Nasional di Indonesia. Kementerian Pemberdayaan Perempuan Republik Indonesia Cooperation with UNDP

Ostergaard L (ed.) (1992) Gender and Development. Routledge. London, UK

Renzetti, C.M., & Curran, D.J. 1995. Women, Men, and Society. Allyn and Bacon. Boston, USA

SEANAFE (2007a) Cashew Marketing in Wonogiri District, Central Java, Indonesia: A Case Study. A Research Report. SEANAFE, Bogor

SEANAFE (2007b) Marketing Curriculum: Case Study Cashew Marketing Chain, Indonesia. SEANAFE, Bogor

SEANAFE-IPB (2008) SEANAFE'S-MAFTP Teacher's Guide. SEANAFE and Institut Pertanian Bogor, Bogor

The robustness of indicator based poverty assessment tools in changing environments - empirical evidence from Indonesia

Xenia van Edig[1*], Stefan Schwarze[1], and Manfred Zeller[2]

[1] Department of Agricultural Economics and Rural Development, University of Göttingen, Germany
[2] Institute for Agricultural Economics and Social Sciences in the Tropics and Subtropics, University of Hohenheim, Germany

*corresponding author: X. van Edig, email: xedig@agr.uni-goettingen.de

Summary

Eradicating poverty is one of the highest priorities of development policies. Besides the necessary improvement of peoples livelihoods, the reduction of poverty is believed to have a positive effect on the stability of the rainforest margins. Better-off households are furthermore less vulnerable to shocks caused by natural hazards.

Organisations aiming to reduce poverty need simple and stable tools to detect (very) poor households. To reliably distinguish poor people, such tools need to be easy to apply and robust in time. Using data from Central Sulawesi, Indonesia, this study aims to test first whether two sets of poverty indicators developed in 2005 are still capable in predicting absolute poverty and second, how the national calibrated tool developed by IRIS[1] predicts poverty using the same data-set.

In 2005 and 2007, almost 20% of the rural population of Central Sulawesi was identified as being very poor with individuals living on less than $1 US per capita and day in purchasing power parities. Beside this relatively high poverty incidence compared to Indonesian average of 7.5%, the tropical rainforest in the research area is threatened by smallholder conversion of forest into farmland.

For the analysis, data from two household surveys were used. In 2005 we surveyed 264 households in the vicinity of the Lore Lindu National Park in Central Sulawesi to obtain indicators of poverty and to derive the daily

[1] IRIS is a research and advisory centre at the Department of Economics, University of Maryland

T. Tscharntke et al. (eds.), *Tropical Rainforests and Agroforests under Global Change*, Environmental Science and Engineering, 191–211, DOI 10.1007/978-3-642-00493-3_9,

per capita consumption expenditures. In total 280 indicators were recorded. Two different multivariate regression models were fit to this data-set. One model (Model 1) included all sampled indicators and the other one (Model 2) contained only easily assessable indicators as ranked by local staff. Each of the models yielded a different set of 15 indicators to best predict poverty. In 2007, we conducted an additional survey with the identical questionnaires in the same households. We used the data from 2007 to estimate the poverty status of the households with the indicators derived in 2005. Furthermore, we tested the national calibrated poverty assessment tool developed by The IRIS Centre and USAID[2] with the 2007 household data from Central Sulawesi.

As to the results, we can state that both tools calibrated for Central Sulawesi in 2005 lose accuracy because they tend to over-predict the poor. The *Poverty Accuracy* of both tools declined between about 0.5% and 21%. Only in the case of one-step OLS of Model 2 the Poverty Accuracy increased by about 10%. Instead the accuracy performance of the national calibrated tool provided by IRIS and USAID are overall disappointing.

Keywords: poverty assessment, poverty indicators, time robustness of targeting tools, Indonesia, Central Sulawesi

1 Problem setting

1.1 The need for poverty reduction in economical and ecological terms

Although the first millennium development goal of the United Nations is to reduce extreme poverty and hunger by half until 2015 (United Nations 2008), that goal has yet to be achieved and poverty remains a pervasive problem in many countries. In general, poverty reduction is one of the main goals of development policies, programs and projects (e.g. Zeller et al. 2001, Collier and Dollar 2002). In Central Sulawesi we found that almost 20% of the households are very poor, i.e. the household members live on less than \$1 US purchasing power parities (PPP) per capita and day. This poverty headcount is quite high in comparison to the Indonesian average of 7.5% (HDR 2007/2008).

In Indonesia, the annual deforestation rate rose from 1.2% in the 1990s to 2.0% from 2000 through to 2005 (FAO 2009). Erasmi et al. (2004) give the research area an annual deforestation rate of 0.6% on data from 1972-2001. Schwarze et al. (2005) find a positive correlation between the relative poverty status of a household and forest encroachment. The Nature Conservancy (2005) state that the natural forest in Central Sulawesi is threatened by smallholder conversion of forest into farmland. At the rainforest margin of the Lore Lindu National Park, land use change is mainly driven by a change of strategy from "food crops first" to "cash crops first" (Weber et al. 2007).

[2] United States Agency for International Development

This is in part from the Bugi migrants who brought knowledge of cocoa cultivation into the region and the desire by the indigenous population to also gain the economic benefits of cacao-production. The local ethnic groups often sell their land to the Bugi migrants and clear new plots in the forest for themselves (Weber et al. 2007). This is further evidence that forest degradation is often fostered by poverty combined with internal and external change factors such as population pressure as described by Wunder (2001). Poor people often clear forest areas for short term gains, even if they are aware of the long term negative effects this could have (Eckholm et al. 1984). Hence, poverty reduction in Central Sulawesi could not only improve peoples livelihoods but also contribute to the achievement of conservation goals.

Beside their important ecological role forests are often crucial for peoples livelihoods. They can support subsistence, generate income or act as safety networks. Therefore, forests are often extensively used by rural households. Thus, clearing forest areas to increase arable land can also have a negative impact as it diminishes the availability of the forest as a resource (PEN 2009). In the research area 76% of the households collect forest products, mostly firewood. However for the poorest people in the region the sale of rattan is an important income source (Schwarze et al. 2007).

Moreover, forests provide environmental services - such as coffee pollination (see for example Olschewski et al. 2007).

Furthermore, forests can fill gaps as part of the response to *ex ante* risks, e.g. to overcome seasonal fluctuations in the availability and affordability of goods. Moreover, they can act as a form of insurance for larger *ex post* shocks such as droughts (Wunder 2001). Such shocks can get more frequent and severe due to climate change. It is assumed that climate change, which deforestation contributes to, affects poor people more severely than wealthy people (IPCC 2001, OECD 2009). Therefore, forests are not only important for the global eco- and climatic systems but also important for the livelihood of rural people.

The picture drawn above clearly shows the need for poverty alleviation for both economical and ecological reasons. To better target poor households with poverty reduction programs, organisations aiming to provide these projects need good instruments to detect poor households.

1.2 The need for poverty assessment tools (PATs)

Several attempts in poverty assessment try to meet poverty as a multidimensional phenomenon in contrast to a pure measure of inadequate income or expenditures (Osmani 2003). In his book "Development as freedom" (1999) A. Sen argues that poverty rather is a deprivation of basic capabilities and not only a matter of the lowness of income. Nevertheless, he also admits that low income is one of the major causes of poverty in the sense that it is often a reason for capability deprivation. Hence, it is necessary to use approaches which account for low incomes and other forms of deprivation.

To better target absolute poor households easy applicable tools for poverty assessment are needed. For non-governmental organisations (NGOs) and other stakeholders concerned with poverty reduction, it is particularly important that tools which enable the detection of absolute poor households are low in costs and contain indicators which are robust over time and space.

Most poverty assessments done in the last 25 years referred to relative poverty (Zeller 2004). The concept of relative poverty defines the situation of an individual or the situation of a group of persons in relation to the average living standard of their society (see for example Foster 1998, Witt 1998). For example, Grootaert and Braithwaite (1998) conducted research on correlates of poverty and indicator based targeting in Eastern Europe using a relative poverty line for their analysis. In this study they used multivariate regression analysis to detect determinants of poverty. They found that a strong relationship between poverty incidence and the number of children in a household exists and that poverty in Eastern Europe has, to some extent, an age and gender dimension. Another example of an approach to assess relative poverty is "the construction of a poverty index using a range of qualitative and quantitative indicators" (Zeller et al. 2001, p. 3) as done by Zeller et al. (2001) using Principal Component Analysis (PCA) to derive the indicators.

Until now few attempts have been made to assess absolute poverty. One approach used is to assess a households poverty status via food security scales. Three different scales - non food insecure, moderately food insecure, and severely food insecure - are used to predict daily per capita expenditures. This tool faces the problem that food insecurity is not always identical with (income) poverty (Alcaraz V. and Zeller 2008).

1.3 Objectives of the study presented

The aim of the survey in Central Sulawesi was to test new tools for the assessment of absolute poverty. The methodology used is based on poverty assessment tools (PATs) developed by The IRIS Centre on behalf of USAID. Out of the nine regression models tested by IRIS, two very promising types of regression models were tested in Central Sulawesi (see section 2). The study used two sets of household data. In 2005, we conducted research to identify two sets of 15 indicators each for poverty assessment in Central Sulawesi, Indonesia. We wanted to compare the capability of the models in predicting very poor households with the observed poverty headcounts.

In 2007, we conducted the same survey again to test the identified sets of indicators regarding their capability in poverty prediction and robustness over time. Furthermore, we tested a national calibrated poverty assessment tool developed and provided by The IRIS Centre and USAID with the 2007 household data from Central Sulawesi. Such poverty assessment tools are provided for over 20 different countries at URL: http://www.povertytools.org. They are approved by USAID and can be used by anyone. USAID requires its implementation partners, including organizations that deal with micro-enterprises with the aim of poverty reduction and receiving funds from USAID, to assess their target population with these tools.

2 Indicator based models for the assessment of absolute poverty

2.1 Background

One approach to assess absolute poverty was developed by The IRIS Centre at the University of Maryland in collaboration with the US Agency for International Development (USAID). These organisations developed and tested different poverty assessment tools for targeting poverty reduction projects, especially those dealing with micro-enterprises. These tools were developed in order to meet the requirements of the US Congress. In the year 2000, the US Congress adopted the *Micro enterprise for self-reliance act.* In 2003, an amendment to this act was adopted which made USAID responsible for the development and certification of low-cost poverty assessment tools. Further requirements for these tools were that they should be objective and quantitative. Hence, they should be based on income or expenditure and able to identify individuals who fall short of one of two poverty lines which were

1. the bottom 50% of the national poverty line, or
2. the international poverty line of 1$ US PPP per day

(Zeller 2004). For further information see http://www.povertytools.org.

The approach of indicator based poverty assessment connects indicators of different dimensions of poverty with the commonly used poverty line. Indicators of poverty should - as the word indicator suggests - indicate a persons or households standard of living or income and yield information about the social conditions of the poor (Schubert 1994, Minot 2000).

Poverty indicators can be a constitutive part in developing poverty reduction strategies as they try to measure poverty as a multidimensional phenomenon. While the indicators vary between the subjective and objective perspectives on poverty, they often have the same scale in the relative and absolute approach (Lok 1995). One problem identified is that poverty indicators face difficulties in differentiating chronic from temporary poverty. For example monetary poverty is less persistent than malnutrition or low school enrolment (Baulch and Masset 2003). Therefore monetary poverty indicators are eventually more valid for transient poor whereas indicators dealing with nutrition or education could tell more about the chronic poor. Another difficulty for poverty indicators is the seasonal fluctuation in poverty (Muller 1997). In our study we refer to the latter problem by using recall periods of at least 12 months.

A commonly used approach to assess poverty is the "construction of a poverty line and (the) computation of various measures that take into account the way in which household expenditures fall short of the poverty line" (Zeller et al. 2001, p. 3). In practice, however, total household expenditures are used as a measure to evaluate household living standard. Whether the household income is sufficient to meet food security and other basic needs is used as a criterion. The "basket of basic needs" or a monetary poverty line is applied. This "basket of foods and services corresponding with the local consumption pattern and satisfying a pre-set level of basic needs for one person is constructed and ranked at local consumer prices to compute its minimum costs" (Zeller et al. 2001, p. 3-4). The value of this basket represents the poverty line, mostly in terms of daily per capita expenditure.

2.2 Poverty assessment in Central Sulawesi

In the study conducted, two models for poverty assessment were tested in Central Sulawesi, Indonesia. These models search for sets of 15 poverty indicators to predict daily per capita expenditures of a certain household. For the first model (Model 1), every surveyed indicator could possibly be included. In the second model (Model 2), only indicators which were ranked as "easy to verify" by the Indonesian staff were included. Many of the variables from Model 1 were either difficult to survey or difficult to verify. The following two examples, out of the 15 indicators for Model 1 from 2005, should illustrate this: The subjective indicator "Household feels that its healthcare expenditures are above its needs" is very difficult to verify because of its subjectiveness. "The average clothing expenditures per capita in the last 12 month" instead is an objective indicator. Nevertheless, the required information is difficult to obtain and difficult to verify too. Model 2 only included indicators which were "easy to verify": E.g. "the total number of rooms in a dwelling" an indicator used for the tool, can be obtained and verified easily by the enumerator during the interview.

Why two different models? Although, Model 1 was more likely to achieve a better accuracy performance because it used all variables, Model 2 referred to two categories of problems which might occur in the analysis of indicators. First, information might be difficult to obtain, especially regarding the aspects of time, social costs and money. Second, indicators might be difficult to verify, especially if they are recall-related (Zeller et al. 2005).

A similar approach was used by Benin and Randriamamony (2008) to assess household income via a set of indicators in order to monitor and evaluate public investments in several countries in sub-Saharan Africa. They also used different proxy indicators in the model estimation to develop an econometric prediction model for household income.

As aforementioned, a second part of the study was to test the national calibrated tool of The IRIS Centre and USAID. USAID requires its partners to use for targeting support. In general, development projects use targeting to increase their cost-efficiency (Minot 2000).

2.3 Accuracy measures

For purposes of assessing the prediction power of a regression model (or tool) for poverty assessment, we used the following measures of performance for each model in this study:

Total Accuracy is the percentage of households whose poverty status is correctly predicted by the regression model.

Poverty Accuracy is the percentage of very poor households whose poverty status is correctly predicted by the regression model. It is expressed as a percentage of the total amount of very poor households.

Non-poverty Accuracy is the percentage of not very poor households whose poverty status is correctly predicted by the regression model. It is expressed as percentage of the total number of not very poor households.

Undercoverage represents the error of predicting very poor households as being not very-poor, expressed as a percentage of the total number of very poor households.

Leakage reflects the error of predicting not very poor households as very poor, expressed as a percentage of the total number of very poor households.

Poverty Incidence Error (PIE) is defined as the difference between the predicted and the actual (observed) poverty incidence (here headcount), measured in percentage points.

Balanced Poverty Accuracy Criterion (BPAC) is defined as the *Poverty Accuracy* minus the absolute difference between *Undercoverage* and *Leakage*, each expressed as a percentage of the total number of the very poor. When *Undercoverage* and *Leakage* are equal, the BPAC is equal to the *Poverty Accuracy*. BPAC is measured in percentage points (Zeller et al. 2005 /The IRIS Centre 2005).

3 Data collection and analysis

3.1 Obtaining expenditure data and indicators of poverty

Household surveys are the most important data source for poverty measurement and poverty comparison. They can provide direct information about the distribution of living standards in a society or in a certain region, for example how many households do not attain a certain consumption level. With the availability of such quantitative data, the poor can be assessed and an assessment of poverty policies can be done (Ravallion 1992).

The study used household data from two survey years. In both survey years, data were collected in 13 villages in the vicinity of the Lore Lindu

National Park. In both years the same randomly selected households participated in the survey. In 2005, the models were estimated with data from 279 households. In 2007 data from 282 households were obtained. The intersection of both samples was 264 households.

Two questionnaires were completed in both years. One was a benchmark questionnaire to obtain the daily per capita consumption expenditures of each household and resembled the consumption module of the *Living Standard Measurement Survey* (LSMS) of the World Bank. Thus it had the same purpose of the LSMS which is to "collect information to describe poverty and monitor it over time" (Grosh and Glewwe 2000, p. 30). Thus, the benchmark questionnaire focused on the economic dimension of poverty.

Second, we used a composite questionnaire to derive indicators of poverty in several dimensions like health, education or housing. As poverty is a complex phenomenon, the composite questionnaire tried to capture different dimensions of poverty.

3.2 Analysis of household survey data from 2005 and 2007

Model estimation and the selection of poverty assessment indicators in 2005

In each survey year, almost 280 independent variables were compiled from the composite questionnaire. The amount of independent variables had to be reduced because of a lack of degrees of freedom for the model estimation. For this purpose several steps for the indicator selection were employed. Primarily, for Model 1 all indicators were grouped on different dimension of poverty such as education, food, durables etc. For each of these dimensions we used an ordinary least square regression (OLS) which delivered indicators for the final model estimation. For Model 2, the number of indicators was restricted by the condition of being "easy-to-verify" and there was no need for pre-selection. In each step of the indicator selection, nine control variables were forced in the model estimation: Four demographic variables controlled for demographic factors and five regional dummies controlled for agro-ecological differences. For the variable selection OLS and the MAXR routine implemented in SAS were used. MAXR seeks to maximize the R^2 considering all possible combinations of regressors. For the final selection of indicators various checks and adjustments, especially regarding the sign of the coefficient of each indicator, had to be done. The sign was expected to concur with the direction one would expect from theory. For example the indicator "share of expenditures spend on food consumption" only was included in the model if the sign of the regression coefficient was negative because, as stated in Engels law, the expenditure share on food declines when income increases.

In addition to the control variables, we included 86 regressors in Model 1 and 90 regressors in Model 2.

To improve the accuracy of the models different regression methods, one and two-step OLS as well as one and two-step quantile regression, were tested. For the two step regressions, two steps of indicator selection were employed with the first step identical to the OLS/MAXR- regression described above. The second step included a sub-sample which contained a higher percentage of poor households. Hence, the second step should improve the accuracy of identifying the poor.

Using the models to predict households daily per capita expenditures

Any of the variable sets found can be described as a poverty assessment tool for the purpose of identifying the poverty status of a household. The dependent variable (per capita daily expenditures) was, like any other variable defined in monetary values (as expenditures or values of assets), converted into the natural logarithm of Indonesian Rupiah (IDR). All ordinal variables, such as the type of exterior wall of the dwelling, with lower values indicating inferior materials and higher values indicating superior materials, are converted into a set of dummy variables (Zeller et al. 2005).

To calculate the predicted daily per capita expenditures of household j the equation

$$Y_j = \beta_0 + \sum_{i=l}^{N} \cdot \beta_i \cdot \chi_i \tag{1}$$

was used, where Y_j is the natural logarithm of the daily per capita expenditures; β_0 is the intercept, β_i are the coefficients, and χ_i are the surveyed values of the indicator used in the model.

For the one-step regression the use of this equation is straightforward. For a two-step model, it is necessary to use two steps to calculate the predicted per capita expenditures. In practice this means that the second step includes only those households whose predicted daily per capita expenditures are below the expenditure percentile found during the indicator selection. For the households in our study which were predicted in the first step (one-step OLS/ one-step quantile) as having less expenditures as the 32 percentile (Model 1) or 38 percentile (Model 2), a second indicator set to predict their expenditures was applied. For all other households the predicted values from the one-step regression remained in the model.

In 2007, the indicator sets derived in 2005 were applied to the new dataset. Furthermore, the indicators proposed by IRIS for Indonesia were tested using the data from 2007.

Measurement errors and error term

Any regression model is potentially biased by measurement errors in either the dependent variable or in the independent variable. For the presented regression models the measurement error in the dependent variables is not relevant in a direct sense because the model presented is not a causal analysis but a prediction model. Nevertheless, the accuracy measures (section 2.3) could be affected because the predicted expenditures are compared to the measured expenditures. When we assume the true equation

$$Y = \beta_0 + \beta_1 \cdot X_1 + \varepsilon, \tag{2}$$

where Y is the dependent variable (here per capita expenditures), β_0 is the intercept, β_1 is the regression coefficient, X_1 is the true value, and ε is error term. But what is observed is

$$Y^* = \beta_0 + \beta_1 \cdot X_1 + (\varepsilon + \omega), \tag{3}$$

where Y^* is the observed dependent variable (here per capita expenditures), β_0 is the intercept, β_1 is the regression coefficient, X_1 is the true value, ε is error term and ω is the random error. In this case the estimated coefficients are not biased but the overall fit of the regression is lower and therefore the error term is larger.

A measurement error in the independent variables could be a reason for occasionally underestimating the daily per capita expenditures as we assume a true regression function of

$$Y = \beta_0 + \beta_1 \cdot X_1 + \varepsilon, \tag{4}$$

where Y is the dependent variable (here predicted per capita expenditures), β_0 is the intercept, β_1 is the regression coefficient, X_1 is the true value, and ε is error term. But what is observed is

$$Y = \beta_0 + \beta_1 \cdot X_1^* + (\varepsilon + \beta_1 * \omega), \tag{5}$$

where Y is the dependent variable (here predicted per capita expenditures), β_0 is the intercept, β_1 is the regression coefficient, X_1^* is the observed value, ε is error term and $\beta_1 \cdot \omega$ is the random error in the regression coefficients caused by the measurement error in X_1. In this case the measure of true X is noisy and the estimations of β_1 are biased towards 0.

In OLS regression a normal distribution of the error term, i.e. the residuals, is assumed. For our Models, we can state the residuals of the estimations are distributed normally.

4 Results

4.1 Poverty incidence in Central Sulawesi 2005 and 2007

In the research area, the number of very poor people slightly decreased from 19.3% in 2005 to 18.2 % in 2007 (Table 1). While the number of poor, i.e.

those who live on less then $2 US per capita and day, increased to 59.1%. This concurs with the overall trend observed for Indonesia. The World Bank (2008) state that after the poverty rates sank to the pre-economic crisis level in 2005 and the situation worsened again after 2006 mainly due to increasing food prices.

Table 1. Percentage of poor households in Central Sulawesi using alternative definitions of poverty in 2005 and 2007. Source: own data

Poverty line	Poverty line (IDR per capita and day)		Headcount Index (%) (N=264 [3])	
	2005	2007	2005	2007
$1 US PPP	2723	3436	19.3	18.2
$2 US PPP	5445	6872	47.0	59.1

4.2 Capability of the poverty assessment tools for Central Sulawesi developed in 2005

Model 1

The best accuracy performance in 2005 for Model 1 was achieved with a two-step quantile regression. When using the indicators selected in 2005 as well as their estimated coefficients with the data from 2007, the total accuracy dropped from 92.11% to 81.21%; the poverty accuracy dropped by about 20 percentage points from 79.69% to 58.49%. The non-poverty accuracy declined from 95.11% in 2005 to 86.46% in 2007. As both prediction errors increased - the undercoverage rose from 20.37% to 41.51% and the leakage from 20.37% to 58.49% - the BPAC decreased from 79.69 to 41.59 (Figure 1).

To detect which of the 2005 indicator sets - with their corresponding coefficients - fitted the 2007 data-set best, we calculated the accuracy of every tested regression method, i.e. one- and two-step OLS and one- and two-step quantile. Even if the overall accuracy of two-step quantile dropped, it remained the best way to predict the daily per capita consumption expenditures of the households and therefore poverty status of the households with Model 1.

If we only looked at the poverty accuracy, one-step OLS provided a better result in 2007 (77.36%), but the leakage was very high for this method (107.55%).

In Table 2, the corresponding indicators for the second step of the two step regressions are displayed.

As discussed above, Model 1 faced several difficulties with the included indicators including the fact that rural households in the study area normally do

[3] 264 households were the intersection of both samples

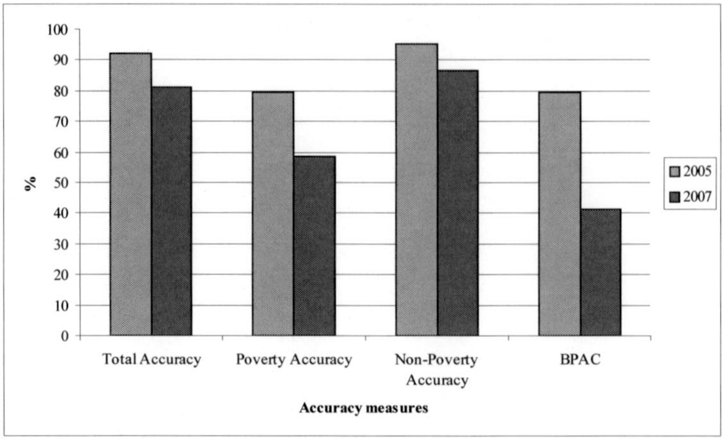

Fig. 1. Comparison of accuracy results of Model 1 2005-2007, two-step quantile regression. Source: own data

Table 2. Indicators for two step regressions from 2005 of Model 1. Source: own data

Household size
Household size squared
Age of household head
Age of household head squared
District (5 district dummies)
Maximum education of female household member is completed secondary level
Dummy: Household member lost weight because of food scarcity in the last 12 month
Food expenditure share of total consumption expenditures in percent (from section C: summary expenditures last 12 month)
Dummy: Household eats rice mixed with maize because of food scarcity in the last 12 month
Age of youngest household member
Percentage of dependents younger than 18 and older than 60 years (in relation to household size)
Dummy: Household head works outside of agriculture
Dummy: Trunk or suitcase ownership
Total value of furniture sets owned by household
Dummy: Household agrees that people in the neighbourhood are basically honest and can be trusted
Dummy: Household borrowed money from informal market in the last three years
LOG of annualised total consumption expenditures from section C (summary expenditures last 12 month)
Total value of transportation assets
Dummy: Household made a recent home improvement
Dummy: Exterior walls are out of brick or stone

not own a scale to monitor their weight. Therefore, the indicator "household member lost weight because of food scarcity" relies on their own impressions. As well, the indicator "food expenditure share of total consumption expenditures in percent" refers to questions in the composite questionnaire (section C) which asks for summaries of expenditures on food and non food categories. This indicator might therefore be biased by wrong guesses of the interviewed person. The dummy indicator "household agrees that people in the neighbourhood are basically honest and can be trusted" was used as a proxy for social capital.

Model 2

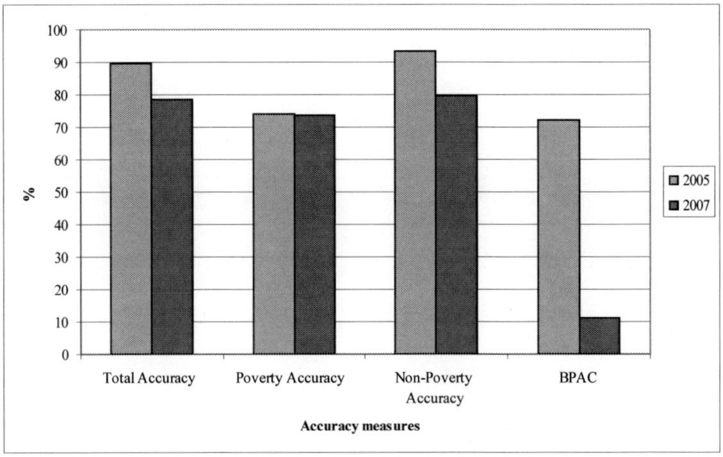

Fig. 2. Comparison of accuracy results of Model 2 (2005-2007), one-step quantile regression. Source: own data

One step quantile regression provided the best accuracy results for Model 2 in 2005. In Figure 2, the accuracy results for this regression method are shown. The total accuracy decreased by 11.24 percentage points, but poverty accuracy only by 0.54 percentage points while non-poverty accuracy declined by 13.85 percentage points. As a result of increased leakage (leakage rose by 61.08 percentage points from 27.78% to 88.86%), the BPAC dropped from 72.22% to only 11.14%.

In 2007, one-step OLS gave the best accuracy results (Figure 4). The increase of the BPAC (from -0.01% in 2005 to 50.94%) in 2007 can be explained with the higher poverty accuracy (44.44% in 2005 and 54.72% in 2007) and the decline of the prediction error undercoverage (from 55.56% in 2005 to 45.28% in 2007). Another reason is the increase in the error leakage from (11.11% in 2005 to 49.0% in 2007) because now both errors cancel each other out.

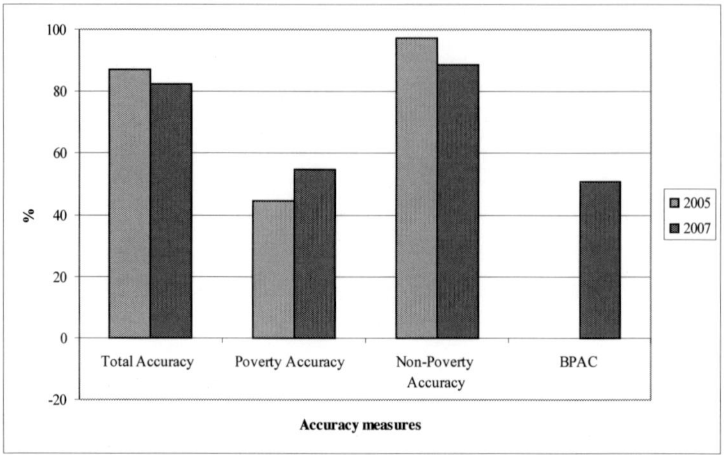

Fig. 3. Comparison of accuracy results of Model 2 (2005-2007), one-step OLS regression. Source: own data

Table 3. Indicators for one step OLS regression from 2005 of Model 2. Source: own data

Age of household head
Age of household head squared
Household size
Household size squared
District (5 dummies)
Total number of rooms in the dwelling
Dummy: Metal cooking pots ownership
Dummy: Clock or watch ownership
Dummy: VCD player ownership
Dummy: Motorcycle ownership
Dummy: Cow ownership
Dummy: Household uses other cooking fuel than collected wood
Dummy: Toilet is own pit toilet
Dummy: Main source of drinking water is water from well in residence yard
Dummy: Household head sleeps in bed with thin mattress out of fibres
Dummy: Household cooks in separate kitchen
Dummy: Household has own or shared electricity (including generator)
Percentage of dependents younger than 18 and older than 60 years (in relation to household size)
Dummy: Household made a recent home improvement
Number of trunks and suitcases owned

The indicators used for the one-step regression of Model 2 are summarised in Table 3.

The indicators in Model 2 are mostly time invariable or change only very slowly over time. The potential problem with this variable set is that it might not capture short term poverty dynamics and therefore rather detects the chronic than the transitory poor.

4.3 Capability of the national calibrated poverty assessment tool for Indonesia provided by IRIS

Table 4. Indicators from national calibrated tool. Source: IRIS (2009)

Household size
Age of household head
Household size squared
Age of household head squared
Region (Dummies for 7 different regions, Central Sulawesi is in region 5)
Dummy: Household live in rural area (the dummy for urban area was omitted because all households live in a rural area)
Dummy: Household head has incomplete secondary education
Dummy: Household head has any university education
Share of household member with incomplete secondary education
Share of household members with any university education
Dummy: Household head can read and write
Dummy: Floor of dwelling is earth
Area of the dwelling
Dummy: Main source of drinking water is bottled water
Dummy: Main source of drinking water is water from tab
Dummy: Main source of drinking water is water from pump
Dummy: Toilet facility is other (i.e. bush etc.)
Dummy: Main source of lightening is oil lamp
Dummy: Household received food aid in past 6 month
Dummy: Any household member bought new set of clothes in the previous year
Dummy: Any household member rent a stall/ shop outside of the households dwelling

IRIS is providing tools for poverty assessment by means of a small set of indicators for several countries (The IRIS Center 2008). These tools consist of the necessary questionnaires, a data entry sheet, and a data analysis tool (http://www.povertytools.org). The tools IRIS provides are nationally calibrated. The indicators used for Indonesia are listed in Table 4.

As raised earlier, IRIS and USAID are using two possible definitions of poverty lines. For the nationally calibrated poverty assessment tool developed by IRIS for Indonesia, the reference poverty line is called the median poverty line and is the bottom 50% of the national poverty line. For Indonesia this was 82,747 IDR per capita per month for rural areas. This number refers to the official national poverty line of 96,512 IDR likewise per capita per month for rural areas at 2002 prices (The IRIS Center 2008). Adjusted to the 2007 price levels by using the Consumer Price Index (CPI) the "median" poverty line per capita per month was 122,760 IDR, or 4092 IDR per day.

To measure the performance of the national calibrated tool using our data the headcount poverty rate for Sulawesi had to be calculated with the poverty line used by IRIS/USAID. (Table 5).

Table 5. Poverty rates in Central Sulawesi 2007. Source: own data/IRIS(2009)

Poverty line	Poverty line (IDR per capita and day)	Headcount Index (%) (N=282)
International poverty line of $1 US	3436	18.79
Median of national poverty line used by IRIS	4092	28.4

When predicting the poverty status of rural households in Central Sulawesi with our household data from 2007, but using the coefficients of the indicators listed in Table 4 provided by IRIS, the predicted headcount was 67.38%. Thus the poverty incidence error (PIE) was 38.98%. In terms of its accuracy, the performance of the tool is disappointing: the total accuracy was only 0.71%, the poverty accuracy was 0% and the non poverty accuracy 1.04%. The prediction errors undercoverage (100%) and leakage (211.11%) were very high and hence the BPAC was incredible low.

In a second calculation we used the $1 US poverty line of 3436 IDR as reference for the predictions of the IRIS tool in order to make the prediction comparable to the models described above (4.2) given the use of different poverty lines. In this analysis, a headcount of 11.7% was predicted by the IRIS tool. In Table 5, the observed and the predicted headcounts as well as PIE are summarized for all regression methods presented.

Table 6 shows that the coefficients obtained by the one-step OLS regression method of Model 2 best predicted the poverty incidence with a PIE of only 0.71%; the second best prediction of poverty headcount by using the two-step OLS coefficients of Model 2. The national calibrated poverty assessment tool of IRIS provided the fifth best prediction of the poverty headcount. Nevertheless, the other accuracy results were disappointing: the total accuracy was only 20.92%, none of the very poor households were correctly predicted as being very poor, the non poverty accuracy was 23.69%. As well, both undercoverage and leakage were again very high.

Table 6. Observes vs. predicted headcount (N=282). Source: own data

Method	Actual headcount in %	Predicted headcount in %	Poverty Incidence Error (PIE) in %
Model 1			
One-step OLS	18.79	24.27	5.48
Two-step OLS		28.36	9.51
One-step quantile		34.75	15.96
Two-step quantile		22	3.21
Model 2			
One-step OLS	18.79	19.5	0.71
Two-step OLS		20.2	1.41
One-step quantile		30.5	11.71
Two-step quantile		35.46	16.67
IRIS tool			
	18.79	11.7	7.09

5 Conclusion and discussion

NGOs, micro enterprises and other organisations or institutions concerned with poverty reduction are in need of a low-cost poverty assessment tool which is able to easily detect their clients. In the research region the role of poverty reduction is not only crucial for the improvement of peoples livelihood but also for the protection of the natural rainforest. As discussed in section 1, there is a strong link between poverty and forest degradation.

In 2005, when the tools for Central Sulawesi were developed, one of the biggest problems was the trade-off between the practicability of a tool and its accuracy (van Edig 2006, van Edig et al. 2007). Johannsen and Zeller (2006) also found that the exclusion of monetary indicators (as done in Model 2) reduces the accuracy of the tool. In addition, Zeller (2004) describes another problem poverty assessment has to face: the trade-off between accuracy and costs. Our results indicate another potential weakness of poverty assessment tools: their stability over time. When predicting the poverty status of households in 2007 with 2005 indicators, the accuracy of both models tested dropped. In all cases, except two-step OLS of Model 2, the leakage increased. Thus, within two years the capability of the tools was limited by errors predicting non-poor households as being very poor. Even so the average decline in the different accuracy measures is not dramatic and, for practitioners, still sufficient.

We expected Model 1 to perform somewhat better because it includes many short-term indicators like the "number of days in last week any superior food (large fish, beef/pork/buffalo meat, chicken/duck or egg) was eaten" or the "natural logarithm of expenditures on other expenditures, social events and leisure in the last 12 months" (both examples from one-step regressions). These indicators tend to change with the same speed as household expenditures. Model 2 instead mostly used long-term variables like "total rooms of the dwelling" which do not change as fast as expenditures. In contrast to our expectations, one-step OLS of Model 2 provided the best overall accuracy. This is opposite to the findings of Zeller et al. (2005) that OLS is less able to predict the poverty incidence when the actual headcount is relatively low.

That one-step OLS of Model 2 provided the best overall accuracy is only true if we use the BPAC as benchmark. The best poverty accuracy was achieved using two-step quantile regression with for Model 2. Even if one-step OLS of Model 2 is providing the best BPAC, the poverty accuracy with this method is comparatively low (44.44% in 2005 and 54.72% in 2007). Nevertheless, the predicted poverty headcount with this method was 19.5%, which is very close to the actual headcount of 18.79%. Finally, we can state that Model 2, which could be easily applied by local organizations for targeting, is still a good choice for poverty assessment. Even if the one-step OLS coefficients provide a better BPAC, we would recommend the use of one-step quantile coefficients because they provided a better poverty accuracy (73.53%) than one-step OLS, but the leakage is not as high as with two-step quantile. This tool could be applied by practitioners straightforwardly.

In general, one could improve the methodology using a bigger sample where an out-of-sample test would be possible. An out-of-sample test would be to split a sample randomly into two parts. One of these parts would be used for the tool calibration and the second part would be used for poverty assessment.

When applying the national calibrated tool of IRIS/USAID to our data set, the accuracy of the tool is very low. This result suggests that it is very difficult, if not impossible, to develop indicator based poverty assessment tools which are applicable in all regions of a country. This seems to be particularly true in such a diverse country as Indonesia. Furthermore, we see what a huge impact the choice of the poverty line has. Therefore, the two definitions the US congress chose as benchmarks for poverty assessment tools are somehow questionable. In general, it is a very good and ambitious idea to develop country-wide applicable PATs but, in reviewing our results, we think it is critical that USAID requires its implementation partners to use these poverty assessmenttools for their targeting in order to receive funds from USAID. That said, we believe that - at least in diverse countries like Indonesia - the development of regional tools would be more effective at accurately targeting the very poor.

References

Aho G, Larivière S, Martin F (editors) (1998) Poverty Analysis Manual - With Applications to Benin. United Development Program, Université Nationale du Bénin, Université Laval (Canada)

Alcaraz V. G, Zeller M (2008) Use of household food insecurity scales for assessing poverty in Bangladesh and Uganda. Discussion paper No. 1/2008. Research in development economics and poverty. University of Hohenheim

Baulch B, Masset E (2003) Do monetary and non-monetary indicators tell the same story about chronic poverty? A study of Vietnam in the 1990s. World development (31) 3: 441-453

Benin S, Randriamamonjy J. (2008) Estimating household income to monitor and evaluate public investment in Sub-Saharan Africa. IFPRI discussion paper 0771

Collier P, Dollar D (2002) Aid allocation and poverty reduction. European Economic Review (46): 1475-1500

David IP (2000) Poverty Statistics and Indicators: How often should they be measured?. Asian Development Bank URL: http://unstats.un.org/unsd/methods/poverty/Sid Aug 2000 Povnote.PDF (14.08.05)

Eckholm E, Foley G, Barnard G, Timberlake L (1984) Fuel wood - the energy crisis that won't go away. Earthscan Paperbacks - International Inst. for Environment and Development (RU)

van Edig X (2006) Measurement of Absolute Poverty and Indicators of Poverty among Rural Households in Central Sulawesi, Indonesia. Staats- und Universitätsbibliothek Göttingen URL: http://webdoc.sub.gwdg.de/master/2006/vanedig/vanedig.pdf

van Edig X, Schwarze S, Zeller M (2007) Indicator based poverty assessment for rural Central Sulawesi. Quarterly Journal of International Agriculture (46) 2: 145-158

Erasmi S, Twele A, Ardiansya M, Malik A, Kappas, M (2004) Mapping deforestation and land cover conversion at the rainforest margins in Central Sulawesi, Indonesia. EARSeL eProceedings 3(3): 288-297

FAO (Food and Agriculture Organization of the United Nations) (2009) State of the Worlds Forests 2009. URL: http://www.fao.org/docrep/011/i0350e/i0350e00.htm (31.03.09)

Foster J E (1998) Absolute vs. Relative Poverty. The American Economic Review (88) 2, Papers and Proceedings of the Hundred and Tenth Annual Meeting of the American Economic Association: 335-341

Grootaert C, Braithwaite J (1998) Poverty correlates and indicator-based targeting in Eastern Europe and the Former Soviet Union. Policy Research Working Paper 1942. The World Bank

Grosh M, Glewwe P (2000). Designing Household Survey Questionnaires for Developing Countries - Lessons from 15 years of the Living Standard Measurement Study Vol. 1, The World Bank . Washington D.C.

HDR (Human Development Reports) (2007/2008) URL: http://hdrstats.undp.org/indicators/23.html (28.04.09)

IPCC (Intergovernmental Panel on Climate Change) (2001) URL http://www.ipcc.ch/ (31.03.09)

Johannsen J, Zeller M (2006) Operational Poverty Targeting by the Means of Proxy Indicators - The Example of Peru. Contributed paper for presentation on the International Association of Agricultural Economists Conference, Gold Coast, Australia, August 12-18, 2006

Lok R (1995) Technical support document: Poverty - Module 1: Poverty Indicators. SEPED/BPPS; UNDP 13.09.1995, URL: http://www.undp.org/poverty/publications/tsd/tsd1/tsd1.pdf (14.08.05)

Minot N (2000) Generating Disaggregated Poverty Maps: An Application to Vietnam. World Development Vol. 28, Issue 2, February 2000, 319-331

Muller C (1997) Transient seasonal and chronic poverty of peasants in Rwanda. The Centre for the Study of African Economies Working Paper series. CSAE 1997, paper 56

OECD (2009) Poverty and climate change - Reducing the Vulnerability of the Poor through Adoption. URL: http://www.oecd.org/dataoecd/60/27/250 2872.pdf (12.01.09)

Olschewski R, Tscharnke T, Benítez P C, Schwarze S, Klein A-M (2007) Economic evaluation of ecosystem services as a basis for stabilizing rainforest margins? The example of pollination services and pest management in coffee landscapes. In Tscharnke, Leuschner, Zeller, Guharrdja, Bidin (Eds.) Stability of Tropical Rainforest Margins - Linking Ecological, Economic and Social Constrains. Springer Heidelberg Berlin

Osmani S R (2003) Envolving Views on Poverty - Concept, Assessment and Strategy. Poverty and Social Development Papers. No 7. 2003 Asian Development Bank

PEN (Poverty and Environment Network), 2009: URL: http://www.cifor.cgiar.org/pen/_ref/links/index.htm (26.03.2009)

Ravaillion M (1992) Poverty Comparison - A Guide through Methods and Concepts. LSMS Working Paper 88. The World Bank. Washington D.C.

Sen A (1999) Development as freedom, Oxford University Press

Schubert R (1994) "Armut in Entwicklungsländern - Begriff, Ausmass, Konsequenzen". In: Schäfer (Ed.) pp. 13-39, Schriften des Vereins für Socialpolitik Bd. 234. Ducker & Humboldt. Berlin

Schwarze S, Nuryartono N, Wardhono A, Zeller M, Sanim B, Anwar C (2005) Deforestation - A household perspective. Presentation at the International Symposium "The Stability of Tropical Rainforest Margins: Linking Ecological, Economic and Social Constraints of Land Use and Conservation", Göttingen, September 2005

Schwarze S, Schippers B, Weber R, Faust H, Wardhono A, Zeller M, Kreisel W (2007) Forest Products and Household Incomes: Evidence from Rural Households Living in the Rainforest Margins of Central Sulawesi. In Tscharnke, Leuschner, Zeller, Guharrdja, Bidin (editors), Stability of Tropical Rainforest Margins - Linking Ecological, Economic and Social Constrains. Springer Heidelberg Berlin

The IRIS Center (2005) Developing Poverty Assessment Tools Project - Note on Assessment and Improvement of Tool Accuracy. URL: http://www.povertytools.org/documents/Assessing and Improving Accuracy. pdf (14.08.05)

The IRIS Center (2008) Developing Poverty Assessment Tools - Accuracy Results for 20 Poverty Assessment Tool Countries. URL: http://www.povertytools.org/other_documents/PAT_20_country_accuracy_results_Dec2008. pdf (12.01.09)

United Nations (2008) URL: http://www.un.org/millenniumgoals/ (24.11.08)

Weber R, Faust H, Schippers B, Mamar S, Sutarto E, Kreisel W (2007) Migration and ethnicity as cultural impact factors on land use change in the rainforest margins of Central Sulawesi, Indonesia. In: Tscharnke, Leuschner, Zeller, Guharrdja, Bidin (Eds.) Stability of Tropical Rainforest Margins - Linking Ecological, Economic and Social Constrains. Springer Heidelberg Berlin

Witt M (1998) Der *Fondo de Inversión Social* - Ein effizientes entwicklungspolitisches Instrument zur Bekämpfung der Armut in Bolivien? Göttinger Studien zur Entwicklungsökonomie 6. Vervuert. Frankfurt am Main

World Bank (2008) URL: http://indopov.org/poverty.html (24.11.08)

Wunder S (2001) Poverty Alleviation and Tropical Forests - What Scope for Synergies? World Development (29) 11: 1817-1833

Zeller M, Sharma M, Henry C Lapenu C (2001) An operational tool for evaluating poverty outreach of development policies and projects. FCDN Discussion Papers No. 111. Food and Nutrition Division. International Food Policy Research Institute. Washington, D.C.

Zeller M (2004) Review of poverty assessment tools. AMAP

Zeller M, Johannsen J, Alcaraz V. G (2005) Developing and testing poverty assessment tools: results from accuracy tests in Peru - Final Report March 2005. USAID, EGAT, and AMAP

Demography, development, and deforestation at the rainforest margin in Indonesia

Stephan Klasen[1*], Heiko Faust[2], Michael Grimm[3], and Stefan Schwarze[4]

[1] University of Göttingen, Departmenf of Economics,Platz der Göttinger Sieben 3,37073 Göttingen, Germany
[2] University of Göttingen, Institute of Geography, Division of Human Geography, Goldschmidtstr. 5, 37077 Göttingen, Germany
[3] International Institute of Social Studies, Erasmus University Rotterdam, Kortenaerkade 12, 2518AX The Hague, The Netherlands
[4] University of Göttingen, Department of Agricultural Economics and Rural Development, Platz der Göttinger Sieben 5, 37073 Göttingen

*corresponding author: S. Klasen, email: sklasen@uni-goettingen.de.

Summary

In this paper, we study the factors influencing both economic development and deforestation at the rainforest margin in Indonesia. We base our analysis on an econometric analysis of the village survey in the STORMA region as well as qualitative evidence generated in the project area. We show that a key driver of economic development is demographic change, particularly immigration into the Lore Lindu area. This in turn affects the emergence of land rights, including formal land titles as well as semi-formal ones. These land rights then spur technological change and economic development. Unfortunately, immigration as well as some aspects of technological change are also major drivers of deforestation, while economic development appears to reduce deforestation.

Keywords: deforestation, migration, land rights, institutions, economic development

1 Introduction

A central challenge for poor countries richly endowed with natural resources is to promote economic development while ensuring environmental sustainability. As a microcosm, this is precisely the challenge faced by the STORMA

T. Tscharntke et al. (eds.), *Tropical Rainforests and Agroforests under Global Change*,
Environmental Science and Engineering, 213–236, DOI 10.1007/978-3-642-00493-3_10,

research program surrounding the Lore Lindu National Park in Central Sulawesi, where protection of the rainforest and the promotion of growth and poverty reduction are the central aims.

The purpose of this paper is to jointly investigate the determinants of economic development and deforestation at the rainforest margin in an empirical framework which includes an analysis of a village survey of over 70 villages in the STORMA research area linked to GIS data on deforestation as well as qualitative evidence. Our results suggest that demography-induced institutional change, which in our case is the establishment of formal and semi-formal titles over agricultural land, are the key drivers of agricultural productivity improvements, which in turn determine economic development outcomes at the rainforest margin. At the same time, several of these drivers of economic development themselves promote deforestation while the conditional effect of economic development (controlling for its drivers) is associated with reduced deforestation.

The paper is organized as follows. Section 2 discusses the conceptual framework. Section 3 describes the data, the setting, and the econometric method. Section 4 focuses on the results, while section 5 concludes.

1.1 Conceptual Background

When studying the literature on determinants of economic development in general, and development in rural settings, seemingly competing hypotheses are invoked. A first strand of the literature has debated the respective roles of geography versus institutions for long-term economic development. One group has argues that "geography", proxied by indicators such as climate, disease environment, topography, market access, and soil quality of the cultivated land area, is the dominant factor in determining long-term economic development, including particularly agricultural development (see e.g. Diamond, 1997; Gallup, Sachs and Mellinger, 1998). The opposing view is that institutions such as property rights and the rule of law are much more important determinants of long-term economic progress (e.g. Hall and Jones, 1999; Acemoglu *et al.*, 2001; Easterly and Levine, 2003; Rodrik, Subramanian and Trebbi, 2004). Those in the latter camp allow, however, for the fact that institutions have evolved endogenously responding to, among other things, geographic conditions. This is done most explicitly in Acemoglu et al. (2001) where geographic conditions, particularly a high disease burden, affected European settlement patterns. This, in turn, led to extractive institutions in non-settler economies and development-friendly institutions in settler economies. Through historical persistence, these institutions still heavily influence the economic fate of nations today.[1] It is noticeable that in both of these views, demographic change

[1] See Grimm and Klasen (2008) for a more detailed exposition of this geography versus institutions argument, and a more detailed econometric analysis of the transmission channels from migration to technological change. In the paper here,

will play an important role. In the pure geography view, regions with unfavorable geography will suffer from high mortality and out-migration, while in the institutions view, immigration and settlement patterns play a crucial role in shaping institutional change.

A closely related but rather independent strand of the literature also emphasizes the role of endogenous institutional change, this time with an explicit focus on improvements in agriculture (North 1990; Hayami and Ruttan 1985). In this literature, the role of formal land titles has received particular emphasis (e.g. Binswanger, Deininger, and Feder 1995; Deininger 2003). According to this argument, land titles would provide security to the land owner, would lower the cost of trading land and could be used as collateral for credit. This in turn would have a positive impact on investment in land improvement as well as new and more productive cultivation technologies. However, the literature also emphasizes that land titles have to be considered as endogenous, responding, among others, to past investment decisions in the land, land scarcity, land quality, as well as the differential power of different rural groups (e.g. Binswanger, Deininger, and Feder 1995; Rozelle and Li 1998; Besley 1995; Brasselle, Gaspart and Platteau 2002). The empirical evidence on the effectiveness of land titles on technological change and agricultural productivity is quite mixed, as will be discussed below.

A third strand of the literature emphasizes population size and density, and associated pressure on land, inducing technological improvements or the adoption of existing technologies (see e.g. Boserup 1981; Kremer 1993; Klasen and Nestmann 2006). This demographic pressure can be significantly enhanced through immigration which can therefore accelerate the induced technological improvements.

These three strands of the literature have evolved quite independently and there are only few studies that explicitly test the relative importance or the inter-relationships between these competing hypotheses regarding the long-term drivers of economic change. In this paper we argue that migration and associated population pressure in turn intensify land pressure and conflict over land in these areas. Land pressure induces communities to opt for formal land titles, which in turn increase the incentive of farmers to invest in agricultural technology. Eventually, agricultural technology enhances agricultural growth which in turns drives economic development and poverty reduction. In short, endogenously generated institutional change is the core element of our transmission channel from demography to economic development.

The causal chain from demography to economic development we have in mind is as follows. Villages with relatively high immigration in turn experience population pressure on increasingly scarce land resources. Land pressure and associated conflicts induces villages to regulate their land market and to opt for formal land titles, for which there is a demand-driven system of land

we focus on the larger causal chain from geography to deforestation and also incorporate qualitative evidence to support our analysis.

titles in place in Indonesia (see below). In line with a large literature on this topic, land titles increase incentives and means for agricultural productivity improvements as farmers are able to capture the returns to long-term investments through higher productivity or higher land values and can use the land as collateral for credit (e.g. Braselle et al. 2002). Tenure security is particularly important in our study area because the main innovation in agriculture is the cultivation of cacao, a perennial crop, which starts to generate returns from the third year onwards. Eventually, agricultural technology enhances agricultural growth and economic development.[2]

The reader might notice some similarity between our hypothesis and the one by Boserup (1981) mentioned above. Boserup (1981) argued convincingly that demographic pressure (which could come about through population growth and/or immigration) and associated food shortages would induce technological improvements or the adoption of existing technologies. A rising population density would force individuals to modify the mode of land use and to employ progressively modern agricultural technologies. The argument put forward in our paper is consistent with Boserup's line of reasoning, but the crucial issue is, that we assume that institutional change is the critical intervening variable between population and technology. Without that institutional change investments in new technologies will not take place for the reasons given above.

Regarding the link between this causal chain and possible deforestation, the literature suggests competing influences (see Maertens et al. 2006; Angelson 1999; Angelsen et al. 2001). On the one hand, deforestation and environmental degradation more generally is seen as a consequence of population pressure, poverty, and poor technologies, where people are pushed onto marginal lands and/or convert forests for agricultural use. Conversely, deforestation can also result from profitable income-earning opportunities in agriculture that induces the expansion of a land frontier, i.e. deforestation is a result of pull factors to the rainforest margin. As has been suggested in the literature (e.g. Maertens et al. 2006, Angelsen et al. 2001), the type of technological change might matter as well, with labor-saving technological improvements favoring forest clearing while land-saving technologies discouraging land conversion. Some of these forces have actually been investigated using the same data from Indonesia. In particular, Maertens et al. (2006) examine the impact of land- and labor-saving technologies in rice production, as well as population pressure on land expansion in a set of Indonesian villages. They find that population pressure and labor-saving technologies promote deforestation while land-saving ones reduce it.[3]

[2] The details of this causal chain as well as the literature on aspects of this causal chain are discussed in Grimm and Klasen (2008).

[3] Our study differs from their approach by explicitly considering the drivers of technological change, by looking directly at deforestation rates (rather than land use changes as a proxy), by considering more covariates, and by including an explicit panel dimension in the analysis.

This literature thus suggests that many of the determinants along our causal change, including favorable geography, immigration, population pressure, land rights, technological change as well as economic development could all have an impact on land-use changes and deforestation rates and that it is therefore important to consider the direct and indirect effects of these determinants in our empirical analysis.

2 Study context, and econometric method

2.1 Data

The data used in this analysis consists of a village survey matched to GIS data on forest cover and its changes. The village survey we use was conducted during March to July in 2001 in the Lore Lindu region. This region includes the Lore Lindu National Park and the five surrounding sub-districts. It is situated south of Palu, the provincial capital of Central Sulawesi/Indonesia. The survey is part of an international and interdisciplinary research program known as "Stability of Rain Forest Margins" (STORMA) which studies the determinants of biodiversity and land use in this region and how such biodiversity can be protected through appropriate socioeconomic mechanisms. For the survey 80 of the 119 villages in the region were selected using a stratified random sampling method (Zeller et al. 2002). The survey collected data on current and past demographics, land use practices and technology adoption, conservation issues, infrastructure and qualitative information on income and well-being. The years for which this information is reported are 1980, 1990, 1995 and 2001. Additional information on geographic features was taken from secondary data sources and added to the data set by Maertens et al. (2006).

This data set was complemented by land cover data derived from satellite images for 1981 and 2001 (for details see Erasmi and Priess 2007). The information on forest cover was extracted from the land cover data, then aggregated from the pixel level with a resolution of 500m x 500m to the village level, and finally merged with the survey data.

To add further detail to our story, we supplement the quantitative analysis with qualitative data that confirm our causal linkages from demography via economic development to deforestation. Our qualitative methods included formal, semi-structured in-depth interviews with the heads of selected migrant and non-migrant households belonging to different ethnic groups, and with village representatives. We also conducted informal talks, discussions, and participant observations. The qualitative data have been conducted in six villages within the study region in 2003-2007 (Weber 2006, Burkard and Fremerey 2008, Koch et al. 2008). In total we conducted more than 200 interviews with foci on social and cultural development (including ethnicity and networking), land tenure and land use change, as well as the role of institutions. The villages are selected based on previous qualitative studies regarding cultural change in the Lore Lindu region during the 20th century (Faust et al. 2003; Kreisel et al. 2004). The villages are assumed to represent different stages on a transition to modernized agriculture, a transition which is for the most part initiated by the villagers' migration patterns. This continuum encompasses a relatively static type of villages (Toro, Rompo), where the number of new migrants is low during the last 15 years, to quite dynamic or post-transitional villages (Bulili, Sintuwu). The latter is characterized by a high share of local migrants who were among the first settlers in the 1970s, and a high share of regional migrants from South Sulawesi who settled down during a relatively short period (1980-2000). The villages in transition (Watumaeta and Lempelero) are situated between the two extremes, as the number of migrants is rising steadily during the past decade. All six villages are located in the uplands of the Lore Lindu region and have similar access to roadways.

2.2 Study context

The role of agriculture

The Lore Lindu region is rural. 87% of the 33,000 households living in the region depend economically on agriculture. 15% of the total areaexcluding the National Parkis used for agricultural production. The rest of the area is mainly grasslands and forests. The principal food crop is paddy rice. Important cash crops are cocoa and coffee. Households mainly operate as smallholders and with very few exceptions there are almost no large plantations in the region (see Maertens *et al.* 2006). Logging is mainly done informally (mainly for the conversion of forest land into cacao plots and not to sell the wood) and has then only a marginal importance for the local population.

Migration

During the past decades a significant part of the immigration into the study region has taken place from the south and middle-west of Sulawesi to the north-east of the Lore Lindu region, in particular to the districts of Palolo, Sigi Biromaru and Lore Utara. Some immigration has also taken place within so called "transmigration programs", organized by the government mainly during the 1960s and 1970s. These programs resettled people in particular from the islands of Java, Bali and Lombok to Central-Sulawesi. The places were chosen according to factors such as soil fertility and land availability (Faust et al. 2003). Most of these migrants have today returned and the programs are seen as having failed. In our sample only three villages were affected by such migration programs during the period 1990-2001. We excluded these three villages from our sample (reducing the sample size to 77). No village was affected by these programs during the 1980s.

Land rights

Formal land titles became more and more widespread over time in the Lore Lindu region. Some villages have had land titles since the early 1980s. Others introduced them only recently and a significant share of the villages is even today without such titles. In the villages where land titles exist they were in most cases established in the framework of the land certification schemes PRONA (*Proyek Operasi Nasional Agraria*) and PRODA (P*rogram Proyek Agraria Daerah*), which can provide ownership rights to land holders. These schemes were created by the Indonesian Government in 1981. However, no central or regional government beyond the village level ever enforced land titling and land redistribution in the study area using these mechanisms. PRONA/PRODA is rather an available scheme which can be used if there is a demand and the willingness to opt for land titling by villagers (Siagian and Neldysavrino 2007). The costs of land titling under these schemes have to be borne by the villagers. The process of land titling needs collective action by the villagers and usually starts with a proposal to the land administration office. This implies that the process of land titling is - consistent with our hypothesis - demand and not supply driven. There is some heterogeneity in the type of land titles issued under these schemes, ranging from a formal land certificate under these schemes, a letter of land transaction, to a letter from the village headman, and some other rarely used options (Nuryartono 2005). The use of these options also depends on the type of land acquisition, i.e. clearing of forests, inheritance/marriage, or purchase (see below).

2.3 Estimation Strategy

First, we show that in the Lore Lindu Region agricultural technology is an important driver of agricultural household income. Although the empirical literature has shown many times that technology drives agricultural development, we think it is important to show that this link is also significant in our case. Therefore, we estimate using ordinary least-squares (OLS) the following equation:

$$Y_i = \mu + A_i^{'}\alpha + X_i^{'}\gamma + \varepsilon_i \tag{1}$$

where the index i stands for the villages. Since the survey does not provide any information on village mean income or alike, we use the percentage of all houses (used for the purpose of human residence) in each village built from stone, bricks or cement. Throughout the Lore Lindu region having a stone house is seen as sign of prosperity and wealth and therefore that variable should be a good measure of the villager's living standard, Y. As can be seen in Table 1, the share of stone houses varies significantly in our data set and therefore should contain enough information about differences in well-being across villages and over time during the period we look at.[4]

[4] It should also be noted that stones and bricks are often made or collected in the surroundings of the villages and hence, no road is necessary to bring them. Also,

Table 1. Descriptive statistics of used variables (continued in table 2)

	Invariant	1980	1990	1995	2001
Development:					
Share of houses built from stone		0.054	0.125	0.214	0.317
		(0.107)	(0.180)	(0.235)	(0.303)
Deforestation Rate	0.073				
(% 1981-2001)	(0.056)				
Geography:					
Share of agr. land on steep slopes	0.156				
	(0.259)				
Number of years to last drought	9.299				
	(10.664)				
Village accessible by car		0.597	0.714	0.727	0.753
Altitude in m (level above sea)	647.3				
	(339.9)				
Demography:					
Net immigration rate of housh.[a)]		0.021	0.021	0.014	
		(0.131)	(0.066)	(0.100)	
Village population size		713.5	913.8	987.2	1101.9
		(694.1)	(821.1)	(857.3)	(876.4)
Change in Household Number (%)	1.34				
	(1.65)				
Land rights		0.091	0.351	0.403	0.636

Source: CRC STORMA A3 Village Survey.
Notes: see table 2

As measures of agricultural technology (A) we use the existence of technical or semi-technical irrigation systems (usually village schemes), the construction of terraces as well as the use of fertilizer, pesticides, and improved seeds in the villages. Irrigation systems are only reported for villages with paddy rice fields. This concerns 70 out of the 77 villages and only those are included in the respective regressions. Likewise, terraces are only relevant for villages which have fields on steep slopes. This concerns 46 out of the 77 villages and again only those are included in the relevant regressions. One should also note that irrigation and terraces are rather long term investments, whereas fertilizer, pesticides and improved seeds are short term investments. In the case of such short-term investments, land rights matter if land can be used as collateral for credit. In the study region this is the case and credits are an

heavy materials including stones are in the Lore Lindu region traditionally and still frequently transported using buffalos, donkeys, horses or motorcycles. Given that labor is very cheap, transport time plays no important role. In 2001, among the 15 villages without any stone house, 11 are not accessible by car and 4 are accessible. Conversely, 8 villages among the 19 villages which are not accessible by car, have a significant share of stone houses.

Table 2. Descriptive statistics of used variables.

	Invariant	1980	1990	1995	2001
Technology use:					
Hand Tractors per HH in 1980	0.039				
Irrigation system available [b)]		0.200	0.329	0.371	0.514
Fertilizer use		0.403	0.584	0.649	0.727
Pesticides use		0.455	0.636	0.753	0.948
Improved seeds use		0.286	0.416	0.545	0.870
Terraces building [c)]		0.065	0.217	0.283	0.523
Other control variables:					
Male population per ha land			0.971	1.053	
			(0.583)	(0.636)	
Share of population 19-45 years					0.380
					(0.088)
Population density (pop per ha)		1.205	1.484	1.652	1.829
		(0.919)	(1.041)	(1.173)	(1.187)
Primary school in village		0.857	0.961		0.987
Newspaper in village					0.052
Drinking water connection		0.416	0.455		0.896
Electricity connection		0.104	0.247		0.922
Doctor available		0.169	0.338		0.442
Credit available [d)].		0.901	0.922		0.909

Source: CRC STORMA A3 Village Survey.
Notes:
Standard deviations in parentheses where appropriate.
[a)] The net immigration rate relates to the periods 1980-1990, 1990-1995 and 1995-2001.
[b)] Information about irrigation is only available in villages cultivating paddy rice (mean computed over those villages, i.e. 70 out of 77).
[c)] Terraces are only relevant for villages with steep slopes (mean computed over those villages, i.e. 46 out of 77).
[d)] The "credit available" variable is here shown for the periods "past 20 years", "past 10 years" and in 2001.

important device to finance inputs such as fertilizer in the pre-harvest period (Nuryartono 2005).

The vector X stands for additional control variables such as the male per agricultural land ratio in the village, the share of the village population between 19 and 45 years old (both measures of labor availability/abundance) and whether the village had connection to drinking water in 1990. We also control for adult education, which we measure by a dummy variable indicating whether the village had a primary school in 1980.[5]

[5] We also used a few other control variables but they did not change the results and were mostly not significant. Due to the relatively small sample size, we cannot include a large set of control variables in a single estimation.

In the basic specification the dependent variable is measured in 2001 and all explanatory variables, including technology, in the mid-nineties to allow for a time lag until these investments translate into higher incomes. [6]Given that we have for most of our variables also retrospective information, we estimate Equation 1 also with a panel fixed-effects estimator to control for all time-invariant unobserved village effects:

$$Y_{it} = \mu_i + A_{it}^{'}\alpha + X_{it}^{'}\gamma + \varepsilon_{it} \tag{2}$$

Clearly, endogeneity bias could be a problem here as income might affect technology just as much as technology affects income. In order to deal with this problem, we will now resort to our causal chain from demographic pressure via land rights to technologies and economic development. Details of this causal chain are examined in Grimm and Klasen (2008). Here we will simply resort to using immigration between 1980 and 1995 and population levels in 1980 as instruments for technologies in 1995. Thus the specification we use is

$$Y_i = \mu + A_i^{'}\alpha + X_i^{'}\gamma + \varepsilon_i, \tag{3}$$

with

$$A_i = \lambda_G + \beta_G D_i + X_i^{'}\gamma_G. \tag{4}$$

Where D stands for demographic pressure and is proxied by immigration rates between 1980 and 1995 and population levels in 1980. We also include a specification where the left-hand side variable is income growth (rather than income levels, again using our proxy for incomes), using the same framework.

To test the impact of the causal chain on our deforestation rates between 1980 and 2001, we need to consider that some of the drivers of economic development (such as population pressure) have a direct impact on changes in the forest cover (FC) while others will have an indirect effect via economic development. The easiest way to capture this effect is to specify an OLS regression of forest losses at the village level on the drivers of economic development as well as economic development itself. This way one can then later identify (and quantify) direct effects of these factors using the OLS regression and consider indirect effects by examining the earlier regressions and the impact of economic development on deforestation.[7]

$$\Delta \mathrm{FC}_i = \mu + G_i^{'}\alpha + \Delta D_i^{'}\beta + \Delta A(\mathrm{ld})_i^{'}\gamma + \Delta A(\mathrm{lb})_i^{'}\delta + \Delta Y_i^{'}\varphi + \varepsilon_i \tag{5}$$

Where the delta-signs stand for change over time and *A(ld)* and *A(lb)* distinguish between land-saving (*A(ld)*) and labor-saving (*A(lb)*) technological

[6] Even in the case of technologies that can have a short-run impact such as fertilizer and improved seeds, it is appropriate to consider some time lag as it takes some time to optimally apply these technologies and, in any case, farmers tend to use them throughout once they have adopted them.

[7] See Klasen (2002) for an example of such a path analysis.

change; G refers to geographical features that might additionally affect deforestation (such as accessibility) and D is again demographic pressure now proxied by the growth rate of households in a village. We use the use of hand-tractors as an indicator of labor-saving technological change (mostly used in rice production) and consider the changes in the package of modern seeds, fertilizer and pesticide use as one indicator of land-saving technological change, and the implementation of terraces as another one.[8] We will consider both the effect of economic growth (as shown in the equation) and, alternatively, the level of economic development as a factor influencing deforestation.

Tables 1 and 2 present the descriptive statistics of all key and the principal control variables in our analysis. As the statistics show, migration, land rights, technology adoption, and income growth show a sizable variation across villages and over time. Both should help to identify the parameters we are interested in. This spatial and intertemporal variation also shows that our region of analysis is a region under substantial transformation. Regarding forest loss, the forest cover in the villages is reduced by about 7 percentage points between 1980 and 2001 (from about 77% forest cover to 70%). This is somewhat smaller than the average for Indonesia and might also be affected by the protective function of the national park (see Schwarze et al. 2008). But also here there is significant variation and thus it is very interesting and important to understand the drivers of this differential in deforestation rates.

3 Results

3.1 Technology and economic development

Tables 3 and 4 report regressions of Equation (1), i.e. of the share of houses built from stone, bricks or cement, our measure of economic development, on various variables of agricultural technology.

Columns (1)-(5) show that all used technology variables have a positive and highly significant impact on economic performance. Note that technology is measured in 1995 and the share of houses in 2001, taking into account the time it takes until new technologies can translate into durably higher incomes. Column (6) shows a regression in which we use irrigation and a dummy variable as technology variables - the latter taking the value one if the village used fertilizer, pesticides and improved seeds simultaneously[9] - and, as an additional control variable, the average share of households who emigrated

[8] As terraces are only relevant in some of the villages, we include a variable "terraces not relevant" as an additional regressor in order to be able to use the full data set. Please note also that regression (6) is focused on those villages where there was some forest cover in 1980; this led to the exclusion of three villages from the analysis.

[9] Note that these techniques are often adopted in a sequence, starting with irrigation, followed first by fertilizer, second by pesticides and last by improved seeds.

Table 3. The effect of technology on development: OLS regressions.

	OLS regressions. Dependent Variable: Share of stone houses in 2001						
	(1)	(2)	(3)	(4)	(5)	(6)	(7)
Irrigation	0.379					0.303	0.191
1995 [a)]	(0.066)***					(0.097)***	(0.069)***
Fertilizer		0.395					
1995		(0.046)***					
Pesticides			0.313				
1995			(0.050)***				
Improved seeds				0.304			
1995				(0.060)***			
F., P. & S.						0.204	0.234
1995 [c)]						(0.097)**	(0.069)***
Terraces					0.265		
1995 [b)]					(0.087)***		
Emigration						0.364	
1990-2001						(1.463)	
Male population							0.039
per ha. 1995							(0.027)
Population							0.361
share 19-45yrs							(0.261)
Primary school							0.080
1980							(0.083)
Drinking water							0.171
connection 1990							(0.060)***
Constant	0.185	0.062	0.082	0.149	0.194	0.101	-0.181
	(0.036)***	(0.022)***	(0.028)***	(0.038)***	(0.044)***	(0.045)***	(0.147)
R^2	0.356	0.394	0.204	0.252	0.185	0.536	0.620
N	70	77	77	77	46	46	70

Source: CRC STORMA A3 Village Survey. *Notes:* see Table 3.

Table 4. The effect of technology on development: FE regressions.

	FE regressions. Dependent Variable: Share of stone houses[d]					
	(8)	(9)	(10)	(11)	(12)	(13)
Irrigation [a]	0.190					0.143
	(0.031)***					(0.127)***
Fertilizer		0.139				
		(0.026)***				
Pesticides			0.097			
			(0.018)***			
Improved seeds				0.145		
				(0.019)***		
F., P. & S.						0.127
						(0.029)***
Terraces [b]					0.072	0.040
					(0.013)***	(0.016)**
R^2	0.238	0.256	0.167	0.236	0.076	0.425
N	280	308	308	308	184	160

Source: CRC STORMA A3 Village Survey.
Notes:
Robust standard errors in parentheses. *** significant at 1%, ** significant at 5%, * significant at 10%.
Variable explanations see Section 4.
[a] Information about irrigation is only available in villages cultivating paddy rice.
[b] Terraces are only relevant for villages with steep slopes.
[c] "F., P. & S." stands for the simultaneous use of fertilizer, pesticides and improved seeds.
[d] In the FE regressions stone houses and technology in 1980, 1990, 1995 and 2001 are used.

from the village between 1995 and 2001. This latter variable is insignificant and thus makes it unlikely that the share of stone houses is strongly related to remittances coming from former villagers who migrated to the city.

Column (7) introduces as additional controls the ratio between the male population and the total size of agricultural land in 1995, the share of villagers between 19 and 45 years old, adult education approximated by the availability of a primary school in the village in 1980, and a dummy variable whether the village had in 1990 a drinking water system (to proxy for good current infrastructure). Using the results of that regression we find that on average in a village with irrigation the share of stone houses is higher by almost 20 percentage points than in a village without irrigation. Using fertilizer, pesticides and improved seeds simultaneously increases this share again by 23 percentage points. This model explains more than 62% of the total variance in the data. We also estimated the models presented in columns (1)-(7) with maximum likelihood using a generalized linear model which might be more appropriate given that our dependant variable is a ratio bounded between 0 and 1. It turned out that the standard errors were nearly identical and hence we decided to stick to the simpler OLS model.

Given the relatively small sample size we are of course constrained by the number of control variables we can introduce in the model. However, given that we have for the relevant variables observations over at least four different points in time (usually 1980, 1990, 1995 and 2001), we can estimate our model also using fixed effects as specified in Equation (2) and thus at least control for the influence of all time-invariant village effects (including the potential role played by geography). Columns (8)-(13) show that all results hold and that most of the technology coefficients have a similar magnitude.[10]

Of course all these results might be affected by a possible endogeneity of technology to income, although we mitigate this problem by using appropriate lags. In principle proper instruments are needed to solve the endogeneity problem satisfactory. We implement this below. At this stage of the analysis we simply conclude, as many other empirical studies have done before, that agricultural technology enhances rural development. This motivates us to look now at the determinants of technology adoption. We will show that technology is driven by migration-induced land rights.

3.2 Revisiting the technology-growth link using demographic change as an instrument

As shown in Grimm and Klasen (2008), it is possible to show that geography-induced demographic pressure indeed affects the emergence of land rights and this in turn affects technological change. Thus we can use demographic pressure as an instrument for technological change to revisit the technology-growth linkage. This is done in Table 5 where we consider regressions investigating the level of income in 2001 as well as the growth of incomes between 1995 and 2001 (both proxied by our housing variable). The first stage regression shows that demographic variables indeed have a significant influence on technological change and thus are clearly relevant instruments. The second stage regression continues to show a large and significant influence of technological change on economic development in the villages, regardless of whether we consider levels or growth rates. While we cannot conclusively show that our instruments fulfill the exclusion restriction, an overidentification restriction test supports the validity of the instruments and the demographic variables no longer have a separate significant influence on economic growth if included in addition in the second stage regression.[11]

---

[10] In Grimm and Klasen (2008) we also estimate growth rather than level regressions with very similar results.

[11] Results are available on request.

3.3 Modeling the determinants of deforestation

We now turn to investigating the determinants of deforestation in our village data set. In Table 6 we show two specifications which only differ in their use of economic growth (column 1) or income levels in 2001 (column 2) as a covariate. The results show that many of the drivers of economic development indeed also promote deforestation. Among the geography variables, good early infrastructure access (access to car in 1980) and low elevation not only help induce economic development, but they also have a direct effect of promoting deforestation. It thus appears that in locations that are accessible and in the lowlands, forest conversion is accelerated. Similar observations relate to population pressure which also helped promote technological change but also has a direct and sizable effect in promoting deforestation. In contrast to these trade-offs, less frequent droughts appear to both promote technological change *and* are associated with reduced deforestation. Here two effects might play a role. First, it could be the case that the absence of severe drought shocks reduce encroachment into the rainforest during such shocks to make up for lower yields.[12] Second, lower drought frequency might alter crop choice and favor irrigated rice over the more drought-resistant but extensively farmed cocoa (see Maertens et al. 2006).

The effects of technology also present some trade-offs. While labor-saving (and thus land-using) technology is, as expected associated with higher deforestation, the package of modern seeds, fertilizers, and pesticides also is associated with higher levels deforestation in the villages. It appears that the processes that generate better technologies and higher incomes encourage further encroachment. Only the implementation of terracing leads to lower deforestation rates. This shows that more labor-intensive paddy rice production indeed has a forest preserving effect, which is in line with Maertens et al. (2006).

Lastly, it is most interesting to see that, conditional on all other included variables, income growth is associated with significantly lower deforestation rates. Thus a "pure" development effect in the sense of rising prosperity appears to lower deforestation. This might be related to a falling dependence on the forest encroachment, and agriculture in general, for a livelihood, thus reducing the need to convert forest land. Interestingly, this effect is only true for economic growth, but not income levels in 2001 (or earlier).[13] Thus it is not a pure "prosperity" effect, but rather an effect of *rising* prosperity.

Given the fact that many drivers of economic development seem to promote deforestation while economic development itself has a forest conserving effect, one can quantify the net effect of these drivers. This has to be done

---

[12] In the time period under investigation, the role of the severe El Nino Southern Oscillation event in 1998 which led to a severe drought in Indonesia might have played a significant role.

[13] We also tried specifications with income levels in 1980, 1990 and 1995. It was never a significant determinant of deforestation.

Table 5. Geography, Technology and Economic Development: IV Regression.

	Growth (95-01) (1)	Income Level (2001) (2)
Second Stage:		
Technology: Seeds,	0.032	0.545
Fert., Pest. 1995	(0.010)***	(0.169)***
Primary school 1980	-0.005	-0.020
	(0.008)	(0.141)
Water access 1980	0.008	0.138
	(0.006)	(0.096)
Constant	0.003	0.009
	(0.007)	(0.130)
First Stage: Technology:		
Primary school 1980	0.044	0.044
	(0.169)	(0.169)
Water access 1980	0.210	0.210
	(0.119)**	(0.119)**
Population 1980	0.0003	0.0003
	(0.0001)***	(0.0001)***
Net migration rate	0.157	0.157
1980-1995	(0.008)**	(0.008)**
Constant	0.150	0.150
	(0.301)*	(0.301)*
N	70	70

Source: CRC STORMA A3 Village Survey.
Notes:
Robust standard errors in parentheses. *** significant at 1%, ** significant at 5%, * significant at 10%.
Variable explanations see Section 4.
Technology in the first stage refers to the share of villages having adopted the package of improved seeds, fertilizers, and pesticides in 1995.

with some caution as such a quantitative assessment is focused on the point estimates (without considering the sizable standard errors) and it neglects the fact that these point estimates are quite sensitive to included and excluded control variables and thus depend on the specification used. It turns out that the forest-clearing direct effect is always larger than the forest-conserving indirect effect of promoting economic development. Thus, on net, population growth, the modern technology package, better accessibility and lower elevation all end up furthering deforestation despite their forest-conserving impact on economic development; but the size of the net effect various considerably and is particularly large in the case of population growth and road access, while much smaller it the case of low elevation and the modern technology package. Thus for these factors, a trade-off remains between clearing the for-

Table 6. Determinants of Deforestation.

	Forest Loss (80-01)* (1)	Forest Loss (80-01) (2)
Geography:		
Altitude above sea	-0.00008	-0.00007
	(0.00003)***	(0.00003)***
Years to last drought	-0.0010	-0.0011
	(0.0004)**	(0.0004)***
Access by car in 1980	0.0426	0.0388
	(0.0130)***	(0.0125)***
Demography:		
Growth of #	0.0123	0.0130
of households	(0.0018)***	(0.0023)***
Technology:		
Labor-Saving: Hand-tractors	0.0436	0.0407
per hh in 1980	(0.0155)***	(0.0154)***
Land-Saving: Change in	0.0155	0.1466
seeds, Fert. Pest. (80-01)	(0.0101)*	(0.0109)*
Land-Saving: Change in	-0.0209	-0.0189
Terraces (80-01)	(0.0113)**	(0.0145)*
Terracing	-0.0077	-0.0123
not relevant	(0.0135)	(0.0145)
Economic Development:		
Economic Growth (81-01)	-1.1214	
	(0.6391)**	
Income (2001)		-0.0211
		(0.0246)
Constant	0.1624	0.1486
	(0.0335)***	(0.0362)***
N	70	70
R2	0.50	0.49

Source: CRC STORMA A3 Village Survey.
Notes:
Robust standard errors in parentheses. *** significant at 1%, ** significant at 5%, * significant at 10%.
Variable explanations see Section 4.
A positive coefficient means that this factor contributed to higher rates of deforestation.

est and promoting economic development; in contrast this is not the case for terracing , where a win-win situation appears to be feasible.

4 Descriptive and qualitative evidence

The econometric story told so far also is confirmed by descriptive and qualitative evidence along this causal change proposed here. We can also draw on results from a stratified random household survey (n=300) in 13 villages surrounding Lore Lindu National Park (Nuryatono et al. 2004), a village census in three villages ($n = 898$; Schippers and Faust 2009) and on our qualitative investigations (Burkard and Fremerey 2008, Weber 2006). Of particular importance is the link between immigration, land rights, ethnicity and technological change.

The migration fostered population growth and induced land pressure, which leads villages to adopt land rights and promotes technological modernization and economic development. (Kreisel et al. 2004). The favourable geographic conditions such as access to roads, regular rainfall, an agricultural-friendly landform, a moderate altitude, the relatively low population density and thereby availability of land attract migration to our research region for instance from the densely populated South Sulawesi (Weber 2006, Weber et al. 2007). Migration and ethnicity emerge with a great influence on land use decisions and on the accessibility of land in the Lore Lindu region. The results show a general land use change in the region, ranging from a strategy that places food first (e.g., rice) to one that places cash first (Schippers and Faust 2009). The special success of agroforestry systems in the research region is based on the intensification of cacao plantations (Sitorus 2002, Abbate 2007).

This change cannot be observed to the same extent in all parts of the study region. It varies from village to village, depending, among other things, on the share of households belonging to the Bugis migrants. Bugis as an ethnic group and as migrants have an enormous effect on the land use decisions of local ethnic groups in their respective villages. These migrants accelerate the issuing of land titles by acquiring land through purchases from villagers. With this process, the formalization of land titles gets under way as such purchases very often involve the issuing of formal land titles, either a certificate, a letter of transaction, or a letter from the village headman (Burkard and Fremerey 2008). These transactions enhance the issuing of land titles to other villagers as well. With these land titles, the propensity to invest in perennial crops and modern technologies rises considerably in the whole village (Weber et al. 2007, Koch et al. 2008). On the other hand, at the end of the process, a substantial share of the autochthonous population finds itself either landless or is forced to cut marginal forest land, even inside Lore Lindu National Park.

Thus, it is the immigrants who get land access primarily through purchase, but in contrast to the general argument of Carr (2004) that immigration directly leads to forest clearing, the Buginese migrants do not clear the forest themselves. Mainly ethnic groups, which call themselves local in the respective villages, clear the primary forest. The land acquisition by Buginese migrants is predominantly done by buying. In this context we can show that the majority of the Buginese migrant farmers are found in the relatively high local income groups (Weber et al. 2007). They have the financial background to purchase land. In contrast the local ethnic groups, which regard themselves as the owners of their village's territory, do not deem it necessary to buy land, but they do realize the opportunity of generating cash by selling parts of their land. The forest, which represents their customary land reserve, is cleared by the local groups to create a balance with the land sold outside the forest or to sell it in order to increase their cash income. This additional income is mainly used for buying status symbols (e.g., motorcycles, TV sets, furniture), purchases that are the results of the improved living standards displayed by the affluent migrants and the modern lifestyles portrayed on television programs (Weber et al. 2007, Burkard and Fremerey 2008).

In conclusion the capital-driven agroforestry intensification the Lore Lindu National Park region results in land titling and economic development, but it also increased social stratification between migrants and local ethnic groups as well as deforestation. On the other hand restricting land ownership to traditional forms of community land rights avoids the formation of a class of newly landless locals, but it comes at costs in terms of social discrimination and lost agricultural income (see Barkmann et al. in this book).

5 Concluding remarks

As we argued in the beginning of our paper there is considerable debate about the main drivers of technological change and economic development in a poor rural economy. The literature emphasizes among other things geography, population pressure and institutional change as important determinants, without however establishing clear transmission channels between these factors. Our hypothesis was that a demographic pressure creates pressure on land and associated conflicts over land use. This provides an incentive for villagers and village leaders to opt for land rights which in turn provide an incentive to invest in agricultural technology, which in turn promotes economic development. We tested this hypothesis empirically using longitudinal data on villages situated on the Indonesian island of Sulawesi.

We also find, however, that many of the factors that promote economic development also appear to promote deforestation. These trade-offs, as well as the identified win-win situations, require more careful scrutiny in future work where we hope to integrate another round of the village surveys with more recent data on deforestation in our analysis.

Acknowledgements
We would like to thank participants International Symposium "Tropical Rainforests and Agroforests under Global Change" in Bali as well as two anonymous referees for helpful comments and discussion. We also want to thank Stefan Erasmi and Jörg Priess for granting access to their data on deforestation. Funding from the German Research Foundation as part of CRC STORMA is gratefully acknowledged.

References

Abbate M (2007) The "Sweet Desire" Cacao Cultivation and its Knowledge Transfer in Central Sulawesi, Indonesia. STORMA Discussion Paper Series Sub-Program A, No. 17, 27 p

Acemoglu D, Johnson S, Robinson JA (2001) The Colonial Origins of Comparative Development: An Empirical Investigation. American Economic Review 91 (5): 1369-1401

Angelsen A (1999) Agricultural expansion and deforestation: Modeling the impact of population, market forces, and property rights. Journal of Development Economics 58: 195-218

Barkmann J, Burkhard G, Faust H, Fremerey M, Koch S, Lanini A (2009) Land tenure rights, village institutions, and rainforest conversion in Central Sulawesi (Indonesia). In: Tscharntke, T et al. (eds.) (2009): Tropical Rainforests and Agroforests under Global Change. Springer, Berlin

Besley T (1995) Property Rights and Investment Incentives: Theory and Evidence from Ghana. Journal of Political Economy 103 (5): 903-937

Binswanger H, Deininger K, Feder G (1995) Power, Distortions, Revolt and Reform in Agricultural Land Relations. In: Behrman J, Srinivasan T (eds.) Handbook of Development Economics, Volume III Elsevier North Holland, Amsterdam

Boserup E (1981) Population and technological change: a study of long term trends. University of Chicago Press, Chicago

Brassselle AS, Gaspart F, Platteau JP(2002) Land tenure security and investment incentives: puzzling evidence from Burkina Faso. Journal of Development Economics 67: 373-418

Broeck K, vd Newman C, Tarp F. (2007) Land Titles and Rice Production in Vietnam. Trinity Economic Papers No. 1207, Trinity College Dublin

Burkard G (2009) Locating Rural Communities and Natural Resources in Indonesian Law. In: Engel C et al. (eds.) Development - Organization - Interculturalism. Supplement No. 91 to the Journal of Agriculture and Rural Development in the Tropics and Subtropics, 25-42

Burkard G, Fremerey, M (2008, eds.) A Matter of Mutual Survival: Social Organization of Forest Management in Central Sulawesi, Indonesia. Münster

Byerlee D, Diao X, Jackson C (2005) Agriculture, Rural Development, and Pro-poor Growth. Country Experiences in the Post-Reform Era. Agriculture and Rural Development Discussion Paper 21, World Bank, Washington D.C.

Carr DL (2004) Proximate Population Factors and Deforestation in Tropical Agricultural Frontiers. Popul Environ 25 (6): 585-612

Datt G, Ravallion M (1996) How important to India's poor is the sectoral composition of economic growth? World Bank Economic Review 10 (1): 1-25

Datt G, Ravallion M (2002) Why has economic growth been more pro-poor in some states of India than others? Journal of Development Economics 68: 381-400

Deininger K (2003) Land Policies for Growth and Poverty Reduction. World Bank Policy Research Report. World Bank/Oxford University Press: Washington

Diamond J (1997) Guns, Germs and Steel: The Fates of Human Societies. W. W. Norton & Company, New York

Easterly W, R Levine (2003) Tropics, Germs, and Crops: How Endowments influence economic development. Journal of Monetary Economics 50 (1): 3-39

Erasmi S, Priess JA (2007) Satellite and survey data: a multiple source approach to study regional land-cover / land-use change in Indonesia. In: Dickmann, F. (ed.): Geovisualisation in Human Geography. Kartographische Schriften 13: 101-114

Faust H, Maertens M, Weber R, Nuryartono N, van Rheenen T, Birner R (2003), Does Migration lead to Destabilization of Forest Margins? STORMA Discussion Paper Series No. 11, STORMA, University of Göttingen

Faust H, Schippers B, Weber R (2009) Cultural Dimensions of rural poverty in Indonesia - a methodological approach focussing on education and origin. In: Engel C et al. (eds.) Development - Organization - Interculturalism. Supplement No. 91 to the Journal of Agriculture and Rural Development in the Tropics and Subtropics, p. 45-65

Gallup JL, Sachs JD, Mellinger AD (1998) Geography and Economic Development. NBER Working Paper 6849, NBER Cambridge, MA

Grimm M, Klasen S (2008) Geography versus Institutions at the Village Level. CESifo Working Paper No. 2259, CESifo, Munich

Grimm M, Klasen S, McKay A (2007) Determinants of Pro-Poor Growth. Palgrave-Macmillan, London

Hall RE, Jones CI (1999) Why Do Some Countries Produce So Much More Output Per Worker Than Others? Quarterly Journal of Economics 114 (1): 83-116

Hayami Y, Ruttan VW (1985) Agricultural Development. 2nd ed. Johns Hopkins University Press, Baltimore

Jacoby HG, Minten B (2007) Is Land Titling in Sub-Saharan Africa Cost-Effective? Evidence from Madagascar. World Bank Economic Review 21 (3): 461-485

Klasen S, Nestmann T (2006), Population, Population Density and Technological Change. Journal of Population Economics 19: 611-626

Klasen S (2002) Low schooling for girls, slower growth for all? World Bank Economic Review 16: 345-373

Koch S, Faust H, Barkmann J (2008) Differences in Power Structures Regarding Access to Natural Resources at the Village Level in Central Sulawesi (Indonesia). Austrian Journal of South-East Asian Studies 1 (2): 59-81

Kreisel W, Weber R, Faust H (2004) Historical impacts on use and management of natural resources in the rainforest margins of Central Sulawesi. In: Gerold G, Fremerey M, Guhardja E (eds) Land use, nature conservation and the stability of rainforest margins in Southeast Asia. Springer, Berlin Heidelberg New York, pp 39-65

Kremer M (1993) Population Growth and Technological Change: 1,000,000 B.C. to 1990. Quarterly Journal of Economics 108 (3): 681-716

Krugman P (1991) Geography and Trade. Leuven University Press, Leuven/Cambridge MA

Maertens M, Zeller M, Birner R (2006) Sustainable agricultural intensification in forest frontier areas. Agricultural Economics 34: 197-2006

Migot-Adholla S, Hazell P, Blarel B, Place F (1991) Indigenous Land Rights Systems in Sub-Saharan Africa: A Constraint on Productivity? World Bank Economic Review 5 (1): 155-175

Mundlak Y, Larson DF, Butzer R (2002) Determinants of Agricultural Growth in Indonesia, the Philippines, and Thailand. Mimeo, University of Chicago and World Bank

North DC (1990) Institutions, Institutional Change and Economic Performance. Cambridge University Press, New York

Nuryartono N, Schwarze S, Zeller M (2004) Smallholders access to credit and its impact on the adoption of agricultural technology: Evidence and lessons from agriculture practices in the Vicinity of the Lore Lindu National Park, Central Sulawesi, Indonesia. Mimeo, Institute of Rural Development, University of Göttingen

Nuryartzono N (2005) Impact of Smallholders' Access to Land and Credit Markets on Technology Adoption and Land Use Decisions. Göttingen: Cullivier Verlag

Place F, Hazell P (1993) Productivity Effects of Indigenous Land Tenure Systems in Sub-Saharan Africa. American Journal of Agricultural Economics 75 (1): 10-19

Ravallion M, Chen S (2007), China's (uneven) progress against poverty. Journal of Development Economics 82 (1): 1-42

Rodrik D, Subramanian A, Trebbi F (2004) Institutions Rule: The Primacy of Institutions over Geography and Integration in Economic Development. Journal of Economic Growth 9: 131-165

Rozelle S, Li G (1998) Village Leaders and Land-Rights Formation in China. American Economic Review 88: 433-438

Sachs JD (2003) Institutions don't role: The effects of Geography on Per Capita Income. NBER Working Paper 9490, NBER, Cambridge

Schippers B, Faust H (2009), Migrants as cash crop-"pioneers"? - Sociocultural change and land use in Central Sulawesi. STORMA Discussion Paper Series Sub-Program A, No. 29, 22 Sp

Schwarze S, Erasmi S, Priess JA, Zeller M (2008) Has the Lore Lindu National Park been working? An analysis of the effectiveness of the LLNP in protecting the forest. Paper presented at the International Symposium

"Tropical rainforests and agroforests under global change", Bali, Indonesia, October 5 - 9, 2008

Siagian Y, Neldysavrino Y (2007) Collective Action to Secure Land Management Rights for Poor Communities. Governance Brief, May, No. 35, Center for International Forestry Research, Bogor, Indonesia

Sitorus F (2002) "REVOLUSI COKELAT" - Social Formation, Agrarian Structure, and Forest Margins in Upland Sulawesi, Indonesia. STORMA Discussion Paper Series Sub-Program A, No. 9, 19 Sp

Stock JH, Yogo M (2004) Testing for Weak Instruments in Linear IV Regression. Mimeo, Department of Economics, Harvard University

Thurlow J, Wobst P (2007) The Role of Agriculture in Pro-Poor Growth: Lessons from Zambia. In: Grimm M, Klasen S, McKay A (eds.) Determinants of Pro-Poor Growth. Palgrave-Macmillan, London

Weber R (2006) Kulturlandschaftswandel in Zentralsulawesi: historisch-geographische Analyse einer indonesischen Bergregenwaldregion. Göttingen.(= Pazifik Forum Bd. 12)

Weber R, Faust H (2006), Kulturelle Aspekte der Landnutzung in Indonesien. Einflüsse von Migration, Ethnizität und Wissen auf bäuerliche Entscheidungsprozesse im Hochland Zentralsulawesis. Geographica Helvetica 61 (4): 237-245

Weber R, Faust H, Schippers B, Mamar S, Soetarto E, Kreisel W (2007) Migration and ethnicity as cultural impact factors on land use change in the rainforest margins of Central Sulawesi, Indonesia. In: Tscharntke T et al. (eds.) The stability of tropical rainforest margins, linking ecological, economic and social constraints of land use and conservation, Springer Verlag Berlin, pp. 417-436

World Bank (2007) World Development Report 2008. Agriculture for Development. World Bank, Washington D.C

Zeller M, Schwarze S, van Rheenen T (2002) Statistical Sampling Frame and Methods Used for the Selection of Villages and Households in the Scope of the Research Program on Stability of Rainforest Margins in Indonesia. STORMA Discussion Paper Series No. 1, STORMA, University of Göttingen

Part II

Climate change effects on tropical rainforests and agroforests

Functional biodiversity and climate change along an altitudinal gradient in a tropical mountain rainforest

Jörg Bendix[1*], Hermann Behling[2], Thorsten Peters[3], Michael Richter[3], and Erwin Beck[4]

[1] Laboratory for Climatology and Remote Sensing, Faculty of Geography University of Marburg, Deutschhausstr. 10, D 35032 Marburg, Germany
[2] Department of Palynology and Climate Dynamics, Albrecht-von-Haller Institute of Plant Sciences, University of Goettingen, Untere Karspuele 2, D 37073 Goettingen, Germany
[3] Institute of Geography, University of Erlangen, Kochstr.4/4, D 91054 Erlangen, Germany
[4] Dept. of Plant Physiology, University of Bayreuth, Universitaetsstr. 30, D 95440 Bayreuth, Germany

*corresponding author: J. Bendix, email: bendix@staff.uni-marburg.de

Summary

This article investigates possible consequences of climate change for a hotspot of tropical biodiversity, exemplified by a comparatively small area of a neotropical mountain rain forest in the eastern range of the South Ecuadorian Andes. In the introduction, several approaches for such predictions are evaluated with respect to their applicability to the eco-region. After a short presentation of the research area and its biodiversity, climate and vegetation development during the Holocene is described showing the range of possible fluctuations between Puna-like grassland and tropical mountain forest. Data of climate dynamics during the past 50 – 60 years, covering several ENSO (El Niño Southern Oscillation) events, suggest a significant increase in temperature but no dramatic changes in the precipitation regimes of the region. Due to the altitudinal span of the area, the rise in temperature will shift the ecothermic belts by several hundred meters uphill and thus increase the distribution ranges of ectothermic organisms. To assess the consequences of this shift on biodiversity two model approaches were applied, namely the species-area-approach and the energetic-equivalence rule, using the extremely diverse insect group of moths. Combining both approaches the consequences of a climate change

T. Tscharntke et al. (eds.), *Tropical Rainforests and Agroforests under Global Change*, Environmental Science and Engineering, 239–268, DOI 10.1007/978-3-642-00493-3_11,

can be estimated for the various scenarios of greenhouse gas emissions published by the International Panel on Climate Change (IPCC 2007). Applying the most realistic scenario A1B a reduction of moth species by 31% until the year 2100 can be predicted for the RBSF area. Due to their greater life-span, woody plants are much more resilient to climate change, especially so in a megadiverse forest with usually small population sizes. Therefore impacts of global warming on the local vegetation can only be assessed on the basis of qualitative data of the forest structure rather than of the floristic composition. The most conspicuous trait of the tropical mountain rain forest in the region is the low elevation of the upper tree-line which on average is by 1000 to 1500 m lower than in the other parts of the tropical Andes. Due to the particular orographic situation, the extreme environmental conditions (quasi-permanent easterly storms carrying a tremendous load of precipitation) do not allow growth of trees in the peak regions of the mountains. Only if the prevailing trade-wind system dampened and the mountain range received less precipitation, a change of the environmental conditions could be expected resulting in a situation like in the early and mid Holocene, when a forest covered the entire mountain range. However, up to present, symptoms for such a change are lacking. A final outlook comments on the priority ranking of climate change vs. direct anthropogenic impacts with respect to conservation measures. For the investigated tropical forest ecosystem a reduction of human impact is more urgent than ever.

Keywords: Andes, South Ecuador, Biodiversity, climate change, extreme weather events, timberline, species range shift

1 Introduction

Organisms react in different ways to climate change, (i) by thriving in a modifying climate through genetic adaptation, phenotypic plasticity, or ecological buffering, (ii) by migrating to areas with more proper climates, or (iii) by going extinct (Theurillat and Guisan 2001). Thus, the impact of climate change on biodiversity can result in species enrichment as postulated for many biomes of cooler climates (Thuiller 2004) as well as in species losses as predicted for most dry biomes and for parts of the wet tropics as well (Colwell et al. 2008). Williams et al. (2007) argued that climate conditions which favoured some of the biodiversity hotspots during the past century may entirely disappear during the 21th century. In that respect the biodiversity hotspot area of the tropical Andes will be most endangered (their Figs. 2 E and F). Climate changes can result in range-shifts and/or extinction of species, but may also be without substantial effects on species distribution patterns. For the first two assumptions the formation of novel organismic communities is obvious. Several approaches could be worth considering to predict effects of climate change on the biological diversity of an area:

- Modelling the species or group-specific responses to altered climatic conditions is one possibility to forecast future climatic impacts on biodiversity, though such predictions should be considered as first approximation, rather than as reliable prognoses (e.g. Pearson and Dawson 2003).
- Niche-theory models based on species – habitat relations represent a more sophisticated option for biodiversity perspectives (e.g. Thuiller et al. 2005). However, such Species Distribution Models (SDM) are based on the assumption that the observed distribution of a species is in equilibrium with its current habitat (Guisan and Thuiller 2005).
- Coupling the biosphere and pedosphere with Regional Atmospheric Modelling Systems Models (RAMS) may be of higher accuracy for developing biodiversity scenarios. Problems occur, if ecological systems comprise poorly quantifiable processes rendering the applicability of RAMS questionable (Botkin et al. 2007).
- Finally, species–area curve models that consider all species or at least large aggregates of species are also used in biodiversity research (Thomas et al. 2004). Nevertheless, various limitations of this approach for predicting climate change effects must be critically assessed (Lewis 2006).

Most of the mentioned models were developed in Europe and northern America and have been implemented for predicting biodiversity changes in cooler and temperate zones. Nearly all predictions for these comparatively simple-structured habitats anticipate partially dramatic species losses already during the current century. However, traits and processes that could preserve species diversity or even result in diversity enrichment remain largely unconsidered. Examples for that are biotic interactions, speed-up effects by fragmentation or (unpredictable) inputs by invasive species. Apart from the above mentioned particular limitations, the quoted approaches seem less useful for tropical biomes and especially not for ecosystems harboring biodiversity hotspots like a tropical mountain rain forest.

In contrast, the approach used in this paper for a tropical mountain rainforest attempts to integrate existing data and well investigated ecological processes into present theories and to combine them with own findings. Using established models, the species-area approach considers the elevational area attenuation in mountains as one of the major problems of the altitudinal shift of species distribution ranges. Another approach, the energetic-equivalence rule regards the possible expansion or attenuation, respectively, of a well investigated ectothermic insect group upon future warming. Possible changes in plant diversity patterns upon climate change will be discussed on the basis of ecotonal structures, biotic interrelations, triggers for species area shifts and extinction, disturbance regimes, and invasion of pioneer species. Since Botkin et al. (2007) state that more frequent use of fossil records would improve forecasting methods, pollen and charcoal records have also been included in this work.

2 The Study area and its biodiversity

The study site in southern Ecuador covers an area of around 100 km^2 in the valley of the Rio San Francisco, a tributary of the Rio Zamora. The site which partly belongs to the Podocarpus Nacional Parque has been studied since 1997 and a first comprehensive description of it has been published by Beck et al. (2008a). It encompasses a core area of 11 km^2 (termed Reserva Biológica San Francisco, RBSF, Fig. 1) and some smaller satellite areas. As part of the eastern range of the South Ecuadorian Andes, the entire region belongs to the tropical humid ecozone. From the North it reaches into an important ecological transition as well as border zone, the so-called Andean Depression. Separating the central and northern Andes, this orographic depression stretches 500 km N-S from the Girón-Paute drainage basin around Cuenca in southern Ecuador to the Rio Chicama-Rio Huallaga intersection around Cajamarca in northern Peru (Weigend 2002). Here, the Andes barely reach 4000 m and the tree line is as low as 3000 to 3400 m asl. Despite or just because of its low elevation, the Andean depression is considered an important biogeographical barrier. Especially its eastern escarpment harbours an important and fascinating centre of endemism (Young and Reynel 1997).

Global censuses of vascular plant biodiversity (Barthlott et al. 2007) identified five centres with diversity maxima of over 5000 vascular plant species per 10,000 km^2. In addition to the Costa Rica-Chocó area, south-eastern Brazil, northern Borneo and the New Guinea mountain range, the tropical Andes-Amazonia transition zone in Ecuador belongs to these five top megadiverse plant diversity hotspots. Within this large biogeographical zone, mountain rainforests because of their extraordinary biodiversity deserve particular attention (Beck and Richter 2008). Inventorying is still going on in the core and the total study area, but already until 2008 (Liede-Schumann and Breckle 2008, Brehm et al. 2008) 1208 seed plant species, 257 ferns and fern allies, 515 Bryophyta, 323 lichens, 83 Glomeromycota, and 96 Basidiomycota had been identified. Animal species inventories resulted in 21 bats, 379 birds, 243 butterflies (Papilionididae), 2396 moths (which is a world record), and representatives of several other arthropod groups.

3 Landscape history: Past vegetation, fire and climate dynamics

Knowledge of the landscape history is a prerequisite for the understanding of the emergence of a biodiversity hotspot. Comprehending the dynamics of the landscape is also necessary for appropriate management and conservation measures. Paleoecological studies in the Podocarpus Nacional Park (PNP) region in the southern Ecuadorian Andes provide interesting insights into the dynamics of climate, vegetation, plant diversity, and the role of fire during the late Quaternary.

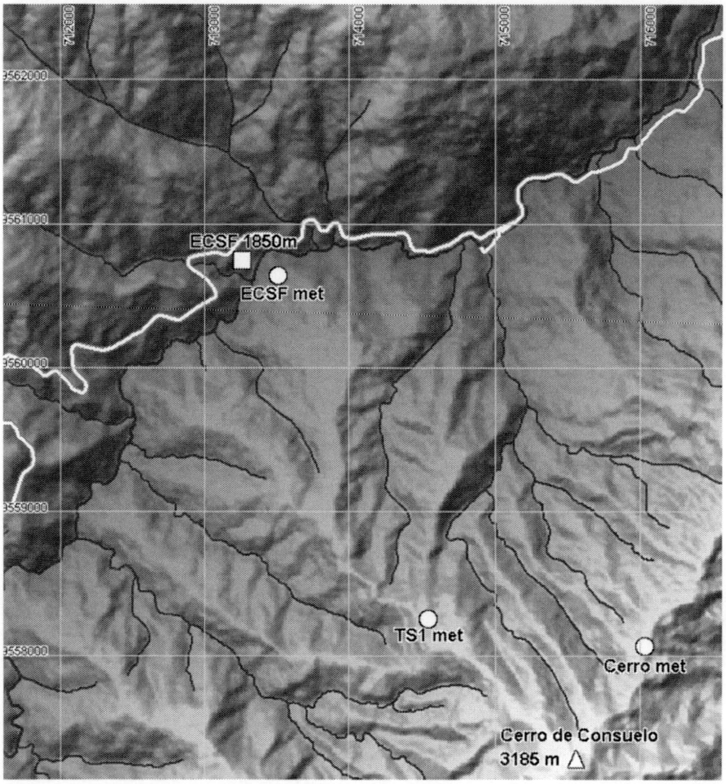

Fig. 1. Digital hill-shading image of the entire core study area that was used for the climate simulations. The connection road between Loja and Zamora is displayed in white, the river system in black, and "met" shows the sites of the meteorological stations. ECSF marks the location of the research station (Estación Científica San Francisco, lat. 3°58'18" S, long. 79°4'45" W, alt. 1,860 m a.s.l.). Cerro del Consuelo is the highest peak of the area. The river separating the north- and south-facing slopes is the Rio San Francisco. The research area corresponds to site # 4 in Fig. 10.

Studies have been carried out on a sediment core of a small bog at El Tiro pass (S 03°50'25.9", W 79°08'43.2", 2810 m elevation) in the eastern Cordillera, next to the road from Loja to Zamora near the ECSF research area (Niemann and Behling 2008).

The study site is today covered by a shrubby subpáramo vegetation. The 17,000 yr BP (uncalibrated radiocarbon years before present) pollen record (Figs. 2 and 3) documents that grass páramo covered the region during the last glacial maximum (LGM) and Late-glacial periods. According to that record, only small populations of wind-pollinated tropical mountain forest trees occurred in the region, probably further down in the valley of the Rio San

Francisco. The tree line must have been markedly lower. The high frequency of *Plantago rigida* suggests wet climatic conditions in the Late-glacial period. From about 8000 to 3000 yr BP, i.e. in the early and mid Holocene, the former grass páramo was almost completely replaced by upper mountain rainforest, indicating drier and warmer climatic conditions than today. Woody genera, like *Hedyosmum, Symplocos*, and later *Myrsine*, *Ilex* and members of the Podocarpaceae were much more common than at present. The warmer conditions are also reflected by the stronger decomposition of the peat deposits compared to the late Holocene. Since the last 3000 years the area above 2780m a.s.l. is mainly covered by a shrubby subpáramo, dominated by species of the Melastomataceae.

The charcoal record shows that fires were rare during last glacial period and the early Holocene, until 8000 yr BP. Later on, fires became common on the slopes of El Tiro and further down. Three periods with increasing fire intensity could be identified: 8000 – 3000 yr BP, 2700 – 1800 yr BP and 1000 – 600 yr BP. During the last ca. 600 years fire frequency decreased again. The increased fire frequency during the wetter late Holocene suggests that fires were mostly of anthropogenic rather than of natural origin, e.g. lightning. It is assumed that the increasing use of slash-and-burn by humans settling in the drier valleys (e.g. in the Loja area) caused the spreading of fire into the mountains especially during the drier phases of the year. The decrease of common fern taxa including tree ferns can be interpreted as resulting from the effect of fire, in particular in the drier crest regions like El Tiro. The results underpin the notion that mountain ecosystems are quite sensitive to natural (climate) and anthropogenic impacts (fires).

In the El Tiro region as well as in the Cerro Toledo region (Brunschön and Behling, submitted) in the northern part of the Podocarpus Nacional Parque, mountain rainforest taxa were very rare or absent during the LGM. Sizable populations occurred only at lower elevations. Climatic conditions were cold and relatively wet and valleys above 3000 m elevation were probably glaciated. Present páramo islands were then connected to larger areas. Mountain rainforests accrued during the Late-glacial and in particular during the Holocene period, were markedly reduced and fragmented during cooler glacial periods. Connection and disconnection of various plant populations may have been an important component triggering speciation during the Quaternary (e.g. Hughes and Eastwood, 2006).

During the Late-glacial and, especially the early Holocene mountain rainforests shifted up-mountain and developed, replacing the former grassland vegetation. The composition of the early to mid Holocene vegetation was different from that of the late Holocene. During early-mid Holocene, upper mountain rainforest was predominant at the coring site and the ample decomposition of organic material during that time suggests relatively warm and also somewhat drier conditions.

Tree populations have expanded their areas to higher elevations and replaced the existing páramo vegetation. However, in the Andean Depression, uplifting of habitat boundaries was limited by the relatively low elevation of the mountains. Nevertheless the areas covered now by subpáramo must have supported growth of trees at the early and middle Holocene. Today, the crest areas are subjected to nearly permanent strong easterly winds which allow growth of small trees only in hollows and small depressions. It must be concluded that during the early mid Holocene the wind regime and speed was different from the recent conditions.

Present-day vegetation became established after ca. 3000 yr BP, when the climate again became somewhat cooler and wetter. Major vegetation changes after 3000 yr BP have also been recorded in the Amazon lowland, where Amazonian rainforest expanded northwards and southwards of the equator (Behling and Hooghiemstra 2000, 2001, Mayle et al. 2000). This supports the hypothesis that changes in the climate regime of the Amazon lowland influenced the mountain rainforest of the eastern Andes.

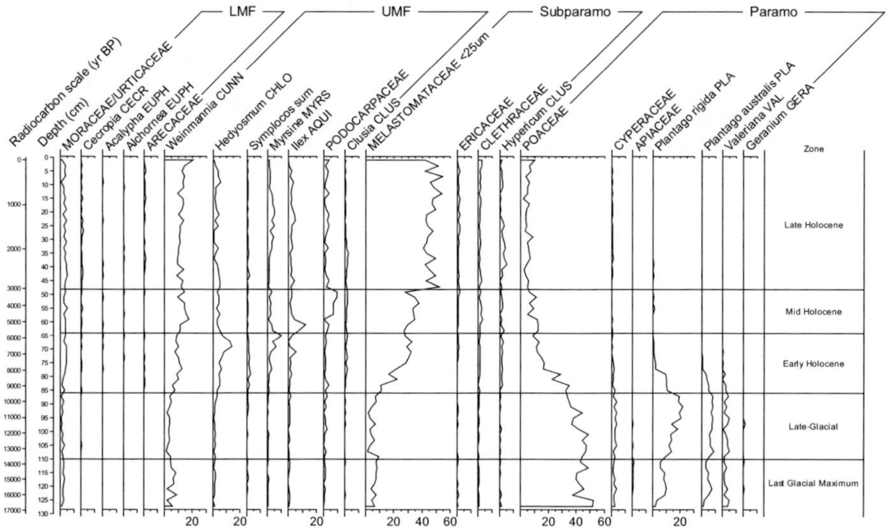

Fig. 2. Percentage pollen diagram of El Tiro in the southeastern Andes of Ecuador including selected taxa of lower mountain forest (LMF), upper mountain forest (UMF) subpáramo and páramo.

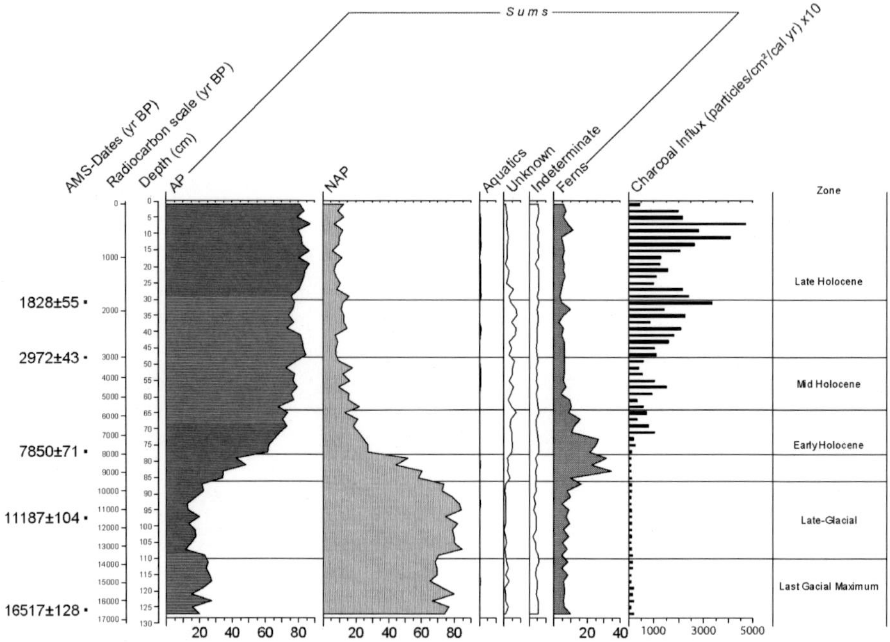

Fig. 3. Summary pollen diagram of El Tiro at 2810 m a.s.l. also showing the sum of arboreal pollen (AP) and non-arboreal pollen (NP) of El Tiro.

4 Recent climate changes and area impacts

4.1 Recent climate dynamics

Figure 4 reveals the recent climate dynamics (1948-2005) for the grid cell of southern Ecuador using the NCEP/NCAR[1] reanalysis data, which encompass the RBSF study area (for data refer to Kistler et al. 2001). The plot shows a clear increase of the annual average air temperature over the last 57 years which runs parallel to the globally observed increase in atmospheric CO_2-concentration. The warming trend since 1948 yields 1.3°C, which means +0.22°C per decade. Quasi-periodic oscillations in air temperature are related to El Niño (EN, positive temperature anomalies) and La Niña (LN, negative temperature anomalies) events. No clear trend in rainfall exists in the data set. Also EN and LN years are not characterised by marked rainfall anomalies.

Generally, meteorological data from stations in the wider region confirm the reanalysis results for southern Ecuador (Fig. 5). The time series of air temperature at the station Loja (1950-2006), ~30 km west of the ECSF meteoro-

[1] *N*ational *C*enters for *E*nvironmental *P*rediction / *N*ational *C*enter of *A*tmospheric *R*esearch

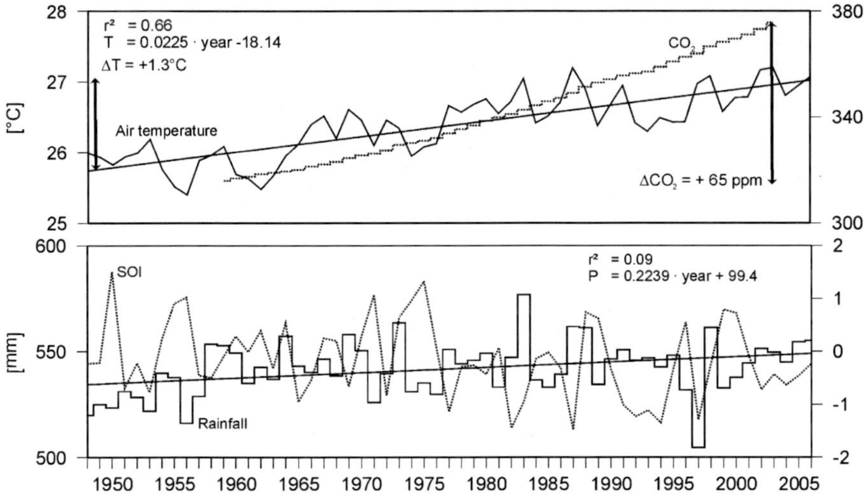

Fig. 4. NCEP/NCAR reanalysis data, average air temperature and rainfall totals for the grid cell of southern Ecuador (lat. 3-6°S; long. 76-79°W); CO_2-concentrations at NOAA-Mauna Loa observatory, Hawaii; Southern Oscillation Index SOI (source: NOAA); negative SOI represent warm phase conditions (El Niño like), positive SOI indicate cold phase conditions (La Niña like).

logical station on the western side of the main Cordillera, reveals a warming trend of ~0.6°C over the last 43 years (0.14°C per decade), somewhat lower than in the NCEP/NCAR data set. EN and LN show the same tendencies as in the NCEP/NCAR data set.

Also in the wider region no significant trend in rainfall can be recognized. However, a slight positive trend for the region west of the main cordillera (Loja) opposes a minimal decrease on the eastern escarpment at Zamora for the period 1963-1993. These tendencies on a first glance reflect the strong difference in climate dynamics along the short horizontal distance east and west of the main cordillera (refer to Bendix et al. 2008b). Considering only one of these sites, e.g. the closest rainfall station (San Ramon) ~1 km down-valley the ECSF with a longer record, reveals no trend in rainfall between 1980 and 2000.

Only 10 years of meteorological data are available from the core study site recorded by the meteorological stations shown in Fig. 1 (see Bendix et al. 2008a). Indeed, the years from 1999-2006 show generally the same warming trend in the upper part of the study area at the Cerro met station as in Loja ($r^2 = 0.56$ with Loja temperature data). This means that the warming trend is most probably the clearest local signal in climate development of global climate change.

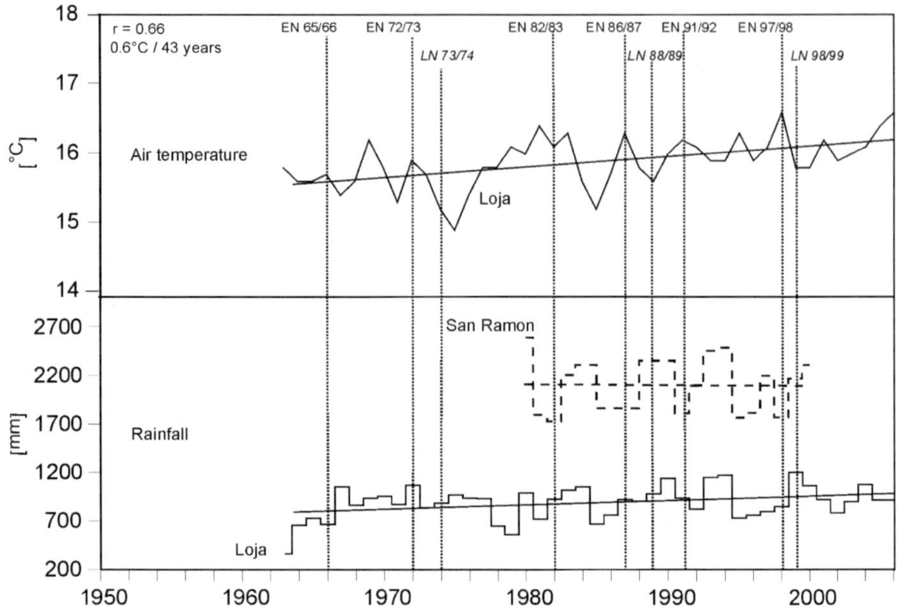

Fig. 5. Climate trends from observational data in the wider study area of southern Ecuador (data source: INAMHI (INAMHI: Instituto Nacional de Meteorología y Hidrología del Ecuador))

4.2 Areal impacts of a projected climate change

The most likely forecast of the climate towards the end of the current century can be derived from the A1B scenario of greenhouse gas emission development until the year 2100 (see appendix). In spite of an expanding use of renewable sources of energy and a moderate economic growth, a fourfold increase of greenhouse gas emissions by 2100 relative to the pre-industrial level is predicted by this model. For the grid cell of southern Ecuador, only a slight increase of rainfall (+8%) and cloud fraction (+4%) but a marked increase of air temperature by +3°C relative to the average of 1980-1999 is expected (Meehl et al. 2007). The decadal increase of +0.3°C is higher than the observed trends of the recent past (+0.22°C NCEP/NCAR; +0.14°C at Loja, see above). Nevertheless, data as well as model calculations point to a clear warming trend in the study area. Ecological importance of such a thermal shift is obvious. By assuming a stationary average lapse rate of –0.61°C per 100 m of elevation increase (Bendix et al. 2008a) until 2100, the increase of temperature must result in an altitudinal shift of ecothermal belts in the study area (Fig. 6). As a consequence of a temperature increase by 3°C in the course of the 21st century the area with an average air temperature equal or higher than that of the ECSF meteorological station today (1860 m a.s.l.; 15.5°C) would increase by a factor of 4.4 and the current thermal conditions of the ECSF met station would shift from 1860 to 2300 m a.s.l..

This must effect changes in ecosystem functional components and processes, e.g. the soil respiration rate and in turn the carbon balance and deposition of organic material. Iost et al. (2008) showed that soil carbon efflux is highly correlated to soil temperature and moisture, with flux rates of around 9.32 Mg C ha^{-1} a^{-1} at 1890 m asl (the elevation of the ECSF met station) and 3.83 Mg C ha^{-1} a^{-1} at around 3000 m. By assuming that soil temperature will linearly rise by ~3°C (concomitantly with a decrease of moisture), the thermal area increase by 60% (Fig. 1) with high efflux rates could boost the carbon efflux by additional 12.6 t C a^{-1}.

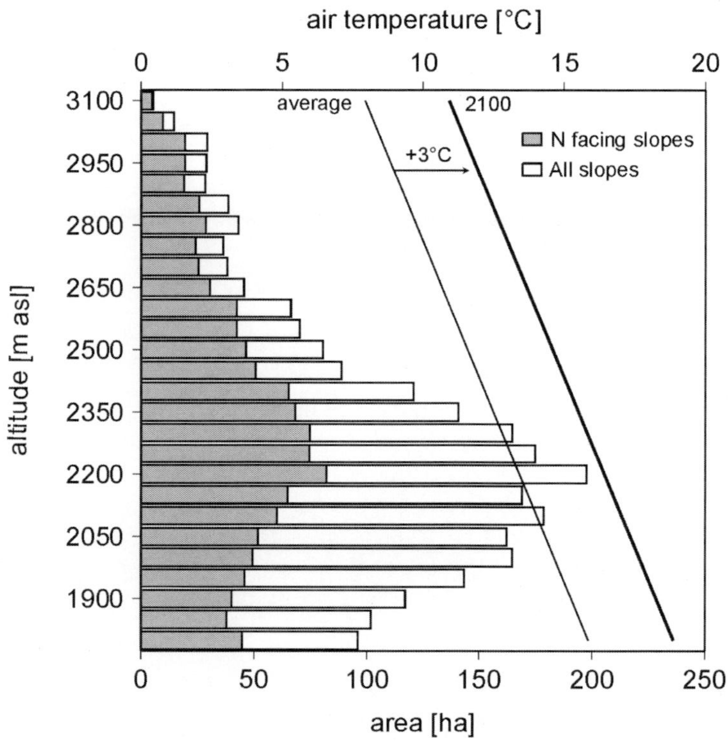

Fig. 6. Areas of altitudinal belts of the study area (Fig. 1), average air temperature (1998 - 2006) and linear forward projection based on IPCC A1B scenario for 2100. Total bars represent the terrain height distribution respective of the entire area of figure 1, grey-shaded bars the area distribution of the north-facing slopes (south of the Rio San Francisco), which up to ~2900 m are covered by natural forest. Note that the altitudinal distribution is displayed only for the landscape subset presented in figure 1, which implies that the center point of altitudes is not in the lower but in lower-mid elevations (2200 m a.s.l.)

5 Climate change and biodiversity in the RBSF

5.1 A case-study: Impact of climate warming on an ectothermic faunal group

Not much is known about the impact of climate warming on the tropical lowland rain forests. In mountains, an uplift of the altitudinal ranges of species must be expected and has been shown for European high mountain regions (Walther et al. 2005). Due to this uplift a general increase in plant species richness has been observed in temperate mountains (Klanderud and Birks 2003) but at the same time high altitudinal species which, because of the limited orographic height, can not escape to higher regions will go extinct (Grabherr et al. 2001, Theurillat and Guisan 2001). However, the reactions of plant populations to a temperature increase are slow compared to e.g. ectothermic insect groups. Fast changes in altitudinal ranges, species numbers and biodiversity can be expected for such type of organisms like butterflies and moths, if they find appropriate host plants in the new altitudinal range. In this regard, Chen et al. (2009) recently provided evidence from Mount Kinabalu in Borneo that 102 montane tropical moth species (Geometridae) have already extended their habitats by about 67 altitudinal meters over the past 42 years.

The species-area approach: Colwell et al. (2008) could show for high neotropical mountains, that a temperature-triggered upwards shift of species ranges is usually coupled with a reduction of the available space. One possible consequence could be that species which are more competitive (usually generalists) might cause the extinction of standing, less competitive species (most likely specialists) in the respective altitudinal belt. Of course, this simplistic approach should be refined by including interaction networks but especially in megadiverse regions like the study area, such data are very difficult to compile.

Nonetheless, Thomas et al. (2004) applied a simple power law relation to estimate the proportion (E) of species in a region going extinct due to an altitudinal shift and range reduction as a result of climate change (species-area approach):

$$E = 1 - \left(\frac{\sum A_{new}}{\sum A_{original}} \right)^2 \qquad (1)$$

where A_{original} is the area initially occupied by a species and A_{new} is the future area projected for the same species.

To demonstrate a possible effect of climate change in the study area on ectothermic organisms, we exemplarily have applied the equation using published data of moths ensembles (Tab. 11.3.1 in Fiedler et al. 2008), bearing in mind that moth species counts are almost necessarily incomplete (Brehm et al. 2005). The calculation for the topographic situation of the north-facing slopes (Fig. 1) assumes that all species ranges will finally shift 500 m uphill to match their optimal thermal range upon the increase of air temperature by 3°C. Figure 7 reveals that species numbers may rise if the available area increases simultaneously with altitude, as indicated by the simulation for the altitudinal range areas up to 2200 m a.s.l. (Fig. 6). This situation implies that additional species are still present at lower altitudes, and that the natural forest as the basic resource remains intact in the target altitude. However, it should be kept in mind that the presented distribution of altitudinal range areas applies only to the study site and not to the entire eastern Cordillera. For elevations above 2200 m the species-area approach predicts losses of species as the available area decreases with altitude. According to the area distribution the number of moth species might decrease by ~6% in the altitudinal range between 2000 and 2500 m, and by ~68% between 2200 and 2700 m. If the suitable temperature range would shift from 2600 m to 3100 m, extinction of up to 92% of the species are predicted by the model. It should be stressed that the straightforward species-area model assumes that an increase of resources due to a warmer climate would necessarily end in an increase of species numbers, rather than of individual numbers of some of the respective species. This assumption is a serious constraint of the theory.

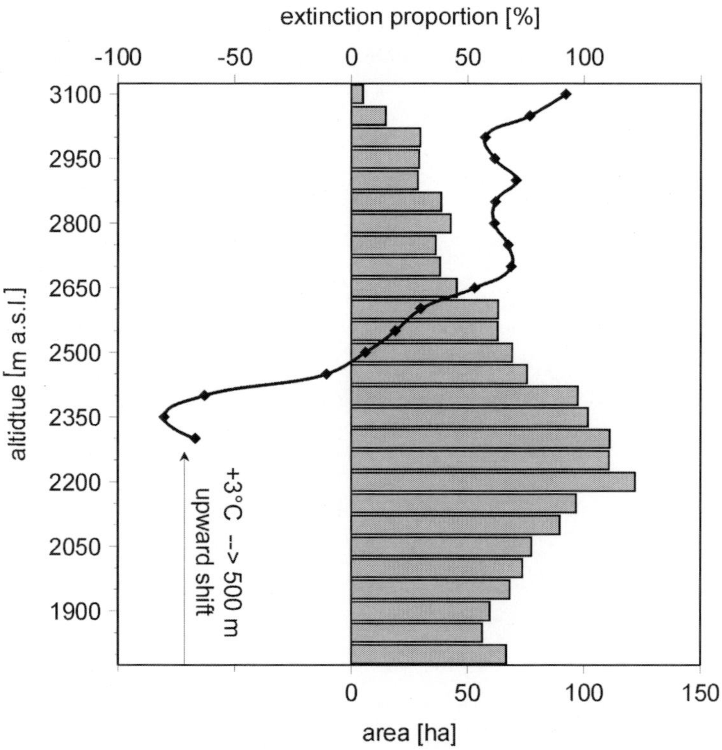

Fig. 7. Actual area of altitudinal belts for the north facing slopes of the study area (Fig. 1) and potential proportion of species going extinct based on the IPCC A1B scenario for 2100.

A comparison with current moth counts (Arctiidae, Geometridae, Pyraloidae combined) at different altitudes reveals that calculated species losses might be overestimated by the species-area approach. Fiedler et al. (2008) could show by light trap captures that the reduction of species numbers between 2180 and 2671 m a.s.l. in the ECSF forest ranges between 22% and 38% (depending on the selected location), in contrast to calculated -68% between 2200 and 2700 m a.s.l. by applying the area approach (Fig. 7). One reason of this overestimation could be that the species-area relation of individual moth groups differs (as e.g. the Geometridae, Brehm et al. 2003) resulting in changes of the quantitative composition patterns rather than in ultimate extinction.

Application of the energetic-equivalence rule: Regarding the prospected future temperature increase, the energetic-equivalence rule (Allen et al. 2002) in contrast to the species-area approach anticipates a general increase of species diversity, particularly along altitudinal gradients. The universal energetic relationship for ectothermic organisms is:

$$ln(S) = \left(\frac{-E}{1000 \cdot k}\right) \cdot \left(\frac{1000}{T}\right) + C_l \tag{2}$$

where S is the number of species, E the activation energy of metabolism (= 0.78 V), k the Boltzmann constant (= 8.62 10^{-5} V K^{-1}), T the environmental temperature (K), and C_l is

$$C_l = ln\left[\left(\frac{B_0}{B_T}\right) \cdot \left(\frac{J}{A}\right)\right] \tag{3}$$

where A is the range area of the community, J the number of individuals, B_T the total energy flux of a population (~80 W km^{-2}) and B_0 is derived from the body size distribution.

$$B_0 = b_0 \cdot M^{\frac{3}{4}} \tag{4}$$

with $b_0 \sim 2.65 \cdot 10^{10}$ W $g^{-3/4}$ and

$$M^{\frac{3}{4}} = \frac{B_T}{xn \cdot b_0 \cdot x} \tag{5}$$

with xn is the population density (km^{-1}) and x is the Boltzman factor

$$x = e^{\frac{-E}{k \cdot T}} \tag{6}$$

To test the possible warming-induced effect on ectothermic individuals as predicted by the energetic-equivalence rule in the study area (Fig. 1) we exemplarily used published data on geometrid moths (Brehm et al. 2005, refer to table 1 of this publication). It should be kept in mind that the samples are error-prone snapshot figures, most likely not representing the "real world". Nevertheless, moths are by far the best investigated ectothermic organismic group in the study area and thus, to date the only meaningful group to develop different climate change scenarios. The authors of the above mentioned study found 32,845 individuals (J in equation 3) and 1,075 species (S in equation 2) in the altitudinal range between 1800 and 2677 m a.s.l.. Restrictively it must be mentioned that parameter A in equation 3 is hitherto unknown due to a lack of sufficient observations in the surrounding area above and below the altitudinal range of 1800-2677 m a.s.l.. Simplistically, A is assumed to correspond to the area of the respective altitudinal range as obtained from the digital elevation model (DEM) for the north-facing slopes (Fig. 1, $A = 14.86\ km^2$). We further suppose that individual and species numbers are representative for this altitudinal segment, as we know that species numbers of Geometrids are fairly constant over this altitudinal range (Brehm et al. 2003, Fiedler et al. 2008). For calculations of the population density we suggest that the light traps used by the authors are representative for an area of $\sim$50 by 50 m^2 (Brehm et al. 2005). Further, we apply a bulk approach where we consider the altitudinal range as a homogenous entity without any differentiation into altitudinal levels, which is characterised by the actual average annual air temperature of 13.1°C (based on meteorological data between 1998-2006). Finally the average air temperature shall increase stepwise by 1°C in the range of the expected future warming which is the parameter T in equations (2) and (6).

The prospected extent of warming depends on the chosen emission scenario, where a worst case scenario (SRES-A1Fl, see appendix) points to a maximum possible temperature increment of +6.4°C by the year 2100 while the most optimistic scenario family (B1, see appendix) predicts a change of +1.1°C (Solomon et al. 2007). Assuming a linear correlation between the average temperature of the investigated slope section and global warming rates according to the IPCC scenarios, Fig. 8 reveals that almost a doubling of species numbers could be expected for the altitudinal range of 1800-2677 m a.s.l. from the energetic point of view under the worst case warming rate.

However, captures of Geometrid moths in lower altitudes under currently warmer temperatures do not reveal significant higher species numbers in contrast to the moth group of Pyraloideae (Fiedler et al. 2008). These differences might be due to the snapshot character of the sampling, to effects of site selection and of the inhomogeneity of the area which might blur a potential effect of a warmer climate on species numbers. Furthermore, a species range uplift would need additional species from warmer altitudinal belts below. In this respect, immigration of novel species could be constrained by a limited local/regional species pool, preventing an increase of species even if warming would promote such a process.

Whereas the energetic-equivalence rule would generally allow for more ectothermic species at any altitudinal level, the species-area approach would at the same time diminish the number of species if a taxonomic group would be shifted upward to regions with decreasing area size.

It is obvious that communities of ectotherms have a specific altitudinal range to which they are thermally adapted. In Fig. 9 the results of the differ-

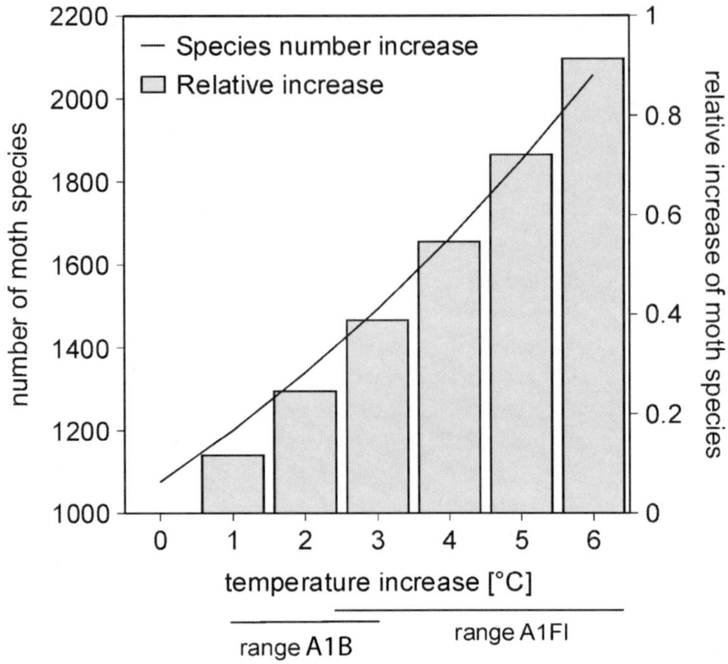

Fig. 8. Thermally induced increase of species numbers by 2100 for two different IPCC scenarios. (A1B represents a moderate and A1F1 the worst case scenario, see appendix). Relative increase is to the actual species number of 1075.

ent approaches are shown for the geometrid moths, presuming a parallel and synchronous uplift of the lower and the upper boundary of an ectothermic faunal group upon warming without range extension. Note that the increase of species numbers predicted by the energy-equivalence rule upon warming is the same as given by the line in Fig. 8. The upper and lower boundary of the moth community after warming is calculated as follows:

$$\Delta hb = \frac{\Delta T}{|\Gamma|} \tag{7}$$

where Δhb is the altitudinal shift of the upper and lower boundary [m] at the temperature increase ΔT [°C] and the actual lapse rate Γ [°C m^{-1}].

The available area after warming is then derived by summing up the pixel area between the height boundaries in the digital elevation model (Fig. 1) for the north-facing slopes.

Regarding the species-area approach, species richness must decrease due to a continuous reduction of available area with increasing altitude, particularly at warming rates >3°C. By combining both approaches

$$S_{com} = S_{EER} - (1975 - S_{SAA}) \tag{8}$$

(where S_{com} is the species number of the combined approaches, S_{EER} of the energy-equivalence rule and S_{SAA} of the species-area approach) a clear reduction of species numbers (from 1,075 to 739 at +3.6°C = -31%) results for the most realistic range of warming (SRES A1B scenario with + 1.7 – 4.4°C by the year 2100). Under more unrealistic warming extremes, temperature-driven species number increase would somewhat overtop the area-driven species losses and consequently attenuate that losses.

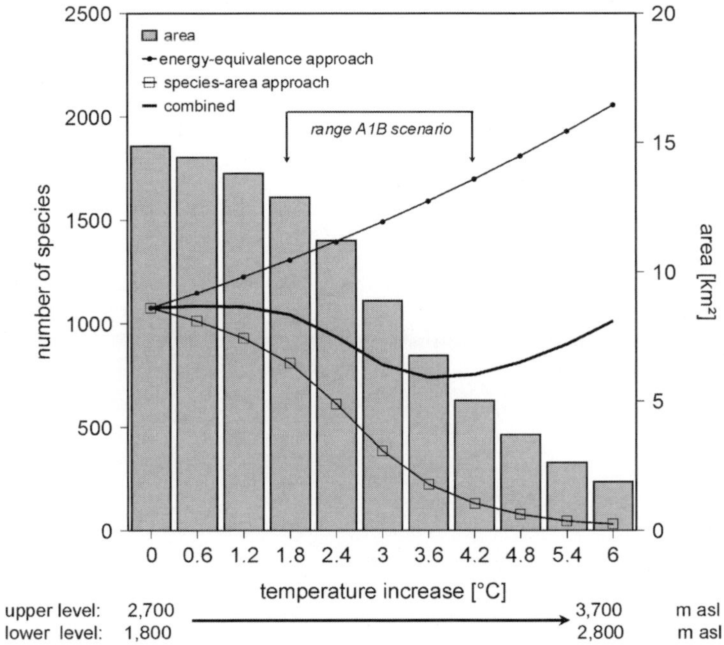

Fig. 9. Possible species losses or gains of geometrid moths on the North-facing slope of the RBSF forest (Fig. 1) under global warming conditions according to the energetic-equivalence rule, the species-area approach, and both approaches combined (equation 8). Note that the highest point in Fig. 1 is ~3200 m a.s.l. so that the available area shrinks dramatically upon warming rates >2.4 °C.

The results of the case study highlight that the straightforward approaches commonly used to estimate global warming effects on species richness must necessarily imply a high degree of uncertainty irrespective of the ecosystem. First of all, the consideration of only one organismic group is unrealistic because altitudinal shifts are always embedded in complex interaction networks and thus, exact knowledge on the reactions of these networks due to global warming is required to properly estimate species extinction or altitudinal shifts of distribution patterns (e.g. Koh et al. 2004, Bascompte et al. 2006). It is obvious that this is a challenging task, particularly for the megadiverse mountain forest of southern Ecuador. Furthermore, human impact along altitudinal gradients can significantly blur the natural elevational gradient of species diversity (Nogués-Bravo et al. 2008) and thus, most likely mask global warming effects. Last but not least it must be stressed that model-required data even in the extensively investigated study area are still incomplete. Thus, intensive future research is necessary to refine the straightforward approaches tested in this section, and to provide sound knowledge for their application.

5.2 Climate change: Possible impacts on vegetation and plant diversity

Both the energetic-equivalence rule and the species-area approach are based on linear changes with altitude. However, in reality and especially on a local scale non-linear ecological processes prevail, as evidenced by the frequent and small-scale changes of the vegetation. One of the non-linearities is the enormous plant species richness itself[2], each species being represented by relatively few individuals. The highly complex taxonomic multiplicity leads to a near-chaotic distribution of tree assemblages, where discrete vegetation zones are lacking, giving rise to a huge continuous ecotone (Richter 2008). Contrasting to this continuous taxonomic turnover, extratropical plant assemblages form a comparatively clear system of characteristic, often monotonous altitudinal vegetation belts. While in the extratropical type of vegetation zonation, climate changes may trigger synchronous responses of more or less all components of a vegetation belt, in tropical mountain forests responses of individual species prevail, maintaining the heterogeneity, however with a changed species composition.

Asynchronous migration trends of plant species upon climate change must be assumed considering the manifold biotic interactions, e.g. between plants and pollinators (e.g. Dziedzioch et al. 2003) or plants and mycorrhiza fungi (Kottke et al. 2008). Clearly delimited local distribution patterns of some endemic taxonomically related species swarms indicate the importance of specific co-evolutionary traits. For example, a high degree of radiation is typical of many miniature orchid genera, of which Pleurothallidinae are prominent members in the RBSF forest. A possible factor for their frequently highly fragmented occurrence might be found in the presence or absence of appropriate mycorrhizal counterparts, as stated by Jost (2004) for similar features in the Rio Pastaza area further north. There is evidence that many of such highly specialised associations are highly fragile and easily dissociate when disposed to migration by substantial climate change. Since rates of range-shift are species-specific and because the members of interaction systems may react differently, niche-based models would be required to predict the effects of climate change on assemblages (Thuiller et al. 2005). Apart from the caveat discussed in the "Introduction" in tropical mountain rainforests of outstanding biological diversity such task appears illusive.

[2] Probably also pertaining to insects

Phenological shifts in plant and animal communities upon climate change are likely, too, as detailed in the chapter on landscape history. On a short-term time-scale flowering and fruiting of 12 tree species showed in spite of the area´s perhumid climate a high extent of inter- and intraspecific synchronisation following the quasi-periodic oscillations of precipitation and cloudiness (Bendix et al. 2006). In the drier inner regions of Loja and Vilcabamba seasonal changes in rainfall and perhaps minimum temperature are most likely triggers which induce flowering (Stimm et al. 2008). The precipitation patterns during the last decades suggest slightly increasing rates for Loja on the western escarpment of the Cordillera Real (Vuille et al. 2008), which agrees with the aforementioned A1B scenario outlook, whereas Zamora on the eastern part may receive slightly less rainfall (Fig. 5). This situation supports the assumption that weakened mid- and upper tropospheric easterlies and strengthened westerlies would cause longer and more dry events. Concurrent clear sky conditions on the eastern side of the Cordillera, by uninhibited nocturnal radiation emission could result in lower minimum temperatures which sporadically might affect flowering and seed production and thus impair propagation. A slight trend towards more droughts is already visible, as only 8 rainless phases for over five days occurred between 1999-2003, while 21 such events were counted between 2004-2008.

Two extreme events of 16 and 18 consecutive days without precipitation in Nov. 2000 and 2005, respectively, resulted in severe desiccation damages and even complete withering of trees and shrubs. Ericaceous species suffered most in the timberline ecotone and the adjacent subpáramos, which is the uppermost vegetation belt, where mountain-top extinction of taxa could occur. All of the peaking nunatakker of the mountain consist of rocky cliffs and thus, species escaping upslope encounter increasingly unfavourable habitats (see "species-area approach", chapter 5.1). Consequently, climate warming would imply that drought sensitive species of an impressively endemic-rich altitudinal belt (Keating 2008) would have "nowhere to go".

The reason for a hampered immigration of montane forest species into the subpáramo is the stable position of the upper tree line of the cloud forest which consists of completely different plant communities. The tree line is located between 2800 and 3530 m a.s.l. (Fig. 10). While on a global scale mean soil temperatures (-10 cm depth) of 6.7 ± 0.8°C during the seasons of plant growth are considered the main reason for limiting growth of trees, Körner and Paulsen (2004) presented a lower value of around 5.5°C for the tropics, and Bendix and Rafiqpoor (2001) measured for even more elevated *Polylepis*-forests at Papallacta (in northern Ecuador) mean soil temperatures of 4.25°C. Environmental factors like abundant precipitation, air humidity and solar radiation can modify the impact of the temperature, and the local topography, soil conditions, natural disturbances and deforestation can add to this variation (Kessler et al. 2007).

Soil temperature measurements from five different tree line sites around the RBSF area show that only the highest patches of forest approach the

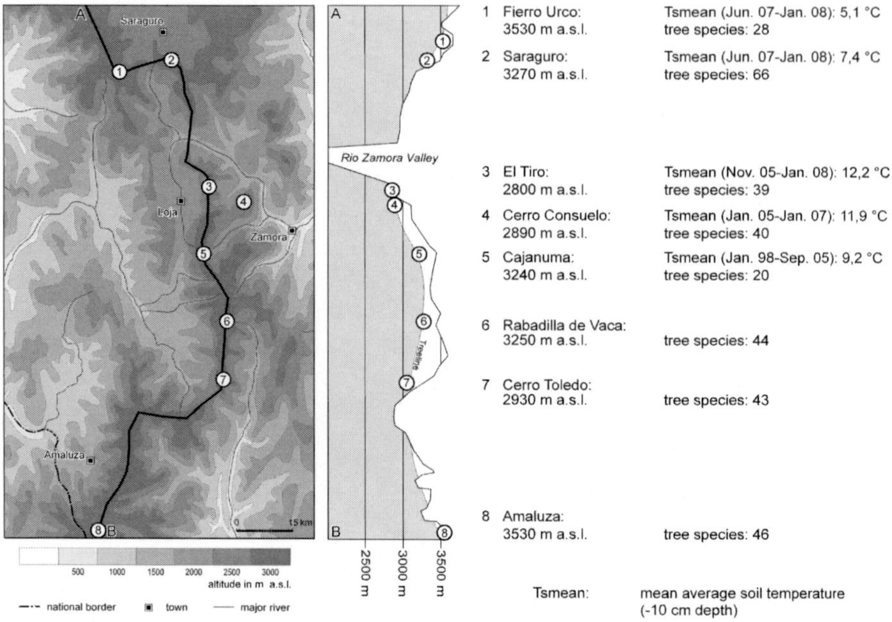

Fig. 10. Current position of upper tree line at the Ecuadorian part of the Andean Depression.

postulated temperature limit of +5.5°C (Fig. 10). In most cases soil temperatures are much higher and thus are not considered a limiting factor for tree growth. Instead, the very strong and quasi-permanent easterly winds prevent development of tree crowns and as a consequence the uplift of the tree line. Furthermore extremely high annual precipitation of up to more than 6000 mm/a (Emck 2007) cause water logging of the soils and leaching of nutrients and thus contributes to the inhibition of tree growth. Another point is the sporadically high global irradiance that by photodamage of the seedlings hinders tree regeneration and thus upslope extension of the forest (Baader et al. 2007). Finally, as in the majority of tropical mountains, livestock grazing and burning (Keating 1998) has a strong impact on the tree line. In summary only bushes and scrubs forming a species-rich subpáramo are encountered on the crests and upper slopes (Richter et al. 2008).

Given that temperature is not the limiting factor for tree growth above the upper tree line, warming fails as a reason for a future uplift of the current timberline. At the moment there are no indications for a forest encroachment into the subparámo. However, it should be kept in mind, that during early and mid Holocene the present-day subpáramo areas were covered by a forest-type vegetation (see chapter 3). An increase of the temperature would presumably allow tree species of lower regions to extend their ranges to higher altitudes thus enhancing tree diversity there. In the Andean Depression the upper tree line fluctuates between 2800 (El Tiro) and 3500 m a.s.l. (Fierro Urco in the north and Amaluza in the south (Fig. 10)). Different from the monotonous tree line in the high Andes the species composition of the tree line in the depression reflects the tree diversity of the upper mountain forest, which is usually an elfin forest. On the eight plots of 800 m^2 each (Fig. 10) between 28 and 66 tree species (DBH $>$ 5 cm) were recorded, but a correlation between altitude and species number could not be detected. Given that the species patterns encountered at the eight sites represent tree diversity saturation situations other factors than thermal must be decisive in competition. Arrival of new species will result in enhanced competition putting more pressure on the recent timberline species, some of which might then completely disappear from the area.

Colwell et al. (2008) stated that suitable habitat corridors to higher areas are a precondition for an uplift of more thermophilous species. In the case of the eastern Cordillera of South Ecuador the numerous valleys from the precordilleran and Amazon forelands could represent such corridors. Many lowland rain forest species have outposts in western Amazonia (Miles et al. 2004) from where they could spread into the valleys and precordilleran ranges.

Another perspective of the effect of climate change on the vegetation leads to the idea that present plant assemblages may disintegrate because of asymmetric elevational range-shifts of the individual species (Williams et al. 2007). Asymmetric range-shifts can be due to species-specific reactions to changes of environmental factors which accompany the increase of temperature, e.g. the above mentioned dry spells. The logical consequence is not only the formation of novel species assemblages but due to the narrow ranges of many taxa also a more frequent formation of species gaps. An uplift of the condensation level as a consequence of warming (Still et al. 1999) implies an increased evapotranspiration especially in the cloud-rich elfin forest where evaporation is low. Species may not be able to adapt to drier conditions and because of the non-transgressibility of the tree line would be prone to extinction. Species-rich cloud and elfin forests with a high proportion of endemics might therefore suffer in particular from depletion, although they represent centres of speciation due to the high proportion of genetically plastic genera such as *Anthurium, Cavendishia* and *Miconia.*

Drought- and heat-tolerant species are most likely gaining terrain and upon reshuffling of communities input from anthropogenic habitats can play a significant role. Sources are pastures, abandoned former cultivated lands, roadside vegetation, and exotic tree plantations close to the RBSF and further downstream in the valley. Elements of these man-made habitats are powerful invaders of gaps in the mountain rainforest, where landslides and mud- or debris-flows form widespread entrance portals for newcomers. *Cortaderia jubata* and ferns like *Sticherus* spp. from roadsides, *Melinis minutiflora* from active and *Baccharis trinervis* and *Erica erecta* from abandoned pastures are examples of such invasive herbaceous and bushy plant species, all of which are light demanding. Therefore, they disappear upon canopy closure and usually can not gain a foothold in the dim subcanopy light climate of the natural forest. This situation is different from that in tropical lowland forests which are more prone to burning (e.g. East Borneo) and where fire is a natural trait of the ecosystem (Goldammer and Seibert 1989). However, several herbaceous road-side followers will benefit from climate warming and might invade the unshaded páramo, among them several highly competitive exotic grasses, such as *Poa annua*, *Dactylis glomerata*, or *Holcus lanatus*.

6 Conclusions

The question which meteorological element of a changing climate is the most important factor for a potential change of species ranges is difficult to answer even for a small mountainous area like the study site. From an energetic viewpoint, ectothermic organisms like moths or other arthropods could react to an increase of the air temperature by expanding their ranges uphill (Fiedler et al. 2008). Such reaction implies that the plants on which the caterpillars feed either already inhabit the enlarged area or that they expand their ranges in a comparable way as the insects, or that the herbivores are flexible in their diet. With respect to herbivorous insects no clear conclusion on the specificity of the insect – host plant binding has been elaborated so far (Stork 2007), but it is not unlikely in general that in an extremely diverse habitat like the study area, biotic interactions are commonly weaker, i.e. that organisms have several partners for the same kind of interaction. This is part of an insurance strategy as population sizes and densities of the species are usually low and the probability to find a specific partner in a short time slot is rather low. This also holds for other plant-animal interactions such as pollination or seed dispersal (Beck et al. 2008b). Nevertheless there are and will be thermal upper limits for ectothermic organisms as well as for some tree species (e.g. *Piptocoma discolor* and *Vismia tomentosa*) whose flowering and seed production suffered from a sequence of only a few consecutive cold days (Stimm et al. 2008). Such not very rare weather irregularities – spells of drought are another example – play a more crucial role in limiting the habitat of a sessile organism than the average temperature.

For tree species, irrespective of their thermal adaptation, the perpetual strong easterlies carrying extraordinary amounts of precipitation constitute another limitation of penetrating into today's subpáramo regions. This results in a significantly lower upper tree line than in other parts of the eastern Andean chains in central and northern Ecuador which are better sheltered against high wind speeds (Richter et al. 2008). At sheltered depressions even at El Tiro, small patches of a dwarf forest composed of species of the upper mountain or elfin forest survive which can be interpreted as outposts of the present or remnants of a former high mountain rain forest. Given a climate change to drier and less windy conditions these islets could be effective as nuclei for the natural reforestation of El Tiro's current subpáramo.

This perspective points to the findings of pollen analysis which revealed the occurrence of forest at the elevated sites of the area (El Tiro) during the most probably warmer and drier early-mid Holocene. Ongoing global warming could result in similar uplifting of upper habitat boundaries of tree populations, causing a stronger competition especially in the upper mountain forest region of the Andean Depression, as during early to mid Holocene. Some plant populations might become then very rare or locally extinct. Additionally, (and this might be even more serious) anthropogenic impact by fire in upper mountain regions strongly affects the subpáramo and upper mountain rainforest populations.

Warming, in general may be associated with changes in the atmospheric circulation. Future global warming simulations show for instance a poleward shift of the South Pacific and South Atlantic subtropical anticyclones, which generally cause alterations in the position and intensity of the tropical easterlies and the Walker-Circulation (Christensen et al. 2007). Similar changes in atmospheric circulation over South America were also suggested for the warmer period in mid Holocene by Grosjean and Nuñez (1994). The change in circulation is expected to weaken particularly the equatorial zonal wind over the Pacific (Vecchi et al. 2006). Polissar et al. (2006) stressed that a reduction of moisture transport from the Atlantic to the eastern tropical Andes in the Holocene is related to an increase of ocean temperature in the tropical eastern Pacific, a higher frequency of westerly winds and, as a consequence, the weakening of easterly stream flow to the tropical Andes.

An increasing frequency of drought events would be another consequence of such an altered circulation. Drought, as discussed in chapter 5.2, can substantially affect plant diversity patterns, in particular of the very sensitive timberline and subpáramo ecotones. Abundant tree line species with a high demand of moisture such as *Axinaea macrophylla*, *Hedyosmum racemosum* and *H. scabrum,* or *Weinmannia rollottii* may not be able to survive under such conditions and may be replaced by more drought tolerant species from lower elevations. Drought, in addition to its direct ecophysiological effects promotes the incidence of wildfires, as is also evident from the charcoal records from the Mid and Late Holocene (see chapter 3). As it is known from tropical Andean ecotones frequent fires can dramatically change species composition giving rise to a monotonous vegetation composed of species which can cope with the recurrent destruction of their aerial plant parts (e.g. Beck et al. 1986).

In that context questions arise concerning the emergence of "novel communities". Identifiable and stable "communities" of coexisting populations (in the sense of the American "community ecology" approach) or "associations" (in the synsystematic sense) can hardly be found in the undisturbed mountain rainforests of southern Ecuador. These forests can be characterized by their structures rather than by the floristic composition. As mentioned above coexistence of organisms has more degrees of freedom which is one of the prerequisites of the high operative dynamics of these forests. As a consequence, newcomers with adequate environmental demands have a realistic chance to enter and integrate into the assemblage, the more so as dominant species are widely missing. While this concept of a "flexible community" seems to hold for the low- and mid-level mountain rainforests, the situation between elfin forests and páramos with their relatively clear boundaries rather corresponds to "fixed communities" which upon climate change could transform into "novel communities".

7 Outlook

Thomas et al. (2004) predicted that species extinctions due to climate change will exceed those caused by direct human impacts like intensification of land-use and hence, claimed to re-evaluate conservation priorities. They argue that further measures to reduce the magnitude of anthropogenic climate change are now the major priority for conservationists. This way of priority ranking is dangerous because endeavours for conservation in particular of the tropical forests may suffer from redirecting resources (Lewis 2006) to technical solutions of the CO_2 problem. Rather, the activities to install a REDD-Regime[3] should be strongly supported in a country, which suffers the highest deforestation rate of the continent (Mosandl et al. 2008).

Acknowledgements
The authors gratefully acknowledge financial support of the work by the German Research Foundation (DFG) in the scope of the Research Units FOR 402 and FOR 816. They would also like to thank the Foundation "Naturaleza y Cultura Internacional" in Loja and San Diego for outstanding logistic assistance. They are grateful to Prof. T. Tscharntke (Göttingen) for the encouragement to write this contribution and for his patience with the delivery of the manuscript.

[3] REDD: Reducing Emissions from Deforestation and forest Degradation<http://www.cbd.int/lifeweb/>

References

Allen AP, Brown JH, Gillooly JF (2002) Global biodiversity, biochemical kinetics, and the energetic-equivalence rule. Science 297: 1545-1548

Baader MY, Geloof I, Rietkerk M (2007) High solar radiation hinders tree regeneration above the alpine tree line in northern Ecuador. Plant Ecol 191: 33-45

Bascompte J, Jordano P, Olesen JM (2006) Asymmetric coevolutionary networks facilitate biodiversity maintenance. Science 312: 431 - 433

Barthlott W, Hostert A, Kier G, Küper W, Kreft H, Mutke J, Rafiqpoor D, Sommer JH (2007) Geographic patterns of vascular plant diversity at continental to global scales. Erdkunde 61: 305-315

Beck E, Makeschin F, Haubrich F, Richter M, Bendix J, Valerezo C (2008a) The ecosystem (Reserva Biológica San Francisco). In: Beck E, Bendix J, Kottke I, Makeschin F, Mosandl R (eds), Gradients in a tropical mountain ecosystem of Ecuador. Ecol Stud 198: 1-13

Beck E, Kottke I, Bendix J, Makeschin F, Mosandl R (2008b) Gradients in a tropical mountain ecosystem – a synthesis. In: Beck E, Bendix J, Kottke I, Makeschin F, Mosandl R (eds), Gradients in a tropical mountain ecosystem of Ecuador. Ecol Stud 198: 451-463

Beck E, Scheibe R, Schulze E-D (1986) Recovery from fire: Observations in the alpine vegetation of western Mt. Kilimajaro (Tanzania). Phytocoenologia 14: 55-77

Beck E, Richter M (2008) Ecological aspects of a biodiversity hotspot in the Andes of southern Ecuador. In: Gradstein SR, Homeier J, Gansert D (eds.): The tropical mountain forest – Patterns and Processes in a Biodiversity Hotspot. Biodiv Ecol Ser 2: 197-219

Behling H, Hooghiemstra H (2000) Holocene Amazon rain forest - savanna dynamics and climatic implications: high resolution pollen record from the Laguna Loma Linda in eastern Colombia. J Quaternary Sci 15: 687-695

Behling H, Hooghiemstra H (2001) Neotropical savanna environments in space and time: Late Quaternary interhemispheric comparisons. In: Markgraf V (ed), Interhemispheric Climate Linkages. Academic Press, New York, 307-323

Bendix, J, Rafiqpoor MD, Daud M (2001) Studies on the thermal conditions of soils at the upper tree line in the Páramo of Papallacta (Eastern cordillera of Ecuador). Erdkunde 55: 257-276

Bendix J, Homeier J, Cueva Ortiz E, Emck P, Breckle S, Richter M, Beck E (2006) Seasonality of weather and tree phenology in a tropical evergreen mountain rain forest. Int J Biometeor 50: 370-384

Bendix J, Rollenbeck R, Richter M, Fabian P, Emck P (2008a) Climate. In: Beck E, Bendix J, Kottke I, Makeschin F, Mosandl R (eds), Gradients in a tropical mountain ecosystem of Ecuador. Ecol Stud 198: 63-73

Bendix J, Rollenbeck R, Fabian P, Emck P, Richter M, Beck E (2008b) Climate variability. In: Beck E, Bendix J, Kottke I, Makeschin F, Mosandl R (eds),

Gradients in a tropical mountain ecosystem of Ecuador. Ecol Stud 198: 281-290

Botkin DB, Saxe H, Araújo MB, Betts R, Bradshaw RHW, Cedhagen T, Chesson P, Dawson TP, Etterson JR, Faith DP, Ferrier S, Guisan A, Skjolborg HA, Hilbert DW, Loehle C, Margules C, New M, Sobel MJ, Stockwell DRB (2007) Forecasting the effects of global warming on biodiversity. BioScience 57: 227-236

Brehm G, Süssenbach D, Fiedler K (2003) Unique elevational diversity patterns of geometrid moths in an Andean montane rainforest. Ecography 26: 456–466

Brehm G, Pitkin LM, Hilt N, Fiedler K (2005) Montane Andean rain forests are a global diversity hotspot of geometrid moths. J Biogeogr 32: 1621-1627

Brehm G, Homeier J, Fiedler K, Kottke I, Illig J, Nöske NM, Werner FA, Breckle S-W (2008) Mountain rain forests in southern Ecuador as a hotspot of biodiversity – limited knowledge and diverging patterns. In: Beck E, Bendix J, Kottke I, Makeschin F, Mosandl R (eds), Gradients in a tropical mountain ecosystem of Ecuador. Ecol Stud 198: 15-23

Brunschön C, Behling H (submitted). Late Quaternary vegetation, fire and climate history reconstructed from two cores at Cerro Toledo, Podocarpus National Park, southeastern Ecuadorian Andes. Quaternary Res

Chen I-C, Shiu H-J, Benedick S, Holloway JD, Chey VK, Barlow HS, Hill JK, Thomas CD (2009) Elevation increases in moth assemblages over 42 years on a tropical mountain. PNAS 103: 10334-10339

Christensen JH, Hewitson B, Busuioc A, Chen A, Gao X, Held I, Jones R, Kolli RK, Kwon W-T, Laprise R, Magaña Rueda V, Mearns L, Menéndez CG, Räisänen J, Rinke A, Sarr A, Whetton P (2007) Regional Climate Projections. In: (Solomon S, Qin D, Manning M, Chen Z, Marquis M, Averyt KB, Tignor M, Miller HL (eds). Climate Change 2007: The Physical Science Basis. Contribution of Working Group I to the Fourth Assessment Report of the Intergovernmental Panel on Climate Change. Cambridge University Press, Cambridge, United Kingdom and New York, NY, USA

Colwell RK, Brehm G, Cardelius CL, Gilman AC, Longino JT (2008) Global warming, elevational range shifts, and lowland biotic attrition in the wet tropics. Science 322: 258-261

Dziedzioch C, Stevens A-D, Gottsberger G (2003) The hummingbird -plant community of a tropical mountain rainforest in southern Ecuador. Plant Biol 5: 331-337

Emck P (2007) A climatology of South Ecuador. With special focus on the major Andean Ridge as Atlantic-Pacific Climate Divide. PhD Thesis, Univ Erlangen
http://www.opus.ub.uni-erlangen.de/opus/frontdoor.php?source_opus=656

Fiedler K, Brehm G, Hilt N, Süßenbach D, Häuser CL (2008) Variation of diversity patterns across moth families along a tropical altitudinal gradient. In: Beck E, Bendix J, Kottke I, Makeschin F, Mosandl R (eds), Gradients in a tropical mountain ecosystem of Ecuador. Ecol Stud 198: 167-179

Goldammer JG, Seibert B (1989) Natural rain-forest fires in Eastern Borneo during the Pleistocene and Holocene. Naturwissenschaften 76: 518–520.

Grabherr G, Gottfried M, Pauli H (2001) Long-term monitoring of mountain peaks in the Alps. Tasks Veg Sci 35: 153-177

Grosjean M, Nuñez L (1994) Late glacial, early and middle Holocene environments, human occupation, and ressource use in the Atacama (northern Chile). Geoarcheology 9: 271-286

Guisan A, Thuiller W (2005) Predicting species distribution: Offering more than simple habitat models. Ecol Lett 9: 993-1009

Hughes C, Eastwood R (2006) Island radiation on a continental scale: Exceptional rates of plant diversification after uplift of the Andes. PNAS, 103, 10334-10339

Iost S, Makeschin F, Abiy M, Haubrich F (2008) Biotic soil activities. In: Beck E, Bendix J, Kottke I, Makeschin F, Mosandl R (eds), Gradients in a tropical mountain ecosystem of Ecuador. Ecol Stud 198: 217-227

Jost L (2004) Explosive local radiation of the genus *Teagueia* (Orchidaceae) in the Upper Pastaza Watershed of Ecuador. Lyonia 7: 42-47

Keating PL (1998) Effects of anthropogenic disturbances on páramo vegetation in Podocarpus National Park, Ecuador. Phys Geogr 19: 221-238

Keating PL (2008) Floristic composition and biogeographical significance of a megadiverse páramo site in the southern Ecuadorian Andes. J Torrey Bot Soc 135: 554-570

Kessler M, Böhner J, Kluge J (2007) Modelling tree height to assess climatic conditions at tree lines in the Bolivian Andes. Ecol Modelling 207: 223-233

Kistler R, Kalnay E, Collins W, Saha S, White G, Woollen J, Chelliah M, Ebisuzaki W, Kanamitsu M, Kousky V, van den Dool H, Jenne R, Fiorino M (2001) The NCEP–NCAR 50-Year Reanalysis: Monthly Means CD-ROM and Documentation. Bull Am Meteorol Soc 82: 247-267

Klanderud K, Birks HJB (2003) Recent increase in species richness and shifts in altitudinal distributions of Norwegian mountain plants. The Holocene 13: 1-6

Körner C, Paulsen J (2004) A world-wide study of high altitude tree line temperatures. J Biogeogr 31: 713-732

Koh LP, Dunn RR, Sodhi NS, Colwell RK, Proctor HC, Smith VS (2004) Species coextinctions and the biodiversity crisis. Science 305: 1632 - 1634

Kottke I, Haug I, Setaro S, Suárez JP, Weiß M, Preußing M, Nebel M, Oberwinkler F (2008) Guilds of mycorrhizal fungi and their relation to trees, ericads, orchids and liverworts in a neotropical mountain rain forest. Basic Appl Ecol 9: 13-23

Lewis OT (2006) Climate change: Species-area curves and the extinction crisis. Phil Trans Roy Soc B 361: 163-171

Liede-Schumann S, Breckle S-W (eds.) (2008) Provisional checklist of flora and fauna of the San Francisco Valley and its surroundings (Reserva Biológica San Francisco, Province Zamora-Chinchipe, southern Ecuador). Ecotrop Monogr 4, 256 p

Mayle F, Burbridge R, Killeen TJ (2000) Millennial-scale dynamics of southern Amazonian rain forests. Science 290: 2291-2294

Meehl GA, Stocker TF, Collins WD, Friedlingstein P , Gaye AT, Gregory JM, Kitoh A, Knutt, R, Murphy JM, Noda A, Raper SCB, Watterson IG, Weaver AJ, Zhao Z-C (2007) Global Climate Projections. In: Solomon S, Qin D, Manning M, Chen Z, Marquis M, Avery, KB, Tignor M, Miller HL (eds) Global Climate Projections - Climate Change 2007: The Physical Science Basis. Contribution of Working Group I to the Fourth Assessment Report of the Intergovernmental Panel on Climate Change. Cambridge University Press, Cambridge, United Kingdom and New York, NY, USA, 996 p

Miles L, Grainger A, Philips O (2004) The impact of global climate change on tropical forest biodiversity in Amazonia. Glob Ecol Biogeogr 13: 553-565

Mosandl R, Günter S, Stimm B, Weber M (2008) Ecuador suffers the highest deforestation rate in South America. In: Beck E, Bendix J, Kottke I, Makeschin F, Mosandl R (eds), Gradients in a tropical mountain ecosystem of Ecuador. Ecol Stud 198: 37-40

Niemann H, Behling H (2008) Late Quaternary vegetation, climate and fire dynamics inferred from the El Tiro record in the southeastern Ecuadorian Andes. J Quaternary Sci 3: 203-212

Nogués-Bravo D, Araújo MB, Romdal T, Rahbek C (2008) Scale effects and human impact on the elevational species richness gradients. Nature 453: 216-220

Pearson RG, Dawson TP (2003) Predicting the impacts of climate change on the distribution of species: are bioclimate envelope models useful? Glob Ecol Biogeogr 12: 361-371

Polissar PJ, Abbott MB, Shemesh A, Wolfe AP, Bradley RS (2006) Holocene hydrologic balance of tropical South America from oxygen isotopes of lake sediment opal, Venezuelan Andes. Earth Planetary Sci Lett 242: 375–389

Richter M (2008) Tropical mountain forests - distribution and general features. In: Gradstein SR; Homeier J, Gansert D (eds): The Tropical Mountain forest – Patterns and Processes in a Biodiversity Hotspot. Biodiv Ecol Ser 2: 7-24

Richter M, Diertl K-H, Peters T and Bussmann RW (2008) Vegetation structure and ecological features of the upper timberline ecotone. In: Beck E, Bendix J, Kottke I, Makeschin F, Mosandl R (eds) Gradients in a Tropical Mountain Ecosystem of Ecuador. Ecol stud 198: 123-135

Solomon S, Qin D, Manning M, Chen Z, Marquis M, Averyt KB, Tignor M, Miller HL (eds) (2007) Summary for Policymakers - Climate Change 2007: The Physical Science Basis. Contribution of Working Group I to the Fourth Assessment Report of the Intergovernmental Panel on Climate Change. IPCC 2007, Cambridge University Press, Cambridge, United Kingdom and New York, NY, USA, pp 18

Still CJ, Prudence N, Foster PN, Stephen H (1999) Stimulating the effects of climate change on tropical montane cloud forests. Nature 398: 608-610

Stimm B, Beck E, Günter S, Aguirre N, Cueva E, Mosandl R, Weber M (2008) Reforestation of abandoned pastures: Seed ecology of native species and production of indigenous plant material. In: Beck E, Bendix J, Kottke I, Makeschin F, Mosandl R (eds), Gradients in a tropical mountain ecosystem of Ecuador. Ecol Stud 198: 417-429

Stork NE (2007) World of Insects. Nature 448: 657-658

Theurillat J-P, Guisan A (2001) Potential impact of climate change on vegetation in the European Alps: A review. Climatic change 50: 77-109

Thomas CD, Cameron A, Green RE, Bakkenes M, Beaumont L, Collingham YC, Erasmus BFN, Ferreira de Siqueira M, Graininger A, Hannah L, Hughes L, Huntley B, Jaarsveld AS van, Midgley GF, Miles L, Ortega-Huerta MA, Townsend Peterson A, Phillips OL, Williams SE (2004) Extinction risk from climate change. Nature 427: 145-148

Thuiller W (2004) Patterns and uncertainties of species range shifts under climate change. Glob Change Biol 10: 2020-2027

Thuiller W, Lavorel S, Araújo MB, Sykes MT, Prentice IC (2005) Climate change threats plant diversity in Europe. Proc Nat Acad Sci USA 102: 8245-8250

Vecchi GA, Soden BJ, Wittenberg AT, Held IM, Leetmaa A, Harrison, MJ (2006) Weakening of tropical Pacific atmospheric circulation due to anthropogenic forcing. Nature 441: 73-76

Vuille M, Francou B, Wagnon P, Juen I, Kaser G, Mark BG, Bradley RS (2008) Climate change and tropical Andean glaciers: Past, present and future. Earth Sci Rev 89: 479-496

Walther G-R, Beissner S, Burga CA (2005) Trends in upward lift of alpine plants. J Veg Sci 16: 541-548

Weigend M (2002) Observations on the biogeography of the Amotape-Huancabamba Zone in northern Peru. Bot Rev 68: 38-54

Williams JW, Jackson ST, Kutzbach JE (2007) Projected distributions of novel and disappearing climates by 2100 AD. Proc Nat Acad Sci USA 104: 5738-5742

Young KR, Reynel C (1997) Huancabamba Region, Peru and Ecuador. In: Davis SD, Heywood VH, Herrera-MacBryde O, Villa-Lobos J, Hamilton AC (eds.) Centers of plant diversity. A guide and strategy for their conservation 3: 465-469

Appendix: SRES-Scenarios

Future climate change is globally estimated by the ***I***ntergovernmental ***P***anel on ***C***limate ***C***hange (IPCC) with global circulation models. The basis of model runs is the radiative forcing due to alternating future greenhouse gas concentrations in the atmosphere. The concentrations are defined by socio-economic scenarios which are constructed with special reference to the production and evolution of greenhouse gases and aerosol precursor emissions during the 21st century. The scenarios are published in the third IPCC Assessment Report as a ***S**pecial **R**eport on **E**missions **S**cenarios* (SRES) and subdivided in four narrative storylines, labelled A1, A2, B1 and B2 (Nakicenovic and Swart 2000). The following scenarios/scenario families are addressed in the current chapter:

B1: This scenario family assumes a convergent world where global population peaks in mid-21 century and declines thereafter. With respect to greenhouse gas emissions it optimistically assumes rapid changes in economic structures toward a service and information economy, with significant reductions in material intensity, and the introduction of clean and ressource-efficient technologies.

A1: The future world in this scenario family is characterised by a very rapid economic growth (Business as Usual, most likely), global population that peaks in mid-century and declines thereafter, and rapid introduction of new and more efficient technologies. In contrary to B1, A1 is more prone to emissions where the A1B scenario is moderate by balancing across energy sources while the A1FI is fossil intensive, generating highest emission rates.

Reference

Nakicenovic N, Swart, R (eds, 2000) Special Report on Emissions Scenarios: A Special Report of Working Group III of the Intergovernmental Panel on Climate Change, Cambridge University Press, Cambridge, U.K., 599 pp

Spatiotemporal trends of forest cover change in Southeast Asia

Stefan Erasmi[1*], Muhammad Ardiansyah[2], Pavel Propastin[1], and Alfredo Huete[3]

[1] University of Göttingen, Institute of Geography, Goldschmidtstr. 5,37077 Göttingen, Germany
[2] Bogor Agricultural University, Department of Soil Sciences and Land resources, Bogor, Indonesia
[3] University of Arizona, Department of Soil, Water and Environmental Science, Tucson, AZ, USA

*corresponding author: S. Erasmi, email: serasmi@uni-goettingen.de

Summary

The current state of tropical forest cover and its change have been identified as key variables in modelling and measuring the consequences of human action on ecosystems. The conversion of tropical forest cover to any other land cover (deforestation) directly contributes to the two main environmental threats of the recent past: 1) the alteration of the global climate by the emission of carbon to the atmosphere and 2) the decline in tropical biodiversity by land use intensification and habitat conversion. The sub-continent of Southeast Asia exhibits one of the highest rates of forest loss and comprises one of the regions with the highest amount and diversity of flora and fauna species, worldwide.

The knowledge of the spatial and temporal trends in the variation of forest cover in tropical regions is a prerequisite for the development and establishment of mitigation strategies from the global to the regional level. However, there is considerable disagreement in recent estimates of tropical forest cover change ranging from continuing and intensified decline in forest loss to a distinct decrease in deforestation rates and up to stagnation in other cases. Against this background, the present study aims at a review and comparison of recently available global forest cover estimates for the region of Southeast Asia. In a case study, the results at the national level will be compared to an analysis at the regional level for the island of Sulawesi, Indonesia. The outcome of the study provides recommendations for future remote sensing based forest assessments in tropical regions.

T. Tscharntke et al. (eds.), *Tropical Rainforests and Agroforests under Global Change*, Environmental Science and Engineering, 269–291, DOI 10.1007/978-3-642-00493-3_12,

Keywords: Tropical deforestation, tropical forest cover, GlobCover, Landsat, Southeast Asia

1 Introduction

Tropical forests are under significant threat from combined effects of long term climate change and short term, human induced disturbance (e.g. deforestation). It is unquestionable that tropical forests account for large amounts of annual carbon sequestration and evapotranspiration. They contain ~25% of the carbon in the terrestrial biosphere and account for ~33% of terrestrial net primary production (NPP) and hence, deforestation directly causes carbon release to the atmosphere (Bonan 2008). Current estimates suggest that deforestation accounts for one fifth of human induced emissions of carbon dioxide (IPCC 2007). Since deforestation is almost exclusively taking place in tropical regions, the major part of the emission from deforestation can be related to the exploitation of tropical forest. The necessity of developing appropriate tools and providing spatially explicit base data for monitoring deforestation and forest degradation has been underlined during the past United Nations Framework Convention on Climate Change (UNFCCC) Cop13 meeting in Bali (Miles and Kapos 2008). The methods and procedures for monitoring, measuring and reporting the efforts for reducing emissions from deforestation and forest degradation (REDD) in developing countries have recently been summarized by GOFC-GOLD (2008).

Including the current focus on climate change and REDD, the overall scope of tropical forest monitoring is much broader (Holmgren et al. 2007). Three major groups of research can be summarized that deal with the development of tools and involve the analysis of spatially explicit data with the overall aim of explaining spatio-temporal processes related to human-environment-interactions: (1) land cover and land use change (LCLUC) and carbon dynamics (e. g. Olander et al. 2008, Achard et al. 2007, DeFries et al. 2002, Herold and Johns 2007); (2) LCLUC and biological conservation (e. g. Sodhi et al. 2009, Gillespie et al. 2008) and (3) vegetation activity and climate variability (e. g. Sarkar et al. 2007, Malhi and Wright 2004, Nagai et al. 2007).

The sub-continent of Southeast Asia features the highest deforestation rates in the tropics (Achard et al., 2002, Sodhi et al., 2004). It covers a total area of 4.8 Mio km^2 whereof 2.8 Mio km^2 (57.6 %) have been forest in 1990 (FAO 2006). Since 1990, the forest area decreased continuously until the most recent FAO forest resource assessment (2005) with an average deforestation rate of 1.2 %, resulting in a forest loss of 214,000 ha every year. Within Southeast Asia, Indonesia and the Philippines have been facing the highest deforestation rates (1.8 % and 2.5 % respectively). In general, deforestation in these countries is ongoing at alarming rates that are twofold compared to the mean rate for Southeast Asia and significantly above the rates for other tropical regions (South America 0.45 %, Africa 0.65 %) (FAO 2006).

The trend of rapid forest loss in wide areas of Southeast Asia during the past two decades has also been confirmed by recent remote sensing based estimates that reveal an average annual deforestation rate of 0.71 % for the time period of 1990 to 1997 (Achard et al. 2002). However, the forest change rates from remote sensing surveys, in general are significantly below the FAO estimates. More recent remote sensing based estimates at the regional level confirm that deforestation is still the major threat to natural resources in the region (Hansen et al. 2008). Up-to-date reliable and objective information about the spatially explicit dimension of forest loss is not yet available but is expected from the next global forest resources assessment of the FAO (FRA 2010). In the meanwhile, a number of studies have investigated different remote sensing data sources and sampling schemes in order to provide regional to cross-continental assessments of tropical forest extent and land cover change for different applications (e.g. Achard et al. 2007, Stibig et al. 2007, Mayaux et al. 1995).

Remote sensing data provide the most reliable data source for accurately and objectively estimating changes in forest cover over large areas, especially in remote areas like the vast and difficult to access regions of mature tropical forest. Global land cover maps created from coarse resolution remote sensing data provide an indispensable means of assessing large-area, comparable and objective information about the state of land cover and forest resources at the national level for a variety of applications (Bartholome and Belward 2005). However, only few studies have dealt with the evaluation of these global land cover maps for tropical regions (Herold et al. 2008, Erasmi et al. 2007) and there is large uncertainty about the quality and accuracy of existing land cover change and deforestation estimates especially in Southeast Asia (Herold et al. 2008).

Against this background, the present study provides a summary of past, present and planned satellite Earth remote sensing missions, the land cover products that have been produced from these data and the main research initiatives and organizations that are involved in tropical forest cover assessment. Finally, an estimate of forest cover and net forest loss for the most recent available data period is carried out within a case study for the sub-continent of Southeast Asia. The aim of the analysis is to evaluate the usefulness of recent satellite based global land cover maps for nationwide to trans-continental tropical forest assessment within the scope of the above mentioned goals of tropical forest monitoring.

2 Products for regional to trans-continental land cover mapping and deforestation monitoring in Southeast Asia

Satellite remote sensing data provide a unique base for the description of the spatial arrangement and composition of (land cover) objects on the land surface. Information about land cover is of utmost importance for the description and study of the environment (Watson et al. 2000).

Forest coverage and forest cover change may be derived from global land cover maps. However, before generating deforestation estimates and interpreting results based on global land cover products, a number of open issues have to be accounted for:

- Is the spatial and temporal resolution of the underlying data appropriate for my investigation?
- Is the definition of (forest) classes consistent in my data sets?
- What exactly is deforestation and how can it be measured from my data?

2.1 Quality and availability of Global Land Cover products for tropical forest assessment

The spatial resolution of global land cover maps has significantly increased since the mid-1980s when the first continental land cover maps were available at 4 km pixel spacing. During the 1990s a number of (pan-) continental to global maps were produced at a spatial resolution of 1 km from the AVHRR sensor on board the US NOAA satellites (e.g. UMD Global Land Cover Characterization, see Table 1). At the end of the 1990s and the beginning of the new millennium, some new global land cover datasets evolved at similar resolution, but produced from advanced spaceborne sensor systems. Despite the same spatial resolution, these products (GLC2000, MODIS) allowed for a spatial and thematic refinement of the previous global maps due to the greater stability of the platforms and the improved spectral characteristics (number of channels, spectral bandwidth) of the sensors. Recent initiatives of the Earth observation community strengthened the efforts to further increase the spatial resolution of global land cover products. As a result, two new global datasets were produced based on Terra MODIS and Envisat-MERIS respectively. The MODIS-derived vegetation continuous fields (VCF) product goes beyond a qualitative land cover characterization and provides spectrally based sub-pixel estimates of percent vegetative cover at a resolution of 500 m (Table 1). The main expected advantage of the continuous classification scheme of the VCF product is that it may depict areas of heterogeneous land cover better than traditional discrete classification schemes (Hansen et al. 2002).

The MERIS-based product GlobCover (Arino et al. 2007) comprises a familiar map of discrete land cover units at the highest available spatial detail so far (300 m). The GlobCover product is designed to update and complement to other existing global or continental products, especially to those of the GLC2000 initiative (e. g. Stibig et al. 2007 for Southeast Asia).

2.2 Consistency of land cover legends

International panels such as GTOS, GOFC-GOLD and the UN Global Land Cover Network (GLCN) have been involved in developing standards on land

Table 1. Area-wide remote sensing based land cover products and programs for Southeast Asia

Product	Sensor	Reference year	Spatial resolution	Coverage	Classification scheme (legend)	Scientific reference	Data access / information
GLCC	AVHRR	1992	1 km	Global	IGBP (17 classes)	Loveland et al. (2000)	http://edc2.usgs.gov/glcc/
UMD land-cover	AVHRR	1992	1 km	Global	Simplified IGBP (14 classes)	Hansen et al. (2000)	http://glcf.umiacs.umd.edu/data/landcover/
TREES I	AVHRR	1992	1 km	pan-tropical	TREES (9 classes)	Achard et al. (2001)	http://www-tem.jrc.it/
Vegetation Continuous Fields (VCF)	AVHRR	1992	1 km	Global	continuous (% tree cover)	DeFries et al. (2000)	http://glcf.umiacs.umd.edu/data/treecover/
GLC2000	SPOT-VGT	2000	1 km	Global	LCCS	Bartholomé and Belward (2005)	http://www-tem.jrc.it/
MODIS land-cover	MODIS	2000	1 km	Global	simplified IGBP	Friedl et al (2002)	http://edcimswww.cr.usgs.gov/pub/imswelcome/
TREES II	SPOT-VGT	2000	1 km	Insular SE Asia	TREES (9 classes)	Stibig et al. (2003)	http://www-tem.jrc.it/
Vegetation Continuous Fields (VCF)	MODIS	2000-2005	500 m	Global	continuous (% vegetation cover)	Hansen et al. (2002)	http://glcf.umiacs.umd.edu/data/vcf/
GlobCover	MERIS	2005	300 m	Global	LCCS	Arino et al (2007)	http://ionia1.esrin.esa.int/index.asp
TREES III	MERIS, Landsat	1990 / 2000 / 2005	30 m (stratified irregular sample)	pan-tropical + Eurasia	N/A	N/A	http://ies.jrc.ec.europa.eu/
FRA 2010	Landsat	1990 / 2000 / 2005	30 m (systematic sample)	Global	FAO (8 classes)	N/A	http://www.fao.org/forestry/44375/en/
NASA LCLUC	MODIS, Landsat	2000 / 2005	30 m (stratified block sampling)	Global	N/A	N/A	http://lcluc.umd.edu/

cover characterisation (Herold et al. 2008). As a consensus, the UN Land Cover Classification System (LCCS) has been accepted as a common land cover legend and language in building land cover maps and translating between different existing land cover products (Herold et al. 2008). The LCCS is not based on a predefined list of class names, but instead uses a simple set of attributes and classifiers that are organized in a modular hierarchical manner (Bartholomé and Belward 2005). This means that maps based on the LCCS may have to be aggregated to a lower level of detail before they can be compared but at this level, a full consistency of class definitions is guaranteed. The LCCS has also been decided as the main classification scheme for an international initiative to harmonize other existing and future global land cover maps in order to support the operational earth observation of land (Herold et al. 2006). The major problem of such harmonization initiatives is that in some cases a direct translation of classes from one legend to another is not possible. As an example, the IGBP legend (MODIS, GLCC, UMD Land Cover) considers forest when the tree cover is above 15 %. On the other hand, LCCS based land cover maps (GLC2000, GlobCover) use a threshold of 65 % tree cover. The consequences of these inconsistencies for forest assessment and land cover change analysis at the national level have been documented by Erasmi et al. (2007) for Indonesia. They show a significant overestimation of forest cover based on IGBP legends and clearly reveal the problems of long term change trend analysis of forest cover based on existing global land cover products.

Besides differences in legends, inconsistencies between land cover products are also a factor of cartographic standards (minimum mapping unit), generalization of land cover objects, geolocation accuracy, cartographic projection and thematic accuracy. An attempt to compare thematic mapping accuracy of different, harmonized global land cover products has been undertaken by Herold et al. (2008). The results of the study show general patterns of agreement for broadleaved evergreen forest but underline the limited ability of global products to discriminate between classes especially in heterogeneous landscapes.

All in all, numerous global land cover products from optical, coarse resolution (>250 m) remote sensing data exist and the quality of these maps in terms of spatial resolution and thematic accuracy considerably increased. The maps have been developed in response to the need for information about land cover and its dynamics but they basically exist as independent datasets that are not comparable. It is expected from the recent agreements on harmonization that future datasets will account for this deficiency. A first attempt in this direction was made with the release of the GlobCover product. The case study in section 4 will examine the consistency of the GlobCover product compared to the GLC2000 product for the region of Southeast Asia.

2.3 Measurement of deforestation

Besides the technical aspects of remote sensing based land cover products, an important semantic issue has to be clarified. The analysis of forest cover change is mostly related to as deforestation. However, the definition of deforestation is not always congruent with the actions of forest and land cover change that are being observed from remote sensing data.

As a synthesis of many approaches Lund (1999) restricts the term deforestation to "the long-term or permanent removal of forest cover and conversion to a non-forested land-use". Considering remote sensing based land cover maps, deforestation would only account for one of four possible processes of land cover conversion with the involvement of forest: deforestation, reforestation, afforestation and regrowth. Where the first three terms describe human induced changes in land cover, the latter describes natural regeneration of a formerly degraded or deforested area. The sum of these processes is the net deforestation or net forest loss. Net forest loss is the key parameter of interest in forest assessments at the national to global level. Hence, it is considered as the target variable in the following investigations. An extensive overview of definitions related to forest is given by Lund (2008).

3 Ongoing and planned science programs and initiatives for tropical land cover assessment

In general, land cover change should refer to a reference and be measured globally over multiple (at least two) years. In recent past, three baseline years have been established in global land cover monitoring (1990, 2000 and 2005).

Global land cover maps provide the basis for national to pan-tropical estimates of general trends in land cover change in tropical regions. They are the only source to establish a spatially explicit area-wide coverage of the land surface but they are limited in use mainly due to the coarse spatial resolution that does not directly allow for change estimations because most changes occur at the sub-pixel level (Achard et al. 2007). On the other hand, they offer a useful tool to stratify the land surface into regions of low to high forest cover or forest change likelihood. This information is particularly useful in developing sampling approaches for more detailed, medium resolution (250-30 m) satellite based forest change analysis. Several approaches for sampling within forest areas have been applied in recent studies (see Czaplewski 2003, Stehman 2005, Hansen et al. 2008).

The following section briefly introduces three initiatives that make use of such sampling procedures for a reliable, consistent and regular mapping of forest cover and change analysis.

3.1 FAO global forest resources assessment (FRA)

FAO has been monitoring the world's forests at 5 to 10 year intervals since 1946. So far, the Global Forest Resources Assessment (FRA) has been based on data that countries provide to FAO in response to a common questionnaire.

There have been many discussions about the objectivity and quality of the FAO forest estimates. Especially, difficulties in monitoring long-term trends in tropical forest change have been revealed by Grainger et al. (2008). They compared the results from the FRA of the reporting frames of 1980, 1990, 2000 and 2005 and figured out, that forest decline between two assessment periods is always a multiple of the decline when the long term trend between the first and the last assessment is evaluated.

To overcome these problems and to ensure long-term consistency in forest resource assessments, FAO will undertake a global remote sensing survey of forests for the first time within the Global Forest Resources Assessment 2010 (FRA 2010). The survey will primarily be based on the use of available Landsat imagery, but will incorporate auxiliary information including other remote sensing images (e.g. ASTER, ALOS-PRISM), local knowledge and results from existing and past field inventories. A systematic sampling design will be used based on a 1° longitude by latitude grid as illustrated in Figure 1 for the Island of Sulawesi.

For each sample plot (20 x 20 km), four Landsat images dating from around 1975, 1990, 2000 and 2005 will be interpreted and classified. The approach is expected to deliver estimates of forest cover change at the regional level. The general legend for the remote sensing survey includes 8 land cover classes. A minimum mapping unit (MMU) of 5 ha (or 50 pixels at 30 m spatial resolution) will be considered (Achard et al. 2007).

3.2 NASA Land Cover and Land Use Change

Another initiative that aims at monitoring global forest and forest cover change is undertaken as part of the Land Cover and Land Use Change program (LCLUC) of NASA. The project uses a block sampling strategy (18.5 x 18.5 km) based on change indicator maps from coarse resolution satellite imagery (MODIS). The coarse resolution maps deliver an estimate of the most likely regions for forest cover change. These samples are subsequently analysed based on medium resolution image pairs (Landsat ETM+) for the years 2000 and 2005. The datasets are taken from the global land surveys (GLS) that have been acquired based on Landsat imagery in recent years, namely "Geocover" (baseline 2000) and the Mid Decadal Global Land Survey (MDGLS) for the baseline year 2005. Together with the existing dataset of the Tropical Rainforest Information Centre (TRFIC), this compilation at present makes the largest single Landsat archive in the world outside the US federal government. The efficacy of the global sample approach to deforestation monitoring is currently tested.

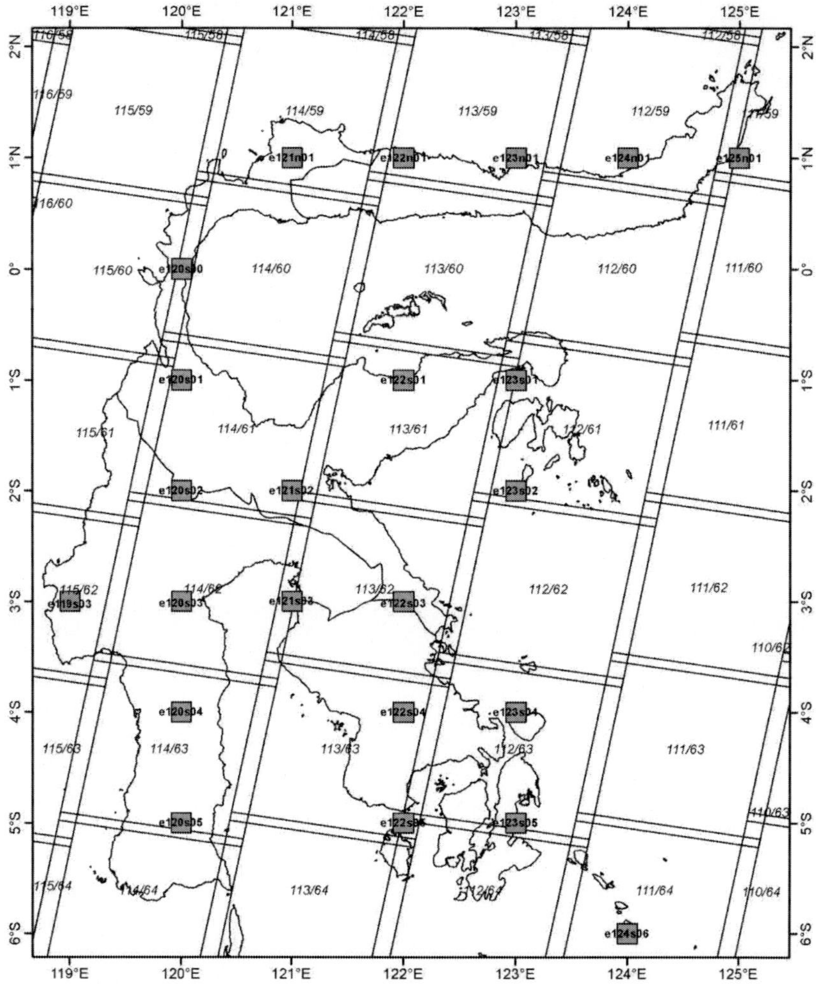

Fig. 1. Distribution of FRA2010 sampling tiles over the island of Sulawesi, Indonesia, superimposed by Landsat ETM+ WRS-2 data frames

3.3 Tropical Ecosystem Environment Observation by Satellites - TREES 3

The TREES-3 action provides quantitative measurements and mapping of changes in forest resources for the EU policies related to global environmental and forestry issues, with a focus on Eurasian boreal forests and tropical forests, including the Caribbean and Pacific regions.

The program will generate regional forest maps, track areas of rapid forest change and produce statistically valid estimates of forest cover change. The TREES-3 products will be used as input for future climate change impact scenarios and will provide a basis for country input into the Kyoto Protocol process. In addition to the forest cover maps, biomass maps and carbon emission estimates will be produced for selected forest ecosystems.

TREES will document forest cover changes for baseline years comparable to the other two mentioned programs (1975, 1990, 2000, 2005) using a similar block sampling approach as NASA LCLUC but based on ENVISAT-MERIS data for coarse resolution global coverage.

4 Forest cover change in Southeast Asia between 2000 and 2005

4.1 Comparison of estimates at the national level

With the release of the GlobCover product, ESA for the first time provided a bi-temporal global coverage of the Earth land surface based on a consistent classification system (LCCS). This fact presumes that the two products, GLC2000 and GlobCover comprise comparable information for a direct and reliable wall-to-wall mapping and change analysis at the global to national level. It is further assumed that a change analysis based on global data should provide comparable results to the global assessments of the FAO. For an evaluation of these three data sources and in order to provide comparability to the forthcoming FRA2010, the global land cover maps were reclassified to the eight proposed land cover classes of FAO FRA (see Tables 2, 3). The aggregation scheme that is shown in the table illustrates the general ability of the LCCS based land cover products to be merged at a broader thematic level. It also documents the higher thematic detail of the GlobCover legend compared to GLC2000.

The reclassification of the global land cover products yielded a single class of "forest" (ID 1 in Tables 2, 3) that is in line with the definition of forest by the FAO: "Land [...] with trees higher than 5 metres and a canopy cover of more than 10 percent [...]" (FAO 2006). This class builds the basis for all subsequent analysis steps of net forest change analysis. The forest cover assessment and calculation of annual net forest change rates has been carried out at the national level for all eleven states of Southeast Asia.

The results of the change analysis from the satellite based global land cover products indicate a mean annual forest loss of 0.2 % compared to an annual rate of -1.3 % for the FAO FRA data (Table 4). Subsequently, the total amount of net forest loss considerably differs between the two estimates (~23 T km2 for GLC/GlobCover and ~144 T km2 for FRA). The comparison of net forest loss rates at the national level reveals considerable differences in some countries (Cambodia, Vietnam) up to opposite trends in change rates for a number of nations (Laos, Malaysia, Myanmar, Philippines). Comparable net forest loss estimates in terms of strength and direction of forest change

Table 2. Workflow for the generalization of the GLC2000 legend to the FAO scheme.

Label GLC2000	Class-ID GLC2000	Class-ID harmonized GLC/ GlobCover	Class-ID FRA*
Tree cover, broadleaved, evergreen, closed and closed to open	1	40	
Tree cover, broadleaved, deciduous, mainly open (incl. Dry Dipterocarpus)	3	50	1
Tree cover, regularly flooded, Mangrove	4	160	
Tree cover, regularly flooded, Swamp	5		
Mosaic: Tree cover / Other nat. vegetation or Cropland (incl. very degraded and open tree cover)	2	30	
Mosaics & Shrub Cover, shrub component dominant, mainly evergreen	6	110	2
Mosaics & Shrub Cover, shrub component dominant, mainly deciduous	7		
Cultivated and managed, irrigated (flooded, rice, shrimp farms)	13	11	
Cultivated and managed, non irrigated (mixed)	12	14	
Mosaics of Cropland / Other natural vegetation (Shifting cultivation in mountains)	9	20	3
Shrub cover, mainly deciduous, (Dry or burnt)	8	130	
Herbaceous Cover (incl. alpine grassland)	10	140	
Sparse herbaceous cover > 3000m	11	150	
Artificial surfaces	16	190	
Bare Areas (Rock: Lime stone)	14	200	4
Snow and Ice	15	220	
Water Bodies	17	210	6
Sea	0		
No Data	18	230	8

* 1=forest; 2=mosaic of trees and other land cover; 3=other vegetated land cover; 4=other non-vegetated land cover; (5=burnt areas); 6=water; (7=clouds); 8=no data

Table 3. Workflow for the generalization of the GlobCover legend to the FAO scheme (for explanation of FRA classes see table 2).

Label GlobCover	Class-ID GLC2000	Class-ID harmonized GLC/ GlobCover	Class-ID FRA*
Closed to open (>15%) broadl. evergr. or semi-deciduous forest (>5m)	40	40	
Closed (>40%) needlel. evergr. forest (>5m)	70		
Closed to open (>15%) mixed broadl. and needlel. forest (>5m)	100		
Closed (>40%) broadl. decid. forest (>5m)	50	50	1
Open (15-40%) broadl. decid. forest/woodland (>5m)	60		
Open (15-40%) needlel. decid. or evergreen forest (>5m)	90		
Closed to open (>15%) broadl. forest regularly flooded (semi-permanently or temporarily) - Fresh or brackish water	160	160	
Closed (>40%) broadl. forest or shrubland permanently flooded - Saline or brackish	170		
Mosaic vegetation (grassland/shrubland/forest) (50-70%) / cropland (20-50%)	30	30	
Mosaic forest or shrubland (50-70%) / grassland (20-50%)	110	110	2
Mosaic grassland (50-70%) / forest or shrubland (20-50%)	120		
Post-flooding or irrigated croplands (or aquatic)	11	11	
Rainfed croplands	14	14	
Mosaic cropland (50-70%) / vegetation (grassland/shrubland/forest) (20-50%)	20	20	
Closed to open (>15%) (broadl. or needlel., evergr. or decid.) shrubland (<5m)	130	130	3
Closed to open (>15%) herbaceous vegetation (grassland, savannas or lichens/mosses)	140	140	
Closed to open (>15%) grassland or woody vegetation on regularly flooded or waterlogged soil - Fresh, brackish or saline	180		
Sparse (<15%) vegetation	150	150	
Artificial surfaces and associated areas (Urban areas >50%)	190	190	
Bare areas	200	200	4
Permanent snow and ice	220	220	
Water bodies	210	210	6
No data	230	230	8

Table 4. Forest cover assessment and net forest loss (2000 to 2005) for Southeast Asia based on global land cover products (pixel level) and FAO data.

		Net forest loss/gain (GLC/GlobCover)[1]		Net forest loss/gain (FRA2005)*		Net forest loss/gain (GLC/GlobCover)[2]	
Country	Total area (km^2)	(km^2)	(%/a)	(km^2)	(%/a)	(km^2)	(%/a)
Brunei	5932	16	0.1	-100	-0.7	-208	-0.8
Cambodia	187110	-18631	-5.1	-10940	-2.0	-51650	-8.3
Indonesia	1921030	-50487	-1.0	-93570	-2.0	-86402	-1.5
Laos	243951	54691	21.8	-3900	-0.5	-64426	-6.8
Malaysia	332765	4734	0.5	-7010	-0.7	-6429	-0.6
Myanmar	720977	18699	1.5	-23320	-1.4	-197475	-7.8
Papua New Guinea	468688	-47290	-2.9	-6950	-0.5	-37997	-2.2
Philippines	304869	7689	2.0	-7870	-2.1	11175	2.5
Singapore	598	13	8.4	0	0	7	3.5
Thailand	537542	-8684	-1.7	-2940	-0.4	-97817	-8.8
Vietnam	345070	15758	5.3	12060	2.0	-29613	-3.4
Total / Mean	5068532	-23492	-0.2	-144540	-1.3	-560834	-3.7

* Data source: FAO (2006); [1] FRA ID1: forest; [2] FRA ID1+2: forest + of trees and other land cover

processes can only be observed for Indonesia and Thailand. However, the patterns of forest change rates are divers and irregular.

These facts make it difficult to evaluate the results in terms of plausibility and consistency. The problem of reliability concerning the FRA estimates is known and has been discussed above. On the other hand, the visual interpretation of the global land cover products also reveals the problems of different spatial resolution and class inconsistency between the GLC2000 and the GlobCover product. As an example, Figure 2 shows a comparison of a subset of both products for the South-eastern part of Borneo (Kalimantan, Indonesia). It is obvious that the enhanced spatial resolution improves the detection and evaluation of the heterogeneity in a landscape. It can also be seen that the increase in spatial detail results in a closer distinction of quasi homogenous areas (e. g. forest or other vegetated land cover). As a consequence, the class mosaic of trees and other land cover has been reduced significantly and the areas have been split into the two homogenous classes forest and vegetated land cover. This means that formerly non-forest areas might now appear as forest, even though they did not change at all, and hence, these pseudo-change processes might overestimate reforestation processes in the region. Thus, in a control study we tested the change statistics at the national level assuming that the mosaic class is dominated by forest land. Again, the results show substantial ambiguities compared to the other estimates and an even unrealistic increase in the overall annual forest loss rate for Southeast Asia (3.7 %, see Table 4). For this reason, this alternative class comparison method has not been taken into account for further investigations.

4.2 Validation of GLC2000 and GlobCover at the regional level

As a follow up to the analysis at the national level and as a consequence of the observed ambiguities, the second part of the analysis dealt with the validation of forest cover and change from the global products based on regional level satellite data analysis for a case study covering the Island of Sulawesi, Indonesia. The validation study is based on a systematic sample of 20x20 km^2 frames as proposed by the FAO for FRA2010 (see section 3). The sampling grid of the tiles is aligned to a 1° by 1° grid, resulting in a sample of 23 image frames over the land surface of Sulawesi (see Figure 1).

For each of these tiles, land cover statistics were extracted from the global land cover products and an annual net forest loss rate was calculated. These data were compared to land cover classifications that were generated based on Landsat ETM+ data. The Landsat data were taken from the GLS for the years 2000 and 2005 and were processed using a supervised classification (maximum likelihood) based on the same eight classes that were defined for the global products. The results of the Landsat classification are summarized in Table 5 and opposed to the global land cover datasets. Due to high cloud coverage, the GLS scenes were substituted by other Landsat data from the

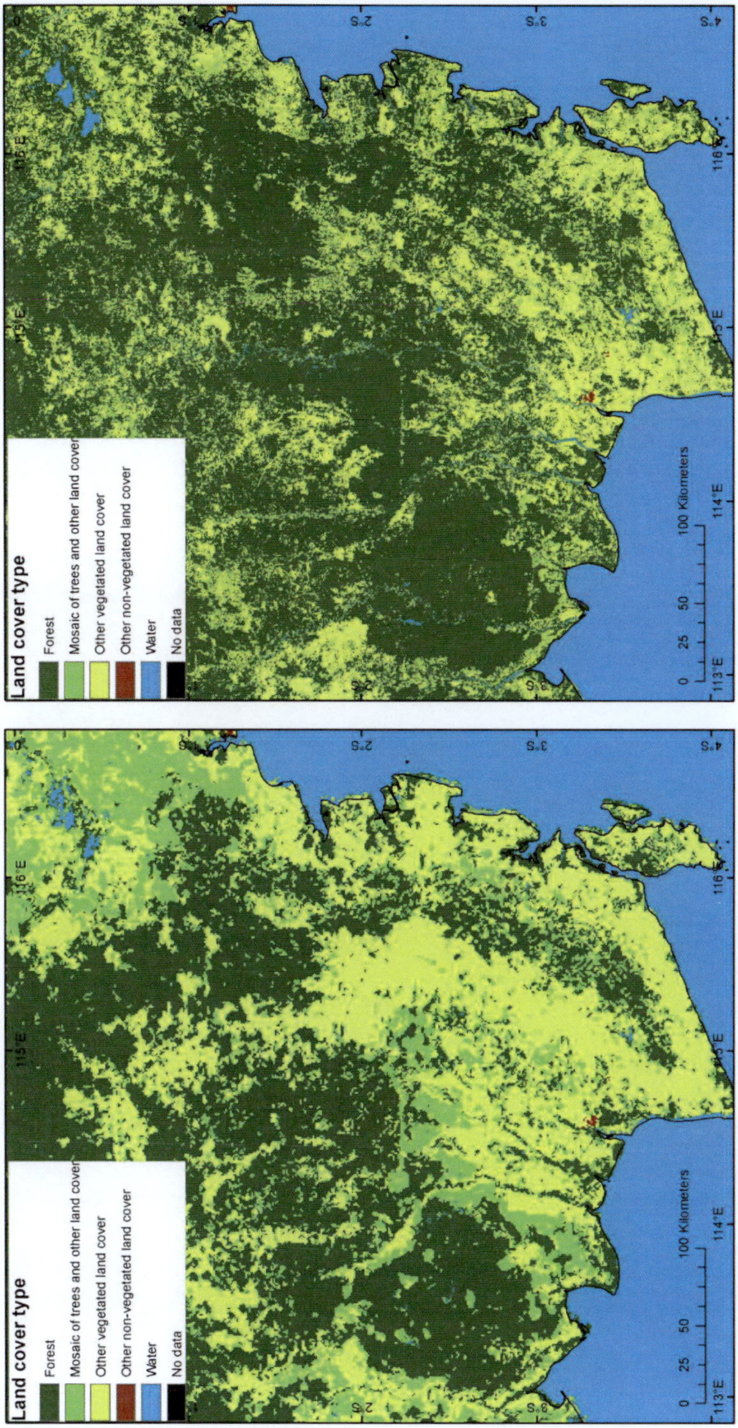

Fig. 2. Land cover variations between two global land cover products, GLC2000 (1 km resolution, left picture) and GlobCover (300 m resolution, right picture), for the South-eastern part of Borneo (Kalimantan, Indonesia). The Land cover legend is based on the aggregated classes as shown in Tables 2 and 3.

USGS archive in two cases, using a time frame of one year before and after the baseline years. Two tiles could not be processed at all due to missing data.

Table 5. Net forest loss estimates (Landsat ETM+ vs. GLC2000/GlobCover) for the island of Sulawesi, Indonesia based on the systematic sampling scheme of FAO FRA2010.

FRA tile ID	Path/row Landsat ETM+		Landsat ETM+ annual net forest loss (%)	GLC2000/ GlobCover annual net forest loss (%)	Land surface per tile (%)
e119s03	115	62	-0.4	-4.2	100
e120s00	115	60	0.0	-10.1	94
e120s01	114	61	-1.8	4.8	100
e120s02	114	61	-7.6	-0.2	100
e120s03	114	62	0.0	-9.9	100
e120s04	114	63	0.0	-15.5	89
e120s05	114	63	0.2	6.6	100
e121n01	114	59	-0.6	-2.9	100
e121s02	114	61	clouds	-3.5	100
e121s03	113	62	1.4	-10.1	11
e122n01	113	59	-8.0	-8.4	79
e122s01	113	61	0.0	-12.1	69
e122s03	113	62	-2.2	-2.0	100
e122s04	113	63	1.5	-5.8	100
e122s05	113	63	-6.2	-11.9	3
e123n01	113	59	-3.0	-14.7	13
e123s01	112	61	-0.4	-8.3	1
e123s02	112	61	0.0	-9.2	1
e123s04	112	63	-2.4	-14.2	33
e123s05	112	63	clouds	-15.9	36
e124n01	112	59	0.0	-15.8	1
e124s06	111	64	-0.2	-14.5	21
e125n01	111	59	-0.4	15.1	9
Mean (area weighted)			-1.8	-5.9	
Standard error			0.54	1.67	

The mean annual forest loss for Sulawesi is estimated at 1.8 % based on Landsat ETM+ data and at 5.9 % using global land cover products. This considerable difference is also expressed in the variability of the estimates ranging from -7.6 % to +1.5 % for Landsat (standard error = 0.54) and from -15.9 to +15.1 % (standard error = 1.67) for global products. In general, both net forest loss rates are significantly above the annual net deforestation as seen from GLC and GlobCover for the entire region of Southeast Asia (-0.2 %) and for Indonesia (-1.0) respectively.

5 Discussion

5.1 Net forest change analysis from global data sources

Net deforestation estimates for Southeast Asia based on a wall-to-wall approach using global land cover products are significantly below FAO estimates for the period 2000 to 2005. In addition, the tendencies in the results at the national level are inconsistent among the two estimates. A reason for these ambiguities is that the FAO FRA2005 estimates are not based on a sole, objective information basis and in many cases only little is known about the data sources. Until today, many countries do not use spatially explicit data sources (e.g. satellite data) for reporting measures within the FAO framework. This process is only just being initialized by the forthcoming FRA2010 and will assure for more consistency and a comparable and harmonized workflow for the generation of net forest change estimates. On the other hand, the evaluated global land cover products do not provide a meaningful data source for the assessment of net forest change at the national to regional level due to differences in spatial resolution and open questions in class assignment (e.g. the mosaic classes in GLC2000 and GlobCover). In particular, the change analysis of the global products revealed extensive forest gain rates for some countries that cannot be explained solely by dedicated reforestation or afforestation programs and that are not in common with the trends reported by the national agencies (e.g. Laos, Vietnam). Summarizing, it can be concluded for this part of the study, that none of the evaluated global data sources yields reliable estimates of net forest change at the country level. Global land cover products provide readily available and easy to use data sources for a spatially explicit mapping of the land surface and its change. They may serve as indicators of large-scale forest change or stratification tools in developing sampling approaches for forest change analysis (Olander et al. 2008, Achard et al. 2007). However, the spatial resolution and ambiguities in class accuracy noticeably restrict the use of such data and call for a comprehensive area-wide validation of the present products especially for those regions that show severe inconsistencies between national estimates and (coarse resolution) satellite based assessments.

5.2 Regional level estimates for net forest change analysis

The regional net deforestation estimate for a selected case study in Sulawesi, Indonesia is significantly higher than the mean net forest loss in Southeast Asia. It is also considerably above that for the entire country of Indonesia which assumes, that at least for the time period under investigation the island of Sulawesi comprises a hot spot of net deforestation in Indonesia. However, the results obtained by the regional validation study with Landsat ETM+ data show net deforestation rates that are more realistic and comparable to findings of other case studies in the region (e.g. Erasmi et al. 2004). This concludes that the GLC2000 / GlobCover net forest loss estimates even though they are significantly below that of FRA2005 still exceed the estimates of regional validation studies. One reason for this is that medium resolution satellite data are more sensitive to gradual changes or modifications in forests (forest degradation) although there is still considerable uncertainty in estimating forest degradation stages even from medium resolution satellite data especially in tropical regions (DeFries et al. 2002, Gibbs et al. 2007).

On the other hand, the analysis of tropical deforestation at the regional scale requires notably more workload than other investigations aiming at trans-continental to cross-continental comparison. A solution to handle the amount of data and analysis workload for regional scale deforestation mapping is to develop an appropriate sampling scheme that provides a balance between data reduction and validity of the sample for the region under investigation. Such a sampling design has been proposed by the FAO and has been applied in the present case study. Besides the sampling scheme, the regional mapping is strongly dependent on the availability of satellite data at sufficient temporal (at least two dates) and spatial (at least Landsat-like) resolution, as well as on the implementation of an adequate land cover classification procedure. In the present study, the 1° by 1° sampling grid of FRA2010 has shown to be capable for generating a systematic, non-weighted subset of land surface tiles for mapping and monitoring land cover and forest changes. Problems have emerged only from the irregular shape of the land surface of Sulawesi which certainly constitutes an extreme case of land demarcation from water areas. In this case, a number of sampling tiles only included small amounts of land cover ($\sim 1/3$ of all tiles with < 25 % land surface).

On the other hand, a systematic sampling precludes the existence of a bias towards deforestation hot spot areas and thus minimizes the problem of overestimation of net deforestation rates, like e. g. in Achard et al. (2002). Another issue of a systematic sampling is the grid resolution and location of the initial cell of the grid (Czaplewski 2003). Duveiller et al. (2008) tested different systematic sampling approaches for the entire region of the Congo basin and figured out that the statistical precision of tropical forest cover change mapping depends on the sampling intensity. Furthermore, they showed that the starting point of the sampling grid considerably influences the estimates of deforestation rates (Duveiller et al. 2008).

Considering the classification methods, a supervised procedure always makes high demands on the interpreter. In most cases it is not feasible for operational mapping purposes due to radiometric inconsistencies and distortions in the underlying satellite data sources. However, supervised classification still provides reliable and meaningful results that can only partly be replaced or complemented by semi-automatic techniques like e.g. segmentation and multi-date object delineation (see Lu et al. 2004 for an overview of change detection techniques).

6 Conclusions

The review of global land cover initiatives and the evaluation of up-to-date land cover products pointed out the achievements and challenges of global products data for pan-tropical mapping of forest cover and monitoring of net forest cover changes. It has been documented in a case study that forest change at the regional level can only reliably be mapped using medium resolution satellite data. The USGS Landsat archive provides an irreplaceable and unique source of information for land cover change analysis and especially for any study related to the baselines 2000 and 2005 through the availability of its GLS collections.

Cloud coverage remains a sustained challenge in any issue related to spatially explicit tropical forest monitoring. According to the data availability within the case study for the island of Sulawesi, it seems feasible to exceed the time frames for the recent baseline years in order to achieve sufficient coverage. However, spatial coverage of other tropical regions may be worse and is strongly dependent on the co-occurrence of cloud frequency and repetition rate of the satellite system. Hence, strategies for future operational monitoring of forest cover change in tropical regions should include strong efforts on the development of a data use policy for existing and planned multi-spectral satellite systems and the development of a multi-sensor (optical, radar) concept for a cross-continental tropical forest assessment.

Acknowledgements
We would like to thank the German Research Foundation (DFG) for funding parts of this work within the collaborative research centre SFB-552 (Stability of Rainforest Margins in Indonesia).

References

Achard F, Eva HD, Stibig HJ, Mayaux P, Gallego J, Richards T, Malingreau JP (2002) Determination of Deforestation Rates of the World's Humid Tropical Forests. Science 297 (9): 999-1002

Achard F, Eva H, Mayaux P (2001) Tropical forest mapping from coarse spatial resolution satellite data: Production and accuracy assessment issues. International Journal of Remote Sensing, 22: 2741-2762

Achard F, DeFries R, Eva H, Hansen M, Mayaux P, Stibig HJ (2007) Pantropical monitoring of deforestation. Environmental Research Letters 2

Arino O, Gross D, Ranera F, Bourg L, Leroy M, Bicheron P, Latham J, Di Gregorio A, Brockman C, Witt R, Defourny P, Vancutsem C, Herold M, Sambale J, Achard F, Durieux L, Plummer S, Weber JL (2008) GlobCover. ESA service for global land cover from MERIS. 2007 IEEE International Geoscience and Remote Sensing Symposium, IGARSS 2007: 2412-15

Bartholomé E, Belward AS (2005) GLC2000: a new approach to global land-cover mapping from Earth observation data. Int J Remote Sens 26 (9): 1959-1977

Bonan GB, (2008) Forests and climate change: Forcings, feedbacks, and the climate benefits of forests. Science 320: 1444-1449

Czaplewski RL (2003) Can a sample of Landsat sensor scenes reliably estimate the global extent of tropical deforestation? Int. J. Remote Sens. 24: 140912

DeFries RS, Houghton RA, Hansen MC, Field CB, Skole D, Townshend J (2002) Carbon emissions from tropical deforestation and regrowth based on satellite observations for the 1980s and 1990s. Proceedings of the National Academy of Sciences of the United States of America 99: 14256-14261

DeFries R, Hansen M, Townshend JRG, Janetos AC, Loveland TR (2000) A new global 1km data set of percent tree cover derived from remote sensing. Global Change Biology 6: 247-254

Duveiller G, Defourny P, Desclée B, Mayaux P (2008) Deforestation in central Africa: estimates at regional, national and landscape levels by advanced processing of systematically-distributed Landsat extracts. Remote Sens. Environ. 112 (5): 1969-1981

Erasmi S, Twele A, Ardiansyah M, Malik A, Kappas M (2004) Mapping deforestation and land-cover conversion at the rainforest margin in Central Sulawesi, Indonesia. EARSeL eProceedings, 3 (3): 388-397

Erasmi S, Kappas M, Twele A, Ardiansyah M. (2007) From global to regional scale: Remote Sensing based concepts and methods for mapping land-cover and land-cover change in tropical regions. In: Stability of tropical rainforest margins: Linking ecological, economic and social constraints (Tscharntke T et al., eds.), Springer, Heidelberg, p 437-462

FAO (2006) FRA 2005 - Global Forest Resources Assessment 2005. (Rome, FAO) 320 pp

Friedl MA, McIver DK, Hodges JCF, Zhang XY, Muchoney D, Strahler AH, Woodcock CE, Gopal S, Schneider A, Cooper A, Baccini A, Gao F, Schaaf C (2002) Global land-cover mapping from MODIS: algorithms and early results. Remote Sens Environ 83 (1): 287-302

Gibbs HK, Brown S, Niles JO, Foley JA (2007) Monitoring and estimating tropical forest carbon stocks: making REDD a reality. Environmental Research Letters 2

Gillespie TW, Foody GM, Rocchini D, Giorgi AP, Saatchi S (2008) Measuring and modelling biodiversity from space. Progress in Physical Geography 32: 203-221

GOFC-GOLD (2008) Reducing greenhouse gas emissions from deforestation and degradation in developing countries: a sourcebook of methods and procedures for monitoring, measuring and reporting, GOFC-GOLD Report version COP13-2, (GOFC-GOLD Project Office, Natural Resources Canada, Alberta, Canada)

Grainger A (2008) Difficulties in tracking the long-term global trend in tropical forest area. Proceedings of the National Academy of Sciences of the United States of America 105: 818-823

Hansen MC, Reed R (2000) A comparison of the IGBP DISCover and University of Maryland 1 km global land-cover products. Int J Remote Sens 21 (6&7): 1365-1373

Hansen MC, DeFries RS, Townshend JRG, Sohlberg R, Dimiceli C, Carroll M (2002) Towards an operational MODIS continuous field of percent tree cover algorithm: examples using AVHRR and MODIS data. Remote Sensing of Environment 83: 303-319

Hansen MC, Stehman SV, Potapov PV, Loveland TR, Townshend JRG, DeFries RS, Pittman KW, Arunarwati B, Stolle F, Steininger MK, Carroll M, DiMiceli C (2008) Humid tropical forest clearing from 2000 to 2005 quantified by using multitemporal and multiresolution remotely sensed data. Proceedings of the National Academy of Sciences of the United States of America 105: 9439-9444

Herold M, Johns T (2007) Linking requirements with capabilities for deforestation monitoring in the context of the UNFCCC-REDD process. Environmental Research Letters 2

Herold M, Latham JS, Gregorio AD, Schmullius CC (2006) Evolving standards in land cover characterization,Journal of Land Use Science 1 (2): 157 168

Herold M, Mayaux P, Woodcock CE, Baccini A, Schmullius C (2008) Some challenges in global land cover mapping: An assessment of agreement and accuracy in existing 1 km datasets. Remote Sensing of Environment 112: 2538-2556

Holmgren P, Marklund LG, Saket M, Wilkie ML (2007) Forest monitoring and assessment for climate change reporting: partnerships, capacity building and delivery. Forest Resources Assessment Working Paper 142. FAO,

Rome, available online at: www.fao.org/forestry/fra (last access 30 January 2009)

Intergovernmental Panel on Climate Change (IPCC) (2007) Climate change 2007: synthesis report. IPCC fourth assessment report (Geneva, Switzerland)

Loveland TR, Reed BC, Brown JF, Ohlen DO, Zhu Z, Yang L, Merchant JW (2000) Development of a global land-cover characteristics database and IGBP DISCover from 1 km AVHRR data. Int J Remote Sens 21 (6+7): 1303-1330

Lu D, Mausel P, Brondizio E, Moran E (2004) Change detection techniques. Int J Remote Sens 25 (12): 2365-2407

Lund G (1999) A "forest" by any other name.. Environmental Science and Policy 2(2): 125-134

Lund H. Gyde (coord.) (2008) Definitions of Forest, Deforestation, Afforestation, and Reforestation. Available online at: http://home.comcast.net/~gyde/DEFpaper.htm (last access: 30 January 2009)

Malhi Y, Wright J (2004) Spatial patterns and recent trends in the climate of tropical rainforest regions. Philosophical Transactions of the Royal Society of London Series B-Biological Sciences 359: 311-329

Mayaux P, Holmgren P, Achard F, Eva H, Stibig H, Branthomme A (2005) Tropical forest cover change in the 1990s and options for future monitoring. Philosophical Transactions of the Royal Society B-Biological Sciences 360: 373-384

Meyfroidt P, Lambin EF (2008) Forest transition in Vietnam and its environmental impacts. Global Change Biology 14: 1319-1336

Miettinen J, Liew SC (2005) Connection between fire and land cover change in Southeast Asia: a remote sensing case study in Riau, Sumatra. International Journal of Remote Sensing 26: 1109-1126

Miles L, Kapos V (2008) Reducing greenhouse gas emissions from deforestation and forest degradation: Global land-use implications. Science 320: 1454-1455

Nagai S, Ichii K, Morimoto H (2007) Interannual variations in vegetation activities and climate variability caused by ENSO in tropical rainforests. International Journal of Remote Sensing 28: 1285-1297

Olander LP, Gibbs HK, Steininger M, Swenson JJ, Murray BC (2008) Reference scenarios for deforestation and forest degradation in support of REDD: a review of data and methods. Environmental Research Letters 3

Sarkar S, Chiu L, Kafatos M, Singh R (2007) Sensitivity of rainfall on land cover change over South East Asia: Some observational results. Advances in Space Research 39: 73-78

Sodhi NS, Koh LP, Brook BW, Ng PKL (2004) Southeast Asian biodiversity: an impending disaster. Trends in Ecology & Evolution 19: 654-660

Sodhi NS, Lee TM, Koh LP, Brook BW (2009) A Meta-Analysis of the Impact of Anthropogenic Forest Disturbance on Southeast Asia's Biotas. Biotropica 41: 103-109

Stehman SV (2005) Comparing estimators of gross change derived from complete coverage mapping versus statistical sampling of remotely sensed data Remote Sens. Environ. 96: 46674

Stibig HJ, Beuchle R, Achard F (2003) Mapping of the tropical forest cover of insular Southeast Asia from SPOT4-Vegetation images. Int J Remote Sens 24 (18): 3651-3662

Stibig HJ, Belward AS, Roy PS, Rosalina-Wasrin U, Agrawal S, Joshi PK, Hildanus, Beuchle R, Fritz S, Mubareka S, Giri C (2007) A land-cover map for South and Southeast Asia derived from SPOT-VEGETATION data. Journal of Biogeography 34: 625-637

Watson R, Noble I, Bolin B, Ravindranath NH, Verardo D, Dokken DJ (eds.) (2000) Land-use, Land-Use Change, and Forestry a special report of the IPCC. Cambridge University Press, Cambridge

Comparison of tree water use characteristics in reforestation and agroforestry stands across the tropics

Diego Dierick*, Norbert Kunert, Michael Köhler, Luitgard Schwendenmann, and Dirk Hölscher

Tropical Silviculture and Forest Ecology, Burckhardt Institute, University of Göttingen, Büsgenweg 1, D- 37077 Göttingen, Germany

*corresponding author: D. Dierick, email: ddieric@gwdg.de

Summary

In the tropics, reforestations and agroforestry become increasingly important and may help mitigate climate change. However, high water use by trees may deplete water resources for associated crops or other purposes. Choice of tree species might reduce water use rates to acceptable levels, but available information on species-specific water use characteristics is scarce. We addressed the following questions: 1) do species differ in xylem sap flux response to fluctuating environmental conditions, 2) are there species-specific differences in quantities of water used, and specifically 3) do universal rules relating tree size to water use apply? This chapter combines data on tree sap flux and water use gathered in Indonesia, Panama and the Philippines. These studies applied the same methods and were conducted in recently established stands (5-12 years old when studied) characterised by small diameter trees and relatively simple stand structure. We analyse data from more than 100 trees belonging to 17 species using a simple sap flux model. Model application suggested species-specific differences in parameters such as maximal sap flux velocity and responses to radiation and vapour pressure deficit. With respect to the quantity of water used per tree, we observed a strong correlation between tree diameter and tree water use, which confirms earlier publications. However, e.g. in the stands in the Philippines where tree diameter explained 65% of observed variation, some species clearly followed distinct trajectories. For a given diameter, up to twofold differences in tree water use among species were observed. Our findings thus support the idea that species selection can be used to control tree water use of future reforestations and within agroforestry

T. Tscharntke et al. (eds.), *Tropical Rainforests and Agroforests under Global Change*, Environmental Science and Engineering, 293–308, DOI 10.1007/978-3-642-00493-3_13,

systems. This will be especially relevant in areas where water resources are limited already or where climate scenarios predict decreasing precipitation.

Keywords: Indonesia , Panama , Philippines , sap flux , tree size , tree species

1 Introduction

Natural forests in the tropics are still being converted at a high rate. Many of the former forest areas have been degraded, fail to produce goods and do not contribute to the protection of climate and biodiversity. For the management of buffer zones around remaining protected forests as well as for the restoration of degraded land, reforestation and agroforestry can be suitable measures. The re-introduction of trees into the landscape and especially reforestations have however been criticised because trees are potentially heavy water users and might deplete water resources (Jackson et al. 2005). From a global synthesis it was concluded that annual runoff was on average reduced by 44% and by 31% when reforesting grass- and shrubland, respectively (Farley et al. 2005). Also trees in agroforestry may be problematic as they may reduce water availability for the main crops and additionally increase stand level transpiration. In Costa Rica, the estimated stand level transpiration by coffee with shade trees was on average twice as high as that of coffee grown without shade trees (van Kanten and Vaast 2006).

It has been suggested that a suitable tree species choice might reduce water use rates to acceptable levels (van Dijk and Keenan 2007). This contention is somewhat theoretical as available information on species-specific water use characteristics is scarce. Furthermore, the effectiveness of species selection is questionable as studies in diverse old-growth forests and a single-species tree plantation showed repeatedly that tree size is the main factor influencing tree water use (Cienciala et al. 2000; Meinzer et al. 2004; McJannet et al. 2007). Meinzer et al. (2005) suggested that, as a result of functional convergence, plants operating within given biophysical limitations, develop common patterns of sap flux and water use in relation to size characteristics across taxa. This would leave little room for species selection to serve as a tool to influence stand water use, at least if wood production or carbon fixation is a major goal.

A thorough analysis of the problem is difficult as single studies of sap flux and tree water use often lack sufficient replicates within a species and/or analyse a limited number of species. Comparison among studies is additionally hampered by differences in applied methods. Here we bring together data from three case studies namely from the Philippines (Dierick and Hölscher 2009), Indonesia (Köhler et al. in press) and Panama (Kunert, unpublished data), which were conducted in relatively young reforestation and agroforestry stands. Our compilation comprises 106 individual trees belonging to 17 species, which were studied and analysed using the same protocol. In this data compilation, we concentrate on time periods with ample soil water supply and on trees fully in leaves. As a diagnostic tool, we apply a model that predicts sap flux density based on meteorological parameters. The following questions were addressed: 1) do species differ in xylem sap flux response to fluctuating environmental conditions, 2) are there species-specific differences in quantities of water used, and specifically 3) do universal rules relating tree size to water use apply?

2 Methods

2.1 Study sites

In the Philippines, we worked on the island Leyte in the Eastern Visayas. Two study sites were located near the villages Marcos (10.765°N, 124.790°E) and Patag (10.736°N, 124.804°E) at an elevation of 30 and 40 m asl, respectively. Average rainfall in the region amounts to 2753 mm y^{-1} and rain is relatively evenly distributed throughout the year (PAGASA 2007). Average annual air temperature is 27.5 °C. The natural vegetation in the region is species-rich lowland dipterocarp forest (Langenberger 2006). After deforestation and intermittent cultivation, the degraded sites were reforested following the so called rainforestation approach (Margraf and Milan 1996). This means reforesting with a mixture of native species, promoting the incorporation of fruit trees and combining fast growing trees with shade tolerant species. At the time of the field study, from June to August 2006 at Marcos and July to September 2007 at Patag, both stands were 12 years old. Stem density was 796 and 1367 stems ha^{-1} in Marcos and Patag, respectively. We selected ten tree species, eight of which were native to the region (Table 1). Each species was represented by five individuals selected to have well exposed crowns. Only for *Hopea plagata* S. Vidal, a species with an in general smaller stature, some individuals had little exposed crowns.

In Indonesia, our study site was located in Central Sulawesi in the vicinity of the village of Marena (1.552°S, 120.020°E) at 560 m asl. Measurements made between 2002 and 2006 at Gimpu (5 km south of Marena, 471 m asl) show that the mean air temperature is 25.5 °C and annual rainfall in the region is around 2092 mm y^{-1}. Rainfall shows a weak bimodal pattern with rainy seasons (>100 mm per month) from March to June and October to December. The structure and species composition of surrounding natural forest, which is usually only remaining at higher elevations, was described by Gradstein et al. (2007). The studied agroforest was located on former agricultural land used for cultivating annual crops such as maize and was 6 years old at the time of study. Cacao (*Theobroma cacao* L.) was growing as crop under *Gliricidia sepium* (Jacq.) Kunth ex Steud. shade trees. The stem density was 1030 stems ha^{-1} for cacao and 325 stems ha^{-1} for *Gliricidia*. Both tree species were represented by 18 individuals and studied for a one-month period in February 2007.

Table 1. Characteristics of the study trees. Scientific names are listed with the respective abbreviations used.

Species abbreviation and scientific name		Family	Study Location	Native or Exotic (-)	Trees studied (n)	DBH (cm)		Tree height (m)		Crown area (m^2)	
						Mean	SD	Mean	SD	Mean	SD
Sc	*Shorea contorta* S. Vidal	Dipterocarpaceae	Philippines	N	5	18.2	7.0	16.1	3.5	14.4	11.7
Pm	*Parashorea malaanonan* Merr.	Dipterocarpaceae	Philippines	N	5	12.0	0.4	13.1	1.6	12.9	3.1
Hm	*Hopea malibato* Foxw.	Dipterocarpaceae	Philippines	N	5	11.6	2.4	13.3	1.8	10.4	6.0
Hp	*Hopea plagata* S. Vidal	Dipterocarpaceae	Philippines	N	5	6.6	1.0	9.4	1.2	6.7	1.0
Sm	*Swietenia macrophylla* King	Meliaceae	Philippines	E	5	14.6	1.3	14.2	1.5	14.5	6.4
Vp	*Vitex parviflora* A. L. Juss	Verbenaceae	Philippines	N	5	20.4	5.5	12.7	1.6	33.2	15.8
Mj	*Myrica javanica* Blume.	Myricaceae	Philippines	N	5	22.1	3.7	11.2	0.6	25.5	11.6
Sk	*Sandoricum koetjape* (Burmf.) Merr.	Meliaceae	Philippines	N	5	16.3	2.7	13.2	1.1	18.0	5.0
Dz	*Durio zibethinus* Murray	Bombacaceae	Philippines	N	5	19.8	7.3	13.8	3.0	34.1	22.6
Ga	*Gmelina arborea* Roxb.	Verbenaceae	Philippines	E	5	21.9	4.0	18.1	2.4	18.6	9.5
Tc	*Theobroma cacao* L.	Malvaceae	Indonesia	E	18	10.1	1.6	4.5	0.8	23.4	9.7
Gs	*Gliricidia sepium* (Jacq.) Kunth ex Steud.	Fabaceae	Indonesia	E	18	15.0	2.5	10.9	2.1	47.3	31.5
Ls	*Luehea seemannii* Triana & Planch	Tiliaceae	Panama	N	4	11.8	1.6	8.7	1.0	11.9	1.7
Ae	*Anacardium excelsum* Beberto & Balb. ex Kunth	Anacardiaceae	Panama	N	4	10.1	0.6	6.4	0.4	9.4	5.1
Hc	*Hura crepitans* L.	Euphorbiaceae	Panama	N	4	18.0	2.3	5.4	1.0	13.7	4.7
Co	*Cedrela odorata* L.	Meliaceae	Panama	N	4	12.0	0.6	11.7	1.1	9.7	4.0
Tr	*Tabebuia rosea* (Bertol) D.C.	Bignoniaceae	Panama	N	4	11.5	1.3	7.4	0.3	13.4	2.1

Our site in Panama was located near the village of Sardinilla, Central Panama (9.317°N, 79.633°W), which is approximately 50 km north of Panama City. The elevation of the site is 70 m asl. Mean annual precipitation measured at Barro Colorado Island (at 30 km distance) is 2627 mm, with 25-50 mm per month during peak dry season (January-March) and 250 mm per month during the rainy season (May to November). The mean annual temperature of the region is 25.9 °C (STRI 2009). The original forest vegetation at the Sardinilla site was probably a tropical moist forest, similar to that of the Barro Colorado National Monument (Leigh et al. 1996).The study site was clear-cut in the 1950s and later used for cattle ranging. At the time of our study, the stands were 6 years old and stem density was about 1100 stems ha^{-1}. Five tree species growing in monocultures were studied with four replicates each during the rainy season between June and September 2007.

2.2 Sap flux measurements

Sap flux density J_s (g cm^{-2} h^{-1}) was measured using 25 mm long thermal dissipation probes (Granier 1985). Per study tree, two sensor pairs were installed on opposite sides of the tree trunk in the outermost xylem. Sensors were shielded and protected by styrofoam boxes, reflective foil, and plastic foil. The thermocouple output from the thermal dissipation probes was measured every 30 seconds and 5 or 30 minute averages were stored using dataloggers and attached multiplexers (CR1000 and AM16/32, Campbell Scientific Inc., Logan, UT, USA). Sap flux density was calculated from raw temperature data using the calibration equation determined by Granier (1987).

2.3 Radial sap flux profiles and water use rates

Radial profiles of sap flux density J_s were used to determine tree water use. Therefore J_s was measured at one or two additional depths below the cambium and expressed relatively (%) to concurrent measurements at the outer reference depth. Sap flow of ring-shaped stem cross-sections was then calculated, taking into account the cross-sectional area of the ring corresponding with the respective installation depth, J_s as measured at reference depth at the outer xylem, and the normalised profile of J_s for the species considered (Hatton et al. 1990; Meinzer et al. 2005). Contributions of the different cross sections were added to determine total tree sap flow (g h^{-1}) and summed over a day to give daily tree water use rates WU (kg d^{-1}). Water use rates were also expressed as transpiration rates (T, mm d^{-1}) by dividing WU by the crown projection area of the respective tree (m^2).

2.4 Sap flux density model

We used a model to capture species characteristics of sap flux density and its responses to environmental conditions in analogy with work of O'Brien et al.

(2004). The sap flux model used is a modification of the Jarvis-type model (Jarvis 1976) which was originally developed to describe stomatal responses to environmental drivers. The model takes the form of a multiplication of non-linear response functions, each depending on a single environmental factor. Each individual response function takes a value between zero and one, thus limiting the overall response if one or more environmental factors become sub-optimal. The fact that environmental variables appear isolated in the model enhances the interpretation of model parameters. This multiplicative type of model has been widely used in a number of variations to describe canopy conductance (Herbst et al. 1999; Granier et al. 2000; Harris et al. 2004), stand transpiration (Oren and Pataki 2001; Whitley et al. 2008) and to describe sap flow patterns in individual trees (Cienciala et al. 2000).

We opted for a sap flux density model with radiation R_g ($W\ m^{-2}$) and vapour pressure deficit VPD (kPa) as explanatory variables. Additional factors such as soil moisture conditions were not included because a preliminary exploration of the data indicated that this would not improve model fit much. The model form used was

$$J_{s\ model} = a \ \frac{R_g}{b + R_g} \ \frac{1}{1 + \exp(\frac{c - \mathrm{VDP}}{d})} \tag{1}$$

with $J_{s\ model}$ = modelled sap flux density ($g\ cm^{-2}\ h^{-1}$), a = maximum modelled sap flux density ($g\ cm^{-2}\ h^{-1}$), b = parameter describing R_g response ($W\ m^{-2}$) and c, d = parameters describing VPD response (kPa).

Modelled sap flux density $J_{s\ model}$ reaches a maximum value a when all environmental conditions are optimal. To account for the influence of radiation we introduced a commonly used hyperbolic response function which asymptotically approaches a value of 1 at high R_g. Parameter b can be interpreted as a measure of the light saturation level. Assuming VPD is non-limiting, the sap flux reaches just over 90% of a if radiation levels equal ten times parameter b. The response function used to describe the influence of VPD was taken from O'Brien et al. (2004). In this response function parameter c equals the vapour pressure deficit for which $J_{s\ model}$ rises to half of the maximum value (R_g non-limiting), whereas d is related to the slope i.e. the increase in $J_{s\ model}$ for a given increase in VPD. Model parameters a, b, c and d in the response functions were estimated by minimising the residual sum of squares using a Gauss-Newton algorithm.

2.5 Statistical analyses

Although the model is nonlinear we calculated an adjusted R^2 in analogy with linear models. Root mean square error (RMSE) is used to characterise model prediction error. When an analysis of variance indicated a significant effect of tree species on a model parameter of interest, a post-hoc Tukey HSD test was carried out to assign species to statistically different groups. The relationships between maximum tree water use rates and tree diameter were established by simple linear regression. All statistical analyses were performed with R version 2.6.2 (R Development Core Team 2008).

3 Results and discussion

3.1 Sap flux densities and environmental controls

Maximum sap flux densities between 13.2 and 52.1 g cm^{-2} h^{-1} were measured in the course of this study. This is in line with values published for tropical forest tree species which are mostly situated between 5 and 50 g cm^{-2} h^{-1} (Granier et al. 1996; Meinzer et al. 2001; Dünisch and Morais 2002), although values of up to 70 g cm^{-2} h^{-1} have been reported (Becker 1996; O'Brien et al. 2004).

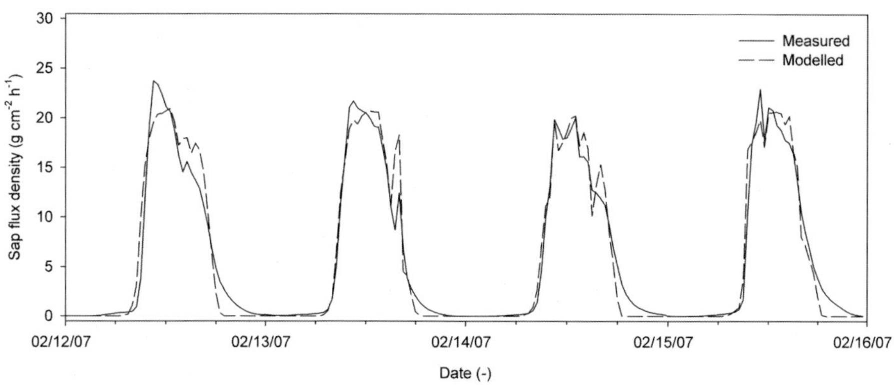

Fig. 1. Measured and modelled sap flux in a selected cacao tree (*Theobroma cacao*) for four days in February 2007, Indonesia. Note that the model form is such that modelled sap flux at night is set to zero.

Performance of the model describing sap flux density was quite satisfying, which is illustrated by high R^2_{adj} values for almost all study trees (R^2_{adj}, ranged from 0.62 to 0.97, average 0.92) and low root mean square errors (RMSE, range 0.8 to 6.9, average 2.0 g cm^{-2} h^{-2}). A visual impression of the agreement between modelled and measured sap fluxes is given for a representative cacao tree in Indonesia (Figure 1). Model performance suggests that the model form was appropriate and that vapour pressure deficit and radiation indeed exerted strong influence on sap flux of the trees in the study periods. We acknowledge that at other times of the year this could be different. Especially when rainfall is limited, soil moisture could become more influential as an explanatory variable and the need may arise to include it in the model. In particular under seasonal climatic conditions such as at our Panamanian site, this is expected to be the case.

For the three study sites combined, the model parameter a differed three-fold among species (14.4 g cm^{-2} h^{-1} for *Vitex parviflora* A. L. Juss and 47.2 g cm^{-2} h^{-1} for *Anacardium excelsum* Beberto & Balb. ex Kunth) and accurately reflected measured maximal sap flux densities in studied species. Within each of the three study sites significant species-specific differences were observed (Figure 2). Model parameter b, which describes the sap flux response to radiation, differed among species at the Philippine and the Indonesian site, but not among the five species studied in Panama. High values for parameter b were found in species which did not have full sun exposed crowns such as *Hopea plagata* in the Philippines and *Theobroma cacao* in Indonesia. A possible explanation is that in trees growing under shade or having a layered crown structure sap flux continues to rise with additional light even at high radiation.

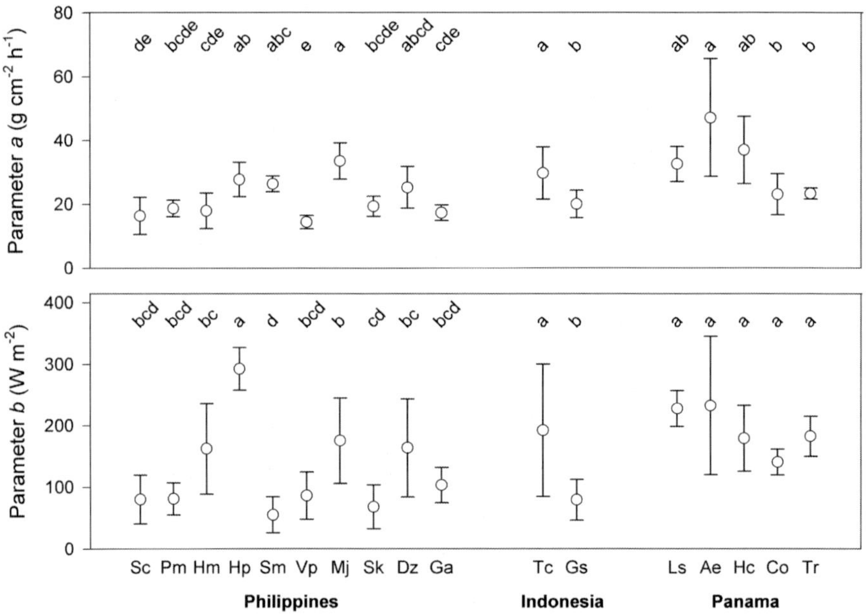

Fig. 2. Estimated model parameters a (upper panel) and b (lower panel) for the species studied (means and standard deviations, n depends on study site (see Table 1)). Significant differences ($p < 0.05$) between species within a study site are indicated by different small letters. The complete species names and abbreviations are listed in Table 1.

3.2 Sap flux density in relation to tree size

At first view our data suggests, despite a very low R^2_{adj}, a negative relation between maximum sap flux density $J_{s\ max}$ and tree diameter (Figure 3). This finding is explained by the significant decline in sap flux density with increasing diameter we observed in the dataset from Indonesia (Figure 3 inset). Rather than by a real effect of diameter on $J_{s\ max}$, the latter is caused by the different diameter range of sampled cacao and *Gliricidia* trees in combination with the different sap flux densities observed for these species. The complete lack of a diameter-related decline of sap flux density with diameter at the two other study sites supports this interpretation.

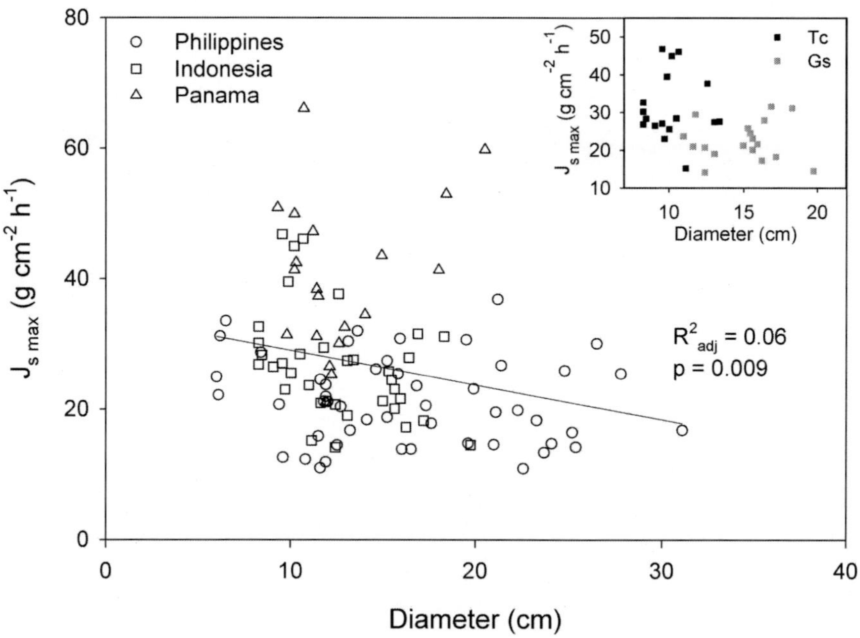

Fig. 3. Maximum sap flux density $J_{s\ max}$ for individual trees in the three studies in relation to tree diameter. The inset shows the data from the Indonesian study site.

The absence of a decline of sap flux density with diameter apparently contradicts the hypothesis of functional convergence (Meinzer et al. 2001; Meinzer et al. 2005). Their work in Panama revealed a strong inverse relationship ($R^2 = 0.85$) between maximum sap flux density and tree diameter in 24 co-occurring species in a Panamanian old-growth forest. Similarly, in a patch of advanced secondary forest in Vietnam, a weak but significant ($R^2 = 0.23$) decline of mean J_s with tree diameter existed (Giambelluca et al. 2003). At the same time however, that study acknowledged the presence of large differences in J_s across and within species and the decisive role of factors such as tree exposure and environmental conditions. That our data did not reveal a similar pattern of declining $J_{s\ max}$ with tree diameter could be due to the limited diameter range encompassed in comparison with the original work (Meinzer et al. 2001). The data presented by Meinzer et al. (2001) revealed considerable scatter, in particular in the lower diameter range where the steepest decline in $J_{s\ max}$ is predicted. A possible implication is that general patterns resulting from functional convergence may remain unnoticed if the range of tree diameters covered is too narrow. In our opinion the species differences we observed in $J_{s\ max}$ and responses of J_s to environmental variables do not necessarily contradict the hypothesis of functional convergence (Meinzer 2003). We hypothesise that this finding merely reflects that considerable variation remains within general patterns observed across species. This can in part be due to species-specific adaptation to a given set of growth conditions and possibly also due to stand structural differences (see also Cienciala et al. 2000).

In stands where functional convergence can be demonstrated, it would be an essential tool in assessing stand transpiration. This is particularly true in species-rich and highly structured natural forest stands (McJannet et al. 2007). However, we argue functional convergence to be less relevant in relatively young mixed reforestation or agroforest stands characterised by limited diameter ranges and stand structure. In our opinion the species-specific differences in sap flux density we observed in co-occurring trees would become relatively more important under such conditions and become a determining factor for stand water use.

3.3 Tree water use, tree diameter, and transpiration

Our data revealed a strong positive correlation between maximum daily tree water use and tree diameter (Figure 4A). This is true for the pooled data, as well as for data from the individual study sites (Figure 4B-D). Also Meinzer et al. (2004) found for different co-occurring species in Panama, that tree size rather than tree species was a determining factor for tree water use. Follow-up studies encompassing more species and replicates revealed similar allometric relationships between tree water use rates and tree diameter (and biomass) that were shared by species belonging to distinct species groups (Meinzer et al. 2005). The relation between tree water use and tree diameter was best described by a sigmoid relationship in 18 mainly tropical angiosperm tree species. That a simple linear regression between tree diameter and tree water use explains most of the variance in our data set, should be seen in relation to the small diameter range covered. Some theoretical relationships describing water use versus tree size characteristics might be inappropriate from a biological point of view, even though they provide an adequate fit in certain datasets (Meinzer et al. 2005). The same appears to be the case in our study.

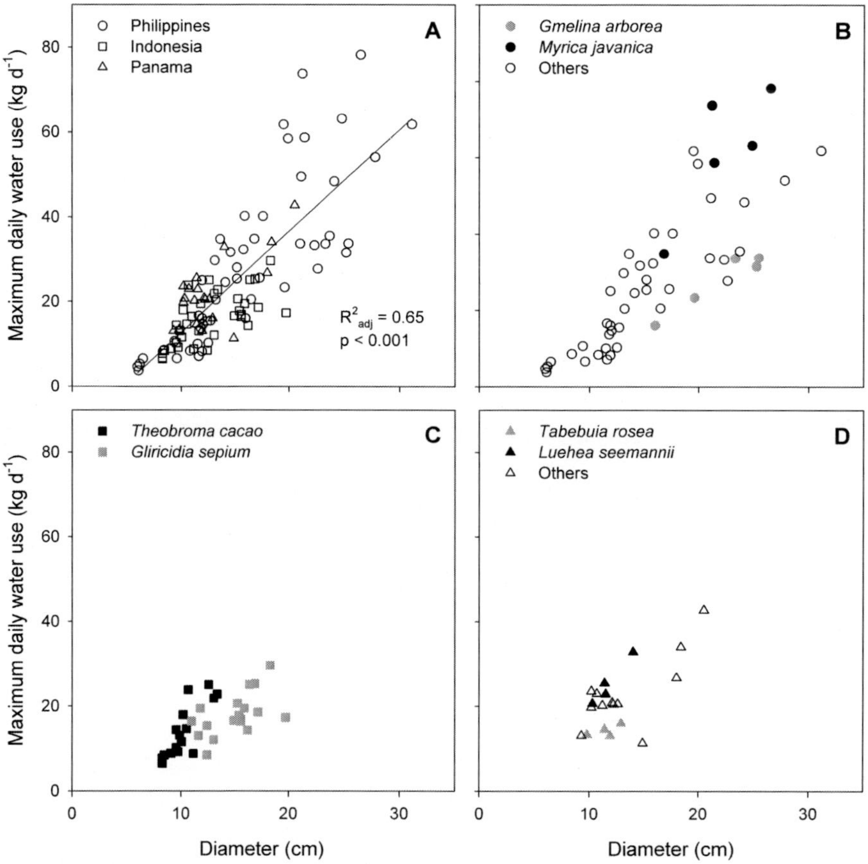

Fig. 4. Maximum daily water use rates WU_{max} for each individual in relation to tree diameter for all studied trees (A). Analogue plots show the data for species in the Philippines (B), Indonesia (C) and Panama (D). In each of the latter graphs two contrasting species are shown with a different symbol.

Despite the strong dependence of maximum water use on diameter for the individuals studied, considerable scatter was present around the fitted curve with species located on distinct trajectories (Figure 4B-D). For a given tree diameter, tree water use differed up to twofold between some of the species studied. This resulted from a combination of both species-specific J_s and species differences in hydroactive xylem depth. Apart from these factors, the relationship would be determined strongly by cross correlation between estimated water use rates and tree diameter. This can be expected to be most pronounced in the lower diameter ranges (Meinzer et al. 2001), making the deviations from a common relationship even more significant.

Maximum transpiration rates observed in individual trees varied from 0.4 to 4.9 mm d^{-1}, except for a single individual of *Swietenia macrophylla* King which had a maximum transpiration rate of 7.5 mm d^{-1}. A small crown projection area and associated larger relative error in this individual could explain the high value. Excluding this single tree, results are comparable with findings from Giambelluca et al. (2003) who measured daily transpiration rates in individual trees from 0.5 to 4.6 mm d^{-1} in a patch of advanced secondary forest in Vietnam during the wet season. In our study, mean transpiration rates over the study periods differed fourfold across species from 0.6 to 2.4 mm d^{-1} (Table 2). For the mentioned study in a forest patch in Vietnam, transpiration rates were also found to be different among species and were, in addition, influenced by crown exposure, seasonality and enhanced transpiration at the forest edge (Giambelluca et al. 2003). It is unlikely that differences in transpiration rates observed between species can be translated directly to transpiration at the stand level as different stand development and stand structure of contrasting species (or species mixtures) are likely to partly reduce species-specific transpiration rates. Nevertheless, we assume that the remarkable differences across tree species in both tree water use and tree transpiration rates could be exploited to manage stand water use.

4 Conclusions

We conclude that water use and transpiration rates found in trees of tropical reforestation and agroforestry stands showed considerable variation across species. Despite the strong dependence of maximal tree water use on tree diameter for the individuals studied, considerable scatter was present around the fitted curve with species located on distinct trajectories. At a given diameter, up to twofold differences between trees of different species were observed. Species selection might thus indeed be an effective tool to control water use in reforestation and agroforestry. It can be used to optimise the balance between wood production or carbon sequestration and the use of water resources.

Acknowledgements
This study was funded by the German Research Foundation in the framework of the SFB 552 (sub-project B4) and a further grant for the study in Panama (Ho-2119/3-1).

Table 2. Maximum (WU_{max}) and mean water use (WU_{mean}) and maximum (T_{max}) and mean (T_{mean}) transpiration rates (means and standard deviations, n depends on study site (see Table 1)). Significant differences ($p < 0.05$) are indicated by different small letters.

Species	WU_{max} (kg d^{-1}) Mean	SD	WU_{mean} (kg d^{-1}) Mean	SD	T_{max} (mm d^{-1}) Mean		SD	T_{mean} (mm d^{-1}) Mean		SD
Shorea contorta	25.6	19.5	18.4	14.4	2.03	abcd	0.93	1.45	abc	0.70
Parashorea malaanonan	15.3	1.2	10.6	1.1	1.24	bcd	0.28	0.85	bc	0.17
Hopea malibato	13.1	8.5	9.1	6.7	1.33	bcd	0.43	0.89	bc	0.29
Hopea plagata	5.7	1.9	4.0	1.3	0.85	cd	0.18	0.59	bc	0.12
Swietenia macrophylla	33.7	4.1	25.5	3.6	3.14	a	2.46	2.38	a	1.90
Vitex parviflora	30.7	14.6	20.7	9.3	1.02	bcd	0.32	0.69	bc	0.20
Myrica javanica	61.7	17.0	43.2	12.5	2.60	ab	0.60	1.80	ab	0.38
Sandoricum koetjape	32.8	16.5	23.4	12.6	1.90	abcd	0.98	1.36	abc	0.77
Durio zibethinus	44.7	18.5	32.9	14.8	1.52	abcd	0.81	1.11	bc	0.62
Gmelina arborea	27.6	7.8	19.8	6.1	1.67	abcd	0.46	1.18	abc	0.32
Theobroma cacao	13.4	6.2	10.0	4.5	0.63	cd	0.35	0.47	c	0.25
Gliricidia sepium	17.9	5.0	13.9	4.1	0.54	d	0.36	0.42	c	0.28
Luehea seemannii	25.5	5.3	13.1	3.6	2.14	abcd	0.26	1.11	bc	0.22
Anacardium excelsum	19.8	4.8	10.5	2.8	2.64	ab	1.58	1.37	abc	0.79
Hura crepitans	28.7	13.3	14.6	7.6	2.01	abcd	0.38	1.01	bc	0.23
Cedrela odorata	20.5	0.3	9.9	2.2	2.34	abc	0.74	1.09	bc	0.29
Tabebuia rosea	14.2	1.4	7.9	0.6	1.09	bcd	0.27	0.60	bc	0.11

References

Becker P (1996) Sap flow in Bornean heath and dipterocarp forest trees during wet and dry periods. Tree Physiology 16: 295-299

Cienciala E, Kucera J, Malmer A (2000) Tree sap flow and stand transpiration of two Acacia mangium plantations in Sabah, Borneo. Journal of Hydrology 236: 109-120

Dierick D, Hölscher D (2009) Species-specific tree water use characteristics in reforestation stands in the Philippines. Agricultural and Forest Meteorology 149: 1317-1326

Dünisch O, Morais RR (2002) Regulation of xylem sap flow in an evergreen, a semi-deciduous, and a deciduous Meliaceae species from the Amazon. Trees-Structure and Function 16: 404-416

Farley KA, Jobbagy EG, Jackson RB (2005) Effects of afforestation on water yield: a global synthesis with implications for policy. Global Change Biology 11: 1565-1576

Giambelluca TW, Ziegler AD, Nullet MA, Truong DM, Tran LT (2003) Transpiration in a small tropical forest patch. Agricultural and Forest Meteorology 117: 1-22

Gradstein SR, Kessler M, Pitopang R (2007) Tree species diversity relative to human land uses in tropical rain forest margins in Central Sulawesi. In: Tscharntke T, Leuschner C, Guhardja E, Zeller M (Eds.), The Stability of Tropical Rainforest Margins: Linking Ecological, Economic and Social Constraints of Land Use and Conservation. Springer, Berlin, p. 321-334

Granier A (1985) A new method of sap flow measurement in tree stems. Annals of Forest Science 42: 193-200

Granier A (1987) Evaluation of transpiration in a Douglas-fir stand by means of sap flow measurements. Tree Physiology 3: 309-319

Granier A, Huc R, Barigah ST (1996) Transpiration of natural rain forest and its dependence on climatic factors. Agricultural and Forest Meteorology 78: 19-29

Granier A, Loustau D, Breda N (2000) A generic model of forest canopy conductance dependent on climate, soil water availability and leaf area index. Annals of Forest Science 57: 755-765

Harris PP, Huntingford C, Cox PM, Gash JHC, Malhi Y (2004) Effects of soil moisture on canopy conductance of Amazonian rainforest. Agricultural and Forest Meteorology 122: 215-227

Hatton TJ, Catchpole EA, Vertessy RA (1990) Integration of sapflow velocity to estimate plant water-use. Tree Physiology 6: 201-209

Herbst M, Eschenbach C, Kappen L (1999) Water use in neighbouring stands of beech (Fagus sylvatica L.) and black alder (Alnus glutinosa (L.) Gaertn.). Annals of Forest Science 56: 107-120

Jackson BJ, Jobbagy EG, Avissar R, Roy SB, Barrett DJ, Cook CW, Farley KA, le Maitre DC, McCarl BA, Murray BC (2005) Trading water for carbon with biological carbon sequestration. Science 310: 1944-1947

Jarvis PG (1976) The interpretation of the variations in leaf water potential and stomatal conductance found in canopies in the field. Philosophical Transactions of the Royal Society of London B- Biological Sciences 273: 593-610

Köhler M, Dierick D, Schwendennman L, Hölscher D (in press) Water use characteristics of cacao and Gliricidia trees in an agroforest in Central Sulawesi, Indonesia. Ecohydrology, DOI: 10.1002/eco.67

Langenberger G (2006) Habitat distribution of dipterocarp species in the Leyte Cordillera: an indicator for species-site suitability in local reforestation programs. Annals of Forest Science 63: 149-156

Leigh EG, Rand AS, Windsor DW (1996) The ecology of a tropical forest, 2nd edn, Smithsonian Press, Washington, DC

Margraf J, Milan PP (1996) Ecology of dipterocarp forests and its relevance for island rehabilitation in Leyte, Philippines. In: Schulte A, Schöne D (Eds.), Dipterocarp Forest Ecosystems: Towards Sustainable Management. World Scientific, Singapore, p. 124-154

McJannet D, Fitch P, Disher M, Wallace J (2007) Measurements of transpiration in four tropical rainforest types of north Queensland, Australia. Hydrological Processes 21: 3549-3564

Meinzer FC (2003) Functional convergence in plant responses to the environment. Oecologia 134: 1-11

Meinzer FC, Goldstein G, Andrade JL (2001) Regulation of water flux through tropical forest canopy trees: Do universal rules apply? Tree Physiology 21: 19-26

Meinzer FC, James SA, Goldstein G (2004) Dynamics of transpiration, sap flow and use of stored water in tropical forest canopy trees. Tree Physiology 24: 901-909

Meinzer FC, Bond BJ, Warren JM, Woodruff DR (2005) Does water transport scale universally with tree size? Functional Ecology 19: 558-565

O'Brien JJ, Oberbauer SF, Clark DB (2004) Whole tree xylem sap flow responses to multiple environmental variables in a wet tropical forest. Plant, Cell and Environment 27: 551-567

Oren R, Pataki DE (2001) Transpiration in response to variation in microclimate and soil moisture in southeastern deciduous forests. Oecologia 127: 549-559

PAGASA (2007) Philippine Atmospheric, Geophysical and Astronomical Services Administration, Philippines.

R Development Core Team (2008) R: A language and environment for statistical computing. R Foundation for Statistical Computing, Vienna, Austria, ISBN 3-900051-07-0, http://www.R-project.org (last visited 28/05/2009)

STRI (2009) Temperature and rainfall data Barro Colorado Island, Panama 1987-2006.
http://striweb.si.edu/esp/physical_monitoring/download_bci.htm (last visited 28/05/2009)

Van Dijk AIJM, Keenan RJ (2007) Planted forests and water in perspective. Forest Ecology and Management 251: 1-9

Van Kanten R, Vaast P (2006) Transpiration of arabica coffee and associated shade tree species in sub-optimal, low-altitude conditions of Costa Rica. Agroforestry Systems 67: 187-202

Whitley R, Zeppel M, Armstrong N, Macinnis-Ng C, Yunusa I, Eamus D (2008) A modified Jarvis-Stewart model for predicting stand-scale transpiration of an Australian native forest. Plant and Soil 305: 35-47

A comparison of throughfall rate and nutrient fluxes in rainforest and cacao plantation in Central Sulawesi, Indonesia

Carsten Gutzler[1*], Stefan Koehler[2], and Gerhard Gerold[1]

[1] Institute of Geography, Department of Landscape Ecology, University of Göttingen, Goldschmidtstr. 5, 37077 Göttingen, Germany
[2] Landscape Ecology and Land Evaluation, Faculty for Agricultural- and Environmental Sciences, University of Rostock, Justus-von-Liebig Weg 6, 18059 Rostock, Germany

*corresponding author: C. Gutzler, email: cgutzle@gwdg.de

Summary

This study investigates throughfall rates and nutrient fluxes in two cacao plots and one natural forest plot in Central Sulawesi, Indonesia. Throughfall was collected over a 12 month period on each plot with bulk precipitation samplers and compared with open area precipitation of nearby reference sites. Samples of both throughfall and gross precipitation were analysed for chemical composition and nutrient enrichment.

Throughfall rates (based on total sums for the whole measurement period) differed strongly between cacao and rainforest sites: throughfall was 89 - 91% for the cacao plots and 81% in the natural forest. In-plot variance was high both between samplers and between sampling dates. Our findings imply an increase of yearly water input by 10% with the conversion of natural forests to cacao plantations as a result of reduced interception. Additional increases through reduced transpiration are documented in the literature. Higher runoff, probably mainly in low flow situations, is to be expected.

For the volume-weighted average concentrations of sodium (Na), calcium (Ca), magnesium (Mg), silicon (Si) and potassium (K) a moderate (Na: 0.9 – 1.5 times) to very high (K: 5.1 – 18.9 times) enrichment in throughfall compared with open area precipitation was determined. Volume-weighted average concentrations were similar on both cacao plots. As expected, enrichment factors in the natural forest were much higher than on the cacao plots.

A reduction of nutrient input with rainfall into the soil will undoubtedly affect soil chemistry and biology.

T. Tscharntke et al. (eds.), *Tropical Rainforests and Agroforests under Global Change*, Environmental Science and Engineering, 309–326, DOI 10.1007/978-3-642-00493-3_14,

Keywords: cacao agroforestry, interception, nutrients, Indonesia

1 Introduction

The disappearance of natural forests worldwide is an ever-increasing cause of concern. Although public awareness seems to centre on the Amazonian rainforest, nowhere are loss rates nearly as high as in South East Asia.

According to the FAO (2007) Indonesia has by far the highest rate of natural forest loss worldwide, having lost as much as 13% in the 5 year period from 2000 to 2005. A study for the World Bank and the Indonesian Department for International Development (PEACE 2007) also considers Indonesia to be the third biggest producer of greenhouse gasses, mainly due to forest conversion. Forests are not only logged (legally and illegally) for timber, but also in order to obtain arable land. Price development on the world market has made cocoa a profitable export commodity and particularly on the island of Sulawesi great areas of land are being converted into cacao plantations. Indonesia has now become the third largest producer of cocoa, supplying 15% of the world market (ICCO 2008). The centre of production is the island of Sulawesi, which provides more than 50% of the national production (ICB 2008). The effects of these changes on a landscape level are difficult to predict.

The quantity of rainfall intercepted by the canopy is of major importance for the hydrological cycle and the water yield of the area. The fraction that never reaches the ground and subsequently evaporates is termed interception, whereas the fraction reaching the ground can be divided into free throughfall, in which the rain falls through canopy gaps without touching the foliage, non-free throughfall, in which the rainwater touches the leaves causing physical and chemical interaction, and stemflow, in which the rainwater flows down along the stems of the trees (Gash 1978). Despite greater biomass and leaf area per square meter (LAI), the percentage of rainwater intercepted by tree canopies is lower in tropical rainforests than in temperate forests. This is primarily due to the characteristics of tropical rainfall: short convective rain showers with rainfall quantities exceeding canopy storage capacity, most rainfall occurring during the afternoon when water vapour deficit and wind speed have decreased, and rain falling with large drop size which negatively affects storage capacity (Crockford & Richardson 2000, Calder 2001).

Cacao agroforests replacing natural rainforests usually have leaf area index (LAI) values lower than that of natural forests. However, even if LAI values are similar in cacao agroforest and natural rainforest, this does not mean that quantity of rainwater reaching the ground will be the same. Although canopy interception is positively correlated with LAI, it may differ strongly depending on leaf characteristics (Hall 2003). Conversion of natural forest to cacao plantation will probably result in an increase in river discharge due to reduced interception and transpiration (Imbach et al. 1989, Kleinhans 2003). This increase is believed to primarily affect base-flow situations. However, if the infiltration capacity of the soil is affected by compaction or trail-building, peak flows and erosion will also increase (Bruijnzeel 1990, Bruijnzeel 2004). Even though a great number of throughfall studies in the inner tropics have already been published, there is still a need for data comparing cacao agroforests with natural forest under identical soil and climatic conditions. In particular, chemical data is required on throughfall composition. The present study makes a contribution to filling this gap.

The elements and nutrients analysed in our study fall into two categories: on the one hand, we analysed the nutrients that are necessary for plant growth and have to be monitored when determining sustainability of land use. On the other hand, we analysed elements such as sodium and silicon, which may help to separate sources of input (biogenic or pedo-/geogenic) and can be used to determine different water flowpaths when incorporated into a watershed model.

The objectives of this study were to assess the differences in throughfall rate and throughfall composition between natural forest, a 7-year-old and an 18-year-old cacao plantation under the same climatic conditions, as well as to calculate the annual input of elements and nutrients with throughfall for both land use types.

2 Study Site and Measurements

2.1 Site Description

The area surrounding the village Bulili in the Palolo Valley (1° 11.1' South; 120° 5.0' East) has experienced a steady land conversion of natural forest into cacao plantations by smallholders. To predict the effects of forest conversion on water balance and nutrient cycle, an experimental catchment site has been instrumented there (Kleinhans 2003) and plot level studies have been implemented.

The plots are located close to the village. At this location, cacao plantations cover the valley floor and extend up the slope of the first hills of the national park. The local community does not respect the official border of the national park and so the actual border between agricultural land and natural forest has constantly moved upwards since 2000. At present the upper part of the hills is still covered with forest. Conversion is carried out by smallholders, who slash and cut trees. Slash and burn is still practiced here. Before planting cacao trees, the ground is usually prepared for one or two years by planting annual crops (mainly maize).

2.2 Research Plots

Gross precipitation and throughfall, the main hydrological input parameters, were measured in a natural forest and on two cacao plantations. Since none

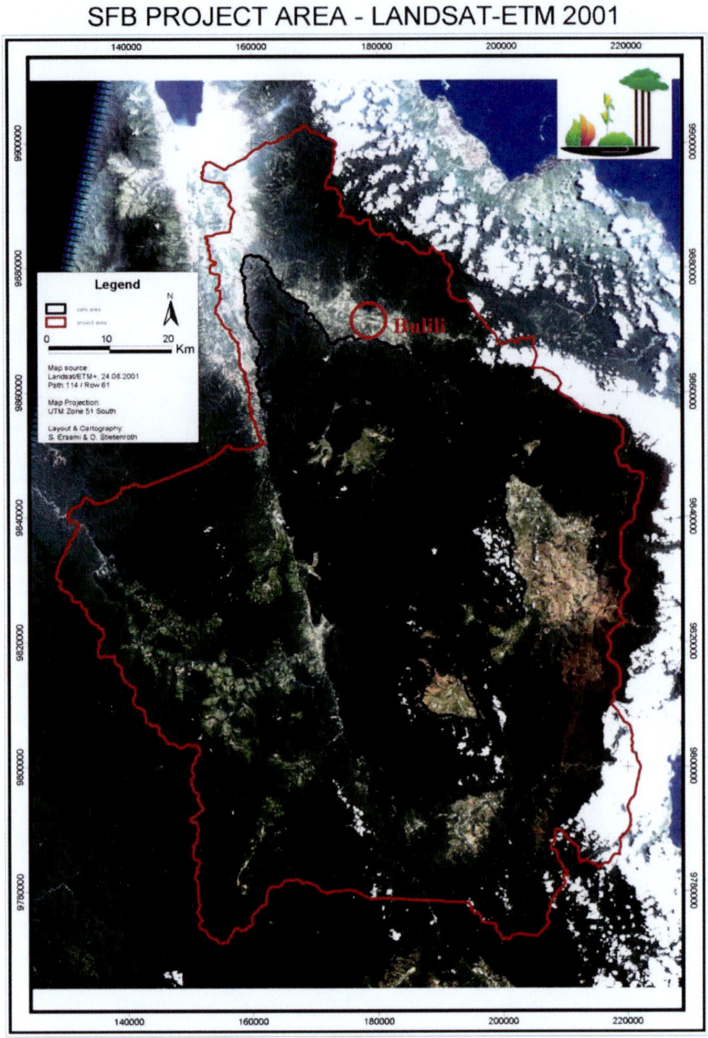

Fig. 1. Research area, modified after Erasmi (2003).

of the plots has palm trees or a high proportion of understory trees (DBH <5 cm) high quantities of stemflow are very improbable (Manfroi et al. 2004). Stemflow was not measured in our study, but values were assumed to be smaller than 1%, in accordance with the results of previous research in the STORMA area by Dietz (Dietz et al. 2006, Dietz 2007; see also Chappell et al. 2001, Marin et al. 2000).

Each throughfall plot measures 10 x 10 m. At the end of the experiment the cacao plots were 7 and 18 years old. The older plot lies on the bottom of the valley at an elevation of 639 m a.s.l. The younger plot is a cacao-site on a moderately inclined mountain slope at an elevation of 721 m a.s.l. This site was natural forest until October 2000. After cutting, maize (January 2001) and cacao (February 2001) were both planted next to each other on the same plot. In June 2003 maize cultivation was discontinued and from that time on the plot was exclusively used for cacao. Both cacao plots are monocultures without shade trees. Trees are cut twice annually; however the younger plot is better managed than the older one. On the former, dead branches are quickly cut off and removed, which results in a higher degree of canopy closure (lower gap fraction) and a lower dry mass of branches in the litter fraction.

The natural forest plot lies on a knoll at an elevation of 1082 m a.s.l. The leaf cover comprises bamboo and evergreen trees. Estimations of leaf coverage by two independent researchers at 20 positions (sampler positions) within the plot gave an average of 40% bamboo and 75% evergreen trees.[1] There are no palm trees or trees with DBH <5 cm on the plot.

2.3 Reference Plots

To account for small scale differences in rainfall distribution within the research area, an open area site was established at the nearest possible location (less than 250 m distance from plot centre) for each of the three throughfall plots. Bulk precipitation was collected at the open area sites on the same dates and with the same type of samplers as on the research plots. A fourth open area plot (Open Area D) was set up on a level piece of land at 837 m a.s.l. The rainfall samples taken there were used as chemical reference for open area precipitation and compared with the samples taken at the three research plots.

2.4 Plot Position and Properties

Leaf area index and openness of the research plots were calculated by means of hemispherical photography using a Nikon Coolpix S3 digital camera with fisheye lens and Caneye Software (Caneye; Jonckheere et al. 2004; Weiss et al. 2004). Photos were taken at 20 positions on each plot (rainfall sampler positions) in May 2007.

All plots exhibited full canopy closure, and the average LAI values were very similar. Openness was highest on the 18 year old cacao plot, despite the higher tree density. This is a result of less intensive cutting of dead branches.

[1] The sum's exceedance of 100% is a result of a bamboo layer beneath the branches of evergreen trees.

[2] Only trees with a breast height diameter (DBH) >5cm were counted. Cacao plot values were extrapolated to give values per hectare, natural forest values were estimated from a 30m x 30m area with the plot at the centre.

Table 1. Position and properties of research- and reference plots.

Plot-Name	Position S	Position E	Altitude [m a.s.l.]	Trees/ha^2	Openness	Average LAI
Cacao 7y	01°11.489'	120°05.044'	721	1100	10%	4.7
Cacao 18y	01°11.081'	120°05.076'	639	1500	14%	4.9
Nat. Forest	01°12.157'.	120°05.472'	1082	311	10%	5.1
Cacao 7y Reference	01°11.457'	120°05.054'	710		100%	0
Cacao 18y Reference	01°11.099'	120°05.025'	641		100%	0
Nat. Forest Reference	01°12.067'	120°05.439'	958		100%	0
Open Area D	01°11.778'	120°05.176'	837		100%	0

2.5 Rainfall Pattern

Total rainfall in the research area is about 2500 mm/year. Thirty years of meteorological measurements conducted by the Dutch Colonial Meteorological and Geophysical Service (Berlage 1949) show high variation in monthly averages. Quantity of rainfall and degree of seasonality depend on the elevation of the site. For the sites at higher elevations a bimodal rainfall pattern exists with absolute maxima in April/May and local maxima in November and December.

In our study we found an increase of rainfall quantity with the elevation of the site. Highest quantities were measured in the natural forest and the Open Area D plot.

Table 2. Rainfall quantity during research period.

Location	Rainfall (05.06.2007 – 04.06.2008)
Open Area D	3039 mm
Cocoa 7 years	2863 mm
Cocoa 18 years	2754 mm
Natural Forest	3062 mm

Seasonal patterns may shift considerably between years. Figure 2 shows monthly rainfall amounts from 06.2007 – 03.2009 measured in the open area plots.

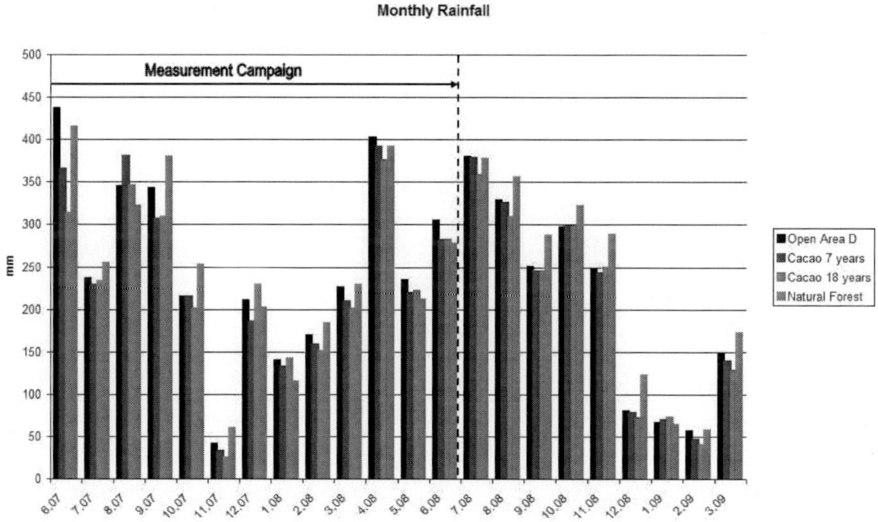

Fig. 2. Monthly rainfall at reference plots.

3 Methods

3.1 Instrumentation

Each throughfall plot was equipped with 20 bulk deposition samplers (BDS) set up at a height of 1 meter. Samplers were distributed randomly on the plot, with a minimum distance of 1 meter between all samplers. Each open area plot was equipped with 3 BDS set up in close proximity to one another.

The bulk deposition samplers consist of a funnel with an 18.8 cm opening (diameter) and a 4 mm rim. Rainwater flows through a PVC tube into a 5 litre plastic canister. Funnels have a 7.5 cm high rim to prevent rainwater from splashing out and an inner diameter of 3 cm to minimize evaporation. Evaporation from samplers between sampling dates was tested and no detectable evaporation (>0.5 mm) was found. To prevent coarse organic material from entering the samplers, a plastic net (mesh size 1 mm) and fine gauze were placed inside the funnel opening. Funnels were cleaned of fallen plant material at each sampling date (twice a week); canisters and funnels were washed out once a month to clean them from organic coating.

3.2 Research Period

From 05.06.2007 until 04.06.2008 all BDS were emptied twice a week. The quantity of collected rainwater was determined manually, using a household measuring jug. To obtain a volume-weighted mixed sample for chemical analysis, the samples of each throughfall plot were mixed in a plastic barrel, and one sample of this mixture was subsequently taken.

The samples of the Open Area D plot (used for chemical reference) were not mixed but all three samples were analysed separately.

3.3 Fixed and roving sampling design

A roving sampling design is recommended for long-term measurements of throughfall (Lloyd et al. 1988). The sampler positions were fixed from June 2007 to January 2008. After that, the sampling scheme was changed to roving sampling to allow a comparison of the two methodologies. From that time on the positions of five random samplers were changed in each throughfall plot after each sampling.

3.4 Chemical Analysis

Collected rainfall and throughfall were analysed for chemical composition. The following parameters were analysed:

Table 3. Parameters analysed in rain samples.

Key Parameters	pH; electrical conductivity
Elements [mg/l]	Ca; K; Mg; Na; Si; Fe
Nitrogen & Phosphor [mg/l]	PO_4-P; NO_3-N; NH_4-N; total nitrogen bound (tNb)

All chemical analysis was done in the STORMA-laboratory at Tadulako University, Palu, Central Sulawesi, Indonesia. The samples were stored in a refrigerator until transport, transported to the laboratory in Palu, filtered with 45 μm cellulose acetate filters (Whatmann, Schleicher & Schell), and stored in refrigerator or freezer until final analysis. Ca, K, Mg, Na, Si and Fe were analysed using an ICP-OES Optima 2000 DV (Perkin Elmer). Minimum concentration for calibration was 0.1 mg/l; detection limit was 0.05 mg/l. PO_4-P, NO_3-N and NH_4-N were analysed using an AA3-Autoanalyzer (Bran & Luebbe). The detection limit was 0.1 mg/l for NO_3-N and NH_4-N. For PO_4-P it was 0.05 mg/l. tNb was analysed using a Dima-N Analyzer (Dimatec); the detection limit was 0.05 mg/l.

4 Raw data management

On a number of sampling dates we recorded throughfall rates greater 100%, which means negative interception. This is quite typical for throughfall measurements in the tropics (Lloyd & Marques 1988, Crockford & Richardson

2000), and considered to be due to one or more of the following: measurement errors in throughfall or open area precipitation, micro scale local differences in rainfall amount between throughfall plot and open area site (possibly caused by the trees themselves), cloud harvesting by forests or dew effects upon leaves. Throughfall values greater than 100% were incorporated into this study, since we believe that possible measurement uncertainties of rainfall will cancel themselves out in the 12 month measurement period and that dew, fog and cloud effects have to be accounted for as part of the hydrological cycle.

Chemical analysis results that were more than three standard deviations above or below the average were excluded. This was done to eliminate sampling- and analysis errors and to account for untypical samples that could be affected by random effects such as bird droppings.

5 Definition of terms in this study

We defined the rainfall collected on a research plot at one sampling date (in mm) as the average of the quantity collected in all the samplers on that plot. The average was used because only the combination of the sampled quantities can adequately represent the plot: the canopy presents itself as a mixture of points open to the sky, shielded points with reduced leaf drip and dripping points. Therefore it is impossible for any single sampler to give a plot typical value.

The opposite is true for rainfall collected on a reference plot. All samplers are placed close to one another under the open sky and collect the quantity of rainfall typical for the site. The median value best represents this typical value and eliminates the influence of outliers.

The throughfall rate was calculated by comparing the total sum of throughfall on a research plot over all sampling events (12 month period) with the sum of rainfall collected in the reference plot.

Calculating throughfall rate in this manner will automatically weight the importance of each sampling event according to the quantity of rainfall collected. Light rainfalls will therefore influence the throughfall rate far less than heavy rainstorms. The rate thus calculated is a useful parameter for water balance modelling.

6 Results

The following table gives the throughfall rates calculated for the 12 months of our study. An obvious difference exists between the rates calculated in both cacao plots and the rate calculated in the natural forest.

Because there is a limit to the storage capacity of the canopy, throughfall rates are generally lower for small rainfall events and higher for strong

Table 4. Throughfall rates calculated in this study.

Location	Gross Precipitation 12 month	Throughfall T 12 month	Throughfall Rate 12 month
Cacao 7 years	2863 mm	2534 mm	89%
Cacao 18 years	2754 mm	2502 mm	91%
Natural Forest	3062 mm	2473 mm	81%

rainstorms. We expect the storage capacity of the natural forest canopy to be higher than that of the cacao plot. This would lead to stronger differences in throughfall rates for weak rainfall. To test this, we divided the sampling events into three categories with an equal number of events in each category. Events with less than 3.5 mm of rainfall were not considered because measurement errors are believed to be disproportionally high if less than 100 ml (~3.5 mm) is collected in the BDS. Note that BDS collection events and not single rainfall events were categorized. The quantity of rain collected during a sampling may represent the sum of more than one rainfall event.

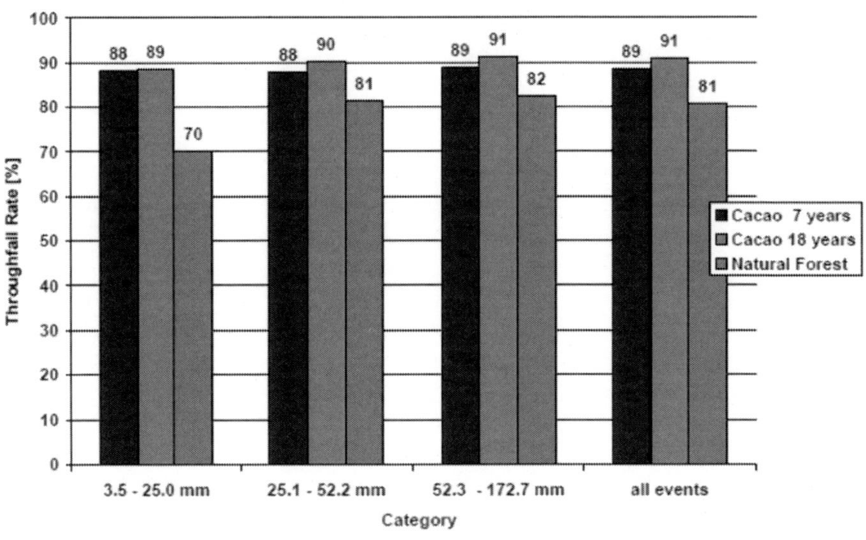

Fig. 3. Throughfall within categories.

Throughfall rates of the two cacao plots are similar, but differ greatly from the natural forest plot. Differences between the natural forest plot and the cacao plots are highest for the smallest rainfall category.

The standard deviation of throughfall rates (calculating and comparing throughfall rates for single events within a category) is highest in the lowest rainfall category and lowest in the highest rainfall category.

Chemical composition of throughfall

During canopy passage rain takes up elements and nutrients. These originate either from a remobilisation of former depositions (wet and dry depositions) or from the plant material itself (leaching). To quantify this effect, rainfall and throughfall samples were analysed for their chemical properties.

Volume-weighted average concentrations

The following table gives the volume-weighted average of chemical parameters analysed in our study. pH values were transformed into H^+ activities before calculating average; then the average was transformed back into pH. For electrical conductivity volume-weighted averages were calculated in the same way as for concentration values.

Table 5. Volume-weighted average concentrations, pH and conductivity.

Location	Ca [mg/l]	K [mg/l]	Mg [mg/l]	Na [mg/l]	Si [mg/l]	Fe [mg/l]
Open Area D	0.37	0.43	0.07	0.52	0.07	0.03
Cacao 7 years	0.43	2.48	0.20	0.66	0.18	0.03
Cacao 18 years	0.44	2.20	0.18	0.46	0.15	0.03
Natural Forest	1.34	8.09	0.37	0.77	0.56	0.03
Location	**PO_4-P [mg/l]**	**tNb [mg/l]**	**NO_3-N [mg/l]**	**NH_4-N [mg/l]**	**pH**	**Conductivity (μS/cm)**
Open Area D	0.01-0.05	0.43	0.00-0.10	0.05-0.10	5.80	7.88
Cocoa 7 years	0.09-0.10	0.60	0.01-0.10	0.07-0.11	6.25	14.09
Cocoa 18 years	0.09-0.10	0.41	0.00-0.10	0.05-0.10	6.38	12.73
Natural Forest	0.65-0.65	0.80	0.13-0.20	0.09-0.12	6.48	32.16

The highest concentration on all throughfall plots was that of K, for which a high degree of leaf leaching has been identified by several studies. For those elements and nutrients, which showed enrichment with canopy passage, the concentrations were highest in the natural forest plot. pH and electrical conductivity also increased during canopy passage, again showing highest values in the natural forest plot.

Fe concentrations were below detection limit on all plots and therefore excluded from further analysis.

In the open area rainfall concentrations for calcium, potassium, sodium and tNb are in a similar range (0.37 – 0.52 mg/l). Values of both cacao plots were very similar despite the age difference. For Ca, K, Mg, NO_3-N and NH_4-N volume-weighted average values differed less than 15%. For all other elements and nutrients (Na, Si, PO_4-P, tNb) differences were less than 40%.

To better compare the concentrations in the plot types, enrichment factors were calculated. The enrichment factor in our study was determined by dividing the volume-weighted concentration in the throughfall plots by the respective concentration of the Open Area D plot.

For PO_4-P, NO_3-N and NH_4-N all values below detection limit were set to the detection limit to allow comparison. For the electrical conductivity the factor was calculated by dividing conductivity values instead of concentrations.

Table 6. Enrichment factors for selected elements.

Location	Ca	K	Mg	Na	Si
Cocoa 7 years	1.2	5.8	3.0	1.3	2.5
Cocoa 18 years	1.2	5.1	2.7	0.9	2.1
Natural Forest	3.6	18.9	5.5	1.5	7.8
Location	PO_4-P *	tNb	NO_3-N*	NH_4-N*	Conductivity**
Cocoa 7 years	2.1	1.4	1.0	1.1	1.8
Cocoa 18 years	1.9	0.9	1.0	1.0	1.6
Natural Forest	12.9	1.9	1.9	1.2	4.1

* Values below detection limit were set to the detection limit ** Factor given for conductivity represents relation of open area and throughfall plot values

Enrichment with canopy passage was identified for most elements and nutrients (Ca, K, Mg, Si, tNb, PO_4-P*, NO3-N* and NH_4-N*). The values in the table above highlight the similarity in the throughfall composition of the two cacao plots, while underlining the difference to the natural forest plot.

Input with rainfall

The input with rainfall (in our study including input with throughfall) is a necessity for creating nutrient budgets and indispensable when evaluating whether a certain type of land use will deplete nutrient stocks. Input was calculated for the 12 month data set. Gaps caused by the exclusion of outliers or missing data were closed by linear interpolation.

The total rainfall during the 12 months of our study was higher than the long term average of 2500 mm for the area. Furthermore, the rainfall in the natural forest plot was 7 - 11% higher than that in the cacao plots, making direct comparison difficult. To facilitate comparison, we calculated both the real input during the 12 month study period and the theoretical input for a standardized 2500 mm gross precipitation on all plots.

Table 7. Annual input during the research period [kg*ha^{-1}].

Location	Ca	K	Mg	Na	Si	PO_4-P	tNb	NO_3-N	NH_4-N
Open Area D	11.36	13.01	2.06	15.73	2.19	0.17-1.53	13.03	0.06-3.06	1.51-3.10
Cocoa 7 years	10.96	62.96	5.16	16.76	4.53	2.19-2.48	15.16	0.16-2.61	1.70-2.72
Cocoa 18 years	11.02	55.04	4.50	11.52	3.75	2.21-2.59	10.18	0.00-2.50	1.31-2.54
Natural Forest	33.22	200.04	9.18	19.12	13.95	15.99-16.13	19.87	3.25-4.83	2.11-3.03

Differences in input with throughfall between the natural forest and the cacao plots are highest for potassium, for which a strong influence on soil biology is to be expected. Strong differences (at far lower input rates) also exist for phosphor which is considered a critical nutrient in many tropical soils.

Table 8. Annual input for 2500 mm gross precipitation on all plots [kg*ha^{-1}].

Location	Ca	K	Mg	Na	Si	PO_4-P	tNb	NO_3-N	NH_4-N
Open Area D	9.35	10.70	1.69	12.94	1.80	0.14-1.26	10.72	0.05-2.52	1.24-2.55
Cocoa 7 years	9.57	54.98	4.50	14.64	3.95	1.91-2.16	13.24	0.14-2.28	1.48-2.37
Cocoa 18 years	10.00	49.97	4.08	10.45	3.40	2.01-2.35	9.24	0.00-2.27	1.19-2.31
Natural Forest	27.12	163.32	7.49	15.61	11.39	13.05-13.17	16.23	2.65-3.95	1.72-2.47

7 Discussion

7.1 Throughfall rates

Studies on throughfall rates in rainforests in the humid tropics differ in size, geographical position of the research area and climatic conditions. The methods applied for sampling and statistical analysis also vary. As a result, there is a wide range of published data. In mature tropical rainforest 80 - 95% of the gross precipitation reaches the ground as throughfall or stemflow (Bruijnzeel 1990; Dykes 1997; Manfroi et al. 2004; Marin et al. 2000; Asdak et al. 2000; Lloyd et al. 1988; Chappell 2001). Throughfall values for tropical cacao plantations are still scarce. In Central Sulawesi, Nicklas (2006) found throughfall rates of 88%. De Miranda (1994) calculated a rate of 85% in Brazil, whereas the findings of Hetzel & Gerold (1998) in Côte d'Ivoire were at 91%.

The values calculated in our study fit well in the context of data published in the literature. It is one of the few studies comparing throughfall rate of cacao and natural forest under the same (micro-) climatic conditions. Throughfall rates obtained by fixed sampling design differed only little from those calculated with roving samplers. This is in line with the results of Holwerda et al. (2005) who found that fixed or roving sampling affected only their standard error of throughfall rates but not the rates themselves (at 0.5 significance level).

Conversion of natural rainforest to cacao plantations in the study area will increase rainwater input to the soils by 10% due to reduced interception. Additional increases resulting from reduced transpiration are to be expected. Since the conversion by smallholders in the project area is done without the use of heavy machinery strong soil compaction is unlikely. However, the creation of new tracks along the steep slopes may favour erosion and surface runoff. If soil infiltration properties remain more or less unchanged, it is to be expected that the major part of the increased water input will infiltrate, adding to ground water recharge and base flow. A scenario simulation with WaSiM-ETH by Kleinhans (2003) for the study area, simulating the effect of conversion from natural forest to cacao plantation on a watershed level, predicted an increase in baseflow by 8%.

Table 9. Enrichment factors of throughfall samples.

	a) in cacao plantations				
Element	this study Cacao 7 years	this study Cacao 18 years	Hetzel (1999)	Lanfer (2003)	Nicklas (2006)
Ca	1.2	1.2	1.4	2.7	2.1
K	5.8	5.1	4.8	9.0	7.6
Mg	3.0	2.7	2.3	9.0	3.4
Na	1.3	0.9	0.9	1.1	1.0
Si	2.5	2.1			3.9

	b) in tropical rainforests			
Element	this study	Burghouts et al. (1998)	McDowell (1998)	Hetzel (1999)
Ca	3.6	2.8	3.0	7.0
K	18.9	19.3	19,2	19.0
Mg	5.5	4.6	2,7	10
Na	1.5	1.2	2.2	0.9
Si	7.8			

7.2 Chemical Composition of Rainfall and Throughfall

Table 9 summarises the enrichment factors for the elements analysed in this study and compares them with the results of Burghouts et al. (1998), McDowell (1998), Hetzel (1999), Lanfer (2003) and Nicklas.

The findings of our study agree well with the results for cacao plantations published by Hetzel (1999). For calcium and magnesium Lanfer (2003) and Nicklas (2006) give higher values. For the natural forest plot the enrichment factors found in our study are very similar to the findings of Burghouts et al. (1998). For calcium Hetzel (1999) reports a far stronger enrichment than the other two studies. For magnesium our study takes, together with Burghouts et al. (1998), a middle position between the extremes of McDowell (1998) and Hetzel (1999). However, all studies agree on the very strong enrichment of potassium in natural forest throughfall.

8 Conclusion

This study was able to show differences between the quantity and chemical composition of throughfall in the natural forest and the cacao plots.

Even though the age of the cacao plots differed (7 and 18 years), the throughfall rates and chemical composition of throughfall were very similar, but differed strongly from those of the natural forest plot. Conversion of natural forest to cocoa plantation is expected to increase river discharge. As long as soil infiltration is not negatively affected this will occur mainly in the form of increased low flow.

The study was able to show how the input of elements and nutrients with throughfall is reduced if natural forests are converted to cacao plantations. Potassium was identified as the element most strongly affected. These changes will influence soil biology. Since our method is not able to distinguish between internal nutrient cycling (wash out) and external nutrient input (wash off) further research is desirable to determine the effect of land use change on nutrient stocks in the soil.

Acknowledgements
The authors wish to thank the following persons and institutions: DFG (Deutsche Forschungsgemeintschaft) for funding the STORMA project, Mr. Dudin, Mr. Kemyl Latupono, Mr. Risman and Mr. Rickson Tiranda for intensive data collection in the field, as well as the staff of the STORMA laboratory at Tadulako University for their effort in analysing the collected samples.

References

Asdak C, Jarvis PG, van Gardingen P, Fraser A (1998) Rainfall Interception Loss in Unlogged and Logged Forest Areas of Central Kalimantan, Indonesia. Journal of Hydrology 206: 237-244

Burghouts TB , van Straalen NM, Bruijnzeel LA (1998) Spatial heterogeneity of element and litter turnover in a Bornean rain forest. Journal of Tropical Ecology 14: 477-506

Bruijnzeel LA (1990) Hydrology of Moist Tropical Forests and Effects of Conversion: a State of Knowledge Review. Humid Tropics Programme, IHP-UNESCO, Paris and Vrije Universiteit, Amsterdam, 224 pp.

Bruijnzeel LA (2004) Hydrological functions of tropical forests: not seeing the soil for the trees? Agriculture Ecosystems and Environment 104: 185-228

Berlage HP (1949) Rainfall in Indonesia. Mean rainfall figures for 4339 rainfall stations in Indonesia, 1879 - 1941. Department van Verkeer, Energie en Mijnwezen, Meteorologische en Geophysische Dienst, Koninklijk

Calder IR (2001) Canopy processes: implications for transpiration, interception and splash induced erosion, ultimately for forest management and water resources. Plant Ecology 153: 203-214

CANEYE: CAN-Eye software, under: http://www.avignon.inra.fr/can_eye (30.04.2009)

Chappel NA, Bidin K, Tych W (2001) Modelling rainfall and canopy controls on net-precipitation beneath selectively-logged tropical forest. Plant Ecology 153: 215-229

Crockford RH, Richardson DP (2000) Partitioning of rainfall into throughfall, stemflow and interception: effect of forest type, ground cover and climate. Hydrological Processes 14: 2903-2920

de Miranda RA (1994) Partitioning of rainfall in a cacao (Theobroma cacao Lour.) plantation. Hydrological Processes 8: 351-358

Dietz J (2007) Rainfall partitioning in differently used montane rainforests, Central Sulawesi, Indonesia. Dissertation, University of Göttingen, online: http://johannes.dietz-pino.com/docs/dietz/2007/dissertation.pdf

Dietz J, Hölscher D, Leuschner C, Hendrayanto (2006) Rainfall partitioning in relation to forest structure in differently managed montane forest stands in Central Sulawesi, Indonesia. Forest Ecology and Management 237: 170-178

Dykes AP (1997) Rainfall interception from a lowland tropical rainforest in Brunei. Journal of Hydrology 200: 260-279

Erasmi S (2003) Maps of study sites and project area. SFB 552 internal database

(FAO) Food and agriculture organisation of the United Nations (2007) The State of World's Forests 2007. Rome

Gash JH (1978) An analytical model of rainfall interception by forests. The Quarterly Journal of the Royal Meteorological Society 105 (443): 43-55

Hetzel F, Gerold G. (1998): The water cycle of a moist deciduous rainforest and a cacao plantation in Côte d'Ivoire, IAHS Publ. 252, 411-418.

Hetzel F (1999) The water and nutrient cycle in a tropical rain forest and a cocoa plantation in Côte d'Ivoire. EcoRegio 2, Göttingen

Hall R (2003) Interception loss as a function of rainfall and forest types: stochastic modelling for tropical canopies revisited. Journal of Hydrology 280: 1-12

Holwerda F, Scatena FN, Bruijnzeel LA (2006) Throughfall in a Puerto Rican lower montane rain forest: A comparison of sampling strategies. Journal of Hydrology 327: 592-602

(ICCO) International cocoa organisation (2008) Assessment of the movement of Global Suppy & Demand - April 2008. http://www.icco.org (30.04.2008)

(ICB) Indonesian cocoa farmer association, Indonesian cocoa board (2008) Indonesian Cocoa Beans: current situation. http://www.icco.org/ (30.04.2008)

Imbach AC, Fassbender HW, Borel R, Beer J, Bonnemann A (1989) Modelling agro-forestry systems of cacao (Theobroma cacao) with laurel (Cordia alliodora) and poro (Erythrina poeppigiana) in Costa Rica. IV. Water balances, nutrient inputs and leaching. Agroforestry Systems 8: 267-287

Jonckheere I, Fleck S, Nackaerts K, Muys B, Coppin P, Weiss M, Baret F (2004) Review of methods for in situ leaf area index determination Part I. Theories, sensors and hemispherical photography. Agricultural and Forest Meteorology 121: 19-35

Kleinhans A (2003) Einfluss der Waldkonversion auf den Wasserhaushalt eines tropischen Regenwaldeinzugsgebietes in Zentral Sulawesi (Indonesien). Dissertation, University of Göttingen

Lanfer N (2003) Landschaftsökologische Untersuchungen zur Standortbewertung und Nachhaltigkeit von Agroökosystemen im Tieflandregenwald Ecuadors. EcoRegio 9, Shaker Verlag, Aachen

Lloyd CR, de Marques A (1988) Spatial variability of throughfall and stemflow measurements in Amazonian rain forest. Agricultural and Forest Meteorology 42: 63-73

Lloyd CR, Gash JH, Shuttleworth WJ (1988) The measurement and modelling of rainfall interception by Amazonian rain forest. Agricultural and Forest Meteorology 43: 277-294

Manfroi OJ, Koichiro K, Nobuaki T, Masakazu S, Nakagawa M, Nakashinzuka T, Chong L (2004) : The stemflow of trees in a Bornean lowland tropical forest. Hydrological Processes 18: 2455-2474

Marin CT, Bouten W, Sevink J (2000) Gross rainfall and its partitioning into throughfall, stemflow and evaporation of intercepted water in four ecosystems in western Amazonia. Journal of Hydrology 237: 40-57

McDowell WH (1998) Internal nutrient fluxes in a Puerto Rican rain forest. Journal of Tropical Ecology 14: 521-536

Nicklas U (2006) Nährstoffeintrag durch Bestandsniederschlag und Streufall in Kakao-Agroforstsystemen in Zentral-Sulawesi, Indonesien. Diplomarbeit, University of Göttingen

(PEACE) Pelangi energi abadi citra enviro (2007) Indonesia and Climate Charge: Current Status and Policies.

http://siteresources.worldbank.org/INTINDONESIA/Resources/ Environment/ClimateChange_Full_EN.pdf (15.01.2009)
Weiss M, Baret F, Smith GJ, Jonckheere I, Coppin P (2004) Review of methods for in situ leaf area index (LAI) determination Part II. Estimation of LAI, errors and sampling. Agricultural and Forest Meteorology 121: 37-53

Effects of "ENSO-events" and rainforest conversion on river discharge in Central Sulawesi (Indonesia)

Gerhard Gerold[1*] and Constanze Leemhuis[2]

[1] Deptartment of Landscape Ecology, Geographical Institute, University of Göttingen, Goldschmidtstr. 5, 37077 Göttingen, Germany
[2] Center for Development Research, University of Bonn, Walter-Flex-Str. 3, 53113 Bonn, Germany

*corresponding author: G. Gerold, email: ggerold@gwdg.de

Summary

Forest conversion and natural interannual climate fluctuation as ENSO-events are key determinants of water balances in tropical catchments. Distributed hydrological modelling that relates land cover changes and climate changes with river discharge are rare for humid tropical catchments at the mesoscale. In the present paper we present a hydrological modelling approach to describe the impact of land cover changes and ENSO-events on the water resources in a mesoscale humid tropical catchment. These are based on hydro-meteorological measurements that were performed in the Gumbasa catchment from 2002 to 2005 in the frame of STORMA.

The distributed hydrological model WASIM-ETH was calibrated and validated for the current land use (2002/2003, Landsat/ETM+ scene). Model results were generally consistent with observed discharge data and reproduced seasonal discharge dynamics well. The implications of possible future climate and land use conditions on the water balance of the Gumbasa River were assessed by scenario analysis. These clearly demonstrate a strong relationship between deforestation rates and increasing discharge variability. In particular, a significant increase of high water discharges was simulated for the applied land use scenarios. Forest conversion by smallholders practicing traditional land use alters the discharge dynamics more than a change in the total annual discharge.

The main results of the scenario analysis are:

1. ENSO anomalies of precipitation lead to an increase in discharge variability.

T. Tscharntke et al. (eds.), *Tropical Rainforests and Agroforests under Global Change*, Environmental Science and Engineering, 327–350, DOI 10.1007/978-3-642-00493-3_15,

2. Strong ENSO-events (El Nino) lead to a 30% reduction in the water yield.
3. ENSO-events (El Nino) decrease the potential (flooding) area of paddy rice cultivation in the Gumbasa Irrigation Scheme by two thirds in the second half of the year.
4. Annual crop scenarios up to 1,200 m a.s.l. showed a 42% and the perennial crop scenario (cacao) a 23% increase in river discharge with high increase in overland flow and flooding risks.

With regard to the high deforestation rates of the research catchment and the proposed increase of El Nino frequency, it is highly likely that the negative changes for the water resources of Central Sulawesi will continue.

Keywords: Hydrological Modelling, ENSO, Tropical Deforestation, River discharge development, Central Sulawesi

1 Introduction

Tropical catchment areas are primarily located in developing countries and are predominantly ungauged or poorly gauged. The vast majority of tropical catchment studies have been conducted at the lower meso spatial (<10 km) and time scale (<5 years) (Bruijnzeel 1996), because a dense experimental set-up can be implemented within its spatial and temporal limits . Costa et al. (2003) confirm that studies relating changes in land cover with changes in river discharge in tropical catchments are abundant especially on an experimental scale (< 1 km). These experimental scale catchment studies contribute to a better hydrological knowledge of humid tropical catchment dynamics. Conversely, there are a number of studies in tropical regions which focus on the macro-hydrological scale using global data sets. However, to investigate the impact of land cover or climate changes on the water resources, a mesoscale catchment unit represents an ideal research scale. The global and local scale, which is employed in the vast majority of tropical hydrological studies (e.g. Güntner 2002; Döll et al. 2003; Godsey et al. 2004; Fleischbein et al. 2005) is far from the regional reality where decisions on water resource management are developed and implemented (Falkenmark and Rockström 2004). Data availability is one of the main difficulties for the analysis of tropical catchment processes on a mesoscale level. An appropriate database, which is urgently needed to enable water resources to be developed and managed, is lacking (IAHS 2003). Therefore, long term records, which are required to study hydrological trends, are normally not available (Manley and Askew 1993). Considering the increasing stress on water resources in humid tropical developing countries due to rapid deforestation rates and climate change, there is an urgent global research need in humid tropical hydrology and its associated mesoscale catchment processes. According to Abbot and Reefsgard (1996), the spatially-distributed and time-dependent hydrological model has become the conditio sine qua non for investigations in this area.

The El Niño/ Southern Oscillation (ENSO) phenomenon is the strongest known natural interannual climate fluctuation. The two most recent extreme ENSO events during 1982/83 and 1997/98 had severe impacts on major parts of Indonesias economy. Understanding of the hydrology of humid tropical catchments is an essential prerequisite for the investigation the impact of climate variability. Knowledge of the hydrological consequences of ENSO-events for the quantitative assessment of future water resource changes is an essential factor in the development of mitigation strategies on a catchment scale. The results can be implemented into long term Integrated Water Resource Management (IWRM) strategies. Additionally, land use changes by forest conversion alter the components of the water and nutrient cycle in catchments of Indonesia. Rainforest conversion into predominately annual cultures and cacao-systems has been intensified in Central Sulawesi during the last decade. The general objectives of STORMA (DFG-project) and IMPENSO (BMBF-project) are to investigate and quantify the impact of ENSO induced climate variability and forest conversion effects on the water balance of a mesoscalic tropical catchment

2 Methods and catchment characteristics

2.1 Study area Gumbasa river catchment

The Gumbasa River catchment (Fig. 1) is a sub-basin of the Palu River catchment (2694 km^2), which is located in Central Sulawesi, Indonesia (1°S, 120°E). The Gumbasa River catchment has a total drainage area of 1275 km^2 and is characterized by a gross annual areal precipitation of 2000 mmy^{-1} and a mean annual discharge of 22 m^3s^{-1}. The seasonal hydrological regime of the catchment is described by two peaks and strongly corresponds with the bimodal monsoonal pattern of the yearly precipitation distribution of Central Sulawesi, Indonesia. Similar to other river basins that are located in monsoon regions the Gumbasa River also shows a great variation of seasonal and annual flow (Oyebande and Balek, 1987). ENSO years are characterized by reduced precipitation from July to October, which represents the dry period of the monsoonal setting of Central Sulawesi. The elevation ranges from 99 to 2,491 m a.s.l. with a mean elevation of 1,119 m a.s.l. In accordance with this steep gradient the slope averages to 10.4°. About two-thirds of the catchment area is classified as slope with moderate and strong inclination (69%), which emphasizes the mountainous character of the watershed. The irrigated paddy fields and other agricultural land use systems are primarily situated in the valleys. Mountainous tropical forest with a total percentage of 86% forms the main land cover type in the watershed (Leemhuis et al. 2007, Fig. 4).

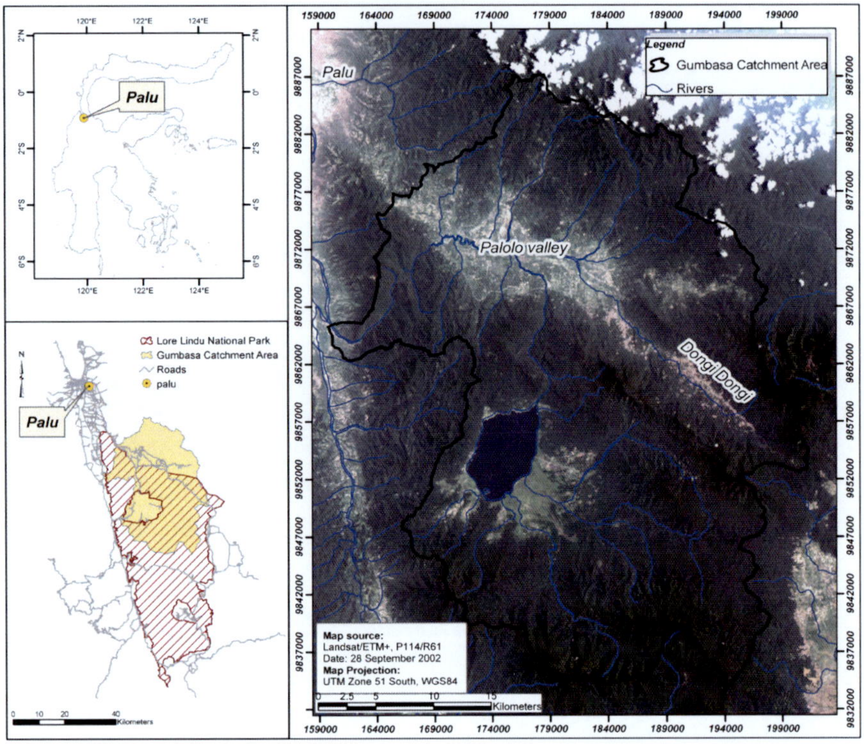

Fig. 1. Project area in Central Sulawesi, Indonesia (left) and the Gumbasa River catchment (right, Landsat ETM 2002).

2.2 Methods

Temporal and spatial data base

In order to study the impact of forest conversion and ENSO effects on the water balance of the Gumbasa River catchment we applied the Water Flow and Balance Simulation Model WASIM-ETH (Schulla and Jasper 1999). WASIM-ETH is a process-based fully distributed catchment model. The spatial resolution is given by a grid, and the time resolution can vary from minutes to days. The main processes of water flux, storage and the phase transition of water are simulated by physically-based simplified process descriptions (Schulla 1997). The meteorological input data of the model are interpolated for each grid cell and are followed by the simulation of the main hydrological processes such as evapotranspiration, interception, infiltration and the separation of discharge into direct flow, interflow and base flow. These calculations are modularly structured and can be adapted to the physical characteristic of the catchment

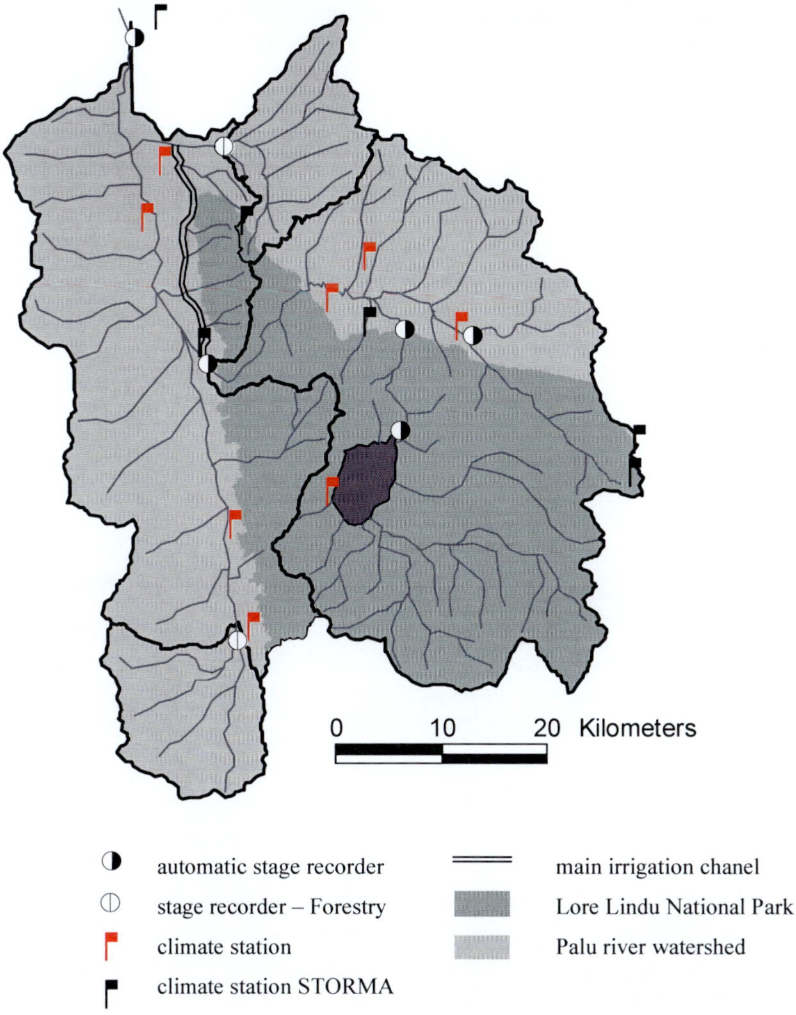

Fig. 2. Location of the climate and hydrological stations within the Palu river watershed, Central Sulawesi, Indonesia.

area. For the set-up of the Gumbasa River catchment a 500 m x 500 m raster spatial, and a daily temporal resolution was chosen. WASIM-ETH requires a Digital Terrain model, a land cover map and a soil map (Fig. 2) as spatial input data. All spatial data is required in a raster data format (grid) with unique spatial resolution and extension. Starting from the primary grid of the topography, secondary grids were generated with a topography analysis program. Spatial data was generated with SAGA software from topographic maps (TK 50) for the Gumbasa catchment (1,275 km^2, 50 m grid, digital terrain model).

Table 1. Land cover classes and its vegetation parameters.

Land cover	α	R_c (ms^{-1})	LAI	v	z_o (m)	z_w (m)	p	Ψg (hPa)
Rainforest	0.13	150	7	1	24		0	3.55
Submontane rainforest (< 1200 m a.s.l.)	0.13	200	6	1	20	1.0	0	3.55
Open forest	0.15	150	6	0.9	20	1.0	0	3.55
Water	0.05	20	1	0.1	0	1.0	-1	3.55
Paddy rice	0.12	100	1	0.2	0.5	0.1	-1	3.55
Annual crops	0.18	100	1.5	0.3	2.0	0.3	-1	3.55
Coconut	0.32	200	4	0.6	10	0.3	-0.5	3.55
Perennial crops (cacao, coffee)	0.15	270	5	0.8	5	1.0	-0.5	3.55
Grassland	0.26	200	1.5	0.6	0.5	0.2	-1	3.55
Reed	0.18	250	2	0.2	2.0	0.4	-1	3.55
Settlement	0.15	150	1	0	0	0.1	-1	3.55

α albedo, R_c (ms^{-1}) surface resistance, LAI leaf area index, v vegetation cover, z_o (m) effective crop height, z_w (m) root depth, p root distribution, Ψg (hPa) minimum suction for reducing pET to aET

In order to obtain a feasible temporal data set of meteorological forcing and hydrological calibration data for the performance of a distributed hydrological model, the catchment was equipped with four automatic stage recorders and eight automatic climate stations monitoring precipitation, temperature and humidity. Additionally, five automatic climate stations measure the meteorological parameters including precipitation, air temperature, relative humidity, wind speed, global radiation at 10 minute intervals (Fig. 2). We measured hydrological and meteorological data for the period from Sept. 2002 to Sept. 2004. For the calibration (01.09.2002 31.08.2003) and validation (01.09.200331.08.2004) of the hydrological model WASIM-ETH we chose the land cover classification of the Landsat/ETM+ scene from 2002. The land cover class concept is based on a global hierarchical class definition system (LCCS) and has been adapted to the project region in Central Sulawesi. Each land cover type refers to corresponding physical vegetation parameters such as stomata resistance, Leaf Area Index (LAI), albedo, vegetation height, rooting depth and vegetation cover. The parameterization of the vegetation physical properties was mainly derived from a global vegetation guide by Matthews (1999), single studies (Mo et al. 2004; Dijk a. Bruijnzeel 2001; Körner 1994) and local studies (Bohmann 2004; Falk 2004, Twele et al. 2008: LAI, minimal stomatal conductance, effective crop height) which were conducted in the research area (table 1). The root depth was derived from a global study of root distributions for terrestrial biomes (Jackson et al., 1996). In WASIM-ETH the root density distribution p describes the geometrical shape of the root system and is defined by: $-1 < p < 0$ (convex shape). Main values of root density distribution were taken from Landon (1984).

For the Gumbasa catchment the only available soil map is the soil map of Indonesia (1 : 1,000,000). Hodenett and Tomasella (2002) showed that most of the pedotransfer functions had been developed with databases from soils of temperate regions and are not transferable to tropical soils. Therefore, a topographic nested scale approach with SAGA was developed to parameterise soil physical properties with an experimental hillslope soil analysis (soil catenas from Nopu and intramontane basin of Toro Häring et al. 2005, Kleinhans 2003) and a classification of the catchment area in potential homogeneous soil texture areas (PHA) (Gerold et al. 2008). Field data, for the verification of the results for the van Genuchten soil water retention parameters of saturated hydraulic conductivity, was obtained with an intensive measurement campaign using a Guelph-Permeameter in the Nopu experimental catchment in 2004. K_{sat} was measured at 10 randomly selected replications for each different land use type and for each soil depth (0.15, 0.30, 0.60 m). Transect length in accordance with the extent of the land use type (one farmer) with same exposition and inclination was 10 to 50 m. This parameterisation of soil hydraulic properties with the PHA approach serves as a rough estimate of soil hydraulic parameters (Gerold et al. 2008). Table 2 lists the required soil hydraulic parameters for the soil model WASIM-ETH according to the PHAs classes.

Model set-up, calibration and validation

In accordance with the existing water level stations, the Gumbasa catchment was divided into four sub-catchments: Takkelemo, Lake Lindu, Sopu, Nopu. For the model set up, WASIM-ETH was applied to the Nopu experimental test catchment with a 30 m x 30 m resolution (Kleinhans 2003) and for the Takkelemo test catchment with a 250 m x 250 m resolution (Leemhuis 2005). Schulla (1997) recommends a grid resolution of 500 m x 500 m for the application of WASIM-ETH in mountainous mesoscale catchments. For the Takkelemo catchment (79 km^2) model efficiency on a daily and weekly time resolution was tested by Leemhuis (2005) for grid resolutions from 250 m x 250 m to 500 m x 500 m. With a coarser grid the model performance declines only slightly for both temporal resolutions. To reduce the running time of the simulations, the raster grid resolution of 500 m x 500 m was chosen for the whole Gumbasa watershed (grid cells reduce from 21,250 to 5,100) with a daily temporal resolution. In Central Sulawesi, WASIM-ETH was applied for the first time in the Inner Tropics.

For the Gumbasa watershed the calibration period was set for the period from 01.09.2002 to 31.08.2003, which corresponds to the first year of

Table 2. Determined PHA classed and its associated soil physical parameters.

PHA	Pot. texture	θ_s	θ_r	α	n	ψ	k_s	k_{rec}
I	sand	0.45	0	7.36	1.23	385	9.0E-4	0.1
II	sand	0.45	0	7.36	1.23	385	5.0E-4	0.1
III	loamy sand	0.41	0	1.86	1.26	375	4.0E-4	0.1
IV	loamy sand	0.41	0	1.86	1.26	375	5.0E-4	0.1
V	sandy loam	0.45	0	4.01	1.2	345	1.5E-3	0.1
VI	loam	0.49	0	4.01	1.2	350	5.3E-3	0.1
VII	sandy clay loam	0.51	0	2.0	1.13	290	5.0E-4	0.1
VIII	sandy clay loam	0.51	0	2.0	1.13	290	4.0E-4	0.1

Morphometric description with mean slope inclination (°) (details see Leemhuis 2005)
I Intramontane basin (low altitude above channel line 1.6), II Intramontane basin (moderate altitude above channel line 2.5), III Valleys outside intramontane basin catchment (large area 9.3), IV Valleys outside intramontane basin catchment (moderate area 12.1), V Slope with moderate inclination (17.4), VI Slope with strong inclination (30.8), VII Summit area with low inclination (5.3), VIII Summit area with moderate inclination (12.7)
θ_s saturation water content, θ_r residual water content, α,n empirical parameter, ψ suction (hPa), k_s saturated hydraulic conductivity (m s^{-1}), k_{rec} recession constant for k_s with increasing soil depth

the measurement period. In order to achieve realistic soil moisture conditions for the initial conditions of the calibration run, the start of the whole model run was predated to the 01.09.2001. The required temporal data for this period of model initialisation was copied from the first year of measurements. The validation period was split into the pure validation period *(01.09.2003 - 31.08.2004)*, and the total period of simulation *(01.09.2002 - 31.08.2004)*. Table 3 lists the complete calibration and validation results for the model efficiency of the Lake Lindu, Sopu and Takkelemo gauging sites with simulation results on a daily and weekly resolution. The coefficient of efficiency R^2 was used (Nash and Sutcliffe 1970). Furthermore, the index of agreement d (Willmot 1981) (sensitive to extreme values), the mean square error (MSE) and the root mean square error (RMSE) were calculated. According to Andersen et al. (2001) the achieved model efficiency based on d (volume error) and R^2 is good (Leemhuis 2005). Moreover, the graphical evaluation of observed and simulated discharge for the Lake Lindu, Takkelemo and Sopu subcatchment (Fig.3) shows a good agreement. The baseflow is simulated in realistic values, but slightly overestimated during periods of low flow conditions. In addition, the discharge fluctuation during the rainy season from March till May is well reproduced.

To test the uncertainty of the parameter variance for the model run, a predictive uncertainty analysis was implemented (Leemhuis 2005, chapter 6.5.5).

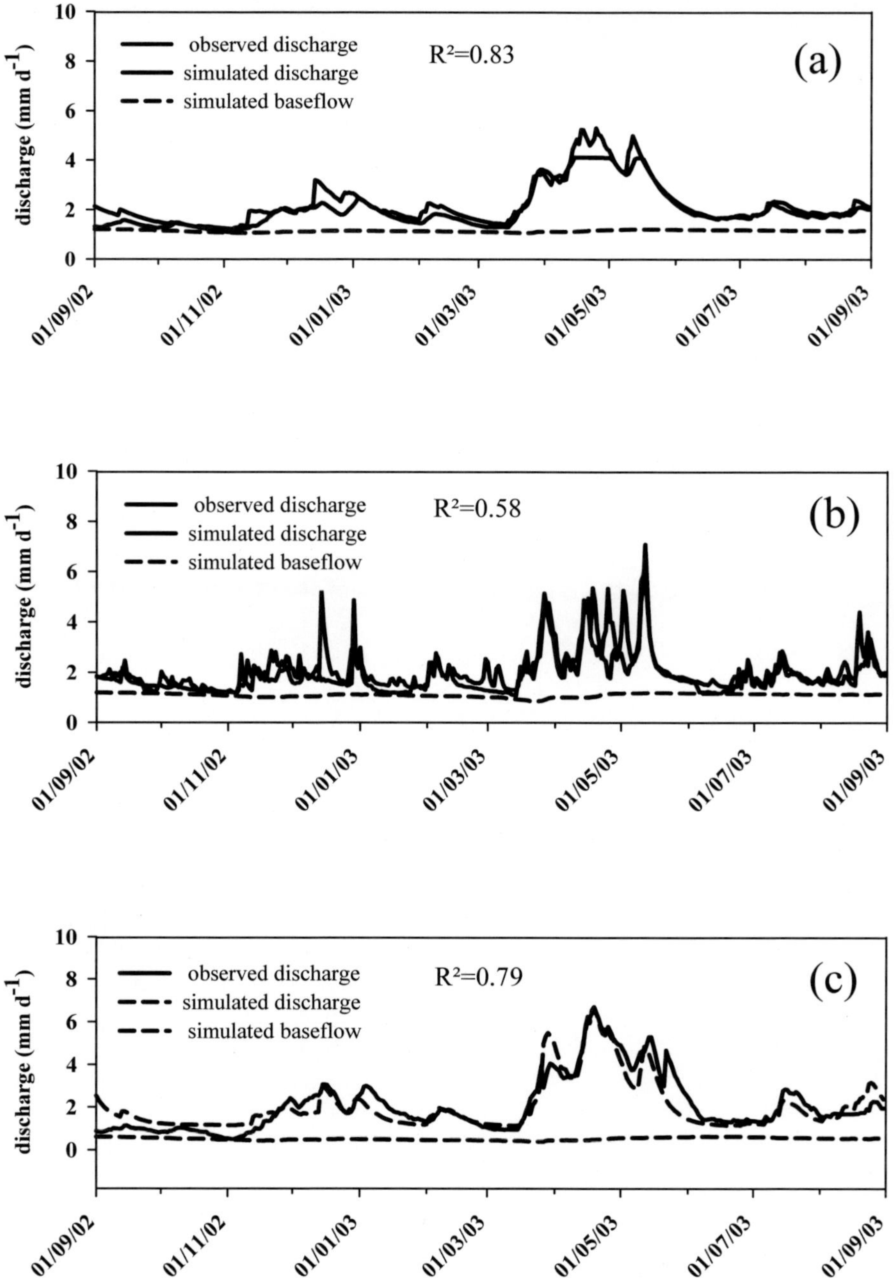

Fig. 3. Results of the calibration (01.09.2002-31.08.2003) of the Danau Lindu (a), Takkelemo (b) and Sopu (c) subcatchment on a daily resolution.

Table 3. List of the coefficient of efficiency R^2 for Gumbasa subcatchments (Lake Lindu, Sopu, Takkelemo, km^2) on a daily and weekly resolution for the calibration, validation-split sample and validation-whole period.

Gumbasa 1275 km^2	Lake Lindu 547		Sopu 592		Takkelemo 79	
R^2	daily	weekly	daily	weekly	daily	weekly
calibration period	0.83	0.86	0.79	0.77	0.58	0.77
validation period - split sample	0.61	0.62	0.58	0.44	0.41	0.50
validation period - whole	0.74	0.76	0.69	0.68	0.45	0.61

With the predictive uncertainty analysis the adjustable parameters were varied in such a way that the variation of these parameters had no effect on model outcomes under calibration conditions. The same adjustable parameter set as for the calibration run was used. The period for the predictive uncertainty analyses was the whole validation period from 01.09.2002 until 31.08.2004. The predictive uncertainty analysis iteratively determines the critical point where maximum model prediction is compatible with calibration imposed constraints on parameter values. Table 4 compares the residual analysis for the calibration run and the predictive uncertainty analysis with PEST (Doherty 2003) for the Danau Lindu subcatchment. The comparison demonstrates that the calibration remains stable within certain limits during the predictive sensitivity analysis when used with the critical point parameter set.

Table 4. Statistical residual analysis for the calibration run and the predictive sensitivity analysis for the Lake Lindu subcatchment.

Residual analysis	Calibration run	Predictive uncertainty analysis
Number of residuals	365	365
Mean value of residuals	- 0.18	- 0.47
Maximum residual	0.38	- 0.06
Minimum residual	- 1.12	- 1.72
Variance of residual	0.11	0.32
Standard error of residuals	0.33	0.56

ENSO and land use scenario generation

Climate variability ENSO: A statistical scenario approach was applied to generate scenarios of spatially and temporally variable ENSO-induced rainfall anomalies as input data for a hydrological scenario run of the Gumbasa

River watershed. Five climate stations with a long term historical record (1981 - 1999: Kalawara, Pandaya, Wuasa, Bora, Kulawi) were located in the area (Fig. 1). The monthly precipitation anomaly of ENSO years was calculated in percent related to the mean monthly precipitation of the observed years. Therefore, two graded ENSO scenarios, one of which reflects the mean anomaly of all observed ENSO events (1987, 1991, 1994, and 1997) and the other, the extreme ENSO event of 1997, were generated. Thereafter, the climate stations used for the interpolation of areal precipitation (8 IMPENSO and 5 STORMA stations, Fig. 2) were allocated to one of the long term stations according to their elevation and topographic position. The allocation is mainly basin-related although each of the long-term climate stations is located within one of the main basins. Aldrians study (2003) has shown that ENSO events have a significant influence on the climate of Central Sulawesi between the dry period from June to October. The ENSO scenario was applied for this time frame, which is affected by ENSO related precipitation anomalies. For the generation of the two graded ENSO scenarios the modelling year 2003 was taken as a base year (Gerold et al. 2008). For the generation of the ENSO scenarios only the precipitation anomalies were considered. Feedback mechanisms of the precipitation anomalies with other climatic parameters were not considered.

We simulated a land-use history scenario with the Landsat/ETM land cover classification of 2001 as input land cover grid for the analysis of the impact of observed land cover changes from 2001 till 2002 in the Gumbasa River catchment (details on land cover changes see Leemhuis et al. 2007). Furthermore, we applied a total change scenario, which assumed that all forested land up to 1,200 m a.s.l. is converted in a first conversion phase into annual crops such as maize or dry rice and in a second step into perennial crops. Cacao is a cash crop in Central Sulawesi and therefore is most likely to be planted as the perennial crop. In Central Sulawesi the cacao plant can be cultivated up to an elevation of 1,200 m a.s.l. since the temperature does not drop beneath 16 C at night up to this elevation (Rehm 1989). For both scenarios the physical vegetation parameters were altered according to the applied land cover type. The soil physical properties were kept constant because the changes in soil physical parameters induced by land cover changes were not investigated for the Gumbasa River catchment.

3 Results

3.1 ENSO effects

For the observed ENSO-events two scenarios which reflect the mean anomaly of all observed ENSO-events (1987, 1991, 1994, 1997; = average ENSO event = EA) and the extreme ENSO event of 1997 (EB) were generated. The actual conditions of 2003 (E0/L0) with rainfall distribution from 2003 (E0) and land cover characteristic 2002/2003 (L0) serve as a control run for the evaluation of the proportional changes. During average ENSO scenario (EA) the total yearly areal precipitation of the Gumbasa River catchment is reduced by 25%, whereas the total yearly evapotranspiration rate is only slightly diminished by 3%. The main differences in the yearly water balance were observed for the total yearly discharge, which is reduced by 32% (Table 5, Fig. 4). With a reduction of the total discharge by 24%, the Lake Lindu subcatchment is least affected by the precipitation anomaly, whereas the small Takkelemo catchment reaches reduction values similar to those in the Gumbasa River catchment (Leemhuis 2005). Even though the strong ENSO event of 1997 is applied by ENSO scenario EB, total yearly changes of the water balance in comparison with ENSO scenario EA are minimal. For example the decrease of the yearly total discharge is only reduced by 0.7% for the Gumbasa River catchment. For both ENSO scenarios the Lake Lindu subcatchment with the highest forest fraction (95%) exhibits the lowest yearly precipitation decline with 22.0% (Scenario EA) and 22.5% (Scenario EB), whereas the Gumbasa River catchment (85% forest) has a yearly average precipitation decline of 24.9% (Scenario EA) and 26.2% (Scenario EB) (Gerold et al.2008).

The simulated yearly precipitation decline is not linear with respect to the simulated yearly discharge decline, which indicates a far wider variance of the different catchment types with 8.4% for ENSO scenario EA and 9.1% for ENSO scenario EB (comparing Lake Lindu and Gumbasa, Table 5). Again for both scenarios the Lake Lindu catchment has the lowest discharge decline with 23.6% for both ENSO scenarios and the Gumbasa River catchment, the highest discharge decline among the observed catchments with 32.0% (ENSO scenario EA) and 32.7% (ENSO scenario EB) (Table 5). The simulated ENSO scenario results for yearly precipitation and discharge decline demonstrate the influence of the overall catchment characteristics on the potency of ENSO caused discharge anomalies. The weakest impact was observed for the Lake Lindu catchment. This demonstrates the buffering effect of the forest with deeper soil moisture storage during the dry season. The low pass flow duration analysis (Leemhuis 2005) generally shows a lower mean discharge for all catchments, whereas the high water discharge is constant for all scenarios.

3.2 Land use change effects

The comparison of the simulated yearly water balance and the applied land cover scenarios gives a first general overview of the impact of land cover change on the hydrological performance of the observed watersheds. Table 5 compares the changes induced by land cover change for the total Gumbasa River catchment and the subcatchments of Lake Lindu (95% forest cover) and Takkelemo (79% forest cover), respectively, of the following water balance components: areal precipitation (P), evapotranspiration (aET) and total discharge (Q). Considered in light of the actual land cover changes from 2001 to 2002, the effect on the yearly simulated water balance is rather insignificant. The hy-

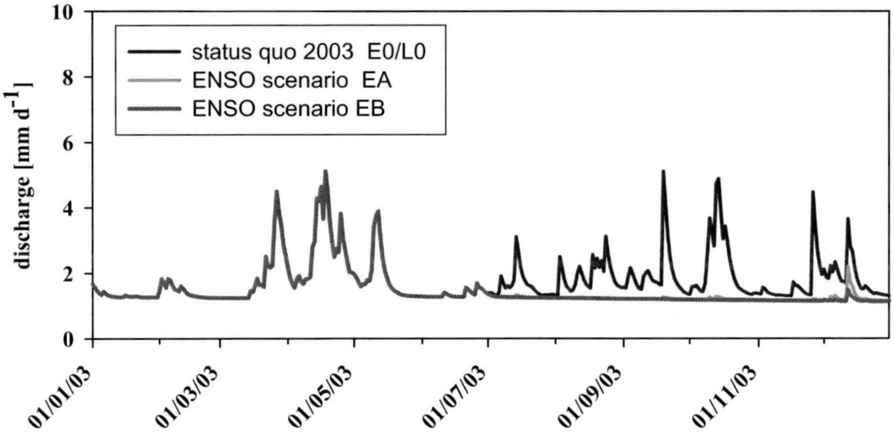

Fig. 4. Scenario results for daily discharge (Gumbasa catchment) with two different ENSO-events (EA = average year with rainfall decrease from 1.07.30.11.2003 with - 60%; EB = average year with rainfall decrease 90%.

drological model calculated a 0.2% of the total water balance decrease of total yearly discharge for the Gumbasa River catchment (Leemhuis et al. 2007).

The annual crop scenario with an increase of 42% in the yearly total runoff for the Gumbasa River catchment and the perennial crop scenario with an increase of 23% in the yearly total discharge indicate a great increase of the yearly total discharge with cumulative deforestation rates (Fig. 5, table 5). All three catchments indicate an increase of 36 - 42% in total discharge (200 - 300 mm), whereas the cacao scenario leads to an increase of 22 - 24% (100 - 200 mm).

Analyses of the temporal discharge variability in mm d^{-1} (monthly mean) with mean low (MLQ), mean (MQ) and mean high water (MHQ) discharges for the Gumbasa River catchment (Leemhuis et al. 2007) demonstrate a moderate monthly rise for both deforestation scenarios. The impact of forest conversion increases for the monthly mean discharge (MQ) and is most significant for the monthly high water discharges (MHQ) with 1 - 3 mm increase during the main rainy season. The results imply that deforestation has a more important impact on the high water discharges than on the low water discharges. The seasonal discharge variation indicates an increase in peak discharge for both scenarios with the maxima for the annual crop scenario (Fig. 5).

4 Discussion

The application of the hydrological model WASIM-ETH on the Gumbasa River catchment in Central Sulawesi, Indonesia is the models first simulation of the water balance of a mesoscale humid tropical catchment. The achieved

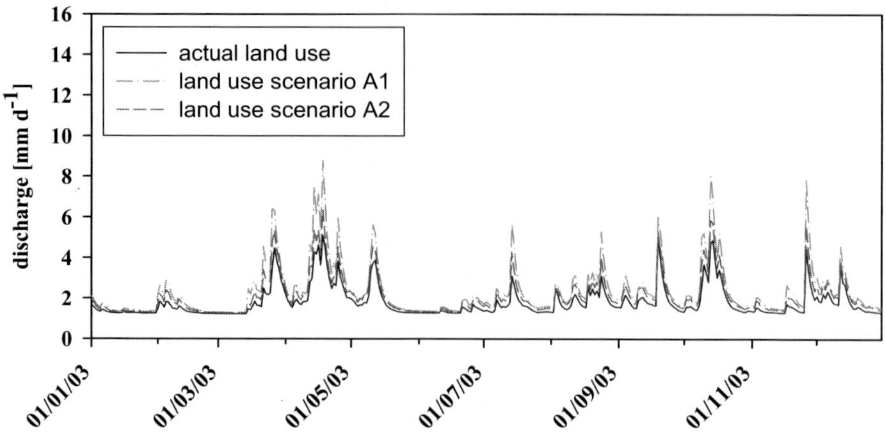

Fig. 5. Scenario results for daily discharge (Gumbasa catchment) with two different land use developments (A1 = conversion of rainforest $\leq$ 1,200 m a.s.l. in annual crops; A2 = conversion of rainforest $\leq$ 1,200 m a.s.l. into cacao plantation.

model efficiencies can be classified among other WASIM-ETH model applications in various catchment studies of different climatic regions (Schulla 1997; Jasper et al. 2002; Piepho 2003). The modularly based character of WASIM-ETH allows a general adaptation of the hydrological model to different climatic regions and catchment data bases in space and time. If the overall model uncertainty is assessed with regard to the available spatial data set, the land use classification and the soil classification generally represent sources of uncertainty. With regard to land use classification derived from satellite images, such as Landsat ETM+, whether different land use classification methods result in different modelling results should be further analysed. Furthermore, the dramatic land use changes within the catchment area are a source of error for the simulation of the water balance of the status quo. The dramatic land use changes which could be observed in the Gumbasa River catchment during the calibration period reflect the aspect of the non-stationary of tropical catchments, which is defined by Klemes (1993) as one major aspect of tropical hydrology. Therefore, for the model application the same static land use classification from one satellite scene was employed over the entire period although in reality the land use was changing continuously during the calibration period, .

Due to the inadequacy of the available soil map of the Gumbasa watershed area, the soil classes were generated on the basis of similar topographic properties. This regionalization concept for hydrological modelling purposes was already conducted by Gerold et al. (2003) and provides promising results if no adequate soil map is available. The parameterisation of soil hydraulic properties was performed with the analysed soil catena in the Nopu catch-

Table 5. Total yearly simulated water balance for the Gumbasa River catchment, compared to an annual crop scenario, perennial crop scenario (cocoa) and 1997 ENSO and average-ENSO-scenarios; Δ represents the changes of precipitation, evapotranspiration and total discharge in percent to the control run (current climate and land use 2002/2003, Leemhuis 2005, Gerold et al. 2008).

Scenario	catchment forest area (%)	precipitation P (mm)	Δ (%)	evapo-transpiration aET (mm)	Δ (%)	total discharge Q (mm)	Δ (%)
current climate a. land use 2003/2002	Gumbasa (82)	2150	–	1543	–	590	–
	L. Lindu (95)	2192	–	1295	–	896	–
	Takkelemo (79)	2259	–	1428	–	804	–
current climate 2003/2002 a. annual crop[1] E0/A1	Gumbasa (45)	2150	–	1265	-20.0	838	42.0
	L. Lindu (65)	2192	–	1112	-14.1	1218	35.9
	Takkelemo (39)	2259	–	1124	-21.3	1113	38.4
current climate 2003/2002 a. cocoa plantation[2] E0/A2	Gumbasa (45)	2150	–	1361	-11.8	724	22.7
	L. Lindu (65)	2192	–	1144	-11.7	1097	22.4
	Takkelemo (39)	2259	–	1222	-14.4	1000	24.4
average ENSO event a. land use 2003 EA/L0	Gumbasa (82)	1614	-24.9	1496	-3.0	401	-32.0
	L. Lindu (65)	1710	-22.0	1281	-1.1	684	-23.6
	Takkelemo (39)	1724	-23.7	1423	-0.4	561	-30.2
ENSO event 1997 a. land use 2003 EB/L0	Gumbasa (82)	1586	-26.2	1487	-3.1	397	-32.7
	L. Lindu (65)	1699	-22.5	1279	-1.2	684	-23.6
	Takkelemo (39)	1709	-24.4	1421	-0.5	557	-30.7
ENSO event 1997 a. annual crop[1] EB/A1	Gumbasa (45)	1586	-26.2	1234	-19.6	606	2.7
	L. Lindu (65)	1699	-22.5	1088	-16.0	827	-7.7
	Takkelemo (39)	1709	-24.4	1110	-22.3	804	0.0
ENSO event 1997 a. cocoa plantation[2] EB/A2	Gumbasa (45)	1586	-26.2	1351	-12.0	515	-12.7
	L. Lindu (65)	1699	-22.5	1167	-9.9	769	-14.2
	Takkelemo (39)	1709	-24.4	1239	-13.2	715	-11.1

[1] forest conversion with annual crop until 1200 m a.s.l.
[2] forest conversion with cocoa plantation until 1200 m a.s.l.

ment (Kleinhans 2003). The Gumbasa River watershed is characterised by steep slopes, which means that in the simulation of the hydrological processes the lateral fluxes prevail and predominately lateral fluxes with interflow generate the total discharge pattern. Using WASIM-ETH (Kleinhans 2003) and digital hydrograph separation (with TSPROC, Doherty 2003) for the Nopu experimental catchment, a fraction of 40 - 50% was calculated for the interflow (Kleinhans and Gerold 2004).

4.1 ENSO effects

Research results of ENSO consequences on river discharge in SE-Asia are rare and mainly address to negative rainfall anomalies with El Niño (Aldrian

2003). Studies from different tropical regions have shown that ENSO events are related to inter-annual variations in stream flow (e.g. Amarasekera et al. 1996, Anderson et al. 2001, Cluis 1998).

ENSO caused precipitation anomalies lead to an overall increase of the discharge variability. Moreover, the ENSO scenario discharge simulations show that mainly the low water and mean discharges are affected in the second half of the year (August till December) with a shift of one to two months. Strong ENSO-events such as the one that occurred in 1997 lead to a 32% reduction in water yield. Catchment characteristics, mainly change in forest area, cause different hydrological response magnitude to the ENSO impact. The relative lower discharge decrease of the Lake Lindu catchment is due to the higher retention capability of Lake Lindu (Table 5). The larger fraction of the forest area is coupled with slower drying out of the soil water reservoir and a reduced evapotranspiration rate for the high altitude rainforest, which results in an overall higher discharge coefficient for the catchment. The Takkelemo headwater catchment showed the lowest increase in discharge variability during ENSO events because under normal conditions it is already characterised by a marginal low water discharge, which is related to its small catchment size. The hydrological system of the total Gumbasa River catchment is more sensitive to ENSO caused precipitation anomalies. This finding is expressed by the highest calculated yearly discharge decline and the greatest simulated discharge variability. It can be concluded that the catchment characteristics have a certain influence on the impact magnitude of ENSO related rainfall anomalies on the water balance of a catchment. But as previously discussed, the high spatial rainfall variability in tropical mesoscale catchment cause uncertainty in the interpretation of the subcatchment differences.

Moreover, alterations in the surface energy balance due to the changed vegetation physics are not considered by the applied methodology. If atmospheric feedbacks were to be included in the scenario analysis, changes in runoff would tend to be buffered (Costa 2005). This is also true for the applied ENSO scenarios, where ENSO caused precipitation anomalies would also alter the meteorological parameters, such as temperature, relative humidity and global radiation. Hence, an increased potential evapotranspiration rate may be assumed during ENSO events, which again would alter the discharge rate. To achieve a complete analysis of the impact of ENSO-caused precipitation anomalies on the hydrology of a catchment, a model which incorporates the atmospheric-vegetation feedback system on a catchment level would further improve analysis.

During ENSO conditions (EB/A1, EB/A2), the applied land use changes (A1, A2) result in a significantly higher yearly discharge variability, which is expressed by an increase of the extreme low water and high water discharge events (Leemhuis 2005, Fig. 7.6; Table 5). The yearly discharge for the coupled ENSO and land use scenario compensate for the high ENSO decrease in the second part of the year by higher peak discharges in the first part of the year. During average ENSO years cacao plantation shows a clear ENSO water yield reduction of 10 - 11% (Leemhuis 2005) and in extreme ENSO years (1997, EB/A2 Table 5) of 11 - 14%.

4.2 Land use change effects

The land use change scenario results from Gumbasa (Leemhuis 2005) agree with the findings of the Nopu tropical catchment study by Kleinhans (2003), who applied the hydrological model WASIM-ETH on a small headwater catchment of the Gumbasa River catchment. Both studies used the same hydrological model, therefore it was likely that similar land use scenario results would be also calculated on a larger scale. The cacao plantation scenario exhibits a comparable result in the magnitude of discharge increase (Nopu + 16.8%), whereas simulation results for the annual crop scenario are quiet different (Nopu +8%). One reason for this is the use of different vegetation parameters. After sensitivity analysis by Kleinhans (Fig. 47, 2003) the parameter albedo (α) and minimal stomatal resistance (r_c) react very sensitively in the estimation of aET and therefore for the discharge (Q). A further reason could be the different calibration period for the model.

All changes in the total discharge were induced by an alteration of the annual evapotranspiration rate, due to a decline of LAI, root depth, vegetation height and stomata conductance. Less water evaporates due to interception; hence, throughfall increases and more water can infiltrate into the soil. Plot research results within STORMA shows a decrease of interception from 21% (rainforest) to 10% (cacao) (Gutzler et al., this book) and a reduction in aET for annual crops (maize) with an order of 15 - 20% and for perennial crops (cacao) with 5 - 12% (Falk 2004). This reduction in the evapotranspiration causes the soil to be wetter and therefore more responsive to rainfall (Bruijnzeel 2004). The graduated impact of annual and perennial crop scenario on yearly runoff is related to the more similar physical vegetation parameters of the cacao tree compared to the broadleaf evergreen forest of the catchment. Results from Oltchev et al. (2008) indicate a decrease of aET (actual evapotranspiration) from rainforest to cacao plantation in the range of 1.0 mm d^{-1}. Regarding the crop scenario for the Lake Lindu subcatchment the lower increase of discharge is due to the lesser decrease of evapotranspiration.

Experimental paired catchment studies confirm the simulated increases in peak flows after forest removal (Bonel and Balek 1993; Chandler and Walter 1998). This is explained by the associated reduction in aET that causes the soil to be wetter and therefore more responsive to rainfall (Bruijnzeel 2004). Lal (1983) reported an increase of 140mm $year^{-1}$ (+12%) with forest conversion to annual crops in Nigeria, but hydrological scenarios for annual crops and agroforestry in the Inner Tropics are rare. Hibbert (1967) reports an increase in discharge of between 2.5 and 4.5 mm per 1% of forest conversion. For the Gumbasa catchment with E0/A1 scenario (Table 5) this would result in an increase of 100 - 180 mm total discharge, compared to the 248 mm simulation value. Bruijnzeel (2004, p.21) wrote: increases in water yield proved to be roughly proportional to the fraction of biomass removed. Detailed experimental measurements with microcatchments (forest and pasture) and process modelling also for the mesoscale (Rio Chiquito in Costa Rica; Bruijnzeel 2006) with FIESTA_CQflow[1] show an overall increase in water yield of 150 mm year^{-1} (+4.3%) for the scenario with total conversion of cloud- and rainforest to pasture. The increase in stormflow and peakflow at the local scale disappear at the larger scale due to the diluting effect of spatial rainfall variability.

The main effect that is with the spatial heterogeneity of rainfall in Central Sulawesi. We compare the measured discharge differences between the subcatchments of Gumbasa, Lake Lindu and Takkelemo (Table 5), but the correlation between forest cover differences in relation to discharge differences is not good.

In order to evaluate the impact of land use changes on the monthly temporal variability, LQ (mean low water discharge), MQ (mean water discharge) and HQ (high water discharge) were calculated for the Gumbasa river catchment (Leemhuis 2005, Fig. 7.6). The diagram of the monthly variation of the low water discharge LQ demonstrate a moderate monthly rise if the land use changes from forest to cacao plantations and annual crops. But the increase is < 0.25 mm d^{-1}. More significant are the increases for the mean discharges (MQ) and for the mean high water discharges (HQ) with +1.0 mm d^{-1} (cacao) and +3.0 mm d^{-1} (annual crop). This implies that the applied land use scenarios have a more important influence on high water discharge and flood generation than on low water discharges.

Bruijnzeel (2004) initiated the discussion about the low flow problem with reference to dry season flow and deforestation. He emphasises the contradiction that, although reduced evaporation associated with the removal of forest should have produced higher baseflow, numerous studies of smaller tropical catchments show a reduced low flow after forest removal, whereas some other studies report an increased baseflow (chapter 3.1 in Bruijnzeel, 2004). Whether the baseflow is raised or reduced after forest removal is strongly linked to the degree of surface disturbance. Costa (2005) argues that as a result of the decreased hydraulic conductivity due to soil consolidation less water can infiltrate. This results in an increased direct discharge and actually leads to a decrease of the low water discharge. Analysis of saturated hydraulic conductivity in relation to different land use types and their age indicate that for the young cacao plantations and annual crops (maize, peanuts, beans) in the Nopu catchment k_{sat} (saturated hydraulic conductivity) is still very high (above 100 mm h^{-1}) and favours infiltration and groundwater recharge with

[1] The FIESTA_delivery model is a process model which simulates the hydrological balance including inputs of wind-driven precipitation and fog and outputs of evapotranspiration. The resulting balance is cumulated along river flow networks to give an indication of runoff (Bruijnzeel 2006).

no decrease in base flow (Leemhuis et al. 2008). But with aging and perhaps mechanizing of crop cultivation this can change in the future.

Furthermore, the scenario results by Leemhuis (2005) show that after the complete conversion of all forest area into cacao plantations the high water discharge is still characterised by enhanced peak flows, which indicates a continued alteration of the water balance for cacao plantations. The forest/cacao experiment with converting forest to cacao plantation in Malaysia (Abdul Rahim 1988) showed an increase in discharge over four years of +25%, whereas with cacao planting under emergent trees in the rainforest no change in the water balance components occurred (Baconquis 1996).

5 Conclusions

The implications of possible future climate and land use conditions on the water balance of the Gumbasa River catchment were assessed by a scenario analysis, which also included severe ENSO events (e.g. 1997) and future land use trends. The results of the hydrological model scenario application clearly demonstrate a strong relationship between deforestation rates and increasing discharge variability. In particular, a significant increase of high water discharges was simulated for the applied land use scenarios. The main results of the scenario analysis are:

1. that ENSO anomalies of precipitation lead to an increase of discharge variability;
2. that strong ENSO-events (El Niño) lead to a 30% reduction in the water yield, and annual crop scenarios up to 1,200 m a.s.l. showed a 42% and the perennial crop scenario (cacao) a 23% increase in river discharge with high increase in overland flow and flooding risks;
3. that ENSO-events (El Niño) decrease the potential (flooding) area of paddy rice cultivation in the Gumbasa Irrigation Scheme by two thirds in the second half of the year.

Forest replacement by annual crops and cacao plantations tends to increase annual streamflow quantities and storm flow events with higher peak discharges. But the hydrological impacts of forest conversion in complex mountainous terrain varies considerably in magnitude and frequency by scale and from catchment to catchment related to land-cover change, and to specific climatic, topographic and soil physical characteristics. Therefore, spatial modelling with climate and land use scenarios contribute to a better understanding of potential change in water yield and flood risks. But these require catchment-specific data, which are often lacking in developing countries. Population pressure by migration from South to Central Sulawesi and the cacao boom with smallholder cash crop potential for income generation will lead to an ongoing increase in deforestation in the future. This will result in elevated risk of high river discharge and flooding in the alluvial plains and river basins. But total runoff and peak discharge depends on the change of pedohydrological characteristics with land use practice in cacao agroforestry systems. Without soil compaction cacao plantation buffer the danger of a large increase in surface runoff, which often follows forest conversion to annual crops.

Acknowledgements
This study was supported by the Federal Ministry of Education and Research (BMBF) and by the German Research Foundation (DFG, SFB 552). Special thanks to the Institut Pertanian Bogor (IBP), Java, Indonesia and the Universitas Tadulako (UNTAD), Palu, Central Sulawesi, Indonesia for the productive research cooperation.

References

Abbott MB, Refsgaard JP (1996) Terminology, modelling protocol and classification of hydrological model codes. In: Abbott MB, Refsgaard JP (Eds.): Distributed hydrological modelling. Kluwer Academic Publishers, Dordrecht, Boston, London, pp. 17-37

Abdul Rahim N (1988) Water yield changes after forest conversion to agricultural land use in Peninsular Malaysia. Journal of tropical Forest Science 1: 67-84

Aldrian E (2003) Simulations of Indonesian rainfall with a hierarchy of climate models. Diss., Universität Hamburg, 23-45

Amarasekera KN, Lee RF, Williams ER, Eltahir E (1996) ENSO and the natural variability in the flow of tropical rivers. Journal of Hydrology 200: 24-39

Andeson ML, Kavvas ML, Mierzwa MD (2001) Probalistic/ensemble forecasting: a case study using hydrologic response distributions with El Niño Southern Oscillation (ENSO). Journal of Hydrology 249: 134-147

Baconquis SR (1996): Impact of agroforestry and gully stabilization on the water budget of a secondary Dipterocarp forest watershed, Philippines. Sylvatrop 3 (2): 1-26

Beven K, Feyen J (2002) The future of distributed modelling. Hydrological Processes 16: 169-172

Bohman K (2004) Functional and morphological diversity of trees in different land use types along a rainforest margin, in Sulawesi, Indonesia. Dissertation, Universität Göttingen, pp. 55-63

Bruijnzeel LA (1996) Hydrology of moist tropical forests and effects of conversion: a state of knowledge review. UNESCO International Hydrological Programme, 20-20

Bruijnzeel LA (2004) Hydrological functions of tropical forests: not seeing the soil for the trees? Agriculture, Ecosystems & Environment 104: 185-228

Bruijnzeel LA (2006) Hydrological impacts of converting tropical montane cloud forest to pasture, with initial reference to northern Costa Rica. Final Technical Report DFID-FRP Project no. R7991, pp. 52

Cluis D (1998) Analysis of long runoff series of selected rivers of the Asia-Pacific region in relation with climate change and El Niño effects. Global Runoff Center, Koblenz, pp. 13-20

Costa MH, Botta A, Cardille JA (2003) Effects of large-scale changes in land cover on the discharge of the Tocantins River, Southeastern Amazonia. Journal of Hydrology 283: 206217

Costa MH (2005) Large scale hydrological impacts of tropical forest conversion. In: Bonell M and Bruijnzeel LA (Eds.): Forests, Water and People in the Humid Tropics. University Press, Cambridge, 590-598

Dijk A, Bruijnzeel LA (2001) Modelling rainfall interception by vegetation of variable density using an adapted analytical model. Part 2. Model validation for a tropical upland mixed cropping system. Journal of Hydrology 247: 239-262

Doherty J (2003) PEST Model-independent parameter estimation. 1-9

Döll P, Kaspar F, Lehner B (2003) A global hydrological model for deriving water availability indicators: model tuning and validation. Journal of Hydrology 270: 105-134

Erasmi S, Twele A (2008) Regional land cover mapping in the humid tropics using combined optical and SAR satellite data a case study from Central Sulawesi, Indonesia. Int. Journal of Remote Sensing (accepted)

Falk U (2004) Turbulent fluxes of CO2, H2O and energy in the atmospheric boundary layer above tropical vegetation investigated by Eddy-Covariance measurements. Diss. Göttingen

Falk U, Ibrom A, Oltchev A, Kreilein H, June T, Rauf A, Merklein J, Gravenhorts G (2005) Energy and water fluxes above a cacao agroforestry system in Central Sulawesi, Indonesia, indicate effects of land use change on local climate. Zschr. Meteorol. 14: 219-225

Falkenmark M, Rockström J (2004) Balancing water for humans and nature. Earthscan, London, pp. 181-199

Fleischbein K, Wilcke W, Goller R, Boy R, Valarezo C, Zech W, Knoblich K (2005) Rainfall interception in a lower montane forest in Ecuador: effects of canopy properties. Hydrological Processes 19: 1355-1371

Gerold G, Sutmöller J, Krüger J-P, Herbst M, Busch G, Peschke G, Zimmermann S, Etzenberg C, Töpfer J (2003) Reliefgestützte und wissensbasierte Regionalisierung in der Hydrologie. EcoRegio 6,, Shaker Verlag, pp. 19-32

Gerold G, Leemhuis C (2008) Modelling the effects of ENSO-events and rainforest conversion on river discharge in Central Sulawesi (Indonesia) problems and solutions for coarse spatial parameter distribution. Int. Congress on Environmental Modelling and Software Integrating Sciences and Information Technology for Environmental Assessment and Decision Making, Barcelona, 553-565

Godsey S, Elsenbeer H, Stallard R(2004) Overland flow generation in two lithologically distinct rainforest catchments. Journal of Hydrology 295: 276-290

Güntner A (2002) Large-Scale hydrological modelling in the Semi-Arid North-East of Brasil. Dissertation, Universität Potsdam

Gutzler C, Koehler S, Gerold G (2009) A comparison of throughfall rate and nutrient fluxes in rainforest and cocoa-plantation in Central Sulawesi, Indonesia. *(this book)*

Häring V, Köhler S, Gerold G (2005) Ergebnisse der Bodenkartierung Toro, Nopu (Sulawesi, Indonesien) im Rahmen des SFB 552. Göttingen, pp. 40

Hibbert A (1967) Forest treatment effects on water yioeld. In Sopper W.E., Lull H.W. (Eds.) Int. Symposium on Forest Hydrology. Oxford, pp. 527-545

Hodnett MG, Tomasella J (2002) Marked differences between van Genuchten soil water-retention parameters for temperate and tropical soils: a new water retention pedo-transfer functions developed for tropical soils. Geoderma 108: 155-180

IAHS (2003) Prediction in ungauged basins. International Hydrology today: 26-28

Jackson RB, Canadell J, Ehleringer JR, Mooney HA, Sala OE, Schulze ED (1996) A global analysis of root distributions for terrestrial biomes. Oecologia 108: 389-411

Jasper K, Gurtz J, Lang H (2002) Advances flood forecasting in Alpine watersheds by coupling meteorological observations and forecasts with a distributed hydrological model. Journal of Hydrology 267: 40-52

Kleinhans A (2003) Einfluss der Waldkonversion auf den Wasserhaushalt eines tropischen Regenwaldeinzugsgebietes in Zentral Sulawesi (Indonesien). Diss. SUB Göttingen

Kleinhans A, Gerold G (2004) Die Abschätzung des Einflusses von Landnutz ungs-Änderungen auf den Wasserhaushalt eines tropischen Einzugsgebietes ein Modellierungsansatz. Forum für Hydrologie und Wasserbewirtschaftung 2, H.05.04, 199-202

Klemes V (1993) The problems of the humid tropics-oppurtunities of reassessment of hydrological methodology. In: Bonell M, Hufschmidt M, Gladwell J (Eds.): Hydrology and water management in the humid tropics. Camebridge University Press, Camebridge, pp. 45-51

Körner C (1994) Leaf diffusive conductances in the major vegetation types of the globe. In: Schulze ED, Caldwell MM (Eds.): Ecophysiology of Photosynthesis. Springer, Berlin, pp. 463-490

Landon JR (1984) Tropical soil manual. Booker Agriculture International, New York, pp. 289-291

Leemhuis C (2005) The impact of El Niño Southern Oscillation Events on water resource availability in Central Sulawesi, Indonesia. EcoRegio 21, pp. 149

Leemhuis C, Erasmi S, Twele A, Kreilein H, Oltchev A, Gerold G (2007) Rainforest conversion in Central Sulawesi, Indonesia: Recent development and consequences for river discharge and water resources. Erdkunde 61: 284-293

Matthews E (1999) Global vegetation guide (1972-1983). http://www.daac.ornl.gov/

Mo X, Liu S, Lin Z, Zhao W (2004) Simulating temporal and spatial variation of evapotranspiration over the Lushi basin. Journal of Hydrology 285: 125-142

Nash JE, Sutcliffe JV (1970) River flow forecasting through conceptual models. Journal of Hydrology 10: 282-290

Oltchev A, Ibrom A, Priess J, Erasmi S, Leemhuis C, Twele A, Radler K, Kreilein H, Panferov O, Gravenhorst G (2008) Effects of land use changes on evapotranspiration of tropical rain forest margin area in Central Sulawesi (Indonesia): modelling study with a regional SVAT model. Journal Ecological Modelling 212: 131-137

Oyebande L, Balek J (1987) Humid warm sloping land. In: Falkenmark M, Chapman T (Eds.): Comparative hydrology. UNESCO, Paris, pp. 224-274

Piepho B (2003) Untersuchungen zum hydrologischen Reaktionsverhalten eines kleinen Einzugsgebietes unter Verwendung des Wasserhaushaltsmodells WaSiM-ETH. Diss. TU Braunschweig

Rehm S (1989) Spezieller Pflanzenbau in den Tropen und Subtropen. Ulmer, Stuttgart, 437-446

Schulla J (1997) Hydrologische Modellierung von Flussgebieten zur Abschätzung der Folgen von Klimaänderungen. Züricher Geographische Schrif ten, Verlag Geographisches Institut ETH Zürich, Verlag Geographisches Institut ETH Zürich

Schulla J, Jasper K (1999) Model Description WASIM-ETH, Zürich

Scurlock JMO, Assner GP, Gower ST (2001) Global Leaf Area Index from Field Measurements, 1932-2000 http://www.daac.ornl.gov

Twele A, Erasmi S, Kappas M (2008) Spatially explicit estimation of leaf area index using EO-1 Hyperion and Landsat ETM+ data: Implications of spectral bandwidth and shortwave infrared data on prediction accuracy in a tropical montane environment. GIScience & Remote Sensing, 45: 1-20. DOI: 10.2747/1548-1603.45.2.1

WRB-FAO (2006) World Reference Base for Soil Resources. FAO, Rom

Adaptation to climate change in Indonesia - livelihood strategies of rural households in the face of ENSO related droughts

Norbert B. Binternagel[1*], Jana Juhrbandt[2], Sebastian Koch[3], Mangku Purnomo[4], Stefan Schwarze[2], Jan Barkmann[2], and Heiko Faust[1]

[1] Georg-August-University Göttingen, Faculty of Geosciences and Geography, Department of Human Geography, Goldschmidtstrae 5, 37077 Göttingen, Germany
[2] Georg-August-University Göttingen, Faculty of Agricultural Sciences, Department for Agricultural Economics and Rural Development, Platz der Göttingen Sieben, 37073 Göttingen, Germany
[3] Georg-August-University Göttingen, Albrecht-von-Haller Institut for Plant Science, Didactics of Biology, Waldweg 26, 37073 Göttingen, Germany
[4] Brawijaya University of Malang, Faculty of Agriculture, Department of Socio-Economics, Jalan Veteran, 65145 Malang, Indonesia

*corresponding author: N. B. Binternagel,
email: Norbert.Binternagel@geo.uni-goettingen.de

Summary

Climate change poses a major threat to the livelihoods of many people, especially those dependent on agriculture. Adaptation to climate change is a necessity to reduce social vulnerability. The adoption of agricultural innovations presents a suitable strategy to reduce negative impacts of increased inter-annual variability in rainfall and temperature. Focussing on the dominant agroforestry system cacao in the Indonesian province of Central Sulawesi, we investigate the adaptation strategies of agricultural smallholders to El Nico-Southern Oscillation (ENSO) related droughts. We apply a triangulation of Rapid Rural Appraisal Methods (RRA), semi-structured in-depth interviews and participatory observations. In addition, results from an intensive cacao-study (n=144) on yields, plot history, structure and location are used.

The analysis shows that all interviewed households have been affected by extreme climatic effects such as ENSO related droughts which lead to a decline in agricultural outputs. Our study identifies various types of adaptation strategies on the micro level. Most common strategies are reactive or

T. Tscharntke et al. (eds.), *Tropical Rainforests and Agroforests under Global Change*, Environmental Science and Engineering, 351–375, DOI 10.1007/978-3-642-00493-3_16,

ex-post adaptations which support the household in coping with the effects of droughts. Further, the study assesses anticipatory adaptation strategies which, if precisely implemented, alter the household's exposure to future droughts, increase the resilience to cope with changes and reduce the degree of sensitivity of affected households. On the basis of these results we examine why different household types do not implement certain strategies, which could enhance their social resilience. In this context we reveal that the adoption of anticipatory adaptation strategies is strongly influenced by social and human capital. The knowledge transfer - which promotes adaptation - is mainly taking place along ethnic lines and local institutions. For the successful adoption of innovations to cope with ENSO related droughts in the study area, membership in certain ethnic groups, local institutions and networks is of utmost importance.

Keywords: adaptation, mitigation, natural hazards, local knowledge, social vulnerability, resilience, adaptive capacity, social capital, innovations, Indonesia

1 Introduction

Climate change may further intensify the frequency and magnitude of El Niño-Southern Oscillation (ENSO) droughts (IPCC 2001). In tropical Indonesia, these ENSO related droughts do not only reduce overall precipitation during several months of effected years, they also increase the inter-annual variability in rainfall and temperature. Even without intensification these droughts have a considerable impact on Indonesia with substantial crop failures, water shortages and forest fires (Cruz et al. 2007, FAO 2008). Agricultural smallholders are particularly affected by ENSO related droughts, because they are often not able to sustain their production assets during a drought event. In such cases, the primary drought impact extends production losses into subsequent years and thus harms the ability to produce and react to changing conditions and new opportunities in the future (Verchot et al. 2007).

According to Adger et al. (2007), there is a long record of social- and technological practices to adapt to ENSO related droughts. These include reactive or ex-post adaptations as well as pro-active or anticipatory measures. The latter have received advanced significance in the last decades since Cane et al. (1986) developed an operational capability to forecast the onset of ENSO events several months in advance. Within an evaluation of the responses to the 1997-98 ENSO across 16 developing countries, Glantz (2001) drew attention to a number of barriers to effective adaptation. However, most of them are on a meta-level focusing on forecast, monitoring and national capacities. Further research highlights that technological solutions - such as seasonal forecasting - are not sufficient to address the underlying social drivers of vulnerability to climate change (Agrawala & Broad 2002). In addition, social inequities in access to climate information severely constrain anticipatory adaptation (Pfaff et al. 1999).

In this study we concentrate on the differentiation of livelihood strategies in the face of ENSO related droughts for different household types. We exclusively focus on three different household types of cacao specialists. The time scale ranges from the present to the last 15 years. Main foci are the processes, determinants and drivers which lead to a successful adaptation of the natural resource systems on which societies depend. In doing so, we first assess micro level adaptation strategies by analysing different types of rural households in Central Sulawesi (Indonesia). Secondly, we conduct a community-based vulnerability evaluation aiming to contribute to practical adaptation initiatives as "research that focuses on the implementation processes for adaptation is still not common [...] and certainly not in the climate change field"(Smit & Wandel 2006). Consequently, we highlight the underlying causes for the processes of adaptation and examine why certain household types do not implement anticipatory- or pro-active adaptation strategies which could reduce their vulnerability against ENSO related droughts. We use the people centred, asset based "Sustainable Livelihood Framework" (Chambers & Conway 1991) to relate household's adaptation strategies to the reduction of vulnerability and an increase in resilience. We show that strategies and institutions which enhance resilience already exist on the village level. However, a widespread diffusion of adaptation strategies is impeded by interethnic barriers in knowledge transfer. Finally we reveal that interventions aiming to reduce the social vulnerability of smallholders should include support in the management of the agroforestry system cacao, transparent and standardised fees to obtain land tenures as well as improved local knowledge transfer.

In the following section the "Sustainable Livelihood Framework" will be explained in detail. A short description of the research area is followed by the methods and data section. Next, the results with respect to the impacts of ENSO related droughts in the research area as well as strategies against such impacts are presented. Finally, in the discussion section, we highlight our findings and confront them with those in the international debate.

2 Conceptual Framework

2.1 Definitions

The concepts of vulnerability, adaptation, adaptive capacity and resilience have a wide application to global change science and the analyses of these concepts range tremendously in scale (Smit & Wandel 2006). Due to continuing debates among scientists in various academic disciplines there is the need to apply these terms coherently with the object of interest and in the most recent and accepted definitions. Consequently, we use the term social vulnerability to emphasize the social dimension of vulnerability which is the focus of this research. Following Smit & Wandel (2006) our community-based vulnerability assessment is neither aiming to count adaptations, measure relative vulnerabilities nor quantify impacts or estimate effects of assumed adaptations. Rather, the focus is to document the ways in which the households experience ENSO related droughts and the processes of decision-making. To avoid misunderstandings the concepts will be defined more precisely in the following.

"Adaptations are manifestations of adaptive capacity, and they represent ways of reducing vulnerability" (Smit & Wandel 2006). Hence, system modifications in order to better cope with problematic exposures and sensitivities finally reflect adaptive capacity. However, adaptation is not similar to adaptive capacity. Households who possess the essential assets to carry out an adaptation strategy are not necessarily implementing these activities. The action itself is determined through the decision-making process of the individual household. Following Adger (2005) adaptation can be separated in "reactive adaptation" and "anticipatory adaptation" due to their timing relative to the event. By reactive adaptation we mean short term coping strategies, triggered by past or current events. By anticipatory adaptation we mean long term strategies that take place before impacts of climate change are observed and which are based on an assessment of conditions in the future. Furthermore, Adger (2005) states that there are three requirements for successful adaptation: reducing the sensitivity of a system to climate change; altering the exposure of a system to climate change; and increasing the resilience of a system to cope with changes.

A general conceptual model of vulnerability has emerged in the climate change literature (Kelly & Adger 2000, Turner et al. 2003, Smit & Pilifosova 2003, Yohe et al. 2003, Adger 2006). Consistent throughout the literature is the notion that the vulnerability of any system is a function of the exposure and sensitivity of that system to hazardous conditions in addition to the resilience or ability of the system to cope, adapt or recover from the effects of these conditions (Smit & Wandel 2006). As a result, households which are more exposed and sensitive to ENSO related droughts are more vulnerable, and households that have more adaptive capacity tend to be less vulnerable.

Adger (2005) stresses that the absolute assessment of successful adaptation is quite problematic. Additionally, Smit & Wandel (2006) highlight that vulnerability is a dynamic process, with all elements and determinants varying over time, by type and place. Furthermore the elements and determinants of vulnerability are household specific, including the adaptive capacity. These assertions, combined with our research focusing on the implementation processes of adaptation, mean that we can not estimate the degree of adaptation of households relative to each other; because all variables show substantial discrepancy in values and are not comparable (e.g. households exposure due to climatic conditions and land use patterns, livelihood assets in time, and adaptive capacity from village to village - over time and among social groups). Accordingly, as outlined above, the goal is not to produce a score or rating of household's current or future vulnerability. Rather, the aim is to attain information on the nature of social vulnerability, its components, and the decision-making processes which lead or hinder the adoption of innovations in order to identify ways in which the adaptive capacity can be increased and exposure-sensitivities decreased.

2.2 Sustainable Livelihood Framework

To relate the adoption of innovation to the reduction of vulnerability and the increase in resilience, the "Sustainable Livelihoods Framework" (Chambers & Conway 1991) is applied. The "Sustainable Livelihood Framework" is a tool to improve the understanding of livelihoods, particularly those of the poor. Based on early studies of Chambers (1989) concerning livelihood strategies and coping mechanisms, the "Livelihood-Research" was developed (Scoones 1998). Shortly after this, the Department for International Development (DFID) of the United Kingdom adopted the framework with great success for its practical implementation of development projects (Bohle & Glade 2008).

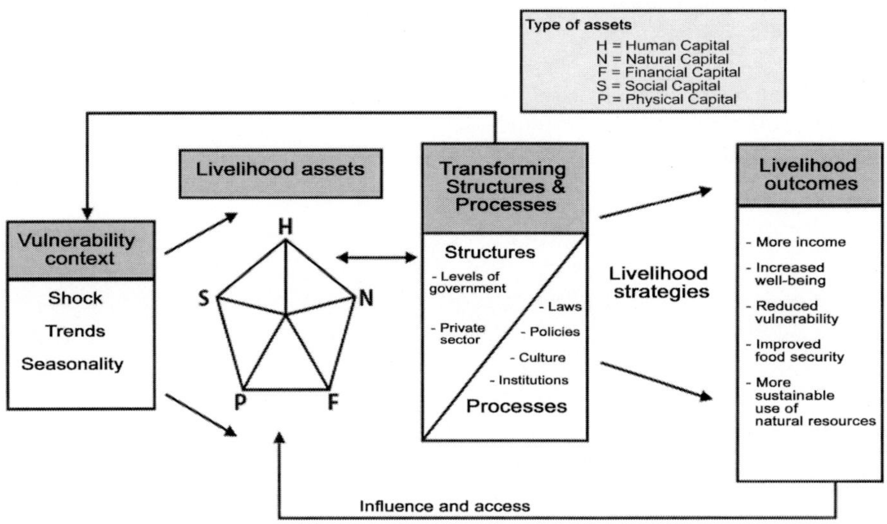

Fig. 1. Sustainable Livelihood Framework. Source: modification according to DFID (1999) and Hamill et al. (2005)

Within the "Sustainable Livelihoods Framework" ENSO related droughts as the threatening hazards would be classified as 'Shock' in the 'Vulnerability Context' (Figure 1). The 'Vulnerability Context' refers to the external environment and is the part of the framework that lies furthest outside people's control. 'Shocks' represent the most extreme and unexpected changes in people's livelihoods, for example through floods, droughts or landslides. 'Shocks' might destroy assets directly. 'Trends', such as economic and population trends influence people's livelihoods, but are usually more predictable and long term. 'Seasonality' refers to seasonal fluctuations in prices and employment opportunities, as well as to the availability of food and resources such as water due to seasonal weather and other factors.

Central to the Sustainable Livelihood Framework are the five different types of 'livelihood assets':

1. Human Capital consists of knowledge, skills, ability to work, and health conditions of the household members. It is required in order to make use of any of the four other types of assets.
2. Social Capital refers to memberships in groups and networks, relationships of trust, reciprocity, common rules, norms and sanctions as well as connectedness in institutions.
3. Natural Capital includes land, forest, wild resources, water, quality of the environment, biodiversity, erosion protection, waste assimilation and air quality.
4. Financial Capital comprises savings in form of cash, bank deposits or liquid assets such as livestock and jewellery as well as the access to credit.
5. Physical capital contains infrastructure such as transport, houses and buildings, water and sanitation systems, energy and access to information (communication) as well as means of production such as tools and equipment (Hamill et al. 2005).

According to DFID (1999) households with more assets tend to have a greater range of options and an ability to switch between multiple strategies to secure their livelihoods. Furthermore, poverty analyses have shown that people's ability to escape from poverty is critically dependant upon their access to assets (ibid. 1999).

3 Data & Methods

3.1 Description of the research area

The research took place in six villages in the vicinity of the Lore Lindu National Park (LLNP) encompassing the subdistricts of Kulawi, Palolo, Sigibiromaru, Lore Tengah and Lore Utarah in Central Sulawesi, Indonesia. The national park covers some 2290 square kilometres of tropical rainforest with an altitude range from about 200-2610 meters a.s.l. It is characterized by rift valleys and the rainfall varies from 500 to 2500 mm per year.

3.2 Data Collection

In 2004, a household census (n=898) in three villages was conducted. The census covered socio-cultural interrelations, environmental perception and socio-economic data. Using a multivariate cluster analysis, the households were classified by land use aspects (Schippers et al. 2007). The aim of the classification was to extract a manageable number of household groups differing in their land-use. The households of each group should be as similar as possible (high internal homogeneity), whereas the groups should differ among each other as far as possible (external heterogeneity). The classification criteria of the cluster (as shown in Table 1) were used to randomly select two households within each cluster for qualitative interviews in 2007 in the three census-villages.

Table 1. Land use cluster and their criteria.

Cluster	Label	Scale	Classification criteria		N	N/Cluster Groups
					30	
1	Cacao specialists	Small		<140 ares	373	
2		Medium	>66% cacao area	≥140ares <300ares	150	588
3		Large		≤ 300ares	65	
4	Cacao-rice combiners	Small		<140ares	95	
5		Medium	Cacao- & rice area: each 33%, OR sum of cacao- & rice area >75%;	≥140ares <300ares	30	134
6		Large		≤ 300ares	9	
7	Wet rice specialists		>66% rice area; all scales		69	
8	Other specialists (coffee, maize, other crops)		>66% coffee area OR >66% area of other cropping types; all scales		21	
9	Other combiners or multi-diverse households		All remaining households with agriculture		56	

Source: modification according to Schippers et al. (2007)

Additionally, three further villages were included in this study because they differ in their development dynamics (migration), composition of population (ethnicity), transformation process (land use and land availability) as well as climatic conditions (altitude) from the census-villages (Weber 2006; Weber & Faust 2006; Faust et al. 2003). In these three additional villages respondents were chosen by applying RRA methods and informal interviews with key informants and village representatives to identify households which fit into the household classification characteristics developed by Schippers et al. (2007).

Consequently, in each village we conducted 18 interviews for nine different household clusters. However, as some of the land use clusters do not exist in certain villages the number of interviews differs from village to village as the number of land use clusters does. The observed villages range from being characterised by high influx of regional migrants, scarcity of land, intensive cacao cultivation and surrounding pre-mountain tropical rainforest to villages dominated by multi-cropping systems, relatively good land availability as well as a semi-arid meso-climate. By investigating the three additional villages, the study seeks to extend the knowledge about livelihood strategies of rural households living in the research area. In total during 2007 and 2008 82 semi-structured in-depth interviews focusing on natural hazards, including questions on the adoption of innovations, land use change and decision making processes were conducted in all six villages. Furthermore, 30 in-depth interviews on natural resource use patterns were conducted in the three census villages. The interviews were structured by a flexible interview guide allowing, e.g., for non-standardized comments and explanations, as well as the incidental coverage of additional aspects when deemed necessary by the interviewer.

All interviews followed a "problem-centred interview" (PCI) approach (Witzel 1989). The interviews were conducted in Bahasa Indonesia supported by an Indonesian research assistant and recorded on tape. In a second step, the complete material was transcribed and later translated into English. To avoid a loss of side information and to immediately clarify misunderstandings we discussed and documented the relevant outcome and circumstances with the research assistant after each interview.

Rapid Rural Appraisal methods (RRA) were used as well. Some of the RRA's were conducted before the in-depth interviews, because they work as an 'icebreaker' in the local communities and they set the right kind of historical perspective for further analysis by in-depth interviews or other RRA-methods (Kumar 2002). Shortly after arriving in the villages, we carried out the 'Timeline' focussing on the chronology of natural disasters. Other methods such as 'Impact Diagrams' are much more sensitive and complex and were therefore conducted beside the interviews or at least after one week actively participating in the community life.

The outcomes proved to be very supportive. Local people may talk about a particular drought but may not recall the exact time period because they possess a different concept of time and therefore do not remember droughts in terms of the Gregorian calendar but in chronology of important local events experienced in their lives. But with ENSO related droughts as the specific research topic, we needed to connect the perceived agricultural drought by the villagers with the specific ENSO years calculated by the climate research centres. With the outcome of the 'Timeline' we could present the respondents, within the in-depth interviews, a local timeframe and they were able to connect the perceived agricultural droughts with years. Furthermore, the events of the 'Timeline', which were chosen out of the people's perspective, had been depicted as well in images and therefore even illiterate people were capable to use this time frame.

To examine the institutional relationship we supported the in-depth interviews by carrying out a 'Venn diagram'. Here, the importance and accessibility of local institutions and individuals in case of a disaster were discussed. The method provides valuable insights and analyses of power structures and decision making processes. Additionally, the need to strengthen the community's institutions can also be ascertained (Kumar 2002). As an outsider there are limitations on what to imagine and thus what to ask. Therefore, we decided to obtain this information by conducting an impact diagram (Figure 2). The visual nature of the impact diagram makes it easy for the local people to depict their complex realties and the cross linkages between the various effects of droughts (ibid 2002).

Finally, an intensive cacao agroforestry study was conducted in 2007 in 12 villages of the project region (Central Sulawesi Cacao Agroforestry Management Study, for short *CAMS*). CAMS villages are located in four different valleys (Palu, Palolo, Kulawi, Napu) covering altitudes from 75 to 1275 m a.s.l. In each of the 12 villages, a sample of one cacao plot of each of 12 cocoa producing households was selected, resulting in a total sample size of 144 cacao plots. The plots were not randomly selected but were systematically chosen to represent the entire intensification gradient of high to low canopy closure values (CC). Per village, three plots were identified for each of 4 *a priori* shading categories: 1 (near) natural forest cover (>85% CC) > 2 dense shade cover (>65% CC) > 3 medium shade cover (>35% CC) > 4 low to zero shade (0-35% CC). 81 CAMS households simultaneously belong to a stratified random household sample used for the generation of socio-economic panel data in the project region (Central Sulawesi Rural Household Survey, for short *RHS*, n=293) (Zeller et al. 2002). Selection biases could not be detected (T-tests: RHS households in CAMS vs. RHS households not in CAMS, CAMS households in RHS vs. CAMS households not in RHS).

Plot history, structure and location were surveyed including the measurement of shade tree cover. Supported by local staff, farmers made weekly entries about their plot management and inputs for the duration of one year. Surveyed parameters include management activities and labour, (changes in) plot structure and intercropping, material input, yield of fresh pods and dry bean marketing. Furthermore, perceptions about soil fertility, patterns of in-

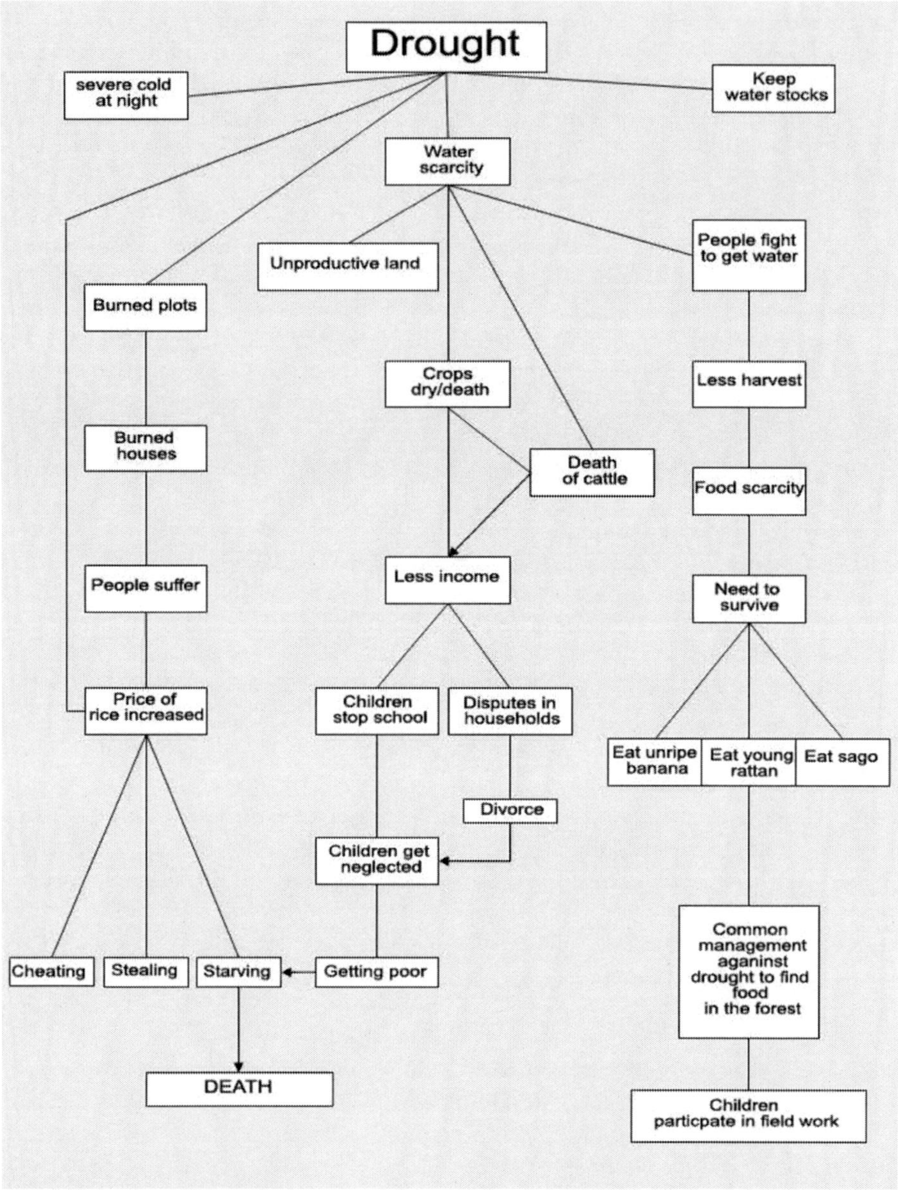

Fig. 2. Outcome of the RRA impact diagram on effects on drought

put use and adoption of agricultural innovations as well as the impact of pests, diseases, droughts and tree age on cocoa production were recorded.

4 Results

4.1 Perceived ENSO-Droughts and their impacts

As Table 1 shows, nearly two thirds (65.5%) of all households are concentrated in the cluster group 'cacao specialists'. For this reason the chapter focuses mainly on the analysis of this cluster. Consequently, the following results are all related to 'cacao specialists'.

All interviewed households in the research area have been affected by abnormal climatic conditions and stated negative effects on agricultural production. As a result of the RRA-Timeline the respondents could merge the perceived agricultural drought[1] with the exact years within the in-depth interviews. The significant highest nomination of agricultural droughts was in 2002. Further, the strongest agricultural droughts were perceived in 1997-98 and 2002, which were also the most severe ENSO events of the last 15 years (NOAA 2008). The farming household's perceptions are in line with findings from Keil (2004) in Central Sulawesi.

However, other major events were also recalled having a direct or indirect relation to ENSO related drought. These include famine, forest fires, rice supply by the government or the construction of new irrigation systems. Everywhere slight deterioration in agriculture, i.e. the drop of leaves and blooms, were found. Production losses (yield decline) due to a lack of water and fire were reported as well as serious damage (death of trees).

The intensive cacao study (n=144) carried out in 12 selected villages shows that the impact on cocoa trees varies between regions. In Palu valley, 56% of the households state that their cacao trees are affected when droughts occur, whereas in Kulawi, it is only 20%, in Palolo 31% and in Napu valley 32% of the households. Yield losses due to droughts as perceived by farmers range between 0 and 60%. Including also the non-effected households, mean yield losses are significantly highest in Palu valley (17%, N=36) whereas Napu (3%, N=49), Palolo (8%, N=36) and Kulawi valley (5%, N=24) show no significant differences (oneway ANOVA and Tukey posthoc test). Moreover, drought related yield loss declines with elevation ($r = -0.344$; $P < 0.01$; Pearson) and increases with average tree age of cocoa trees ($r = 0.302$; $P < 0.01$; Pearson).

4.2 General Adaptation Strategies against ENSO related droughts

We noticed different perceptions of drought by the different household clusters. For large cacao specialists (cluster 3) with older trees, even long-lasting

[1] Agricultural drought occurs when soil moisture is insufficient to meet the requirements of a particular crop at a particular stage of the grow cycle, resulting in a decline in yield (Wilhite & Glantz 1985).

Table 2. Reactive Adaptation Strategies.

Reactive Strategies	Description
Reduce food expenditures & change food patterns	(1) reduce the number of meals (eat twice instead three time a day), (2) replace rice as the main food component through cassava, young rattan or sago, (3) reduce the amount of food within the meals (eat smaller portions)
Paid labour	Mostly weeding and cleaning of other villagers' plots
Raise a loan	Mainly about social networks (local kiosk, own farming-group, neighbours) in cash or rice
Rattan collection	Often illegally done in the forest - usually organised by a (small) trader who collects the rattan in the village and pays in food, goods or cash
Hand irrigation techniques	Carry bamboo sticks (2m long*10cm diameter) filled with water from the river to the plots
	Bucket & hole technique (digging a 20cm hole in diameter around each tree, filling the hole with water carried with a bucket and covering with leaves on the top)
	Bottle/bamboo drops-technique (fill plastic bottle with water- seal it- sting a small hole inside and install it up-side down above the young plant)
	Tent technique (see Figure 3) to sustain soil moisture
Land use change	Temporary abandoning of plots
	Temporary changes in the cultivation system (sawah to corn)
	Change plots to less vulnerable location
	Plant drought resistant crops (cassava, batata, keladi)
Avoidance of weeding and cleaning of the plots (sanitation)	To sustain soil moisture

Source: own compilation

droughts for several months are moderate to cope with. They usually stated that trees older than three years can cope with a scarcity of precipitation. They produce less fruits but still survive. For wet rice specialists (cluster 7) a drought of more than three months means a strong burden. As wet rice is very sensitive to water scarcity they change from rice to corn cultivation in times of a drought because the saturation of soil moisture is insufficient for wet rice but still adequate for one single harvest of corn (usually after three months). If the drought continues for longer than three month, there is no additional harvest from corn as soil moisture is insufficient after that period just as it had been for wet rice before.

The study identifies various types of adaptation strategies on the household level. Following Adger (2005) the strategies are classified into reactive and anticipatory adaptation. Most commonly identified strategies are reactive adaptations which support the household in coping with the effects of drought, but do not reduce the vulnerability of the households to future ENSO related droughts. Table 2 lists the most commonly identified strategies of reactive adaptation.

Fig. 3. Reactive Adaptation Strategies "Tent technique"

Further applied reactive adaptation strategies are the reduction of household's expenditures, replanting of crops or sales of household assets. Moreover, the study identifies anticipatory adaptation strategies (Table 3).

These strategies suggest that households cope with the immediate effects of a drought as, if purposefully implemented, they can alter the exposure of the households to future droughts and increase resilience to cope with changes and reduce sensitivity. In contrast to strategies of reactive adaptation these activities are very rare and depend on knowledge transfer within local institutions or along ethnic lines. The ethnic group of Buginese plays the most important role in cacao agroforestry. This ethnic group predominantly

Table 3. Anticipatory Adaptation Strategies.

Anticipatory Strategies	Description
Land use change	Permanent abandoning of drought susceptible crops e.g. wet rice
Switching to cacao cultivation	Perceived as innovation by local farmers. In this context a tendency of specialisation in cacao agriculture and the expansion of cacao plots emerges.
Shade trees	Are seen as more and more necessary to protect the system. Maintaining the shade trees in cacao plantations improves the ecological resilience of cacao agroforestry systems to drought effects
Food stocks	Usually households possess a stock of unpeeled rice (paddy) lasting for one to six month
Irrigation management	(1) Ditches / drainage systems: digging channels in the cacao plots (see Figure 4), in times of drought the end of the channel is blocked to keep the soil moisture. Thus, no hand irrigation is needed. In times of severe rain protection against soil erosion
	(2) Buying pumps and dig wells at plots to reduce sensitivity from a future lack of precipitation
	(3) Retention basin (to possess a water storage in times of drought)
Drought resistant seeds	Change to more drought resistant rice seeds (obtained from the local agricultural extension agent), which are, however, associated with yield decline
Migration	Migration to less drought exposed regions

Source: own compilation

consists of economically powerful migrant households from South Sulawesi, mostly well educated in cacao cultivation. Their members are well-connected throughout the study area, passing knowledge and support to each other.

In interviews with Javanese migrants, we noticed the existence of an extensive network providing information on soil properties in Central Sulawesi back to Java. Javanese agricultural smallholders who arrived recently in Central Sulawesi are already equipped with the most suitable brand of fertilizer and pump machines for coping with a lack of precipitation in order to increase the production. The local farming group is one of the most important and powerful local social institutions to cope with droughts. For its members the group provides technological innovations free of charge (chemicals and technical equipment), food in case of disaster, and support with the implementation of innovations (e.g. terrace building). The farming group is related

Fig. 4. Anticipatory Adaptation Strategies "channels in cacao plots"

to the cropping system and not every villager is a member of a group. The agricultural smallholder has to apply for membership and the farming group itself decides about acceptance. By utilising a 'Venn-Diagram' the local kiosk was identified as one of the key institutions to provide food (within a lending-system secured by the long term connection of the customers and based on a relationship of trust, reciprocity and exchanges). Another important social institution is the praying group[2]. The kinship in the village of origin presents a further important network. By regularly sending presents or money to relatives the households sustain the possibility to return to their village of origin in case of emergency.

4.3 Comparison of different strategies of adaptation

ENSO related drought effects vary considerably between small and large specialised cacao farmers (cluster 1 and 3). Small cacao farmers use mainly simple reactive adaptation strategies to cope with droughts (reduced food consumption, paid labour and rattan collection to generate additional income). Therefore, their coping strategies are almost entirely based on the household's

[2] Almost every villager is member in a praying group

limited human capital itself. Extended droughts will further decrease human capital (ongoing reduced food consumption combined with more paid labour leads to deteriorated physical health).

In contrast, large specialised cacao farmers mentioned the utilisation of a range of anticipatory strategies (e.g. shadow trees, food stocks, terrace building, irrigation management and crop diversification) along with technological innovations plus their social networks to overcome droughts. Further, in interviews they regularly mention possessing additional sources of income from kiosks, as public servants or from rental. Large specialised cacao farmers are the only household type with more than one anticipatory adaptation strategy. Considering this, their strategies are based not only on human capital, but also to a large extent on social- and financial capital. In summary, we find that households with different land use pattern use different strategies of adaptation to ENSO related droughts and differ in their social vulnerability.

5 The agroforestry system cacao in discussion

As Smith & Wandel (2006) stress practical climate change adaptation initiatives are very unlikely to be taken in the light of climate change alone. These adaptations are usually integrated with other objectives. This is particularly true for land use systems. In the study area we have found an ongoing trend of conversion to cacao (Erasmi et al. 2004, Erasmi & Pries 2007) caused by economic reasons (Ruf et al. 1996). But in the same way the social vulnerability of households to droughts could be decreased via the agroforestry system cacao. Following Verchot et al. (2007) tree based production systems often produce crops of higher market value than annual crops, hence the agroforestry systems buffers against income risks associated with climate change. In addition, out of all analyzed land uses the agroforestry systems offer the highest potential for carbon sequestration (IPCC 2000).

Besides, the labour needed to maintain a cacao plantation is less than in annual cropping systems, which sets human capital free. Consequently, this capital could be used to implement innovations or create independent side incomes. These are necessary, as financial capital in the form of available stocks[3] and regular inflow[4] of money is almost non existent, except in cluster 3. Caused by a lack of financial capital, successfully implemented innovations are still fragile and could be undone immediately, as mentioned by one respondent who adopted more drought resistant seeds to reduce the sensitivity of his crops to droughts. However, the new seeds caused yield losses. Therefore, the household stated to return to its less drought resistant seeds if the yield loss continues.

[3] Savings in forms of cash, bank deposits or liquid assets such as livestock or jewellery

[4] Excluding earned income, most common types are pensions, transfers from the state, remittances

The intensive cacao study (n=144) states drought related mean yield decreases up to 16.6% depending on regions and a positive correlation between increased yield loss and the average cacao tree age. A further study by Keil (2004) in Central Sulawesi even reports an average cacao yield decline up to 62% of the usual level (n=69) during the most severe drought. However, as assured in the qualitative interviews, decreasing cacao yields are not just related to ENSO drought effects but are also connected with an ongoing intensification of cacao production and related increases of pests (Cacao Pod Borer *[Conopomorpha cramerella]* and Black Pod Disease *[Phytophtora sp.]*). Moreover, the positive correlation between increased yield loss and the average tree age needs further interpretation in view of the social vulnerability of cacao specialists. Cacao trees are productive after three years and increase their productivity until they are 10 years old. Along with their productivity cacao trees become much more drought resistant as they age. As stated, respondents lost hundreds of young cacao trees due to their sensitivity against lack of precipitation. From the perspective of a cacao specialist, yield decrease by even 40% represents a less severe impact to the household than the loss of many young trees as households will lose their productive capacity for the subsequent years. This interpretation highlights a frequently mentioned approach to cope with droughts: Keeping the focus on cacao production and the remaining harvest as productive cacao trees are resilient enough to survive drought periods in the study area.

Also, perceptions of local farmers show that a productive agroforestry system cacao is seen as more resistant to drought than other crops, especially non-irrigated rice and maize. These perceptions are consistent with recent scientific findings from Verchot et al. (2007). In addition, Keil et al. (2009) show in a stochastic simulation that cacao farmers in Central Sulawesi are the least vulnerable household class, i.e., the least likely to fall temporarily below a specified poverty line, in comparison to rice producers and mixed-cropping farmers. The same study shows that within a 'Drought impact index' cacao farmers perceive the effects of the most severe drought as lightest by far in comparison to all other household classes.

5.1 Land Tenure and Property Rights

For the adoption of cacao agroforestry systems land tenure and property rights are of particular relevance since agricultural smallholders - and especially the poorer ones - are not willing to invest in innovations without definitive property rights. Therefore, property rights have an important impact on land use and decision-making with respect to local farmers. Well defined property rights - private or common - allow long-term planning horizons for farmers and assure them of capturing the returns from their investments (Toni & Holanda Jr. 2008). Insecurity and a lack of property rights not only affect households by making them more vulnerable to environmental risks or hazards, but also lead to a more unsustainable use of natural resources (Beaumont & Walker 1994).

In the qualitative in-depth interviews the majority of respondents stated the intention to obtain property rights but are unable to realise them due to extra financial claims from government officials for transportation, food and accommodation in addition to standard tax claims.

With respect to the study area, Barkmann et al. (2010) and Koch et al. (2008) emphasise that economically more powerful migrant households tend to cultivate agricultural plots for which they have obtained official land titles, whereas poorer local households cultivate agricultural plots with inferior land use quality and of a highly precarious tenure status. Hence, economically more powerful households - mostly cacao specialized Buginese migrants from South-Sulawesi - are less vulnerable to ENSO related droughts.

However, private property is not necessarily the only form of property rights to secure land tenure. As Toni & Holanda Jr. (2008) point out, common-property farming systems are more diversified, gain higher investments and consequently are vulnerable to drought events to a lesser extent as the risk is spread.

5.2 Knowledge transfer & social and human capital

The study analyses the behaviour of rural households exposed to ENSO related droughts with a special perspective on the regional dominant agroforestry system cacao. We could identify a wide range of adaptation strategies used to cope with their harmful consequences and recognise differences in the adoption of innovations. But why do different household types use different strategies of adaptation?

In all study villages, agricultural smallholders stated in group discussions and qualitative in-depth interviews that the major source of information regarding anticipatory adaptation strategies is either personal information (from relatives or neighbours) or social learning (observation and imitation of strategies). Whereas, for reactive adaptation most respondents mentioned that their own knowledge and skills along with 'trial and error techniques' are the source of success for coping with ENSO related droughts. In particular, the respondents stated that the knowledge transfer which promotes both types of adaptation is mainly taking place along ethnic lines and local institutions. For successful adoption of innovations social capital, such as membership in certain ethnic groups, local institutions and networks is of strong importance. This is in line with Keil et al. (2008), who found that the larger the number of village organisations a household is involved in, the more positively this influences its drought resilience and that households are able to benefit from an extensive social network during crises.

The meaning of change agents (e.g. agricultural extension agent) and impersonal information sources (e.g. TV, radio) is eminent for an increase in agricultural production. However, for coping with ENSO related droughts the 'innovation-decision process' is predominantly influenced by social and personal capital. These findings are consistent with results from Verchot et al.

(2007), who state that externally driven adaptation is more likely in the developed part of the world, while farmers in the less endowed part of the world rely more on innovations from within the system if they are going to adapt to changing climate.

6 Implications

Based on the results and the discussion we conclude that policy implications should concentrate on the following aspects:

(1) Increased support for the accumulation of human capital to reduce social vulnerability.

Human capital has a distributing function. It is required in order to make use of any of the other assets. Most of the mentioned anticipatory adaptation strategies are based on human capital. It could be strengthened by drought- or community based disaster risk management courses on village level. Heads of farming groups will have multiplicator effects spreading the knowledge into their groups. Furthermore, the improvement of medical services by improving the technical equipment of the existing facilities and by training the staff will enhance human capital (health). Another opportunity is the support of education by constructing new school facilities and reducing the costs for education.

(2) Strengthening of social capital to implement innovations.

There are certain social- and ethnic groups existing which implement innovations. On the other hand interethnic barriers in knowledge transfer are incontrovertible. Strengthening local institutions, enforced integration of migrants and non-local ethnics will further support the adoption of innovations. Social capital can be a by-product of other activities (like farming groups, traditional village councils and hamlet memberships, religious groups, village development councils, public work activities etc.). These institutions already exist in the study region but the villages are split in migrants and locals, as well as ethnic groups (Buginese, Javanese or Autochthones). Therefore the integration of these groups should be regarded as a crucial topic for local development. Stronger social participation of marginalized groups in important local decision making structures and institutions (e.g. community development) may reduce barriers in personal knowledge transfer and social learning actions. Simultaneously the ability of households is enhanced to implement further anticipatory adaptation strategies.

(3) Adjustment strategies for small cacao farmers to adapt to climate change.

Fig. 5. Construction of Junior High School in Rompo (Lore Tengah) by an APBN-Grant along with exemption from school fees for the next years

Small cacao farmers tend towards reactive adaptation strategies and large cacao farmers tend towards anticipatory strategies. Caused by land size and economies of scale both cluster groups possess different financial assets. To reduce social vulnerability of small cacao farmers, without extending their productive area, intensification of the agroforestry system cacao might be an option. This could be done by adjusting structural parameters like the right degree of shading related to the age of trees, and enforcement of labour intensity at the cacao plots. These measures could lead to improved benefits per hectare and hence create an opportunity for small cacao farmers to change from reactive adaptation strategies to anticipatory adaptation strategies which would increase the resilience of cacao plots to future ENSO related droughts and decrease the social vulnerability of the smallholders. Additionally, after a period of three years, cacao trees are much less sensitive to drought related stress than other cash crops in the research region.

In conclusion, interventions aiming to reduce the social vulnerability should concentrate on rural smallholders and focus on support for the sustainable management of the agroforestry system cacao. As preconditions a transparent and standardised tenure system and a strengthened integration process of local actors are indispensible to stabilize and improve livelihoods in the face of climate change.

Acknowledgements
The authors thank the people around the Lore Lindu National Park in Central Sulawesi, Indonesia, for supporting their surveys and interviews. Furthermore they gratefully acknowledge financial support by the German Scientific Foundation (DFG), and they thank two anonymous reviewers for their constructive comments.

References

Adger WN, Agrawala S, Mirza MMQ, Conde C, O'Brien K, Pulhin J, Pulwarty R, Smit B, Takahashi K (2007) Assessment of adaptation practices, options, constraints and capacity. In: Climate Change 2007: Impacts, Adaptation and Vulnerability. Contribution of Working Group II to the Fourth Assessment Report of the Intergovernmental Panel on Climate Change, Parry ML, Canziani OF, Palutikof JP, van der Linden PJ, Hanson CE, Eds., Cambridge University Press, Cambridge, UK, p. 717-743

Adger WN (2006) Vulnerability. Global Environmental Change 16: 268-281

Adger WN (2005) Successful adaptation to climate change across scales. Global Environmental Change 15: 77-86

Adger WN (1999) Social Vulnerability to Climate Change and Extremes in Coastal Vietnam. World Development 27: 249-269

Agrawala S, Broad K (2002) Technology transfer perspectives on climate forecast applications. Research in Science and Technology Studies 13: 45-69

Barkmann J, Burkhard G, Faust H, Fremerey M, Koch S, Lanini A (2010) Land tenure rights, village institutions, and rainforest conversion in Central Sulawesi (Indonesia). In: Tscharntke T, Leuschner C, Veldkamp E, Faust H, Guhardja E, Bidin A (eds.) Tropical rainforests and agroforests under global change: Ecological and socio-economic valuations. Springer, Berlin

Beaumont P, Walker R (1994) Land degradation and property regimes. Ecological Economics 18 (1): 55-66

Bohle HG, Glade T. (2008) Vulnerabilitätskonzepte in Sozial- und Naturwissenschaften. In: Felgentreff C, Glade T (Eds.) Naturrisiken und Sozialkatastrophen, Berlin, p. 99-117

Cane MA, Zebiak SE, Dolan SC (1986) Experimental Forecasts of El Niño. Nature, 321: 827-832

Chambers R. (1989) Editorial Introduction: Vulnerability, Coping and Policy. IDS Bulletin 20 (2): 1-7

Chambers R, Conway G (1991) Sustainable Rural Livelihoods: practical concepts for the 21st century. Institute of Development Studies Discussion Papers 296, 1991. University of Sussex, Brighton

Cruz RV, Harasawa H, Lal M, Wu S, Anokhin Y, Punsalmaa B, Honda Y, Jafari M, Li C, Huu Ninh N (2007) Asia. In: Climate Change 2007: Impacts, Adaptation and Vulnerability. Contribution of Working Group II to the Fourth Assessment Report of the Intergovernmental Panel on Climate Change, Parry ML, Canziani OF, Palutikof JP, van der Linden PJ, Hanson CE (Eds.), Cambridge University Press, Cambridge, UK, p. 469-506

Denevan WM (1983) Adaptation, variation and cultural geography. Professional Geographer 35: 399-406

DFID (1999) Sustainable Livelihoods Guidance Sheets, Numbers 1-8, London, Department for International Development

Erasmi S, Pries J (2007) Satellite and survey data: a multiple source approach to study regional land-cover / land-use change in Indonesia. In: Dickmann F (Ed.) Geovisualisierung in der Humangeographie. Kartographische Schriften 13: 101-114

Erasmi S, Twele A, Ardiansyah M, Malik A, Kappas M (2004) Mapping Deforestation and Land Cover Conversion at the Rainforest Margin in Central Sulawesi, Indonesia. EARSeL eProceedings 3: 388-397

FAO - Food and Agriculture Organization (2008) Special Programme for Food Security. URL:http://database.deptan.go.id/saims-indonesia/index.php?files=introduction (27/12/2008)

Faust H, Maertens M, Weber R, Nuryartono N, Van Rheenen T, Birner R (2003) Does Migration lead to Destabilization of Forest Margins? Evidence from an interdisciplinary field study in Central Sulawesi. STORMA Discussion Paper Series No.11, Göttingen and Bogor

Glantz MH (2001): Once Burned, Twice Shy? Lessons Learned from the 1997-98 El Niño. United Nations University Press, Tokyo

Gockowski J, Nkamleu GB, Wendt J (2001) Implications of resource-use intensification for the environment and sustainable technology systems in the central african rainforest. In: Lee DR, Barrett CB (Eds.) Tradeoffs or synergies? Agricultural intensification, economic development and the environment, CAB International, Wallingford, UK

Hamill A, Leclerc L, Myatt-Hirvonen O, Salinas Z (2005) Using the sustainable livelihoods approach to reduce vulnerability to climate change. In: Robledo C, Kanninen M, Pedroni L (Eds.) Tropical forests and adaptation to climate change, Bogor, Indonesia

IPCC (2007) Climate Change 2007: Impacts, Adaptation and Vulnerability. Contribution of Working Group II to the Fourth Assessment Report of the Intergovernmental Panel on Climate Change, Parry ML, Canziani OF, Palutikof JP, van der Linden PJ, Hanson CE, (Eds.) Cambridge University Press, Cambridge, UK

IPCC (2001) Climate Change 2001: Synthesis Report. A Contribution of Working Groups I, II and III to the Third Assessment Report of the Intergovernmental Panel on Climate Change (Watson RT and the Core Writing Team (Eds.) Cambridge University Press, Cambridge, United Kingdom, and New York, NY, USA

IPCC (2000) Land-use, land-use change and forestry. Special report of the intergovernmental panel on climate change. Cambridge University Press, Cambridge, UK

Keil A, Teufel N, Gunawan D, Leemhuis C (2009) Vulnerability of smallholder farmers to ENSO-related drought in Indonesia. Climate Research 38 (2): 155-169

Keil A, Zeller M, Wida A, Sanim B, Birner R (2008) What determines farmers' resilience towards ENSO-related drought? An empirical assessment in Central Sulawesi, Indonesia. Climatic Change 86: 291-307

Keil A (2004) The socio-economic impact of ENSO-related drought on farm households in Central Sulawesi, Indonesia. Shaker, Aachen, Germany

Kelly PM, Adger WN, (2000) Theory and practice in assessing vulnerability to climate change and facilitating adaptation. Climate Change 47: 325-352

Koch S, Faust H, Barkmann J (2008): Differences in power structures regarding access to natural resources at the village level in Central Sulawesi (Indonesia). Austrian Journal of South-East Asian Studies 1 (2): 59-81

Kumar S (2002) Methods for community participation: a complete guide for practitioners, London

NOAA - National Oceanic & Atmospheric Administration (2008) Multivariate ENSO Index. URL: http://www.cdc.noaa.gov/people/klaus.wolter/MEI/rank.html (26/12/2008)

Pfaff A, Broad K, Glantz M (1999) Who benefits from climate forecasts? Nature, 397, 645-646

Ruf F, Eheret P, Yoddang J (1996) Smallholder Cocoa in Indonesia: Why a Cocoa Boom in Sulawesi? In: Clarence-Smith W (Eds.) Cocoa Pioneer Fronts since 1800 - The Role of Smallholders, Planters and Merchants. MacMillan, London, p. 212-231

Schippers B, Weber R, Faust H (2007) Agricultural Household Types in Upland Sulawesi, Indonesia - A Classification Approach. STORMA Discussion Paper Series No.21, Göttingen and Bogor

Scoones I (1998) Sustainable rural livelihoods: A framework for analysis. IDS Working Paper 72, IDS, Brighton

Smit B, Wandel J (2006) Adaptation, adaptive capacity and vulnerability. Global Environmental Change 16: 282-292

Smit B, Pilifosova O (2003) From adaptation to adaptive capacity and vulnerability reduction. In: Smith JB, Klein RJT, Huq S (Eds.) Climate Change, Adaptive Capacity and Development. Imperial College Press, London

Toni F, Holanda JrE (2008) The effects of land tenure on vulnerability to droughts in Northeastern Brazil. Global Environmental Change 18: 575-582

Turner BL, Kasperson RE, Matson PA, McCarthy JJ, Corell RW, Christensen L, Eckley N, Kasperson JX, Luers A, Martello ML, Polsky C, Pulsipher A, Schiller A (2003) A framework for vulnerability analysis in sustainability science. Proceedings of the National Academy of Sciences 100: 8074-8079

Verchot LV, van Nordwijk M, Kandji S, Tomich T, Ong C, Albrecht A, Mackensen J, Bantilan C, Anupama KV, Palm C (2007) Climate change: linking adaptation and mitigation through agroforestry. Mitigation and Adaptation Strategies for Global Change 12: 901-918

Weber R (2006) Kulturlandschaftswandel in Zentralsulawesi: Historisch-geographische Analyse einer indonesischen Bergregenwaldregion. Göttingen, Universitätsverlag Göttingen

Weber R, Faust H (2006) Kulturelle Aspekte der Landnutzung in Indonesien. Geographica Helvetica 61: 237-245

Wilhite DA, Glantz MH (1985) Understanding the drought phenomenon: The role of definitions. Water International 10: 111-120

Witzel A (1989) Das problemzentrierte Interview. In: Jüttemann G (Ed.) Qualitative Forschung in der Psychologie. Grundfragen, Verfahrensweisen, Anwendungsfelder. Asanger, Heidelberg, p. 227-256

UN/ISDR (International Strategy for Disaster Reduction) (2004): Living with Risk: A Global Review of Disaster Reduction Initiatives, UN Publications, Geneva

Yohe G, Strzepek K, Pau T, Yohe C (2003) Assessing Vulnerability in the context of changing socioeconomic conditions: a study of Egypt. In: Smith JB, Klein RJT, Huq S (Eds.) Climate Change, Adaptive Capacity and Development. Imperial College Press, London

Zeller M, Schwarze S, and van Rheenen T (2002) Statistical Sampling Frame and Methods Used for the Selection of Villages and Households in the Scope of the Research Programme on Stability of Rainforest Margins in Indonesia. STORMA Discussion Paper Series No 1. Bogor, Indonesia

Terrestrial herb communities of tropical submontane and tropical montane forests in Central Sulawesi, Indonesia

Daniele Cicuzza[1,2*], Michael Kessler[1,2], Ramadhanil Pitopang[3], Sri S. Tjitrosoedirdjo[4], and S. Robbert Gradstein[1]

[1] Institute of Plant Sciences, University of Göttingen, Untere Karspüle 2, 37073 Göttingen, Germany
[2] Institute of Systematic Botany, University of Zürich, Zollikerstrasse 107, 8008 Zürich, Switzerland
[3] Department of Forest Managment and Herbarium Celebense, Tadulako University, Palu, Indonesia
[4] Department of Biology, Faculty of Mathematics and Science, Bogor Agricultural University, Bogor, Indonesia

*corresponding author: D. Cicuzza, email: daniele.cicuzza@systbot.uzh.ch

Summary

Although the diversity of terrestrial herbs is high tropical forests and although herbs may play important roles, e.g., as competitors of tree seedlings, most tropical botanical research to date has focused on trees. We studied the diversity, taxonomic composition, and biogeographical relationships of terrestrial forest herbs at two sites of tropical mountain forest at different elevations (Pono: 1000 m, Bariri: 1400 m) in Central Sulawesi. The study was conducted in 400 plots of 5 x 5 m^2 (200 for each site). At Pono, we recorded 91 angiosperm herb species in 28 families, and 112 ferns and lycophytes in 15 families whereas at Bariri we found 77 angiosperms in 25 families and 94 ferns and lycophytes in 20 families. At both sites, the most species-rich angiosperm families were Araceae, Orchidaceae, and Zingiberaceae. The species numbers recorded by us are much higher than those reported in any previous tropical forest herb inventories and point to a previously underappreciated richness of plant assemblages on Sulawesi. Biogeographically, significantly more fern species reached their western than eastern distributional limits on Sulawesi, showing that the zoogeographical Wallaces line separating continental Asia and its shelf islands from the Moluccan region also holds for spore-dispersed plants.

T. Tscharntke et al. (eds.), *Tropical Rainforests and Agroforests under Global Change*, Environmental Science and Engineering, 377–390, DOI 10.1007/978-3-642-00493-3_17,

Keywords: herbs diversity, species richness, Indonesia, Sulawesi, tropical mountain forest

1 Introduction

Tropical forests contain the most species-rich plant communities (Jacobs 1988) and the Malasian region is considered among the most diverse worldwide with over 40,000 vascular plants species (Baas et al. 1990; Roos 1993). Sulawesi, the largest island of the Wallacean region located between the Greater Sunda Islands and New Guinea, is generally considered to have intermediate levels of plants species richness (Roos et al. 2004). This can be explained by several factors. First, among the larger Indonesian islands, Sulawesi has the lowest collection rates, with less than 25 collections per 100 km^2, and taxonomic studies have been limited (Cannon et al. 2007). Second, historical isolation from the Sunda Shelf through the Quaternary Period prevented the continental enrichment experienced by Borneo, Sumatra, and Java (Whitmore 1987, Hall and Holloway 1998, Moss *et al.* 1998). Third, Sulawesi has a peculiar geography with four narrow peninsulas radiating from a small central area, so that no location is more than 100 km from the coast, resulting in a unique large island without inland (Cannon et al. 2007). On the other hand, Sulawesi has a complex geology, and the long isolation has allowed the evolution of a characteristic and unique flora and fauna, resulting in some of the highest levels of endemisms in the region (Roos et al. 2004; Cannon et al. 2007).

Most studies on tropical vegetation ecology have focused on trees at the expense of herbs, even though the latter may be a major component of the biodiversity of these forests (Poulsen & Pendry 1995). Herbs may be important competitors of tree seedlings and can thereby impact the diversity and composition of the tree communities. Also, due to their rapid life cycle, they may react rapidly to environmental changes, but to date there are no studies concerning understory herb species as possible indicators of reactions of tropical plant communities to global change. Among the few studies inventoring terrestrial herbs in tropical forests, Poulsen & Pendry (1995) for example found 121 species corresponding to 85 angiosperms and 36 ferns at Bukit Belalong, Brunei, Borneo. In South America, Poulsen et al. (2006) found 123 species (29 ferns, 24 palms, and 70 other angiosperms) in Peru, whereas in Amazonian Brazil, Costa (2004) recorded 35 herb species (11 ferns, 24 angiosperms). These and other studies document a wide range of variation in species richness and taxonomic composition of tropical herb communities that is not yet fully understood. To date, there are no specific studies of forest herbs on Sulawesi (Cannon 2001). In the present study, we therefore present the first inventories of terrestrial forest herbs at two tropical mountain forest sites in Lore Lindu National Park, Central Sulawesi. To study biogeographical affinities of the terrestrial herb assemblages, we focused on the pteridophytes.

2 Material and Methods

Study sites and field sampling

Lore Lindu National Park is one of the largest and most important conservation areas in Sulawesi, containing a unique range of habitats (Cannon 2001). Our study was conducted at two localities within the park (Fig. 1). The Bariri site was located in old-growth montane forest at about 1430 m elevation on the eastern side of the Lore Lindu National Park, near Bariri village, province of Poso (13928.44 S 1201024.6 E). The Pono site was located in submontane forest at about 1000 m near the western side of the National Park, just east of Toro village (1 30 35.383S 120 3 25.169E). Human impact on both sites is slight and limited to hunting and gathering of some forest products, especially rattan.

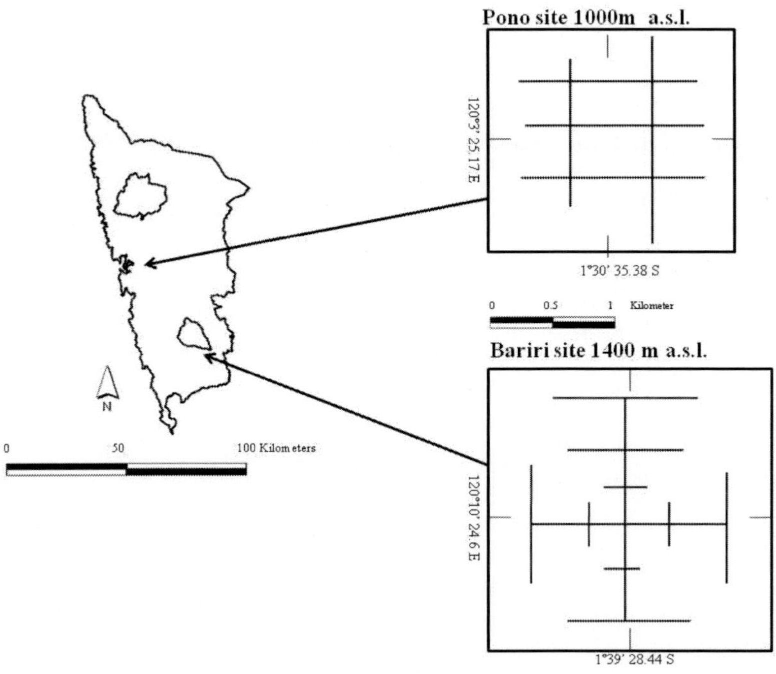

Fig. 1. The outline of Lore Lindu National Park in Central Sulawesi, showing the general location of the two study sites, and details of the plot layout at the two sites, with the individual study plots placed along the transect lines.

At both study sites, we established a network of 200 non-permanent plots of 5 x 5 m^2 each, over an area of about 0.7 km^2 (Fig. 1). Distance between adjacent plots was 20 m for all plots at Pono, whereas at Bariri 120 plots were 20 m apart and the remaining 50 m. Plot layout at each locality was designed to representatively cover the range of topographical conditions (ridges, slopes, valley bottoms) within a given elevational range. Field work was conducted in February 2007 at Bariri and in December-February 2008 at Pono. In each plot, all species of herbs were sampled semiquantitatively recording cover in classes (0-1%, 1-5%, 5-10%, 10-20%, 20-40%, 40-60%, 60-80%, 80-100%). Juvenile individuals less than 10 cm tall and difficult to identify were not recorded.

Voucher specimens were collected for all species within a study area (not in every single plot) with at least seven duplicates. Identification of plant species was done at the herbaria of Göttingen (GOET) and Leiden (L); species difficult to determine were sent to specialists for identification. The collections were deposited in Herbarium Celebense, Palu (CEB), Herbarium Bogoriense, Bogor (BO), and the herbaria of Göttingen (GOET), Leiden (L), Zürich (Z), and Berkeley (UC, ferns only). Information on species distribution ranges was obtained from the Flora Malesiana Series (e.g., 1981, 1991, and 1998) and from the botanical database Tropicos (www.tropicos.org).

Table 1. Environmental characteristics of the study sites. Means and ranges are based on the values of the 200 plots per site, except for climate variables that were measured at one station at each site.

	Pono	Bariri
Elevation mean (m)	1050	1422
Elevation minimum and maximum (m)	900-1100	1400-1500
Mean annual temperature (C)	20.8	19.6
Mean annual precipitation (mm)	3534	1984
pH (mean, min, max)	3.85 (3.01-5.09)	3.91 (3.31-6.84)
C/N (mean, min, max) (of 200 plots)	12.8 (2.71-20.63)	12.73 (8.40-21.3)
C mean (min, max)	3.14 (0.49-7.37)	2.52 (0.70-5.68)
N mean (min, max)	0.25 (0.02-0.48)	0.19 (0.05-0.33)

In each plot, we measured the following environmental variables: elevation, slope inclination, and relative topographical position divided in four categories (shoulder, backslope, footslope and depressional complex). Soil samples were collected in the center of each plot and analyzed for a wide range of parameters (Tab. 1).

3 Results

At Pono, we recorded 91 angiosperm herb species in 28 families, and 112 ferns and lycophytes in 15 families, resulting in a total of 203 terrestrial herb species. Of these, 122 samples were identified to species level, 53 to genus level and the remaining 28 to family level. The most species-rich families were Araceae (20

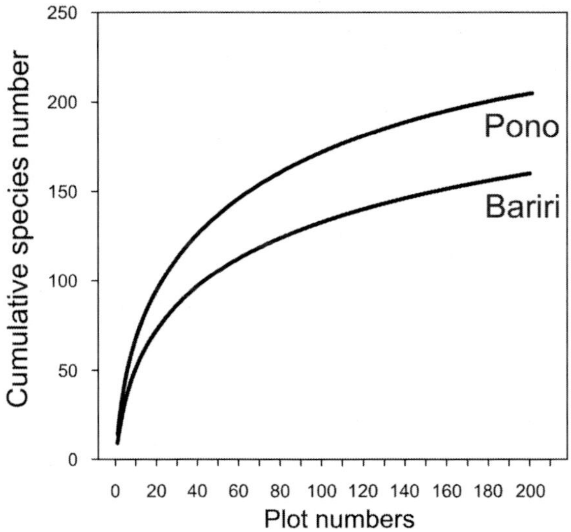

Fig. 2. Species-area curves for the two study sites Pono and Bariri. To calculate the species-accumulation curves the order of the 200 study plots at each site was repeatedly randomized 50 times.

spp.), Orchidaceae (16 spp.), and Zingiberaceae (10 spp.). Among the ferns and lycophytes, the Polypodiaceae had 15 species, followed by Aspleniaceae, Pteridaceae, and Woodsiaceae with 12 species each. At Bariri, we recorded a total of 171 terrestrial herb species, corresponding to 77 angiosperms in 25 families and 94 ferns and lycophytes in 20 families. At this site, 50 samples were identified to species level, 30 to genus, and 91 to family. The most species-rich angiosperm families were Zingiberaceae (16 spp.), Orchidaceae (13 spp.), and Araceae (10 spp.). Among ferns and lycophytes highest species richness was found in Polypodiaceae (16 spp.), Aspleniaceae (11 spp.), and Hymenophyllaceae (9 spp.). Species-accumulation curves for both sites showed signs of saturation, indicating that overall sampling was representative (Fig. 2). This impression was supported by an estimation of the total species richness with the Chao2 estimator (Herzog et al. 2002, Walther & Moore 2005), which predicted total species numbers of 223 for Pono and 221 for Bariri, suggesting that about 91% and 68%, respectively, of all herb species at the sites were encountered.

Terrestrial herbs were recorded in all individual plots, but species richness varied from 1-26 (mean 14.0) species per plot at Bariri and 2-28 (mean 9.7) species at Pono, whereas collective ground cover of all herbs in a plot ranged from 0.01-100 % per plot at Bariri and 0.05-100 % at Pono. These values were not significantly different between sites (t-tests, $P > 0.1$). Ferns and lycophytes were the dominant groups of terrestrial herbs both from the cover index and from the species numbers. Their mean cover index was 48% at Pono and 79% at Bariri, compared to 26% and 7%, respectively, for angiosperms

Table 2. Geographical distribution of the fern and lycophytes species recorded at Pono (55 species) and Bariri (44 species) based on herbarium and literature information. Values in parentheses are percentages.

Distribution	Pono	Bariri
Endemic to Sulawesi	7 (13)	0 (0)
Endemic to Malesia	14 (25)	27 (61)
Malesia + Indochina	19 (35)	15 (34)
Tropical Asia	6 (11)	0 (0)
Tropical Asia + Africa	3 (5)	1 (2)
Pantropical	6 (11)	2 (5)
Reaching W limit in Sulawesi	9 (16)	18 (40)
Reaching E limit in Sulawesi	3 (6)	4 (9)

Because species-level identification was more complete and reliable for ferns and lycophytes (50%) than for angiosperms (23%) (frequently found only in sterile condition), subsequent biogeographical analyses were only based on ferns and lycophytes. At both sites most fern and lycophyte species had distribution ranges covering the entire Malesian Region and often extending to Indochina (Tab. 2). However, at Pono no fewer than 7 species (13%) are currently considered to be endemic to the island of Sulawesi, whereas no such species were recorded at Bariri. The number of geographically widespread species (tropical Asia to pantropical) was limited at both sites. At Bariri, 18 fern and lycophyte species reached their western distributional limits on Sulawesi and extended eastwards to the Moluccas, New Guinea or partly beyond. In contrast, only 4 species found at Bariri reached their eastern limits on Sulawesi. These numbers differed significantly from each other (Chi^2-test, $Chi^2 = 16.70$, $P < 0.01$). At Pono, there were also more species with western (9) than eastern (3) range limits, but this difference was not quite significant (Chi^2-test, $Chi^2 = 3.84$, $P < 0.10$)

4 Discussion

The numbers of terrestrial herb species recorded at both of our study sites was higher than those of any other inventory in tropical forests conducted to date (Tab. 3). This may partly be due to the high number of sampled plots spread over an extensive area (0.7 km^2) in our study, but other studies also covered large areas and this alone can certainly not explain the striking difference

Table 3. Richness and taxonomic composition of terrestrial understory herb assemblages in tropical forests. Number in parentheses are percentages, for the categories ferns and lycophytes and angiosperms these are relative to the total species numbers, for the specific families to the angiosperm number. Continued in Tab. 4.

Locality	Elavation (m)	Mean annual precipitation (mm)	Area (m)	Families	Genera	Species	Fern lycophytes
Pono, Sulawesi	1000	3534	5000	45	93	203	112 (55)
Bariri, Sulawesi	1400	1984	5000	42	80	171	94 (55)
Brunei, Borneo	800	5000	250	27	46	74	15 (20)
Brunei, Borneo	1000	5000	250	22	23	27	5 (19)
Brunei, Borneo	1100	5000	250	31	52	99	35 (35)
SW India	600-660	1600	1200	53	132	155	13 (10)
SW India	900	1014	3200	32	39	42	3 (10)
SW India	900-1300	1014	3200	26	35	39	3 (12)
Budongo, Uganda	1200	1150-1500	2500	22	74	117	32 (27)
Bwindi, Uganda	1450	1400-1900	2500	21	70	101	42 (42)
Merida, Venezuela	2600	2500	400	13	17	20	14 (70)
Manaus, Brazil	80	2478	880	18	24	35	11 (31)
Cuyabano, Ecuador	250-300	---	1000	26	50	99	29 (29)
Los Volcanes, Bolivia deciduous	1050-1100	1200-1500	1000	4	26	31	7 (23)
Los Volcanes, Bolivia, semideciduous	900-1000	1200-1500	1000	6	28	47	17 (36)
Los Volcanes, Bolivia, evergreen	900-950	1200-1500	1000	8	33	63	25 (40)

Table 4. Richness and taxonomic composition of terrestrial understory herb assemblages in tropical forests. Number in parentheses are percentages, for the categories ferns and lycophytes and angiosperms these are relative to the total species numbers, for the specific families to the angiosperm number.

Locality	Angiosperms	Araceae	Orchidaceae	Zingiberaceae	Other families	Reference
Pono, Sulawesi	91 (45)	20 (22)	16 (18)	10 (11)	45 (49)	This study
Bariri, Sulawesi	77 (45)	10 (13)	13 (17)	16 (21)	38 (49)	This study
Brunei, Borneo	59 (80)	10 (17)	7 (12)	20 (34)	22 (37)	Poulsen and Pendry (1995)
Brunei, Borneo	22 (81)	1 (4)	5 (23)	7 (32)	9 (40)	Poulsen and Pendry (1995)
Brunei, Borneo	64 (65)	10 (16)	4 (6)	21 (33)	29 (45)	Poulsen and Pendry (1995)
SW India	124 (90)	2 (2)	6 (4)	8 (6)	108 (79)	Annaselvan and Parthasarathy (1999)
SW India	28 (90)	2 (6)	0	0	26 (84)	Chittibabu and Parthasarathy (2000)
SW India	22 (88)	2 (8)	0	0	20 (80)	Chittibabu and Parthasarathy (2000)
Budongo, Uganda	85 (73)	4 (5)	13 (15)	10 (12)	58 (68)	Poulsen (1997)
Bwindi Uganda	59 (58)	2 (3)	8 (14)	6 (10)	43 (73)	Poulsen (1997)
Merida Venezuela	6 (30)	0 (0)	2 (33)	0 (0)	4 (67)	Kelly et al. (1994)
Manaus, Brazil	24 (69)	2 (8)	0 (0)	1 (4)	21 (88)	Costa (2004)
Cuyabano, Ecuador	70 (71)	28 (40)	0 (0)	3 (4)	39 (56)	Poulsen et al. (2006)
Los Volcanes, Bolivia deciduous	24 (77)	6 (25)	0 (0)	0 (0)	25 (75)	Linares-Palomino et al. (2008)
Los Volcanes, Bolivia semideciduous	30 (64)	7 (16)	2 (7)	0 (0)	23 (77)	Linares-Palomino et al. (2008)
Los Volcanes, Bolivia evergreen	38 (60)	7 (18)	1 (3)	2 (5)	28 (74)	Linares-Palomino et al. (2008)

relative to previous studies. Traditionally, based on the knowledge of animal groups such as snakes (Bosch 1985) and birds (White and Bruce 1986), biotic communities on Sulawesi have been considered to be only moderately rich in species but containing exceptional levels of endemism (Backer and Bakhuizen van den Brink 1963). Studies of plant species numbers across the Sundaic region placed Sulawesi in an average position between less diverse Java and Sumatra on the one hand and more species-rich New Guinea and Borneo on the other (Roos at al. 2004). The moderate diversity and high endemism of Sulawesi have been explained by the limited island size along with the long-term geographical isolation of the island, which has limited colonization and has allowed the evolution and persistence of numerous unique forms (Backer and Bakhuizen van den Brink 1963, Whitten et al. 1987, Roos at al. 2004). However, recent quantitative botanical data suggest that plant communities may be more species-rich on Sulawesi than generally thought. Tree inventories in Lore Lindu National Park show species numbers that are comparable to those of mainland Southeast Asia and the Philippines, or even higher (Schulze et al. 2004, Kessler et al. 2005). Similarly, fern communities on Sulawesi are richer than those on Java, Borneo, and in Peninsular Malaysia (M. Kessler and J. Kluge, unpubl. data), even though they do not approach species numbers from South America (Kessler 2001, Kluge et al. 2006). Bryophyte species richness on Sulawesi is also among the highest ever reported for tropical forests (Sporn et al. 2009). Our study suggests that forest herbs on Sulawesi may also be unexpectedly diverse, although additional surveys from other sites on the islands as well as elsewhere in the tropics are needed to corroborate this impression. The discrepancy between the high diversity of these local studies and the moderate diversity reported based on general collecting activities may well be the low density of botanical collections on Sulawesi, with is less than 5% of that on Java, for example (Cannon 2001).

Taxonomically, 55% of all terrestrial herb species at both study sites were ferns and lycophytes, and 45% angiosperms. This high proportion of ferns contrasts with previous studies in Southeast Asia, Africa and South America, where the percentage of ferns and lycophytes typically ranges from 10% to 25% (Tab. 3). The only comparably high values have been obtained in a montane cloud forest in the Andes of Venezuela (Kelly et al. 1994), suggesting that the high diversity of ferns may be linked to elevation. Indeed, the diversity of terrestrial ferns typically increases with elevation to around 2000 m, and only declines at higher elevations (Kessler 2001, Kluge et al. 2006). The dominance of Araceae, Orchidaceae, and Zingiberaceae among the angiosperms, all belonging to the monocotyledons, appears to be typical of humid tropical forests, but differs strikingly from dry forests in India (Annaselvam and Parthasarathy 1999, Chittibabu and Parthasarathy 2000, Rasingam and Parthasarathy 2009). Among our two study sites, lower-elevation Pono was more species-rich overall than higher-elevation Bariri, suggesting a pattern of decreasing herb diversity within this elevational range. Proctor et al. (1988) and Poulsen and Pendry (1995), however, found increasing herb diversity on low-elevation mountains in Borneo, suggesting an overall hump-shaped elevational pattern of tropical forest herb diversity with elevation, as has previously been documented for ferns (Kessler 2001, Kluge et al. 2006), epiphytic herbs (Cardelz et al. 2006), and vascular epiphytes (Krömer et al. 2005) in tropical America.

Biogeographically, Sulawesi is located just east of Wallaces line, the famous biogeographical boundary separating the Malesian zoogeographical region from the Australasian region (Welzen et al. 2005). Although some plant families also follow this break (e.g., Dipterocarpaceae with over 250 species in Borneo and only 6-7 in Sulawesi; Whitten et al. 1987), van Steenis (1950) proposed that the Wallaces Line is not a sharp demarcation for plants and that it should mainly used to establish faunal provinces. However, Aryanti and Gradstein (2007) recently found that the montane liverwort flora of Mt. Nokilalaki in Lore Lindu National Park was primarily derived from eastern elements and contained few species of western origin, suggesting that the Wallaces Line is also valid for this group of plants. Our study confirmed this pattern for terrestrial ferns, with significantly more species reaching their western than eastern distributional limits on Sulawesi. Because dispersal limitation as such is unlikely to be a limiting factor for spore-dispersed liverworts and ferns, geographical isolation by sea barriers is probably not the main cause for this pattern. Alternative explanations may involve the higher surface area of montane forest habitats in New Guinea as compared to the Greater Sunda islands, differential dispersal determined by prevailing wind streams (e.g., the monsoon), or greater environmental similarity between Sulawesi and New Guinea than to the Greater Sundas (Aryanti and Gradstein 2007).

In conclusion, our study has uncovered an unexpectedly high diversity of terrestrial forest herbs at two sites on the island of Sulawesi. Some of the species in the high number of families found in this study might serve as indicators for global change effects in tropical forests, but this remains to explored with detailed ecological studies.

Biogeographically, there appears to be a more pronounced affinity towards the Moluccas and New Guinea than to the Greater Sundas, although this is based on ferns only and needs to be confirmed for angiosperms. A full understanding of the processes that have generated the high diversity of Sulawesis plant communities will depend on more comprehensive botanical explorations (Cannon 2001) and the development of phylogenetic hypotheses that will provide data on colonization events as well as the modes and timing of plant diversification.

Acknowledgements

We thank Atok Arianto, Sahar Sabir and Fierdaus for help during field work. Soil data was provided by D. Leitner, M. Bealtzik, O. van Straaten, and E. Veldkamp, and climate data by H. Kreilein, A. Oltchev, and G. Gravenhorst.

We are grateful to the STORMA coordination offices in Göttingen, Palu, and Bogor for organisational support, and LIPI for the research permit and supporting field work. This study was funded by the Deutsche Forschungsgemeinschaft (DFG) and the SYNTHESYS Project (http://www.synthesys.info) financed by European Community (NL-TAF-3536).

References

Annaselvam J, Parthasarathy N (1999) Inventories of understory plants in a tropical evergreen forest in the Anamalais, Western Ghats, India. Ecotropica 5: 197-211

Aryanti SN, Gradstein SR (2007) Wallaces line and the distribution of the liverworts of Sulawesi. Cryptogamie, Bryologie 28: 3-14

Baas P., Kalkman K. and Geesink R. (eds) 1990. The Plant Diversity of Malesia. Kluwer, Dordrecht, The Netherlands.

Backer CA, Bakhuizen van den Brink RC Jr (19631968) Flora of Java, Vol. 1-3. Wolters Noordhoff, Groningen, The Netherlands

Cannon CH (2001) The vegetation of Lore Lindu National Park, Central Sulawesi, Indonesia. The Nature Conservancy, Palu, Indonesia

Cannon HC, Summers MHJR, Kessler PJA (2007) Developing conservation priorities based on forest type, condition, and threats in a poorly known ecoregion: Sulawesi, Indonesia. Biotropica 39: 747-759

Cardels CL, Colwell RK, Watkins JE Jr (2006) Vascular epiphyte distribution patterns: explain the mid-elevation richness peak J Ecol 94: 144-156

Costa CF, Magnusson W, Luizao RC (2005) Mesoscale distribution patterns of Amazonia understorey herbs in relation to topography, soil and watersheds. J Ecol 93: 863-878

Costa FC (2004) Structure and composition of ground-herb community in a terra-firme Central Amazonia forest. Acta Amazonica 34: 53-59

Chittibabu CV, Parthasarathy N (2000) Understory plant diversity in a tropical evergreen forest in the Kolli Hills, Eastern Ghats, India. Ecotropica 6: 129-140

Flora Malesia, Pteridophyta (Ferns and Ferns Allies) (1981), Vol. 1. National Herbarium Nederland Univeriteit Leiden branch

Flora Malesia, Pteridophyta (Ferns and Ferns Allies) (1991), Vol. 2. National Herbarium Nederland Univeriteit Leiden branch

Flora Malesia, Pteridophyta (Ferns and Ferns Allies) (1998), Vol. 3. National Herbarium Nederland Univeriteit Leiden branch

Gentry AH, Emmons LH (1987) Geographical variation in fertility phenology and composition of the understory of neotropical forest. Biotropica 19: 216-227

Hall R, Holloway JD (Eds.) (1998) Biogeography and geological evolution of SE Asia. Backhuys, Leiden, The Netherlands

Herzog SK, Kessler M, Cahill TM (2002) Estimating species richness of tropical bird communities from rapid assessment data. Auk 119: 749-769.

In den Bosch HAJ (1985) Snakes of Sulawesi: Checklist, Key and additional biogeographical remarks. Zool Verhandel 217: 1-50

Jacobs M. (1988) The Tropical Rainforest. A First encounter. Springer-Verlag, Berlin,Tokyo, 1-295

Kati JS, Cardenas GG, Tuomisto H (2004) Forest classification in an Amazonian rainforest landscape using pteridophytes as indicator species. Ecography 27: 689-700

Kelly DL, Tanner VJ, Nic Lughadha EM, Kapos V (1994) Floristic and biogeography of a rain forest in the Venezuelan Andes. J Biogeogr 21: 421-440

Kessler M (2001) Pteridophyte species richness in Andean forest in Bolivia. Biodiv Conserv 10: 1473-1495

Kessler M, Kessler PJA, Gradstein SR, Bach K, Schmull M, Pitopang R. (2005) Tree diversity in primary forest and different land use systems in Central Sulawesi. Indonesia. Biodiv Conserv 14: 547-560

Kluge J., Kessler M. Dunn R.R (2006) What drives elevational patterns of diversity? A test of geometric constraints climate and species pool effects for pteridophytes on an elevational gradient in Costa Rica. Global Ecol Biogeogr 15: 359-371

Krömer T, Kessler M, Gradstein SR, Acebey A (2005) Diversity patterns of vascular epiphytes along an elevational gradient in the Andes. J Biogeogr 32: 1799-1809

Linares- Palomino R, Cardona V, Hennig E I, Hersen I, Hoffmann D, Lendzion J, Soto D, Herzog SK, Kessler M (2008) Non-woody life-form contribution to vascular plant species richness in a tropical America forest. Plant Ecol 201: 87-99

Moss SJ, Wilson MEJ (1998) Biogeography implication of the Tertiary palaeogeographic evolution of the Sulawesi and Borneo. Biogeography and geological evolution of Southeastern Asia. Backhuys, Leiden, The Netherlands

Poulsen AD, Pendry CA (1995) Inventories of ground herbs at three altitudes on Bukit Belalong, Brunei, Borneo. Biodiv Conserv 4: 745-757

Poulsen AD, (1997) Plant diversity in forest of western and eastern Zaire (Preliminary results). AAU reposts 36 Department of Systematic Botany, University of Aarhus.

Poulsen AD, Hafashimana D, Eilu G, Liengola IB, Ewango CEN, Hart TB (2005) Composition and species richness of forest along the Albertine Rift, Africa. Biol Skrifter 55: 129-143

Poulsen AD, Tuomisto H, Balslev H (2006) Edaphic and floristic variation within a 1-ha plot of lowland Amazonian rain forest. Biotropica 38: 468-478

Proctor J, Lee YF, Langley AM, Munro WRC, Nelson (1988) Ecological studies on Gunung Silam. A small ultrabasic mountain in Sabah, Malaysia. Environment, forest structure and floristics. J Ecol 76: 320-40

Rasingam and Parthasarathy (2009) Diversity of understory plants in undisturbed tropical lowland forests of Little Andaman Island, India. Biodivers Conserv 18: 1045-1065

Roos M.C. 1993. State of affairs regarding Flora Malesiana: progress in revision and publication schedule.

Flora Malesiana Bulletin 11: 133142.

Roos CM, Kessler PJA, Gradstein SR, Bass P (2004) Species diversity and endemism of five major Malesian islands: diversity-area relationships. J. Biogeogr. 31:1893-1908

Ruokolainen K, Linna A, Tuomisto H (1997) Use of Melastomataceae and pteridophytes for revealing phytogeographical patterns in Amazonian rain forests. J Trop Ecol 13: 243-256

Schulze CH, Waltert M, Kessler PJA, Pitopang R, Shahabuddin, Veddeler D, Steffan-Dewenter I, Mühlenberg M, Gradstein SR & Tscharntke T (2004) Biodiversity indicator taxa of tropical land-use systems: comparing plants, birds and insects. Ecological Applications 14: 1321-1333.

Sporn SG, Bos MM, Hoffstätter-Müncheberg M, Kessler M, Gradstein SR (2009) Microclimate determines community composition but not richness of epiphytic understory bryophytes of rainforest and cacao agroforest in Indonesia. Funct Plant Biol 36: 171-179

van Steenis CGGJ. (1950) The delimitation of Malaysia and its main plant geographical divisions. Pp. Ixx-Ixxv. In: van Steenis CGGJ (ed.), Flora Malesiana. Series 1, 1. Noordhoff-Kolff, Jakarta, Indonesia.

Walther BA, Moore JL (2005) The concepts of bias, precision and accuracy, and their use in testing the preformance of species richness estimators, with a literature review of estimator performance. Ecography 28: 815-829

Welzen JWC van, Slik F, Alahuhta J (2005) Plant distribution pattern and plate tectonics in Malesia. Biol Skrifter 55: 199-217

White CWM, Bruce M (1986) The Birds of Wallacea. BOU Checklist No. 7. British Ornithologists Union, London.

Whitmore TC (1987) Biogeographical evolution of the Malay Archipelago. Oxford University Press, Oxford, UK

Whitten AJ, Muslim M, Henderson GS (1987) The Ecology of Sulawesi. Gadjah Mada University Press, Yogyakarta, Indonesia

The hydraulic performance of tropical rainforest trees in their perhumid environment - is there evidence for drought vulnerability?

Alexandra Zach[1], Bernhard Schuldt[1*], Viviana Horna[1], Soekisman Tjitrosemito[2], and Christoph Leuschner[1]

[1] Plant Ecology, Albrecht von Haller Institute for Plant Sciences, University of Göttingen, Untere Karspüle 2, 37073 Göttingen, Germany
[2] Faculty of Science and Mathematics, Institute Pertanian Bogor, Jl. Raya Pajajaran, Bogor, 16144 Indonesia

*corresponding author: B. Schuldt, email: bernhard.schuldt@googlemail.com

Summary

In a species-rich perhumid tropical rainforest of Central Sulawesi, Indonesia, we studied the hydraulic properties of eight representative tree species and searched for evidence that analyzing hydraulic patterns at the species level provides valuable information to understand the role of abiotic drivers and structural parameters in controlling plant water consumption. We investigated the relationship between xylem hydraulic properties and tree size with the aim to determine possible hydraulic plasticity in response to the vertical variation in environmental conditions in an otherwise constantly humid forest. We found leaf-specific (LSC) and sapwood-area specific (k_S) hydraulic conductivity of twigs to significantly increase with tree height across species. The marked increase of LSC and k_S with tree height was closely coupled with an increase in mean vessel diameters and this trend was consistent for both, stem and twig xylem. Rates of xylem sap flow (J) of all tree species were strongly positively related to atmospheric vapor pressure deficit (VPD) and radiation (R), when evaluated on a daily basis. However, J started to level off at VPD values of about 0.4 kPa in trees of all size classes. We therefore concluded that the stomatal response of tropical moist forest trees is very sensitive to changes in the atmospheric evaporative demand.

The tallest-growing species of our study, the Fagaceae *Castanopsis acuminatissima,* had hydraulic features that were remarkably different from the other species. This species, which is abundant in the study area, showed significantly higher rates of J, and higher hydraulic efficiencies in the root-to-leaf

T. Tscharntke et al. (eds.), *Tropical Rainforests and Agroforests under Global Change*, Environmental Science and Engineering, 391–410, DOI 10.1007/978-3-642-00493-3_18,

flow path and larger xylem vessel sizes than the other co-occurring species. We assumed that this hydraulic adjustment of xylem vessel anatomy is of vital importance for maintaining a sufficient water supply to the canopy under the wet conditions prevailing in Central Sulawesi.

We found a high hydraulic plasticity with variation in tree height. This may be a direct response to changes in ambient conditions along the microclimate gradient from the forest understory to the canopy top. Large trees might be more resistant against drought events in the short run. However, there is evidence that especially tall-growing trees might be more vulnerable to prolonged periods of drought.

Keywords: Castanopsis acuminatissima, hydraulic architecture, hydraulic conductivity, xylem sap flux, premontane rainforest, Sulawesi, tree height, vessel anatomy

1 Introduction

Evergreen tropical rainforests depend on a continuous water supply throughout the year in order to maintain high growth rates and accumulate large amounts of biomass. Permanent high temperatures and high radiation intensities force tropical rainforests to turn over large quantities of water, which can account for 1200 mm (e.g. Calder et al. 1986, Leopoldo et al. 1995, Kumagai et al. 2005) to up to 2000 mm (Larcher 1998, Lösch 2001) per year transpired back to the atmosphere. Such immense water consumption at the canopy level makes the steady replenishment of soil water resources essential.

On the stand level, the bulk of transpiration is mainly contributed by large trees which crowns are located in the upper canopy levels, while suppressed trees have a minor contribution (Granier et al. 1996). In this sense, tree size has been identified as an important structural parameter influencing the hydraulic architecture and hence patterns of water use (Becker 1996, Andrade et al. 1998, Goldstein et al. 1998, Meinzer et al. 2001). Under moist soil conditions, tall trees showed higher rates of xylem sap flux density (J) because their canopies are exposed to a higher evaporative demand and higher radiation than smaller or understory trees (Phillips et al. 1999). Differences in tree stature should result in differences in the hydraulic response of co-occurring tropical trees to changes in the main abiotic drivers of tree transpiration: atmospheric water vapor saturation deficit (VPD) and solar radiation (R). Daytime vertical microclimate in tropical rainforests is often characterized by a decline in VPD and R from the top of the upper canopy towards lower canopy strata (e.g. Fetcher et al. 1985, Dietz et al. 2007). Only a small number of studies investigated morphological and functional differences or similarities in the hydraulic system of tropical tree species and their dependence on important environmental factors (e.g. Granier et al. 1992, Fetcher et al. 1994, O'Brien et al. 2004). Fetcher et al. (1994) found a sensitive stomatal response to VPD for a number of tropical tree species leading to partial stomatal closure at saturation deficits > 1 kPa. A rather conservative water use of tropical moist forest trees was mirrored by a significant reduction in transpiration during periods of soil drought (Granier et al. 1992, O'Brien et al. 2004). Nevertheless, it is still not clear how tree height influences water consumption and water flux control. We suspect that with increasing height, the whole tree hydraulic system exhibits a plastic response to the increasing demand for water release with the higher VPD levels above the upper canopy strata. However, under the aseasonal conditions of perhumid tropical rainforests, a permanent adaptation to water shortage seems less economic.

In the premontane rainforests of Central Sulawesi (Indonesia) with about 3500 mm of annual rainfall, abiotic factors such as VPD, temperature and R vary only slightly throughout the year. Severe dry periods rarely occur in this perhumid environment. The prevailing humid conditions may foster the morphological and physiological plasticity of plant water use and sap flux control of trees along the vertical gradient of the multilayered forest. Consequently, differences in tree hydraulic architecture should be visible with increasing tree height.

In the present chapter, we synthesize key findings of an intensive 1-year measuring campaign studying the hydraulic architecture of abundant tree species in the Lore Lindu National Park of Central Sulawesi and its dependence on stand structural properties and climatic variables. We analyzed patterns of J and studied important anatomical properties of the xylem conducting system likely to determine the axial conductivity along the root-to-leaf flow path. Furthermore, we evaluated the effect of tree height on the relationship between water uptake and environmental parameters under the prevailing conditions of ample soil water supply. Hydraulic plant traits are often species-specific, and thus, species differences are highly relevant for the estimation of stand water use in species-rich tropical forests.

It is still uncertain how tropical moist forest trees growing in a multi-structured, multi-species stand might response to a possible higher drought frequency in a changing global climate. Changes in the hydraulic architecture with tree height often include the development of more and larger vessels with higher conductivity in order to maintain the internal water balance (e.g., Pothier et al. 1989). However, larger vessels bear the risk of xylem embolism (i.e., the cavitation-induced breakdown of the hydraulic water flow in tree vessels, which can cause branch dieback and plant death), especially under conditions of high evaporative demand as found during drought (e.g. Williams et al. 2001). As a consequence, mortality was found to be higher for larger than for smaller trees after increased environmental exposition following forest fragmentation (Laurance et al. 2000). We compared individual trees and tree species to gain insight into the relative importance of specific physiological traits on the one hand and tree height on the other for forest water consumption and its control. We aimed to address the following questions:

1. Do hydraulic properties consistently change with tree size to counteract the effect of high evaporative demand with increasing tree height?
2. Is the environmental control of J directly related to tree height variation?

Based on our results, we finally tried to envisage, if tropical rainforest trees are more vulnerable to drought because of their permanent adaptation to aseasonally wet conditions.

2 Materials and Methods

2.1 Study site and tree selection

The hydraulic measurements were conducted at three study plots of 40 m x 40 m in nearly level terrain between January and December 2008. The plots were situated in a tropical premontane rainforest in the Pono Valley located on the western boundary of the Lore Lindu National Park of Central Sulawesi, Indonesia (UTM 51M, 0172451, 9834650; 1050 m a.s.l). The climate of the study area is aseasonally wet with c. 3500 mm mean annual precipitation. At the most, two months per year receive < 100 mm rainfall classifying this climate as perhumid (Holdridge et al. 1971). Mean annual temperature is 20.8 °C. The soils of this old-growth forest developed on metamorphic rocks, classified as Acrisols (FAO classification). Volumetric soil water content (SWC) within the upper 10 cm, where most of the fine root biomass was found (G. Moser, pers. comm.) was close to saturation throughout our measurement period. In the three study sites, SWC was recorded by M. Köhler (unpubl. data) at half-hour intervals with time domain reflectometry probes (TDR, Campbell Scientific, U.K.).

The average canopy height is about 45 m with a few emergent trees of up to 55 m. The forest harbors about 130 tree species ha^{-1} (Culmsee et al. submitted). For our study, we focused on trees with a DBH > 10 cm from different canopy layers. Our selection included the most common species with different growth strategies in order to represent the variability of tree functional types and tree families in this species-rich premontane forest. Within the three plots we selected tree individuals from the slow-growing mid-story species *Cryptocarya laevigata* (Lauraceae), *Pouteria firma* (Sapotaceae), *Santiria apiculata* (Burseraceae), and two unidentified Myrtaceae species and the fast-growing canopy species *Platea excelsa* (Icacinaceae), *Vernonia arborea* (Asteraceae), *Castanopsis acuminatissima* (Fagaceae). Species identification and growth characterization were provided by H. Culmsee (pers. communication). Due to the high species density in the stand, a given species could not be represented by individuals from all height classes (i.e., 10-20, 20-30 and 30-50 m in our study). Especially taller growing trees > 30 m were mainly represented by *C. acuminatissima.*

For xylem hydraulic conductivity measurements and xylem anatomical studies of trunks and twigs we chose 51 tree individuals of the eight tree species mentioned above. Each species was represented by two to eleven tree individuals differing in tree height (Table 1).

Equipment for permanent sap flux measurements was installed at 39 tree individualsof seven of the eight tree species with each species being represented by three to eight tree individuals of different height (Table 1). Xylem anatomical and the hydraulic studies have not been made from trees where a permanent sap flow set-up was installed on the stem. Rather, we chose tree individuals similar in size for stem wood and twig sampling.

We determined tree height using a laser tree height meter (Vertex III Forester, Haglöf, Långsele, Sweden). The slopes of the sites were taken into account by determining the angle of the stems for calculating total stem length.

2.2 Hydraulic anatomy

Segments of upper canopy twigs and stem cores were used for anatomical studies. Canopy branches were accessed using a pruner (up to 12 m) or by fixed climbing ropes (above 12 m). All branches were at least part of the day sun-lit, either in canopy gaps or in the upper crown. In the case of the stems, we extracted wood cores at breast height with an increment corer (diameter: 5mm, Haglöf, Långsele, Sweden). For microscopic analysis we took samples from the outer 1-2 cm of the core.

For characterizing xylem anatomy and measuring vessel diameter and vessel density of twigs and trunks we used a sliding microtom extracting thin transverse sections (10-20 μm). Prior to microtom cutting, stem wood and twig samples were embedded in polyethylenglycol (PEG 2000, Merck Schuchardt, Germany). The cross-sections were mounted on slides and photographed at 20x or 80x magnification using a light microscope (Photomikroskop III, Carl Zeiss, Germany) equipped with a digital camera (PowerShot A620, Canon, Japan). The images were analyzed with the software ImageJ (v1.36b, http://rsb.info.nih.gov/iJ) using the particle analysis-function to estimate idealized radii (r) from lumen area ($A = \pi r^2$), vessel density (n mm^{-2}) and cumulative vessel lumen area (m^2).

2.3 Hydraulic conductivity

Upper canopy twigs of each tree individual were used to determine empirical axial hydraulic conductivity. Per tree, we sampled 3-6 twigs with a mean diameter of 7.7 mm (± 1.4 SD) and a mean length of 85mm (± 33.3) together with the appending leaf material. To prevent microbial growth we immediately stored the twig segments in polyethylene tubes filled with water containing a sodium-silver-chloride complex (Micropur, Katadyn, Wallisellen, Switzerland). Thereafter, samples were taken to the laboratory and stored at 4 °C. Prior to hydraulic measurement, we recut each segment under water with a razor blade and measurements of k_h were conducted at least three times for each branch following Sperry et al. (1988). Before entering the twig segment,

Table 1. Observed ranges of tree height (m), diameter at breast height (DBH; cm), and Huber value ($HV*10^{-4}$, mean in parentheses) of the sampled tree species. Mean (SD in parentheses) empirical leaf-area specific hydraulic conductivity ($LSC*10^{-4}$, kg $m^{-1}MPa^{-1}s^{-1}$) and branch-sapwood-area specific hydraulic conductivity (k_S, kg $m^{-1}MPa^{-1}s^{-1}$) of exposed, upper-crown twigs of the measured tree species are given. J is the mean daily xylem flux density measured in the outer 2 cm of the active xylem by means of the constant heating method after Granier (1987). Different small letters after the J values indicate significant differences among species ($p < 0.05$). The number of replicates (n) gives the number of trees either measured for hydraulic conductivity or monitored for xylem sap flux density.

Species	Height	DBH	n	$HV*10^{-4}$	$LSC*10^{-4}$	k_S	*n*	J
Santiria apiculata A.W. Benn.	6.5 - 29.4	11.7 - 58.2	11	1.69 - 7.72 (2.80)	6.09 (3.35)	2.43 (1.38)	5	41.87 (11.11) ab
Vernonia arborea Buch.-Ham.	19.5 - 28.7	14.3 - 43.0	8	1.16 - 4.34 (2.06)	3.96 (2.54)	2.11 (1.42)	3	37.18 (14.35) ab
Castanopsis acuminatissima (Blume) Rheder	17.6 - 44.2	11.8 - 131.8	10	0.71 - 2.44 (1.57)	15.90 (7.70)	10.49 (3.47)	8	97.85 (9.04) c
Platea excelsa Bl. var. borneensis (Heine) Sleum.	12.3 - 27.2	10.1 - 45.7	6	1.34 - 2.57 (1.99)	3.26 (2.63)	1.63 (0.89)	6	56.89 (10.11) ab
Cryptocarya laevigata Blume	12.7 - 33.2	11.2 - 27.9	4	1.29 - 2.19 (1.75)	3.95 (2.68)	2.14 (0.87)	6	37.32 (10.24) a
Myrtaceae sp. 8	11.3 - 11.7	15.6 - 16.7	2	1.08 - 1.89 (1.48)	5.15 (9.43)	3.88 (1.51)	–	–
Myrtaceae sp. 10	11.0 - 29.0	11.7 - 18.0	3	1.35 - 2.00 (1.63)	8.79 (2.44)	5.34 (0.31)	–	–
Pouteria firma (Miq.) Baehni	12.8 - 44.3	10.0 - 66.5	7	1.29 - 4.84 (2.50)	4.09 (2.57)	1.68 (0.80)	5	73.19 (11.18) bc
Palaquium luzoniense (Fern.-Vill.) Vidal	17.6 - 44.7	10.4 - 95.0	–	–	–	–	6	48.96 (10.15) ab

the solution was forced trough a 0.20 m membrane filter (Maxi Capsule, Pall, U.S.A.). Water extruding from the distal end of the segment was collected and weighted (precision: 0.1 mg). Hydraulic conductivity (k_h) can be expressed as:

$$k_h = F \cdot \frac{l}{\Delta P} = \frac{\Delta V}{\Delta t} \cdot \frac{l}{\Delta P} \left[\frac{\text{kg m}}{\text{MPa s}} \right] \tag{1}$$

where F is the water flux (kg s^{-1}), l is the length of the twig segment (m), ΔP is the pressure difference applied to the twig segment (MPa), ΔV is the amount of water flowing out of the twig segment (kg) and Δt is the time interval (s). After measuring actual conductivity (k_h^{act}), we determined the maximum axial conductivity (k_h^{max}) by flushing the segments at a pressure of 0.12 MPa. Branch cross-sectional area (m^2) and the supported leaf area (m^2) were used to calculate sapwood-area specific (k_s, kg m^{-1} MPa^{-1} s^{-1}) and leaf-specific conductivity (LSC, kg m^{-1} MPa^{-1} s^{-1}), respectively. We calculated the Huber value (HV) as the ratio of the sapwood cross-sectional area to the supported leaf area distal to the measured branch segment.

2.4 Xylem sap flux density

We measured xylem sap flux density (J, g cm^{-2} d^{-1}) in the stem xylem following Granier (1987). Pairs of 20 mm-long and 2 mm-wide heating probes were inserted in northern and southern trunk directions into the stem sapwood. The upper probe was heated with a constant power of 0.2 W. The temperature difference between the two sensors was recorded with copper-constantan (Cu-Ni) thermocouples placed at the centre of the heating coil. The lower probe, which was not heated, represented the reference for the upper probe. To avoid thermal interference we kept the distance between the two sensor probes constant at about 15 cm. We calculated sap flux velocity according to Granier (1987):

$$J = 119xK^{1.231} \tag{2}$$

where $K = (\Delta T_M - \Delta T)/\Delta T$. ΔT_M is the maximum temperature difference at zero sap flow. Analysis of sap flux data corresponds to the period from May 2007 to January 2008.

2.5 Microclimate and hydrologic measurements

Above-canopy incoming shortwave radiation (R) was monitored with a pyranometer (CS 300, Campbell Scientific, U.K.). We used a combined temperature and humidity probe for air temperature and relative humidity records (CS 215, Campbell Scientific, U.K.). All sensors were installed on a 16 m tall tower located in a large natural forest gap several hundred meters away from the study plots. All values were logged half-hourly by a Campbell CR10X data logger (Campbell Scientific, U.K.). We calculated atmospheric vapor pressure deficit (VPD) from air temperature and relative air humidity according to Goff and Gratch (1946).

2.6 Statistical analysis

We analyzed the relation of hydraulic architectural parameters and tree size by plotting the hydraulic parameters (LSC, k_S, J, vessel diameter, vessel density) against tree height as the independent variable. To analyze stand-level patterns in hydraulic efficiency we applied regression analyses across all measured species in order to represent the tree height distribution in the stand.

Simple regression analysis was used to analyze the relationship between J and VPD as the main driver of J for four tree height classes (10 - 19.9, 20 - 29.9, 30 - 39.9, 40 - 49.9) and across species. For regression analyses we used SigmaPlot 10.0 (Systat software, Inc.).

3 Results and Discussion

3.1 Do hydraulic properties change consistently with tree size?

We found a significant increase in sapwood-area specific (k_S) and in leaf-specific (LSC) hydraulic conductivity of canopy branches with tree height across our measured tree species from a perhumid tropical rainforest of Central Sulawesi, Indonesia (Fig. 1, Table 1). Highest values of k_S and LSC and the steepest increase with height were found in twigs of the tall-growing *C. acuminatissima*. In contrast, lowest values of k_S and LSC were not found in the smallest measured tree; rather, low specific conductivities were found in twigs of various tree species ranging in height between 10 and 20 m. However, due to the dominance of *C. acuminatissima* as tall tree in our sample, this species had a large influence on the height-relationship of hydraulic properties in our study. Rates of tree water supply with increasing size depend on the steepness of the water potential gradient between soil and leaf and the hydraulic resistance of the water flow pathway. To increase the hydraulic performance despite longer flow pathways, an increasing hydrostatic pressure and increasing resistances with tree size, large trees tend to widen their vessel elements (Pothier et al. 1989, West et al. 1999, Burgess et al. 2006). Indeed, across the eight canopy species, we found a significant and positive correlation between tree height and the xylem vessel diameter of stem sapwood, i.e. the taller the tree the larger the vessels at stem base (Fig. 2).

Greater vessel diameters accompanied by high water conductivities usually enhance the risk of xylem embolism and hence xylem dysfunction (e.g. Williams et al. 2001). However, in our study, the xylem of the tallest species *C. acuminatissima* was not found to be more sensitive to embolism than that of smaller trees with thinner vessels (data not shown). This tall-growing species also strikes with surprisingly high hydraulic conductivities allowing for high rates of plant water consumption. We measured mean daily rates of J of up to 97.85 g cm^{-2} d^{-1} in *C. acuminatissima,* more than in any other species; second was *P. firma* (73.19 g cm^{-2} d^{-1}). The lowest mean rate of J was measured for *C. laevigata* (37.32 g cm^{-2} d^{-1}; Table 1).

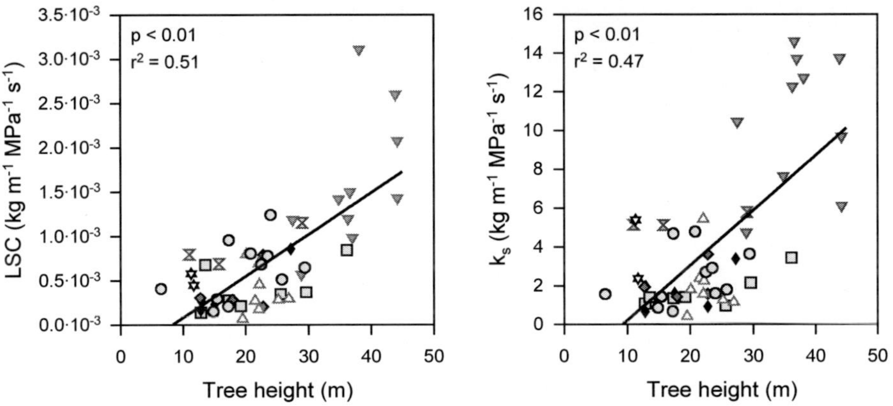

○ = *S. apiculata*, △ = *V. arborea*, ▽ = *C. acuminatissima*, ◆ = *C. laeivgata*, ♦ = *P. excelsa*, ✡ = *Myr. sp.* 8, ⧖ = *Myr. sp.* 10, □ = *P. firma* (see Table 1)

Fig. 1. Empirical leaf-specific (LSC) and sapwood-specific hydraulic conductivity (k_S) of exposed upper-canopy twigs of 8 tree species as related to tree height. Regressions are based on two to 11 tree individuals of each of the eight species measured.

Daily mean rates of J (at 1.3 m height) significantly increased with tree height, though considerable scatter was evident (Fig. 3). The relatively broad correlation resulted from differences in the relationship of J to tree height within single species and from the large influence of *C. acuminatissima* in the height class > 30 m. Plant water use of tropical tree species has often been found to increase with height (e.g., Andrade et al. 1998, Phillips et al. 2001) or DBH (Meinzer et al. 2005). The increase in J with increasing tree height is commonly assumed to result from VPD typically being higher around exposed sun-lit leaves of tall trees compared to the more shaded mid- or understory tree canopies. However, J showed no correlation with tree height in a Panamanian moist forest (Phillips et al. 1999) and two Amazonian forests (Oren et al. 1996). This lack of relationship has been related to the small number of replicates and the large differences in leaf area to sapwood area relationships among the measured tree species (Phillips et al. 1999).

Under conditions of low evaporative demand and high soil moisture, forests tend to hold a higher leaf area per unit sapwood area (Waring et al. 1982, White et al. 1998). This adaptation may help trees growing in humid environments with a high atmospheric humidity to maintain sufficient water flow despite a comparatively small soil-to-leaf water potential difference (Tyree and Ewers 1991). On the other hand, tree twigs reduce their transpiring leaf area in relation to the conductive xylem area with increasing position in the canopy and hence increasing exposure to atmospheric evaporative demand (Vanninen et al. 1996, Phillips et al. 2001, McDowell et al. 2002). This effective strategy helps to prevent excessive leaf water loss and to lower the

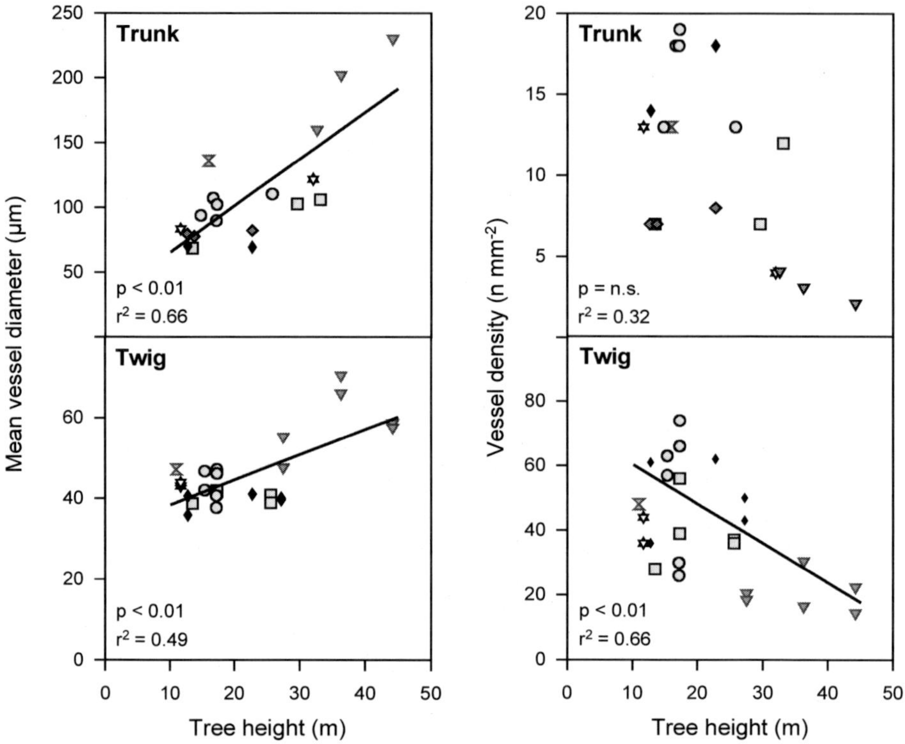

Fig. 2. Mean vessel diameter and vessel density of trunks and twigs as related to tree height. Regressions are based on two to 11 tree individuals of each of the eight species measured (see Table 1). For symbol explanation see Fig. 1.

risk of drought-induced cavitation under conditions of higher VPD in the exposed canopies of tall trees (Waring et al. 1982, Mencuccini and Grace 1995, White et al. 1998, Koch et al. 2004, Burgess et al. 2006). However, major constraints on tree height growth have postulated which are related to increasing hydraulic limitations with increasing water flow path length, eventually resulting in drought-induced reductions of assimilatory carbon gain and hence further growth (Ryan and Yoder 1997). This would imply that an increase in hydraulic efficiency (i.e., conductivity) with increasing tree height is not sufficient to avoid drought stress in the upper canopy of tall trees. On the other hand, Burgess et al. (2006) found branch hydraulic conductivity to increase with increasing position along tall individuals of *Sequoia sempervirens* (tree height range: 60-67 m). They even measured an exponential increase in leaf-specific conductivity with increasing height on the stem of single tree individuals, which was accompanied by a lower leaf area per active sapwood area (i.e., increased HV). Lower resistance to water flow for upper canopy twigs compared to lower positioned twigs has already been noted by Zimmermann

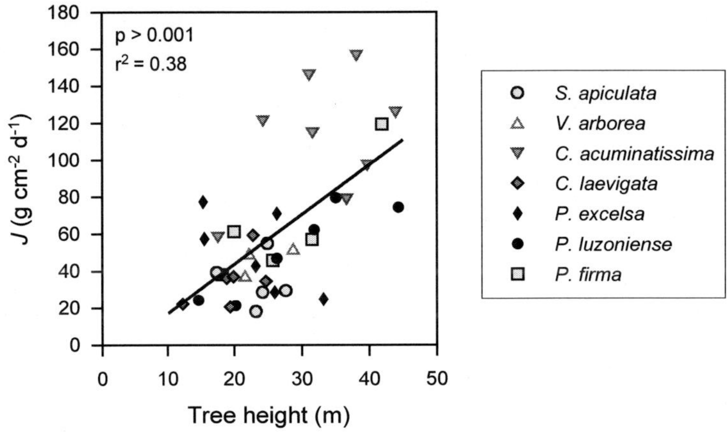

Fig. 3. Daily mean rates of sap flux density (J) for the measurement year 2008 as related to tree height. Regression is based on three to eight tree individuals of each of the seven species measured (see Table 1).

(1978). Comparable with our results, Burgess et al. (2006) found increasing mean conduit diameters in the xylem of branches when approaching the tree top.

Despite considerable debate upon a general safety-efficiency trade-off in the hydraulic system of large trees (Williams et al. 2001, Delzon et al. 2004, Koch et al. 2004, Choat et al. 2005, Woodruff et al. 2007), the overall finding presented here supports the idea that the hydraulic conductivity of tall trees does not suffer from gravitational constraints (Burgess et al. 2006). Rather, adaptive mechanisms such as increasing mean vessel diameters and changes in leaf morphology, which allow for a marked increase in leaf-specific hydraulic conductivity with increasing tree height, seem to be similar across tree species and even across tree functional types (e.g., angiosperms and conifers).

3.2 Is the environmental control of tree sap flux density directly related to tree height variation?

Across all tree species and heights in our study, J was closely related to VPD (Fig. 4). The dominant role played by VPD in the control of J underpins the general finding that VPD is the major abiotic driver determining the intensity of tree water use on a daily basis. VPD and R explained up to 84% of the variation in J in a deciduous temperate broad-leaved forest (Wullschleger et al. 2001). Zeppel et al. (2004) reported values of up to 75% for Australian temperate woodland. Phillips et al. (1999) found J to be slightly closer coupled to R (r= 0.57 - 0.92) than to VPD (r= 0.43 - 0.89) in a tropical moist forest in Panama. We found VPD to explain between 66 and 88% of the variation in J across tree height classes and species (Fig. 4). For all species, J started to level off at about 0.4 kPa (Fig. 4). Further, J clearly followed the diurnal course of R (Fig. 5).

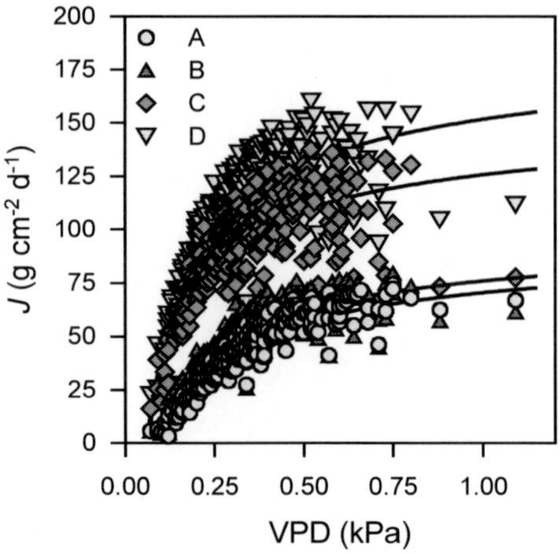

Fig. 4. Rates of sap flux density (J) of different tree height classes (A to D) as related to vapor pressure deficit (VPD). Given are daily mean rates of J for the measurement year 2008. Measured tree individuals are aggregated in four tree height classes and regression analysis is given for each height class separately: A: 10.0 - 19.9 m ($n = 10$, $p < 0.001$, $r^2 = 0.88$); B: 20.0 - 29.9 m ($n = 17$, $p < 0.001$, $r^2 = 0.84$); C: 30.0 - 39.9 m ($n = 9$, $p < 0.001$, $r^2 = 0.66$); D: 40.0 - 49.9 m ($n = 3$, $p < 0.001$, $r^2 = 0.73$). The tree individuals belong to seven co-occurring tropical species as described in Table 1.

However, while the shape of the relationship was the same for all three size classes examined, mean rates of J as a function of VPD varied considerably between trees of different sizes (Fig. 4). The results indicate that under the prevailing favorable conditions of high SWC, trees from all height classes in this perhumid environment are very sensitive to changes in the evaporative demand, because maximum mean rates of J were achieved at VPD below 0.8 kPa. Sap flux density of tropical trees was commonly found to level off at VPD values above 1 kPa (e.g. Granier et al. 1996, O'Brien et al. 2004, Phillips et al. 1999). Larger trees showed a higher level of variation in mean J with increasing VPD, while smaller trees had very uniform rates of J (Fig. 4). This could be a result of the more variable VPD conditions in the upper canopy level in comparison to more constant VPD inside the forest canopy. Unfortunately, the environmental parameters used in our study were measured in a forest opening. These values might better reflect conditions in the upper canopy than the microclimatic conditions inside the forest. However, we found a good correlation between J and VPD even with trees from the lower height classes that are most likely exposed to lower levels of VPD.

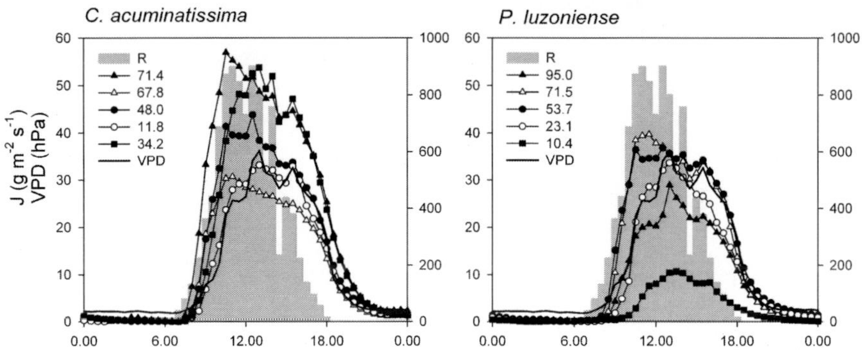

Fig. 5. Diurnal course of solar radiation (R, grey bars), vapor pressure deficit (VPD, solid line), and sap flux density (J, solid lines with symbols) of individual trees from *C. acuminatissima* and *P. luzoniense* during January 9_{th} 2008. The diameter sizes at 1.3 m height (DBH, cm) for each tree are given next to the symbols.

3.3 Are trees from perhumid tropical rainforests more vulnerable to drought?

Global change scenarios commonly expect climate to become more unstable during the 21st century with several regions experiencing more frequent and intense drought events, including the humid tropics of South-east Asia (Hulme and Viner 1998, Timmermann et al. 2004). Although more recent climate change scenarios predict little changes in total precipitation in South-east Asia (IPCC 2007), a future decline in the annual or seasonal amount of rainfall can not entirely be excluded yet. Intensified rainfall seasonality will most likely replace seasonal evergreen forests by more drought-tolerant semi-deciduous forests in the long run (Borchert 1998). On a short-term basis, the response of tropical evergreen trees to severe drought events can include partial leaf shedding, the utilization of stored stem water or the consumption of water from deeper soil layers (Borchert 1998). All these measures can enable tropical trees to maintain their evergreen canopy even under prolonged periods of water limitation, but do, on the other hand, mask how the vitality of tropical trees is dependent on climatic properties.

Changes in leaf area as a result of drought were not reported after an experimental one-year rainfall exclusion in Central Sulawesi, which reduced the incoming precipitation by up to 80% (B. Schuldt, unpublished data). This might allow the conclusion that the foliage in a perhumid environment is less susceptible to drought.

Stem water storage is often estimated from the difference between sap flux near the stem base and at canopy height (Goldstein et al. 1998, Holbrock 1995, Phillips et al. 2003). In this context, Goldstein et al. (1998) and Meinzer (2003) emphasized the role of tree height independent of species in the amount of stored stem water for maximum rates of J when measured near the base of the stem. We found marked differences in the time lag between the increase in radiation and the onset of sap flow in larger compared to smaller trees, which might be interpreted as a tree height effect (Fig. 5). However, the observed patterns have more likely resulted from time lags in the microclimatic changes of R and VPD between the understory, mid-story and upper canopy forest layer over the course of the day. While small and understory trees are shaded by the crowns of larger individuals, tall trees are much earlier exposed to increasing atmospheric demand and incident radiation. However, the atmospheric coupling of J was equally close across all tree height classes in our sample (Fig. 4).

The soils of our study area are characterized by high clay content and a low occurrence of stones (M. Köhler, pers. comm.), hence are rather favorable for deep rooting. Deeper soil water resources might be available for trees of all height classes. Most of the fine root biomass is located in the upper soil horizon in the studied forest, whereas the root densities below 2 m are very low (G. Moser, unpublished data). The role of very deep roots is not known, as are differences in the rooting patterns with tree height. Kume et al. (2007) found the influence of soil water depletion on rates of J a matter of tree size and concomitant differences in the rooting system. Dependent on the soil type and the overall plant water demand, tropical trees are able to build up deep and extensive root systems (Meinzer et al. 1999, Tromp-van Meerveld and McDonnell 2006). For tropical evergreen forests, an average rooting depth of 7.3 m was estimated, with maximum depths of up to 18 m (Canadell et al. 1996). Interestingly, a higher root biomass was found in tropical evergreen than in tropical deciduous or savanna forests in Thailand (Ogawa et al. 1965, Kira et al. 1967).

Extended drought periods should be particularly harmful to tree species of perhumid tropical rainforests with only limited drought experience. This might be especially evident for tall-growing trees. The taller trees in our study showed a higher hydraulic efficiency (i.e., higher rates of k_s and LSC), which allow them to compensate for the longer pathway and concomitant increase in hydraulic resistance. Literature data show that drought-susceptible tree species with a highly efficient hydraulic system are often drought-deciduous and shed their leaves at the onset of a dry period to avoid xylem dysfunction. In contrast, co-existing evergreen species of seasonal tropical forests often have rather narrow vessels, higher vessel densities and lower conductivities, which commonly refer to as drought-tolerant and which allow them to withstand steep pressure gradients without damaging their hydraulic system (Chaot et al. 2005). As a consequence, the tall-growing tropical evergreen species of our study might be rather drought-susceptible. Moreover, tall-growing rainforest trees suddenly threatened by forest fragmentation were found to be more vulnerable than small-stature trees because, among other threats, increased desiccation near forest edges intensified their crown exposure to intense radiation and increased VPD leading to higher drought stress (Kapos 1989, Laurance et al. 2000). Phillips et al. (2009) recently stressed the sensitivity of Amazon rainforests to prolonged water deficits as indicated by elevated tree mortality following the intense drought period in 2005. Increasing dieback of trees with large biomass highly impacts the carbon storage capacity of this forest biome.

4 Conclusions

The trees investigated in our study were very responsive to changes in environmental conditions as indicated by the close relationship found between water uptake and VPD for all tree height classes. While inside the forest, trees from different species converged at similar water uptake rates, taller trees showed a wider range of variation in mean rates of J with increasing VPD. The observed variability might encompass the plasticity of the different species to respond to changes in microclimate. In this context, one might suspect that large moist forest trees might benefit from their physiological plasticity and the xylem anatomical adaptation to tree height, when exposed to prolonged drought. The fact that tall trees generally have to cope with a higher atmospheric evaporative demand compared to the shaded mid- and understory trees could indicate that taller trees might be less vulnerable to drought periods, at least in the short run. Moreover, larger trees will most likely profit from a deeper or at least more extensive root system than smaller trees or saplings to satisfy their water demand. Nevertheless, it has to be kept in mind that natural drought events include atmospheric drought as well as soil water depletion. In our study, we presented the results of species response mainly to changes in atmospheric conditions under well watered soils. In the long run, more frequent and prolonged droughts as predicted for parts of the humid tropics will inevitably induce strong soil water depletion, which might primarily affect the water demanding tall-growing trees. However, because the limits of species-specific and height-related tolerances of tropical moist forest trees to simultaneous atmospheric and pedospheric desiccation are not grasped yet, rate and degree of the concomitant consequences still remain hard to predict. This information is still needed to foresee changes in species composition and tree physiognomy in perhumid tropical rainforests.

Acknowledgements
Collaborative Research Center SFB 552 research program on the stability of rainforest margins (STORMA). We greatly acknowledge financial support by the German Science Foundation (DFG). Sincere thanks to our Indonesian field assistant Atok for invaluable help in tree climbing. We appreciate the data acquisition provided by Sara Brix and Nikolai Brock. We thank Heinz Coners and Hilmar Müller for technical support during the installation, Heike Culmsee for tree species identification and characterization and Wolfram Lorenz for organizing the logistics in Palu. We would like to thank Prof. Dr. Erwin

Beck and one anonymous reviewer for valuable comments on early versions of the manuscript.

References

Andrade JL, Meinzer FC, Goldstein G, Holbrook NM, Cavelier J, Jackson P, Silvera K (1998) Regulation of water flux through trunks, branches, and leaves in trees of a lowland tropical forest. Oecologia 115: 463-471

Becker P (1996) Sap flow in Bornean heath and dipterocarp forest trees during wet and dry periods. Tree Physiology 16: 295-299

Borchert R (1998) Responses of tropical trees to rainfall seasonality and its long-term changes. Climate Change 39: 381-393

Burgess SSO, Pittermann J, Dawson TE (2006) Hydraulic efficiency and safety of branch xylem increases with height in Sequoia sempervirens (D. Don) crowns. Plant Cell and Environment 29: 229-239

Calder IR, Wright IR, Murdiyarso D (1986) A study of evaporation from tropical rain forests - West Java. Journal of Hydrology 89: 13-31

Canadell J, Jackson RB, Ehleringer JR, Mooney HA, Sala OE, Schulze ED (1996) Maximum rooting depth for vegetation types at the global scale. Oecologia 108: 583-595

Choat B, Ball MC, Luly JG, Holtum JAM (2005) Hydraulic architecture of deciduous and evergreen dry rainforest tree species from north-eastern Australia. Trees 19: 305-311

Culmsee H, Moser G, Leuschner Ch, Pitopang R (submitted) Forest aboveground biomass along an altitudinal transect in Sulawesi, Indonesia and the role of extra-tropical Fagaceae. Journal of Biogeography

Delzon S, Sartore M, Burlett R, Dewar R, Loustau D (2004) Hydraulic responses to height growth in maritime pine trees. Plant Cell and Environment 27: 1077-1087

Dietz J, Leuschner C, Hölscher D, Kreilein H (2007) Vertical patterns and duration of surface wetness in an old-growth tropical montane forest, Indonesia. Flora 202: 111-117

Fetcher N, Oberbauer SF, Strain BR (1985) Vegetation effects on microclimate in lowland tropical forest in Costa Rica. International Journal of Biometeorology 29: 145-155

Fetcher N, Oberbauer SF, Chazdon RL (1994) Physiological ecology of trees, shrubs, and herbs at La Selva. In: McDade L.A., Bawa K.S., Hespenheide H.A., and Hartshorn G.S. (Eds.), La Selva: ecology and natural history of a neotropical rainforest. University of Chicago Press, Chicago, p. 128-141

Gebauer T, Horna V, Leuschner Ch (submitted) Atmospheric versus soil water control of sap-flux-scaled transpiration in diffuse- and ring-porous tree species in temperate mixed broad-leaved forests.

Goff JA, Gratch S (1946) Low-pressure properties of water from -160 to 212 °F. In: 52nd annual meeting of the American Society of Heating and Ventilating Engineers. Transactions of the American Society of Heating and Ventilating Engineers. New York, p. 95-122

Goldstein G, Andrade JL, Meinzer FC, Holbrook NM, Cavelier J, Jackson P, Celis A (1998) Stem water storage and diurnal patterns of water use in tropical forest canopy trees. Plant Cell and Environment 21: 397-406

Granier A (1987) Evaluation of transpiration in a Douglas-fir stand by means of sap flow measurements. Tree Physiology 3: 309-319

Granier A, Huc R, Colin F (1992) Transpiration and stomatal conductance of two rain forest species growing in plantations (Simarouba amara and Goupia glabra) in French Guyana. Annals of Forest Science 49: 17-24

Granier A, Huc R, Barigah ST (1996) Transpiration of natural rain forest and its dependence on climatic factors. Agriculture and Forest Meteorology 78: 19-29

Holbrook NM (1995) Stem water storage. In: Gartner B.L. (Ed.), Plant stems: physiology and functional morphology. Academic Press, San Diego, p. 151-174

Holdridge LR, Grenke WC, Hatheway WH, Liang T, Tosi JA (1971) Forest Environments in Tropical Life Zones: A Pilot Study, Pergamon Press, Oxford

Hulme M, Viner D (1998) A climate change scenario for the tropics. Climatic Change 39: 145-176

IPCC (2007) Contribution of Working Groups I, II and III to the Fourth Assessment Report of the Intergovernmental Panel on Climate Change. Pachauri RK and Reisinger A (Eds.), IPCC, Geneva, Switzerland, pp. 104

Kapos VJ (1989) Effects of isolation on the water status of forest patches in the Brazilian Amazon. Journal of Forest Ecology 5: 173-185

Kira T, Ogawa H, Yoda K, Ogino K (1967) Comparative ecological studies on three main types of forest vegetation in Thailand. 4 Dry matter production. Nat. Life S.E. Asia 5: 149-174

Koch GW, Sillett SC, Jennings GM, Davis SD (2004) The limits to tree height. Nature 428: 851-854

Kumagai T, Saitoh TM, Sato Y, Takahashi H, Manfroi OJ, Morooka T, Kuraji K, Suzuki M, Yasunari T, Komatsu H (2005) Annual water balance and seasonality of evapo-transpiration in a Bornean tropical rainforest. Agricultural and Forest Meteorology 128: 81-92

Kume T, Takizawa H, YoshifuJi N, Tanaka K, Tantasirin C, Tanaka N, Suzuki M (2007) Impact of soil drought on sap flow and water status of evergreen trees in a tropical monsoon forest in northern Thailand. Forest Ecology and Management 238: 220-230

Larcher W (1998) Physiological plant ecology. 4th edition, Springer-Verlag, Berlin Heidelberg, Germany, pp. 513

Laurance WF, Delamonica P, Laurance SG, Vasconcelos HL, LoveJoy TE (2000) Rainforest fragmentation kills big trees. Nature 404: 836

Leopoldo PR, Franken WK, Villa Nova NA (1995) Real evapotranspiration and transpiration through a tropical rain forest in central Amazonia as estimated by the water balance method. Forest Ecology and Management 73: 185-195

Lösch R (2001) Wasserhaushalt der Pflanze. Quelle & Meyer Verlag, Wiebelsheim, Germany, pp. 585

McDowell N, Barnard H, Bond BJ, Hinckley T, Hubbard RM, Ihsii H, Köstner B, Magnani F, Marshall JD, Meinzer FC, Phillips N, Ryan MG, Whitehead D (2002) The relationship between tree height and leaf area: sapwood area ratio. Oecologia 132: 12-20

Meinzer FC, Andrade JL, Goldsetin G, Holbrook NM, Cavelier J, Wright JS (1999) Partitioning of soil water among canopy trees in a seasonally dry forest. Oecologia 121: 293-301

Meinzer FC, Goldstein G, Andrade JL (2001) Regulation of water flux through tropical forest canopy trees: Do universal rules apply? Tree Physiology 21: 19-26

Meinzer FC (2003) Functional convergence in plant response to the environment. Oecologia 134: 1-11

Meinzer FC, Bond BJ, Warren JM, Woodruff DR (2005) Does water transport scale universally with tree size? Functional Ecololgy 19: 558-565

Mencuccini M, Grace J (1995) Climate influences the leaf area / sapwood area ratio in Scots pine. Tree Physiology 15: 1-10

O'Brien JJ, Oberbauer SF, Clark DB (2004) Whole tree xylem sap flow response to multiple environmental variables in a wet tropical forest. Plant Cell and Environment 27: 551-567

Ogawa H, Yoda K, Ogino K, Kira T (1965) Comparative ecological studies on three main forest types in Thailand. 2. Plant biomass. Nature and Life in S.E. Asia 4: 49-80

Oren R, Zimmermann R, Terborgh J (1996) Transpiration in upper Amazonian floodplain and upland forests in response to drought-breaking rains. Ecology 77: 968-973

Phillips N, Oren R, Zimmermann R, Wright SJ (1999) Temporal patterns of water flux in trees and lianas in a Panamanian moist forest. Trees 14: 116-123

Phillips N, Bond BJ, Ryan MG (2001) Gas exchange and hydraulic properties in the crowns of two tree species in a Panamanian moist forest. Trees 15: 123-130

Phillips N, Ryan MG, Bond BJ, McDowell NG, Hinckley TM, ermk J (2003) Reliance on stored water increases with tree size in three species in the Pacific Northwest. Tree Physiology 23: 237-245

Phillips OL, Aragao LEOC, Lewis SL et al. (2009) Drought sensitivity of Amazon rainforest. Science 323: 1344-1347

Pothier D, Margolis HA, Waring RH (1989) Patterns of change in saturated sapwood permeability and sapwood conductance with stand development. Canadian Journal of Forest Research 19: 432-439

Ryan MJ, Yoder BJ (1997) Hydraulic limits to tree height and tree growth. Bioscience 47: 235-242

Sperry JS, Donnelly JR, Tyree MT (1988) A method for measuring hydraulic conductivity and embolism in xylem. Plant Cell and Environment 11: 35-40

Timmermann A, Justino F, Jin FF, Goosse H (2004) Surface temperature control in the North and tropical Pacific during the last glacial maximum. Climate Dynamics 23: 353-370

Tromp-van Meerveld, McDonnell JJ (2006) On the interrelations between topography, soil depth, soil moisture, transpiration rates and species distribution at the hillslope scale. Advances in Water Resources 29: 293-310

Tyree MT, Ewers FW (1991) The hydraulic architecture of trees and other woody plants. New Phytologist 119: 345-360

Vanninen P, Ylitalo H, Sieväinen R, Mäkelä A (1996) Effects of age and site quality on the distribution of biomass in Scots pine (*Pinus silvestris* L.). Trees 10: 231-238

Waring RH, Schroeder PE, Oren R (1982) Application of the pipe model theory to predict canopy leaf area. Canadian Journal of Forest Research 12: 556-560

West GB, Brown JH, Enquist BJ (1999) A general model for the structure and allometry of plant vascular systems. Nature 400: 664-667

White D, Beadle C, Worlegde D, Honeysett J, Cherry M (1998) The influence of drought on the relationship between leaf and conducting sapwood area in *Eucalyptus globulus* and *Eucalyptus nitens*. Trees 12: 406-414

Williams M, Bond BJ, Ryan MG (2001) Evaluating different soil and plant hydraulic constraints on tree function using a model and sap flow data from ponderosa pine. Plant Cell and Environment 24: 679-690

Woodruff DR, McCulloh KA, Warren JM, Meinzer FC, Lachenbruch B (2007) Impacts of tree height on leaf hydraulic architecture and stomatal control in Douglas fir. Plant Cell and Environment 30: 559-569

Wullschleger SD, Hanson PJ, Todd DE (2001) Transpiration from a multi-species deciduous forest as estimated by xylem sap flow techniques. Forest Ecology and Management 143: 205-213

Zeppel MJB, Murray BR, Barton C, Eamus D (2004) Seasonal responses of xylem sap velocity to D and solar radiation during drought in a stand of native trees in temperate Australia. Functional Plant Biology 31: 461-470

Zimmermann MH (1978) Hydraulic architecture of some diffuse-porous trees. Canadian Journal of Botany 56: 2286-2295

Zach A, Schuldt B, Brix S, Horna V, Culmsee H, Leuschner C (in press) Vessel diameter and xylem hydraulic conductivity increase with tree height in tropical rainforest trees in Sulawesi, Indonesia. Flora

Part III

Integrated concepts of land use in tropical landscapes

Principle and practice of the buffer zone in biosphere reserves: from global to local – general perspective from managers versus local perspective from villagers in Central Sulawesi, Indonesia

Marion Mehring* and Susanne Stoll-Kleemann

Applied Geography and Sustainability Science, GoBi (Governance of Biodiversity) Research Group, Institute of Geography and Geology, Ernst-Moritz-Arndt-University Greifswald, Jahnstr. 16, 17487 Greifswald, Germany

*corresponding author: Marion Mehring, email: marion.mehring@uni-greifswald.de

Summary

The establishment of buffer zones as part of protected areas – within biosphere reserves but also national parks – is central to conservation strategies designed to safeguard remaining biodiversity, mainly left in tropical forests. However, the real-world situation of protected areas reveals a range of problems. This study seeks to identify the diverse ways of thinking among relevant actors in order to address these issues.

In the present work, characteristics of the global perspective of managers are contrasted with findings from a field study in the forest biosphere reserve Lore Lindu, Central Sulawesi, Indonesia. In this context, the differences between the villagers' point of view regarding the implementation of sustainable-development activities in the buffer zones and the managers' perspective on the sustainability of human uses in the same areas are highlighted. This paper follows two complementary approaches, combining quantitative and qualitative research tools when examining the respective positions of both managers and villagers.

Quantitative analysis of a global survey among biosphere reserve managers revealed that human uses in the buffer zones are generally considered to be sustainable (mean value for evaluation of 6.4 on a scale from 1 to 10). Besides a significant negative correlation (-0.3) between *number of residents in the*

T. Tscharntke et al. (eds.), *Tropical Rainforests and Agroforests under Global Change*, Environmental Science and Engineering, 413–429, DOI 10.1007/978-3-642-00493-3_19,

buffer zone and *accordance to sustainability of human uses*, no meaningful relation was computed either in respect to country income or tropical forest regions.

Because of this lack of conclusive data, we zoomed in on the local level and included the villagers' perspective. Factors influencing the implementation of sustainable-development activities in the buffer zone were the central issue of the field study. A completely different perception was found in Lore Lindu. The villagers expressed the opinion that basic management activities such as *nonexistence of boundary demarcation* and *lack of law enforcement* are the most crucial factors in regard to how sustainable-development activities are implemented. Further barriers are *insufficient awareness of forest conservation needs among groups of villagers* and the ever-increasing *number of immigrants.* Furthermore, the results of this study point to the high relevance of a mediator in the communication process between the management body and local people. Clearly formulated rules and laws and abidance by enforcement of the law seem to play a major role in buffer-zone management. Moreover, the different ethnic groups have to be taken into account by the park management in communication processes. Local people should be considered as partners (in buffer-zone management) rather than as passive objects in this process.

Keywords: biosphere reserve, Seville Strategy, buffer zone, protected area management, law enforcement, boundary demarcation, Central Sulawesi, Indonesia

1 Introduction

Degradation, conversion, and fragmentation threaten the integrity of forested ecosystems worldwide (Achard et al. 2002; Nepstad et al. 1999), with South-East Asia showing the highest rates of deforestation of any major tropical forest region (Sodhi et al. 2004). Developing regions and tropical countries are especially threatened (FAO 2007). Hence in recent years, the establishment of buffer zones as part of protected areas – within biosphere reserves but also national parks – has emerged as a favored nature-conservation tool to conserve the world's remaining biodiversity in tropical forests. However, its implementation is a challenging task that often raises complex challenges at the local level such as rejection by local communities or even conflicts between competing parties. A sound understanding of the diverse ways of thinking among relevant actors is needed in order to address these issues and to better understand current dynamics in land-use change.

The present paper contrasts Biosphere Reserve (BR) managers' perspective on the buffer zone at the global level with villagers' points of view on the challenges raised by its local implementation and management in the context of Lore Lindu BR in Central Sulawesi, Indonesia.

Located on the island of Sulawesi, Lore Lindu BR covers a wide range of vegetation types and is representative of the unique fauna of Sulawesi. Sulawesi is well known as a biogeographical wonder due to its location to the east of Wallace's line and proximity to the Sunda plate (TNC 2002). Its formation and isolation have resulted in a mix of Oriental and Australian fauna with a higher rate of endemic taxa than any other Indonesian island. Most of these endemics are dependent on forest (TNC 2002). Due to its uniqueness, Lore Lindu was recognized as a UNESCO Biosphere Reserve in 1977 and has been nominated as World Heritage Site for its astonishing cultural heritage of ancient stone megaliths from 1300 AD. Since 1993, the core area of the Lore Lindu BR - covering an area of some 200,000 hectares (Shohibuddin 2006) - has been designated a National Park under authority of Lore Lindu National Park management. The buffer zone of the Lore Lindu BR consists of 140 villages of which 64 (with approximately 40,000 inhabitants) are located next to the border of the core area. This buffer zone matches the National Park buffer zone. The transition area has not yet been specified. (Widagdo 2008)

According to the Seville Strategy[1] (UNESCO 1995), BRs should be utilized as models of land management and approaches to sustainable development (goal II) primarily to be implemented in the buffer zone. Hence, the focus of the presented study is the buffer zone of BRs as well as its management and implementation. In order to specify the research objective, two main research questions are raised:

From the managers' perspective: (1) To what extent are human uses in the buffer zone sustainable in such a way that the buffer zone fulfills its target as a protection zone against human influences for the core zone?

To implement goal II on the local BR level, various recommendations for action at the individual reserve level are given in the Seville Strategy, such as developing incentives for the conservation and sustainable use of natural resources and providing alternative means of livelihood for local populations when existing activities are limited or prohibited within the biosphere reserve (UNESCO 1995). This recommended action on the way to implementing sustainable development was investigated in a case study at the Lore Lindu BR, Indonesia. In this context the second research question is raised:

From the villagers' perspective: (2) In the case of the Lore Lindu BR, what influences the implementation of activities for sustainable land use in the buffer zone?

This study follows two complementary approaches. Applying quantitative research tools, a global survey among BR managers was performed to evaluate buffer zones. Adopting qualitative research tools, a case study in the Lore Lindu forest BR was undertaken to investigate the villagers' perspective on implementing sustainable-development activities in the buffer zone.

[1] For background information and detailed explanation, see next chapter *Biosphere Reserve Concept.*

2 Biosphere Reserve Concept

"Biosphere reserves are areas of terrestrial and coastal/marine ecosystems or a combination thereof, which are internationally recognized within the framework of UNESCO's program on Man and the Biosphere (MaB)". (UNESCO 1995:2)

The idea of setting up biosphere reserves resulted from the UNESCO (United Nations Educational, Scientific and Cultural Organization) Conference on Rational Use and Conservation of the Resources of the Biosphere in 1968. The *Man and the Biosphere* (MaB) Program, which was a direct result of this conference, was the first deliberate international effort to identify ways and means of achieving sustainable development of terrestrial ecosystems (Batisse 1993). The BR concept itself was initiated in 1974 with the ambitious idea of reconciling nature conservation with human needs. Worldwide 553 BRs (as of May 2009) form the World Network of Biosphere Reserves (WNBR), which seeks to promote exchange of information and experience. Basically, BRs have three major functions: (1) **conservation** of biological and cultural diversity, (2) **logistic support** through participation in (international) research and monitoring, and (3) **development** in terms of cooperation with local populations living in or around the biosphere reserves to promote sustainable development compatible with conservation objectives.

One way of realizing these basic functions is through zoning BRs in **core areas** (strictly protected areas), **buffer zones** (delimited for management purposes on sustainable land use), and **transition areas** (focus of cooperation with local stakeholders) (Figure 1).

In practice, however, many of the BRs designated in the first decade (before 1995) of the MaB Program (pre-Seville BRs) were selected according to their relevance in regard to biological conservation and potential research interests. Thus, most of these sites had already been declared national parks or equivalent areas where research activities and management facilities could be used or enhanced.

Bearing this in mind, two further milestones in refining and promoting the concept have been accomplished through the International Conferences on Biosphere Reserves in Seville (1995) and Madrid (2008). The conferences drew up the Seville Strategy and the Madrid Action Plan (MAP), respectively. The conference in Seville was organized to enable an evaluation and reflection on the role of BRs in the 21st century (UNESCO 1995) and started a new era for the WNBR. In this context, four goals of major priority are distinguished: (1) to **conserve** natural and cultural diversity, (2) to utilize BRs as models of land management and as approach for **sustainable development**, (3) to use BRs for **research, monitoring, and education**, and (4) to **implement** the BR concept (UNESCO 1995).

The MAP builds on the Seville Strategy and aims to capitalize on its instruments (UNESCO 2008). In attempting to re-orient the MaB and the WNBR towards the challenges of an ongoing global change, the MAP de-

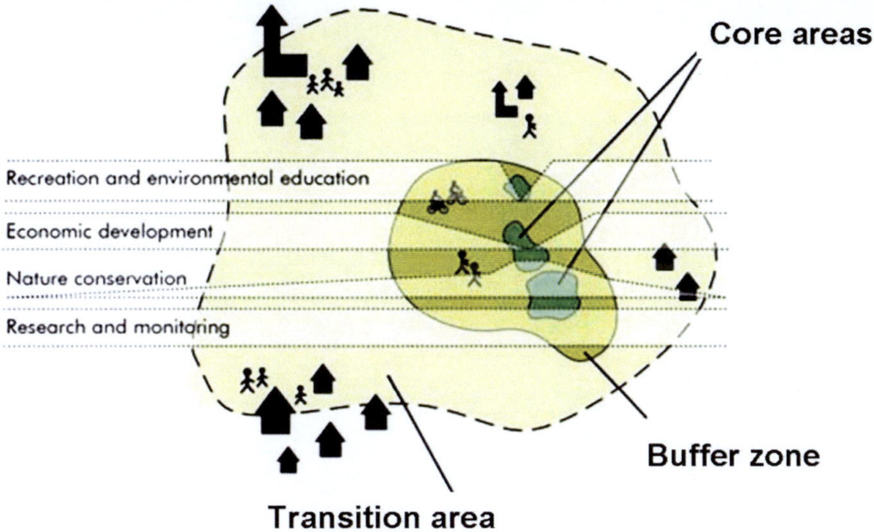

Fig. 1. Biosphere reserve zonation in core area, buffer zone, and transition area (Lange 2005, modified).

fines four main actions: (1) cooperation, management, and communication, (2) zonation – linking functions to space, (3) science and capacity enhancement, and (4) partnership. Actions are to be taken at the local, national, and international level. These documents emphasize participatory management approaches as essential efforts to achieve the above-mentioned targets. Generally, BRs can be seen as forums to integrate people into decision making and generate new ideas for implementing sustainable development. Currently, Indonesia has six BRs that represent various biomes[2] (Figure 2).

3 Research Methods

The study employed a triangulated approach combining quantitative and qualitative research tools to assure validity. Goal II of the Seville Strategy (utilize BRs as models of land management and as approaches for sustainable development), particularly as they are implemented in buffer zones, is the focus of this study. We began the research with a Global Survey on Biosphere Reserve Management. This survey was performed by the Governance of Biodiversity

[2] As of February 2009. On 26 May 2009 the MaB decided to add new sites to the WNBR. One of the new sites includes Giam Siak Kecil – Bukit Batu on the island of Sumatra, Indonesia.

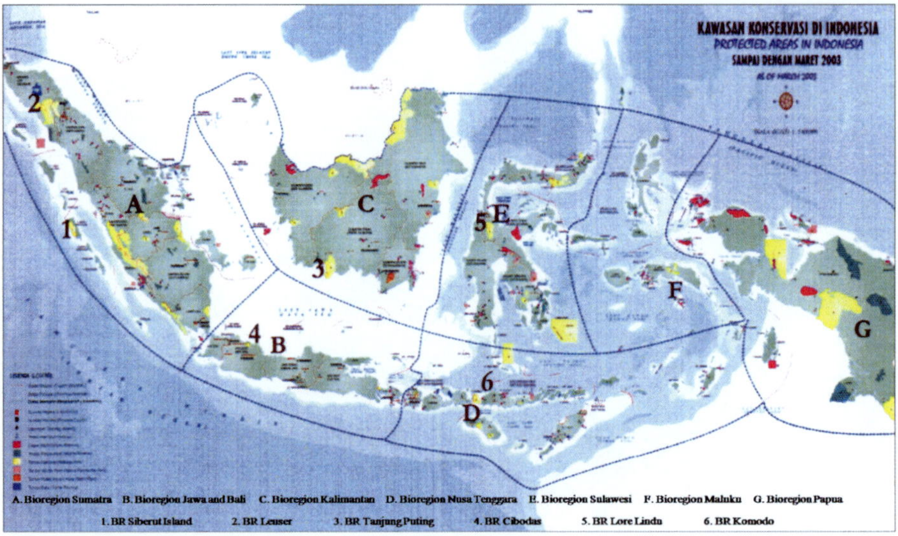

Fig. 2. Location of biosphere reserves in Indonesia (Source: National MAB committee Indonesia, 2008).

(GoBi) Research Project. Among other things, managers were asked to evaluate the buffer zone vis à vis goal II, i.e. to function as protector of the core zone, as related to, e.g., extensive land uses by humans.

To implement goal II at the individual reserve level, specific objectives are recommended. In this context, one tool to ensure sustainable land use is "(sic.) developing incentives for the conservation and sustainable use of natural resources, and (..) alternative means of livelihood for local populations, when existing activities are limited or prohibited within the biosphere reserve" (UNESCO 1995: 6). To assess this type of activity at the local reserve level, a case study was undertaken in the Lore Lindu forest BR. In group discussions, the attitudes of local farmers living in the buffer zone toward factors influencing the implementation of such activities were examined.

3.1 Global Survey

A global telephone survey of BR managers was done from July to December 2006. All elements of the professional population were included, and in total, 225 individuals from 79 countries were interviewed, resulting in an overall response rate of 42% (as of February 2009). The interviews were compiled in a database.

Following a standardized questionnaire, BR managers were asked to express their opinion on the statement "*Human uses inside the buffer zone are sustainable from the point of view of conservation*" as a number. This statement is themed on UNESCO's first definition of the buffer zone (UNESCO

1984) which stipulated that this area of a biosphere reserve may include a larger area where cooperative efforts are emphasized to ensure that its uses are managed in a manner compatible with the conservation and research functions of the biosphere reserve. A metric scale from one to ten was used where "1" denoted total disagreement and "10" absolute agreement. Furthermore, each interviewee was asked about the number of people living inside the BR, as there is a lack of reliable data on BRs in general and in developing countries in particular (cf. Bertzky and Stoll-Kleemann 2009). Due to this prevailing scarcity of data on the BR level, it was not possible to use indicators such as population density or carrying capacity of the buffer zone.

The data used for analysis (number of residents living inside the BR) has to be interpreted as the best available on the level of individual BRs. Regarding the acquisition of additional information concerning ecosystem type, see Mehring and Stoll-Kleemann (2008). A classification into high income and non-high income countries was defined in accordance with the World Bank's gross-national-income-per-capita indicator (World Bank 2007). For specification of climate zones, the classification scheme from UNESCO's MaB Program for individual BRs (see: www.unesco.org/mab) was adopted.

Statistical analysis of the data obtained was performed using SPSS 17.0. A two-tailed t-test was carried out to analyze differences in the mean values of answers on sustainable land use in the buffer zone. Discriminators employed were ecosystem type (forest/non-forest BRs), climate zone (tropical/non-tropical forest BRs), and income status (BRs in high/non-high income countries). As the data was not normally distributed, the Spearman correlation coefficient was computed to analyze the direction and strength of relation between the number of residents and given answers concerning sustainability of human uses inside the buffer zone.

3.2 Field Study

To analyse how sustainable-development activities are perceived by local farmers, a case study was conducted in Lore Lindu applying a qualitative social research approach. The overall objective of this exploratory study was to gain deeper insight into processes and influences of the implementation of sustainable-development activities in buffer zones from the villagers' perspective.

Seven group discussions were conducted with a total of 31 individuals between March and May 2008. A purposive sampling design was applied. The target population of the groups was 1) farmers living in the buffer zone, 2) mixed gender and age composite, and 3) different education levels. Two groups were interviewed per village. Individuals in the villages were chosen by the mayor according to the criteria mentioned. Group size varied from three to eight people. The discussions were conducted with 4 women and 27 men, ranging in age from 20 to 80 years.

Four contrasting villages were chosen with regard to sustainable-development activities undertaken by non-governmental organizations (NGO[3])[4] and the location of the villages within the buffer zone.

Group discussions were based on a semi-structured interview guideline that covered the four goals of the Seville Strategy. This paper discusses only the replies related to goal II of implementing sustainable-development activities. The enquiry included the question of whether the respondents had personally taken part in any of the relevant activities in the village. There was also a discussion of the focus of the activities, the implementing organization (BR management, NGO, government), and the farmers' attitude toward what had been done. The discussions were conducted in the national language (Bahasa Indonesian), with local assistants providing simultaneous translation. Subsequently, the interviews were transcribed by native speakers and translated into English. Applying the content-analysis approach (Punch 2005), computer-based analysis with ATLAS.ti 5.0 and coding of interviews led to the relevant factors.

4 Results

Statistical results of the telephone interviews with BR managers concerning evaluation of the buffer zone are presented below. Next, the villagers' perspective on the implementation of sustainable-development activities in the buffer zone of the Lore Lindu forest BR is described. Furthermore, brief references to particular group discussions are offered where they help explain the findings.

4.1 Global Survey: Management Perspective

Overall, BR managers are in accordance with the statement "*Human uses inside the buffer zone are sustainable from the point of view of conservation*", as expressed by the mean value of 6.4 (see Table 1). No statistically significant difference for given answers was observed with regard to country income, ecosystem type forest, or in combination with tropical climate zone. However, replies on human uses in the buffer zone manifested an inverse correlation to the number of residents of particular BRs: the higher the number of residents in the buffer zone, the lower the value for sustainability of human uses. A significant negative correlation (-0.3) between the two variables emerged.

[3] TNC: The Nature Conservancy (international NGO); YTM: *Yayasan Tanah Merdeka* (the "Independent Earth Foundation" - a national NGO); KARSA ("Initiate" - another national NGO)

[4] NGOs were the first actors to start activities toward sustainable development. Unfortunately, it was not possible to get detailed information about the other two actors (regional government and park management).

Table 1. Valuation of the statement '*Human uses inside the buffer zone are sustainable from the point of view of conservation*'(1: I don't agree, 10: I fully agree). Mean values and correlation coefficient between given answers and number of residents are indicated.

		country income		ecosystem type		climate zone		
	All BRs (n=180)	high income (n=63)	non-high income (n=117)	forest (n=143)	non-forest (n=37)	tropical forest (n=54)	non-tropical forest (n=89)	number of residents
Arithmetic mean value with standard error	6.4 ± 0.2	6.5 ± 0.3	6.4 ± 0.2	6.3 ± 0.2	6.9 ± 0.4	6.2 ± 0.4	6.4 ± 0.3	–
Spearman correlation coefficient	–		–		–		–	-0.3*

* significance level alpha: 0.01

4.2 Field Study: Villagers' Perspective

As a result of the group discussions with farmers residing in the buffer zone of the Lore Lindu forest BR, the local perspective in Central Sulawesi, Indonesia can be presented (see Table 2 and Table 3). We distinguish between positive and negative influences on implementing activities related to sustainable development in the buffer zone in the context of the point of view of the village, internal (villagers), and external actors (NGO, government, and BR management).

It is noteworthy that none of the villagers interviewed associated any positive influence whatsoever with the external actors (NGOs, government, or BR management) who undertook activities toward sustainable development in the villages (see Table 2). However, positive values were attributed to activities of the BR management when they had been implemented in cooperation with the village conservation council (LKD: *Lembaga Konservasi Desa*; see Table 3).

The high relevance of cooperation with this actor becomes clear in this excerpt from a group discussion: *"The one who takes a role here is LKD and the (national park) office." (quotation 1:15)*[5]

Also, in terms of regulations, the village conservation council has strong influence on the success of its implementation: *"So, the public around here is*

[5] Quotations have been retrieved from the hermeneutic unit of ATLAS.ti and are available upon request.

Table 2. Factors and actors influencing the implementation of sustainable development activities in the buffer zone of the Lore Lindu Biosphere Reserve.

Influence on implementation of activities	Internal actor villagers	External actor NGO Government BR management
Positive	• Improved economic situation for people (since cultivation of new land) • Enforcement of law and traditional land-use regulations (rotation system) by themselves • Environmental education program at school by themselves • Awareness of forest conservation (through experience related to flood protection)	
Negative	• Not enough awareness of forest conservation among groups of villagers • Number of immigrants still increasing • Little financial capital • Bad marketing situation (standard price wanted)	• Unclear/lack of boundary demarcation and policy • Lack of law enforcement and compliance with the law • Lack of participation and consideration of traditional knowledge during establishment of National Park • Lack of explanation to the people about National Park (program, purpose) and buffer zone

absolutely not coming to the (core) area because LKD controls it every day." (quotation 1:20)

Furthermore, negative attributes mentioned by the interviewees in regard to external actors are related to basic management activities. One of the most prevalent issues mentioned in all group discussions was *the unclear/lack of boundary demarcation*, as exemplified in a farmer's statement: *"The whole area has no demarcation, so the people come and go there as they want. So there's no agreement between the office and the people about the border of the National Park." (quotation 6:5)*

Table 3. Positive influences on implementation of sustainable-development activities of BR management in cooperation with LKD (local organization), NGOs, and government.

- Agro-forestry activity by National Park, NGO, government, and LKD
- Living boundary activity by National Park, NGO, and LKD for border demarcation
- Law enforcement and controlling by LKD related to border crossing
- Explanation to the people about National Park (program, purpose) together with NGO and LKD

The second crucial factor is the *lack of law enforcement and the failure of the BR staff to abide by the law.* As another local farmer from a different village stated: *"There must be a law, it must be followed and it must be implemented (...). There are documents, but who is going to implement them? Nobody." (quotation 7:19)*

The positive influences mentioned include the improved economic situation in the region (e.g. since cultivation of new land) and factors of the villagers' own initiative such as environmental education programs at school and enforcement of the traditional rotation system applied by villagers.

However, two alarming barriers to implementing activities toward the goal of sustainable development are the *ever-increasing number of immigrants* and the *loss of forest conservation awareness among this group* (quotation 2:4). Another farmer from a different village also mentioned the increasing number of inhabitants as a major problem: *"The destruction of the National Park is caused by the amount of the inhabitants, and it causes the forest destruction like now. Our parents before never destroyed the forest. The impact of the forest destruction could be felt now, silting up of the lake happens. The lake will flatten out four meter if the forest destruction continues." (quotation 7:12)*

5 Discussion

In recent years, the establishment of buffer zones has been used as a part of larger integrated conservation development programs to provide the benefits of ecological buffering of protected areas and socio-economic buffering of neighboring communities.

Our global analysis of buffer zones of biosphere reserves illustrates that according to the managers' answers, human uses inside the buffer zone are sustainable from the point of view of conservation. In their experience, buffer zones achieve their target in buffering the protected core zone. Assuming that national income is related to the income of the farmers living in the vicinity of the BR, we take national income as a proxy for the influence of the farmers' income on sustainable uses in the buffer zone. In our study no such influence can be observed, as no difference was calculated between high income and non-high income countries. Furthermore, differences could neither be detected between different ecosystem types (forest/non-forest) nor climate zones (tropical forests/non-tropical forests). Hence, in our investigation, the success of sustainability of human uses in buffer zones of BRs is not related to the natural resources the villagers depend on, such as forest, or to forests in different climate zones, such as tropical forests that are known to be most threatened (FAO 2007). This finding indicates that BRs with the typical zonation of core, buffer, and transition zone generally seem to be an appropriate instrument in terms of natural (forest) resource conservation.

Stoll-Kleemann (2005) also showed that this model of zonation concerning resource use is the way to go because of its psychological advantage of including instead of excluding local people. The relation concerning number of residents and buffer zone underline the findings from Wittemyer et al. (2008), who highlight a looming threat to protected area effectiveness and biodiversity conservation due to observed higher population growth rates on the borders of protected areas compared to average rural growth. In our study this phenomenon does not seem to be a particular problem in developing countries, as no correlation with income per capita was discovered. However, overall no meaningful differences have been detected among the managers' answers on human uses in the buffer zone. We believe this is due to a combination of two different reasons. On the one hand, the method of using statements rather than asking open questions might simplify issues by restricting answers, and it may also prevent encompassing the complexity of the specific situation. On the other hand, the managers' perception might be different compared to other perspectives from outside the management, such that of as people living in the BR. To complement the managers' perspective, we furthermore discuss the villagers' point of view on factors influencing the implementation of activities devoted to sustainable development in the buffer zone of the Lore Lindu BR.

From the field study in Lore Lindu, good experience was gained through engaging local people to express their opinion. In terms of methodology, group discussions are a beneficial tool to involve local communities and to discover their perspective (Kumar 2006). However, all of the different (e.g. ethnic) groups within a village must be taken into consideration to reflect the true diversity of the villagers and their opinions. Beneficial to the investigation in Lore Lindu was the researcher's independent position vis à vis the biosphere reserve management.

Group discussions with the farmers highlight another perspective. None of the villagers associated any positive influence with sustainable-development activities undertaken by external actors such as NGOs, government, or BR management. Moreover, the most severe problems from the villagers' perspective are fundamental ones such as *lack of boundary demarcation* and *lack of law enforcement*. This result corresponds to Bruner et al. (2001) who identified enforcement and boundary demarcation to be important basic management activities correlating with park effectiveness. Overall, our analysis points out a gap between managers and villagers in the perception of buffer-zone functioning. Mangers hold that human uses are sustainable, while the farmers in Lore Lindu criticize basic management activities on the way to implementing sustainable development activities. We interpret this finding as being due to constrained communication and cooperation between the two parties, as the farmers criticize the *lack of explanation to the people about the purpose and programs of the park and buffer zone*. Furthermore, the management body (head and staff) are regularly (about every other year) moved from one park to another by decisions of the Indonesian Ministry of Forestry to counteract corruption[6]. This also complicates the communication process between management and the local people as it takes a long time for the newly assigned officials to become familiar with the local conditions. In the case of Lore Lindu, most of the villages are remote and only accessible with difficulty.

However, in the same way, our results from Lore Lindu show the high relevance of a mediator. During the interviews, activities from BR management were only positively valued when implemented in cooperation with third parties such as the village conservation council (LKD: *Lembaga Konservasi Desa*). This local council is authorized and accepted by both local farmers and BR management. This mediator can help in situations where cooperation and communication between two parties such as management and local people are hindered. The high conservation potential of regulations monitored by the village conservation council (LKD) is also highlighted in Schwarze et al. (2007), who investigated the link between poverty and collection of forest products in Lore Lindu. The high demand among farmers for regulations and their enforcement become evident from the above quotation (7:19). Examples of facilitating high-quality community participation, where different skills and methods are required, are listed in the study by Stoll-Kleemann and Welp (2008) illustrating various methods of participatory management.

Good communication is particularly important for Lore Lindu as in-migration still plays a major role in this region. The *ever-increasing number of immigrants* and the loss *of forest conservation awareness among this group* have to be taken into consideration to stop forest encroachment and further ensure the integrity of the park. From the local farmers' perspective in Lore Lindu, ongoing in-migration has the most severe impact. This is in line with Faust et al. (2003) who pointed out that spontaneous migration in the Lore Lindu area, which is driven by economic push and pull factors and modified by socio-economic, political, and institutional factors, can have problematic effects with regard to the environment. Furthermore, there is evidence from a meta-analysis on acceptance of protected areas in South East Asia that the most significant factor ascertained is the traditional tie to the ecological environment and identification with it (Richter 2008). Migrants in Lore Lindu

[6] Personal comment from interview with head of management.

are known to be economically strong because they introduce technologies and innovations to the area (Faust et al. 2003). In parallel, though, they appear to lose the relation to the forest and awareness of its conservation needs. This is an important factor for Lore Lindu and has to be considered within the management of the park through e.g. different participation strategies for singular ethnic groups.

Although findings from the case study in Lore Lindu cannot be generalized for other regions, it has become apparent that involvement at the local community level is indispensable for successful buffer zone management, just as Stoll-Kleemann and Welp (2008) showed that participation leads to increased social acceptance of biosphere reserves, and in addition to increased conservation success. Furthermore, Fritz-Vietta and Stoll-Kleemann (2008) conclude from a case study in Madagascar that community participation and incorporation of external actors (in the form of consultants) are the most important success factors for BR management. This has to be considered when managing biosphere reserves in particular, but also protected areas in general. The involvement of local people is an important issue, and they can no longer be considered passive objects in the process.

6 Conclusion

The second goal of the Seville Strategy identifies BRs in the function of models in the realm of land management and approaches to sustainable development (UNESCO 1995). This goal should primarily be realized in the buffer zone.

In the results of the study described in the present work, contrasting the perspective of the managers queried with that of the local villagers in Lore Lindu revealed a clear disparity between the respective points of view on sustainable development in the buffer zone. Our analysis indicates that in the managers' opinions, there is no difference between high-income and non-high-income countries in regard to sustainability of human uses in buffer zones; in both categories, sustainability is ranked as moderate.

From the perspective of the Lore Lindu villagers, as an example of a pre-Seville biosphere reserve, basic management activities like zonation, law enforcement, but also cooperation (lack of participation) and communication are the most important constraints blocking the way to successful implementation of activities for sustainable development. These are also factors the MAP explicitly refers to. Characteristic for pre-Seville BRs (Batisse 1993), Lore Lindu was established without involvement of local people, resulting in conflicts between management and them.

The results of this study also point to the great relevance of involving a mediator such as the village conservation council (LKD) in the communication process. And given the strong influence of migrants, the clear formulation and enforcement of rules and laws by organizations (e.g. LKD) play a major role and are of high value, the exact nature of which needs to be investigated in more detail.

Finally, referring to the Seville Strategy and Madrid Action Plan, our results suggest that biosphere reserves are built upon a concept that offers reasonable solution to addressing difficulties arising in the face of global change. However, success depends on implementation of the Seville Strategy and its acceptance at the local level. Furthermore, detailed studies are essential to explore how the other three goals of the Seville Strategy have been implemented, such as the realization of the BR concept on the national level, to find level-specific recommendations in order to achieve successful BR management.

Acknowledgements

This study was performed in framework of the GoBi (Governance of Biodiversity) research team, which is funded by the Robert Bosch Foundation. We would like to thank our cooperation partner STORMA (Stability of Rainforest Margins in Indonesia, University of Göttingen) for supporting the research in Lore Lindu, Sulawesi. Special thanks go to the GoBi research team for conducting the global survey and to local assistants in Indonesia for transcription, translation, and field assistance.

References

Achard F, Eva HD, Stibig H-J, Mayaux P, Gallego J, Richards T, Malingreau J-P (2002) Determination of Deforestation Rates of the World's Humid Tropical Forests. Science 297: 999-1002

Batisse M (1993) Biosphere reserves: an overview. Nature and Resources 29: 1-4

Bertzky M, Stoll-Kleemann S (2009) Multi-level discrepancies with sharing data on protected areas: What we have and what we need for the global village. Journal of Environmental Management 90 (1): 8-24

Bruner AG, Gullison RE, Rice RE, da Fonseca GAB (2001) Effectiveness of Parks in Protecting Tropical Biodiversity. Science 291: 125-128

FAO (2007) State of the World's Forests. Food and Agriculture Organization of the United Nations, Rome, Italy

Faust H, Maertens M, Weber R, Nuryartono N, van Rheenen T, Birner R (2003) Does Migration lead to Destabilization of Forest Margins? - Evidence from an interdisciplinary field study in Central Sulawesi. Research Project on Stability of Rain Forest Margins (STORMA) 1

Fritz-Vietta NVM, Stoll-Kleemann S (2008) How to Foster Organisational Capacity for Integrated Biosphere Reserve Management - The Biosphere Reserve Mananara-Nord, Madagascar. GAIA 17 (S1): 169-176

Kumar S (2006) Methods for Community Participation. ITDG Publishing, UK

Lange S (2005). Credits for sustainable development. www.credits-for-sustainability.com/images/br/br_zonation.jpg (06-12-2009)

Mehring M, Stoll-Kleemann S (2008) Evaluation of Major Threats to Forest Biosphere Reserves - a Global View. GAIA 17 (S1): 125-133

Nepstad DC, Verissimo A, Alencar A, Nobre CA, Lima E, Lefebvre P, Schlesinger P, Potter C, Moutinho P, Mendoza E, Cochrane M, Brooks V (1999): Large-scale impoverishment of Amazonian forests by logging and fire. Nature 398: 505-508

Punch KF (2005) Introduction to Social Research: Quantitative and Qualitative Approaches. Second ed. Sage Publications, London, U.K.

Richter U (2008) Die Akzeptanz von Schutzgebieten durch die lokale Bevölkerung in Südostasien. Geographisches Institut, Humboldt Universität zu Berlin. Berlin, Germany

Schwarze S, Schippers B, Weber R, Faust H, Wardhono A, Zeller M, Kreisel W (2007) Forest Products and Household Incomes: Evidence from Rural Households Living in the Rainforest Margin of Central Sulawesi. In: Tscharntke T, Leuschner C, Zeller M, Guhardja E, Bidin A (eds.) Stability of Tropical Rainforest Margins - Linking Ecological, Economic and Social Constraints of Land Use and Conservation. Springer, Heidelberg, Germany

Shohibuddin M (2006) Discursive Strategies and Local Power in the Politics of Natural Resource Management: Case of Toro Community in the Western

Margin of Lore Lindu National Park (LLNP), Central Sulawesi. Presented at: Survival of the Commons: Mounting Challenges and New Realities, the Eleventh Conference of the International Association for the Study of Common Property, Bali, Indonesia, June 19-23, 2006

Sodhi NS, Koh LP, Brook BW, Ng PKL (2004) Southeast Asian biodiversity: an impending disaster. Trends in Ecology & Evolution 19 (12): 654-660

Stoll-Kleemann S (2005) Voices for Biodiversity Management in the 21st Century. Environment 47 (10): 24-36

Stoll-Kleemann S, Welp M (2008) Participatory and Integrated Management of Biosphere Reserves - Lessons from Case Studies and a Global Survey. GAIA 17 (S1): 161-168

TNC (2002) Lore Lindu National Park Draft Management Plan 2002-2027. The Nature Conservancy

UNESCO (1984). Action Plan for Biosphere Reserves. *Nature and Resources*, **20** (4), 1-12.

UNESCO (1995) The Seville Strategy for Biosphere Reserves. UNESCO, Paris, France

UNESCO (2008) Madrid Action Plan. UNESCO, Paris, France

Widagdo I (2008) Lore Lindu as one of World's Biosphere Reserve. Palu, Sulawesi, Indonesia

Wittemyer G, Elsen P, Bean WT, Burton ACO, Brashares JS (2008) Accelerated Human Population Growth at Protected Area Edges. Science 321: 123-126

World Bank (2007) World Bank list of economies. World development indicators database. Washington D.C.: World Bank

Institutions for environmental service payment programmes - evidence of community resource management arrangements in Central Sulawesi, Indonesia

Christina Seeberg-Elverfeldt[1*], Stefan Schwarze[2], and Heiko Faust[3]

[1] Natural Resources Management and Environment Department (NRD), FAO, Viale delle Terme di Caracalla, 00153 Rome, Italy
[2] Georg-August University Göttingen, Department of Agricultural Economics and Rural Development, Platz der Göttinger Sieben 5, 37073 Göttingen, Germany
[3] Georg-August University Göttingen, Department of Human Geography, Goldschmidtstraße 5, 37077 Göttingen, Germany

*corresponding author: C. Seeberg-Elverfeldt, email: christina.seebergelverfeldt@fao.org.

Summary

Payments for Environmental Service (PES) schemes are increasingly discussed as a possibility to promote the conservation of natural resources. However, these pilot schemes are frequently small in size and face high transaction costs, leading to the exclusion of smallholders. Solutions could be to use collectively bundled contracts or existing community resource management arrangement structures. Using the example of the institution of the community conservation agreements (CCA) in Central Sulawesi we assess whether a community arrangement can provide the framework conditions to implement a PES project. Four points are necessary: an organisational structure representing the village households; participation of the resource users in the institutional implementation; monitoring and enforcement by the institution of the forest usage regulations; and, finally, the institutions' ability to administer funds. Our findings show that the CCAs are backed up by an organisational structure, the village conservation council. However, in most villages the community members were not involved and did not know of the agreement negotiation. A monitoring entity has been constituted. The awareness for nature conservation has increased only in the recent past, but resource extraction has left its marks, and the participants perceived environmental problems to be growing. Finally,

T. Tscharntke et al. (eds.), *Tropical Rainforests and Agroforests under Global Change*, Environmental Science and Engineering, 431–446, DOI 10.1007/978-3-642-00493-3_20,

compensation payments are regarded on the one hand as a good reimbursement for desisting from using the forest resources. On the other hand, due to the negative experiences with corruption, a clear organisational structure for the administration of such a project is necessary, which does not seem to be given with the current institutional arrangements. Therefore, we recommend using existing community arrangements because established structures can be used. However, these need to be enriched to fulfil the requirements of a PES project. It is of major importance to involve the community members in the management of natural resource projects to increase compliance with regulations.

Keywords: Payments for environmental services, institutions, community resource management, carbon sequestration, focus groups

1 Introduction

Payments for Environmental Service (PES) schemes have been increasingly discussed in the past few years. They are seen as a possibility to support the conservation of natural resources whereby the providers of the service receive compensation payments by the external beneficiaries on the basis of a negotiated agreement. The local landowners and land users in turn have to adopt land use practices which secure the conservation of the environmental service. Changes in land use can be stimulated through compensations (Seeberg-Elverfeldt et al. 2008; Matta and Kerr 2006; Antle and Stoorvogel 2008), whereby the payments raise the private returns of a specific land use system, which generates positive environmental impacts relative to the formerly employed land use. Donors and policy makers regard these schemes as a solution for natural resource management problems since not only environmental objectives are addressed but also social ones. However, experience from PES projects in various countries indicate mixed results (Pagiola et al. 2002; Grieg-Gran et al. 2005; Wunder 2008). These pilot schemes are often small in size and face considerable transaction costs. For smallholders these are effectively a barrier to enter and participate in a PES scheme. A solution can be to build upon existing structures of community resource management arrangements to enhance cost effectiveness and enable smallholder participation.

The Clean Development Mechanism (CDM) of the Kyoto Protocol currently allows only afforestation and reforestation projects in developing countries as emission trading schemes in the Land-Use, Land-Use Change and Forestry (LULUCF) sector. In the voluntary carbon sector the number of avoided deforestation projects is increasing. LULUCF projects focus on carbon sequestration, a forest environmental service. In simple terms the rationale is that northern companies pay farmers in developing countries to plant trees or desist from forest conversion activities. A limitation for smallholders is that to operate at the scale at which the carbon markets work, they are required to have an intermediary body who can negotiate on their behalf with buyers, receive revenues and distribute them to households. The main barriers for smallholder participation in forest PES schemes are institutional ones and the transaction costs incurred (Pfaff et al. 2007; Jindal et al. 2008). Using the example of the institutional framework of the community conservation agreements (CCA) in Central Sulawesi, Indonesia, we have investigated whether the existing structures of a community natural resource management agreement can provide the institutional linkages and framework to implement forest PES projects.

2 Community Environmental Service Payment Schemes

2.1 Institutions for PES Schemes

In this paper institutions will be referred to as the systems of rules, decision-making procedures, and programs that give rise to social practices, and guide interactions among the occupants of relevant roles. Unlike organizations, which are material entities that typically figure as actors in social practices, institutions may be thought of as the rules of the game that determine the character of these practices (Young et al. 1999).

Many carbon sequestration projects are carried out by large-scale plantation forestry and the participation of smallholders is limited. One of the main reasons are the high transaction costs of forest carbon projects (Pfaff et al. 2007). There is wide support for the creation of institutions and financial intermediaries to bundle projects in a portfolio, such that investors are not tied to a particular project (Smith and Scherr 2003). Among local communities technical skills for developing baselines and monitoring plans can be pooled and group contractual arrangements made. Intermediaries for these processes can be different institutions, such as local governments, NGOs, and local community organisations. Consequently, to enhance cost-effectiveness, a strategy is advocated to develop projects whereby smallholders participate in groups, for instance distinguished by local community boundaries, rather than individually. Therefore, local communities act as service providers and obtain a share of the carbon revenues. Experience shows that carbon smallholder projects were often built upon some type of existing community project, particularly community forest plantations or farmer's groups (de Jong et al. 2000). Usually carbon forestry projects entail a variety of transaction costs in its design and implementation. Results from a study by Michaelowa and Jotzo (2005) indicate these to range from US$ 1.48 per tCO_2 for large projects to US$ 14.78 per tCO_2 for small ones. According to experience from a variety of carbon sequestration projects, the monitoring and enforcement activities can be easily integrated into community processes which minimises costs (Cacho et al. 2003). In the International Small Group and Tree Planting Programme (TIST) in Tanzania the monitoring and supervision activities were performed by the local institution, reducing overall transaction costs (Jindal et al. 2008).

2.2 Community Conservation Agreements

Community Conservation Agreements (CCA) have been established in 49 villages located in the surroundings of the Lore Lindu National Park (LLNP) as a co-management strategy. The local communities entered a negotiated arrangement with the Park Authority as a strategy to resolve conflicts between the peoples' needs and conservation demands. The negotiations for the agreements between the LLNP Authority and the villages started in the late 1990s and were promoted and operated by international and local NGOs. In many villages different NGOs have been working either alongside or one after the other. This caused confusion and the community members are not certain which organisation initiated and carried out which activity. The negotiations were usually conducted by the village elders and the Lembaga Adat (LA), the traditional customary institution, which is in charge of the village law. They typically signed the agreement and established a village conservation council (VCC). This new institution looks after the monitoring activities and rule enforcement in the designated CCA area.

The agreements entail rules and sanctions concerning the allowed amount of timber to be harvested, the use and the sale of the timber, forest conversion for agriculture, plantation development, the collection, sale and use of rattan and non-timber forest products, as well as hunting. These are listed in a forest management plan. The village LA has the punishment or sanctioning capacity, but exercising these measures can only be carried out in the presence of the village administration and the village representative body. The sanctions differ between villages but are usually based on the traditional customary rules. The money from the punishment is received by the LA and used for the development of the village.

The majority of households in these villages are agricultural smallholders who cultivate mainly wetland rice as a subsistence food crop, as well as cocoa, the most important cash crop in the region. Central Sulawesi is one of the poorest provinces in Indonesia and in the region almost half of the households fall below the international poverty line of US$2 per capita and day (van Edig 2005). Encroachment on the borders of the LLNP has been recorded and is often driven by agricultural expansion. Consequently deforestation takes place at the fringes of the 220,000 hectares of the National Park which is predominantly covered by mountainous rainforest (Maertens et al. 2006).

2.3 Framework for Institutional Analysis

The objective of the study was to assess whether the institutional arrangement of the CCA in Central Sulawesi can provide a platform for a carbon sequestration project. Figure 1 exhibits the analytical framework which consists of the four focal points: institution, participation, monitoring & enforcement and status of the environment. These are crucial elements in the analysis of a community natural resource management process. The institution (CCA in this case), which provides the framework for a natural resource management project, needs to have an organisational structure representing the village households. The community members should be involved and participate in the resource management process, as the legitimacy of regulatory interventions is increased when the resource users participate in its design and implementation. Furthermore, this institution needs to enable the monitoring and enforcement of the forest usage regulations. A good indicator for the success of these activities is the status of the environment and to which degree resource extraction and environmental impacts are observed. From these focal points several sub-topics were derived such as education, illegal activities and resource extraction.

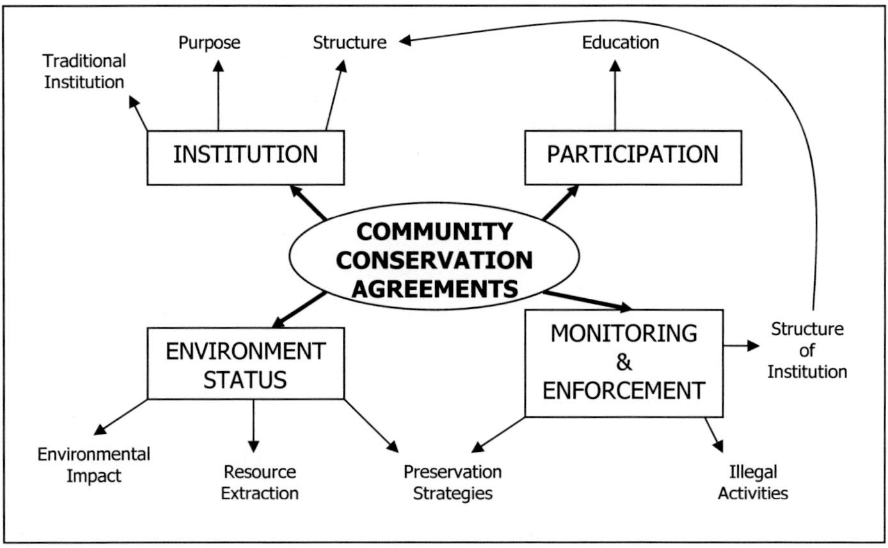

Fig. 1. Framework for Analysis of the CCAs (Source: own elaboration).

3 Methods and Data

We chose a qualitative research design and conducted focus group discussions in 2006 in four villages which border the LLNP. To the extent that we wanted to obtain information about the community members' attitudes, awareness and values pertaining to the impact of the CCA, the focus groups were considered especially appropriate. A focus group is a carefully planned discussion designed to obtain perceptions on a defined area of interest in a permissive, non-threatening environment (Krueger 1994). A focus group can be seen as a combination between a focused interview and a discussion group (Steyaert and Lisoir 2005).

The villages which were chosen for the study are Kapiroe in the Palolo, Wuasa in the Lore Utara, and Salua and Langko in the Kulawi District. We used three main criteria for the selection of these villages. These were the negotiation stage of agreement, the location of the village and the ethnic composition. At the time of research all villages were at different stages of negotiation or execution of the CCA and this allows for obtaining a good impression of the impact of the establishment of the agreements. They are located in different geographical zones, one towards the north-east, one in the east, one in the west and one in the centre of the National Park. Langko differs from the other villages, as it is located in the Lindu enclave in the LLNP and can be only accessed by a narrow path. It is ethnically not mixed, in comparison to Salua and Kapiroe which both have a high proportion of migrants. In Wuasa the indigenous ethnic group still represents the majority of the population.

In every village two focus group meetings were realised. One was carried out with the local village authorities such as the village headman or the village secretary, as well as members of the traditional customary institution and the village conservation council, and the second one with farmers. In each meeting around five people assisted. By organising the meetings with groups from different social strata we acknowledge the importance of their participation, as it allows to involve those affected by, knowledgeable of, or having relevant expertise or experience on the issue at stake in knowledge production and/or decision-making (van Asselt Marjolein and Rijkens-Klomp 2002). In order to assure the validity and quality of the research, we specifically chose to discuss the research topics in two groups in order to separate the village authorities from the farmers, as they could have easily dominated the discussion. The results of the interpretation of the discussion have been presented in workshops 1.5 years later to the same participants. Overall, they agreed with the presented outcome and corroborated the recommendations, an additional assurance for the validity of the results.

The recorded discussions have been assessed with the qualitative content analysis following Mayrings' (2007) approach in order to apply a rule-guided, reproducible assessment of the group discussion interview material. For the analysis an elaborate category system was developed by a theory-driven, deductive approach. The structuring procedure was used for the content analysis as only the text sections with respect to specific topics and content realms were extracted, paraphrased and summarised. Therefore, for each category we obtained evidence from all villages and both groups with respect to their perceptions of positive and negative impacts of the CCA.

4 Results

The in-depth analysis of the discussion contents allows for the detection of the motives and perceptions of the farmers and decision makers with respect to the different identified focal points. The content analysis of the discussion with respect to the community conservation agreements is based on the framework in Figure 1. In addition, the perceptions of the participants with respect to the concept of compensation payments are briefly outlined. The information from all group discussions has been analysed subject to the differences between both groups, as well as to the situation before and after the CCA implementation.

4.1 Community Conservation Agreement Institution

With respect to the traditional customary institution (LA) among the villagers little knowledge and understanding of the institution was observed, as well as decreasing acceptance of the regulatory framework. The purpose behind the agreements is not known by the farmers in two of the villages, whereas the village leadership defined it to have been set up for conservation needs and management. The farmers were familiar with the VCC and they had noticed monitoring activities to take place. A definition of the VCC given in one village was *the village fence to prohibit someone from entering a preserved location* (DM, Kapiroe, 49-52)[1] which points towards its protective function for the forest. The decision makers remarked that the LA gives a good foundation for the CCA and its regulatory framework such as the sanctions. They noted that the monitoring activities are carried out every one to three months, but not on a constant basis. The villagers observed that the participation in the formation of the agreement was restricted to specific people and various participants did not know the date of the beginning of the negotiations.

When we evaluated the change in the institutional setting the farmers across all villages were quite critical towards the LA and its regulatory structure in the past, as rules had not been enforced. However, presently the villagers could see an improvement in the institutional arrangement due to the new monitoring agency; there was only one village in which they still observed rule-breaking. The village leadership remarked that due to the traditional rules of the customary agency, the regulatory framework and its enforcement structure were in place and could be used by the new VCC. However, in one village they noted that the monitoring institution had become an abettor of the government forest rangers.

[1] In brackets is the participant group (DM=decision makers, V= villagers), village, and line number in the English transcript.

4.2 Participation in the Negotiation and Establishment of the CCA

Both groups pointed out that in the past there were hardly any educational activities and information campaigns offered by the LLNP administration with respect to the National Park, the forest and its functions. This has led to wide ignorance among the communities regarding preservation and conservation issues. Sometimes extension programmes were offered to the villages but never put into practice. Following the implementation of the agreements there is still a lack of understanding of the purpose of the National Park among the villagers, whereas the leadership is well informed. Also, the authorities have observed less confrontation between the government forest ranger and the farmers recently. As mentioned above, the participation of the villagers in the past has been very limited in community decision-taking, as well as in the establishment and management of the National Park and the CCAs. The same was remarked by the decision makers who said that there was very little communication by the LLNP authority concerning the rules and regulations of the National Park, as well as their activities and programmes. After the agreement negotiation the farmers still observed a lack of participation in CCA meetings and conservation programmes, whereas the authorities noted an improvement of the community participation in conservation activities.

Decision makers are of mixed opinion, in some villages they note a change and an improvement of knowledge sharing and educational activities by the LLNP administration and other NGOs because of the establishment of the agreements, whereas in other villages they are more critical: *so you have any suggestion for the [government] apparatus that they can have better approaches to the community, not only threatening the villagers. Because it only triggers conflict amongst villagers and forest rangers* (DM, Wuasa, 391-392) and *before the CCA was formed, none of the villagers was willing to support the government to conserve the forest due to shortage of socialisation*[2] (DM, Wuasa, 343). These statements are motivated by the bad collaboration between the community and the National Park forest guards, which have been mentioned in all villages.

4.3 Monitoring and Enforcement

For monitoring and law enforcement it is important to know how illegal activities are defined. In this case they are activities which violate the customary, CCA and the state law with respect to the LLNP. Discussing these is a sensitive issue, since nobody wants to admit their own faults or put the village into a bad light. It is interesting therefore to contrast the opinions between both groups in the same village. In the past the farmers had observed many illegal

[2] This word comes from Bahasa Indonesia and means to make people aware of something through interaction, i.e. meetings.

activities, especially by outsiders. With the new regulations they noticed that less land was cleared for new plantations. The decision makers in Wuasa also declared that illegal logging has taken place inside the CCA area and that they were conscious that communication about the regulations is important to stop these activities. In Langko the decision makers did not observe any deforestation, in contrast to the villagers who did. Some villages have preservation strategies, such as the assignation of specific usage and conservation zones through the customary rules by the LA. However, the traditional boundaries are often different from the official borders of the National Park which causes conflicts.

Monitoring and enforcement activities are mainly looked after by the newly established VCC. The farmers are very critical and remark that in the past the penalties existed only in theory and little implementation and enforcement took place. Timber could be extracted from the National Park without any control. The decision makers ascertain that no monitoring existed and embezzlement of responsibilities occurred frequently. Farmers have different perceptions about the situation after the CCA implementation. In Wuasa and Salua there is apparently no more illegal resource extraction in the forest area which the villagers attribute to the introduction of sanctions. However, in Kapiroe and Langko the farmers say that the existing regulations do not hinder forest conversion, as there is no enforcement of the rules. The decision makers in Wuasa and Salua share the opinion of the farmers in their villages and all ascertain a decrease in deforestation. In Langko the decision makers are of the opinion that the situation has improved in comparison to the past and that the existing customary structures help to support the new CCA regulations. They clarify, however, that their activities are constrained because they do not receive any financial support.

4.4 Status of the Environment

In general, across all villages and both groups the impression was prevailing that the environmental condition was good in the past, that there were plenty of birds and animals and less natural disasters such as flooding and droughts. Both groups perceived the environmental impacts to have become worse after the implementation of the CCA in the recent past. However, they observed less clearing of land taking place now. Obviously, the CCA in itself is not the determining factor for a change in environmental impacts. However the aim is to detect whether it has influenced certain practices which in turn had an impact on the environment. In all villages the farmers as well as the decision makers said that in the past extensive resource extraction such as forest conversion and rattan collection took place. In Langko, the extraction was only for private needs according to the village authorities. Nearly all farmers observed a decrease of natural resource extraction nowadays, whereas in Langko mixed comments were made, in that deforestation still takes place, but less land is opened up for further plantations. The village authorities all perceive a decrease in illegal activities, which corroborates the information by the farmers from all villages excluding Langko. Obviously, forest extraction activities can not just be seen as simply illegal, since people are often also driven by their needs, which have to be satisfied: *in the past, in the age of our ancestor, if the population increased, the land was also extended because they opened up new lands. This is in contrast with the current situation. Nowadays, the number of people increased but the land space is constant. In the past, people were able to open up new land. So, sometimes people break the rules because they have the necessity.* (F, Wuasa, 630).

4.5 Compensation Payments

The fear among villagers is that compensation payments would not be equally distributed, since in general *Indonesia is well known to have corruption* (F, Langko, 629), and very often kin relationships influence the distributional patterns. Payments are seen, however, also as a possibility to stimulate and exert control over forestry extraction activities. Furthermore, since people have to forego a potential income source when they cannot use the forest resources anymore, the compensation is regarded as equitable.

The village authorities argue that if payments would be channelled directly to the communities, the funds could be used efficiently to improve the monitoring system. However, overall the decision makers are quite critical of compensation payments and fear corruption, especially if a variety of institutions are involved. Based on their experience, NGOs intermingle their personal interests with the management of funds, causing embezzlement. Often NGO staff does not fully understand the village realities and therefore use inappropriate targets and objectives in the realisation of projects. They also recognise that the payment cannot compensate for their need of employment, as being idle does not make them happy: *But especially in Suaka Ntodea we disagree [about complete preservation] because we still need the rattan and woods from there. Even if we will be given money, if we do not work anymore, we will be unhealthy.* (Langko, DM, 859-862).

4.6 Differences between Villages

Some of the information has been generalised across all villages, however, in some cases there are also differences, which should be addressed. In Table 1 the main attributes of the four villages and the community conservation agreements are summarised.

In Langko, the decision makers and the villagers are very often of different opinions. This can be attributed to the fact that the decision makers have been much more involved in the negotiation of the agreements and, therefore, do not want to shed bad light on their own actions. Furthermore, a gradient in compliance or acceptance of the CCA which is proportional to the time length the village has been involved in the negotiations is quite apparent.

Table 1. Characteristics of Case Study Villages and CCAs.

	Salua	Langko	Wuasa	Kapiroe
Village established in	1984	1900	1892	1900
No. of households	307	184	648	279
Population	1,244	704	2,644	1,026
Village size (ha)	6,632	7,500	2,839	10,680
Population density (pop/km^2)	19	9	93	10
Ethnic composition	mixed	local	local	mixed
Paddy land (ha)	0	208	330	75
Cacao land (ha)	900	30	430	445
Forest (ha)	5,589	6,738	489	10,015
Start of negotiation process	1996 /2004	2004	1999	2005
Signed by LLNP Authority	Not signed / Not signed in 2006	March 2005	August 2002	Not signed in 2006
Local organization looking after CCA	Lembaga Adat	VCC	VCC	Lembaga Adat

Source: A4 village survey 2007 by Reetz (2008) and own data

First there is Wuasa, followed by Langko, and Kapiroe and Salua have the weakest agreements. Wuasa is probably the most active village in terms of conservation activities and the awareness both among the decision makers as well as the farmers is very high. They have been involved the longest in negotiations with the LLNP administration due to the limited access to the forest caused by the establishment of the National Park. In Langko, even though the views differ between both groups, a base is provided by the customary institution. Some farmers criticise it as not being respected anymore, but it lays the groundwork for the establishment of the new regulatory framework. Among the participants in Kapiroe we detected awareness with respect to the conservation need for the forest and the consequences of deforestation are quite apparent in this area. However, the agreement negotiations are still in process, and in particular the villagers were not very well integrated into the discussions between the different stakeholders. Finally, in Salua the agreements were implemented by different institutions which led to confusion with respect to the responsibilities of the activities. The first negotiations were in 1996 and therefore not well remembered, and the second agreement was never signed. Additionally, the community has not been informed about the purpose of the agreement.

5 Discussion

The CCA is backed up by the organisational structure of the VCC. The community is familiar with this new organisation and aware of its activities. However, in the majority of villages the villagers were not involved in the agreement negotiation and sometimes did not even know of its existence. In contrast, the authorities are familiar with the agreement, indicating a knowledge gap between the different social strata in the village. This finding is corroborated by the results of Mappatoba and Birner (2004) who detected that often persons with functions were among those selected to participate in CCA meetings.

The LA is present in all villages and its regulatory framework provides a good foundation for the rules of the conservation agreement. It can be build upon in order to improve the local population's acceptance of the new regulations. However, the LA has different strength in the four villages, as well as acceptance among the community members which is related to the socio-cultural situation in the village. Both in Wuasa and Langko, the population is still dominated by the indigenous ethnic groups. Thus, the LA and the VCC, especially in Wuasa, have become a voice for the local community to fight for their access to the forest. Burkard (2002) points out that in an ethnically mixed resettlement, such as Kapiroe and Salua, the LA is comparatively weak and does not play a significant role in the management and utilisation of natural resources.

A monitoring entity has also been constituted in most villages and is relatively active. Several cases of law enforcement were recorded; however, restricted or lack of funds constrains the entity's activities considerably and monitoring is not frequent. Similarly, Palmer (2007) demonstrates that the monitoring entities carried out regular checks in just 50 percent of the villages; in 25 percent checks were only carried out when there was a special reason to do so. Approximately two thirds of the monitoring teams did not receive any financial resources for the enforcement activities. Thus, the newly formed monitoring institutions provide a base for the monitoring and enforcement, but it needs to be financially supported and strengthened to be more efficient.

The awareness with respect to nature conservation has become more widespread only in the recent past which can not be attributed purely to the establishment of the CCAs. As considerable resource extraction has left its marks in the region with the participants believing that environmental problems such as flooding and erosion have increased, the villagers are more concerned about protecting the forest.

Finally, compensation payments are regarded on the one hand as a good reimbursement for desisting from using the forest resources. On the other hand, fear of corruption has been expressed, as well as embezzlement of funds, based on previous experience. Indonesia is a country which has considerable problems with corruption, and Transparency International has listed it as Number 143 out of 179 countries on the 2007 Corruption Perception Index (2007). Compensation payments are not seen as a solution for the inherent problem of land scarcity, associated with the need to work, obtain food and pass on land to the villagers' children. This was also mentioned to be one of the main disadvantages of the National Park, that not enough land will be available for their children (Mappatoba and Birner 2004).

6 Conclusions

These findings allow us to make some judgements as to whether the institutional arrangement of the CCA could provide a basis for a carbon sequestration project or more generally for a forest PES project. A resource management project could benefit from the framework of the rules and regulations of the CCA established on the basis of the customary institution, providing an important groundwork for the implementation of a PES project. The given regulatory framework can be used and enriched. However, in the present circumstances the purpose of the agreement has not been communicated to all stakeholders, and the involvement, at least of some villagers or representatives of these villagers, is not given. As argued by Hanna (1995) a resource management process must represent the range of user interests and have a clear purpose and transparent operation, which allows for a better identification of the community with the aims of such a project. Thus, for a PES or forest carbon sequestration project, the participation of all those affected by it can not be guaranteed by the present institutional arrangement of the CCA. For an internationally financed project, the VCC needs to be reinforced and monitoring activities have to be conducted more thoroughly and frequently. More financial support can help to foster these activities. A PES project typically involves payments to the providers of the environmental service at stake. This requires a transparent organisational structure and the objectives and responsibilities have to be clearly defined. The present structure of the CCAs and the VCCs differs between villages, because on the one hand the NGOs used different approaches for the agreements and on the other hand the village structures, due to their ethnic composition and geographic location, diverge. Additionally in Indonesia, due to high levels of corruption, mistrust is engrained in the people; whether projects will be carried out according to their stated objectives and funds handled efficiently and distributed fairly. Thus, with the present institutional arrangement of the CCAs the administration and management of such a PES project is very difficult.

We can conclude from this particular case study in Indonesia, that the structures of existing natural resource management agreements can provide initial institutional linkages and framework conditions to implement a forest PES project. It needs to be assessed on a case to case basis, whether the natural resource management structures are sufficient and the socio-cultural aspects of the specific circumstances need to be taken account of. In addition, it is of major importance to integrate the community members into the management of the natural resource projects. Compliance with regulations increases when they are considered acceptable and legitimate by those whose interests are regulated. Obviously not all community members can participate in these processes, yet an option might be to let the villagers vote on the outcome. Finally, the governance structures in a country are an important factor for the success of development and conservation projects and higher rates influence these positively (Smith and Scheer 2003).

Specifically, for the establishment of PES projects this means that advantage should be taken of intermediary bodies such as traditional community resource management institutions. Using established arrangements can ensure familiarity and trust among participants. Negotiations can be rendered more efficiently, as a contact is given and contracts can be made with the entire group rather than individually. This substantially decreases transaction costs. Additionally, if specific management structures are already established, such as in this case monitoring and enforcement, costs can be reduced even further.

References

Antle JM, Stoorvogel JJ (2008) Agricultural carbon sequestration, poverty, and sustainability. Environment and Development Economics 13 (03): 327-352

Burkard G (2002) Natural Resource Management in Central Sulawesi: Past Experience and Fitire Prospects. Discussion Paper Series 8. Göttingen: STORMA. http://ufgb989.uni-forst.gwdg.de/DPS/index.htm

Cacho OJ, Marshall GR, Milne M (2003) Smallholder Agroforestry Projects: Potential for Carbon Sequestration and Poverty Alleviation. ESA Working Paper 03-06. Agriculture and Economic Development Analysis Division, FAO.

de Jong TR, Tipper R, Montoya-Gomez G (2000) An economic analysis of the potential for carbon sequestration by forests: evidence from southern Mexico. Ecological Economics 33 (2): 313-327

Grieg-Gran M, Porras I, Wunder S (2005) How can market mechanisms for forest environmental services help the poor? Preliminary lessons from Latin America. World Development 33 (9): 1511-1527

Hanna S (1995) Efficiencies of user participation in natural resource management. In Property Rights and the Environment: Social and Ecological Issues, edited by S. Hanna and M. Munasinghe. Washington D.C.: Beijer International Institute of Ecological Economics and the World Bank

Jindal R, Swallow B, Kerr J (2008) Forestry-based carbon sequestration projects in Africa: Potential benefits and challenges. Natural Resources Forum 32 (2): 116-130

Krueger RA (1994) Focus groups: a practical guide for applied research. 2nd Edition ed. London, New Delhi: Thousand Oaks.

Maertens M, Zeller M, Birner R (2006) Sustainable agricultural intensification in forest frontier areas. Agricultural Economics 34 (2): 197-206

Mappatoba M, Birner R (2004) Co-Management of Protected Areas - The Case of Community Agreements on Conservation in the Lore Lindu National Park, Central Sulawesi, Indonesia. Eschborn: Gesellschaft für Technische Zusammenarbeit.

Matta J, Kerr J (2006) Can Environmental Services Payments Sustain Collaborative Forest Management? Journal of Sustainable Forestry 23 (2): 63-79

Mayring P (2007) Qualitative Inhaltsanalyse: Grundlagen und Techniken. 9. Aufl., Dr. nach Typoskript ed, UTB für Wissenschaft ; 8229 Pädagogik. Weinheim u.a.: Beltz

Michaelowa A, Jotzo F (2005) Transaction costs, institutional rigidities and the size of the clean development mechanism. Energy Policy 33 (4): 511-523

Pagiola S, Bishop J, Landell-Mills NE (2002) Selling forest environmental services market-based mechanisms for conservation and development. London: Earthscan Publications Ltd

Palmer C (2007) Background paper: Community forest use and conservation agreements in Lore Lindu, Indonesia. Zürich: ETH.

Pfaff A, Kerr S, Lipper L, Cavatassi R, Davis B, Hendy J, Sanchez-Azofeifa GA (2007) Will buying tropical forest carbon benefit the poor? Evidence from Costa Rica. Land Use Policy 24 (3): 600-610

Reetz S (2008) Data from A4 village survey 2008.

Seeberg-Elverfeldt C, Schwarze S, Zeller M (2008) Payments for Environmental Services - Incentives through Carbon Sequestration Compensation for Cocoa-based Agroforestry Systems in Central Sulawesi, Indonesia. University of Hohenheim. http://www.uni-hohenheim.de/i490a/dps/2008/dp022008.pdf

Smith J, and Scherr SJ (2003) Capturing the Value of Forest Carbon for Local Livelihoods. World Development 31 (12): 21432160

Smith RJ, Muir RDJ, Walpole MJ, Balmford A, Leader-Williams N (2003) Governance and the loss of biodiversity. Nature 426 (6962): 67-70

Steyaert S, Lisoir H (2005) Participatory Methods Toolkit. A practitioner's manual; Method: Focus Group. King Baudouin Foundation and the Flemish Institute for Science and Technology Assessment.

Transparency International 2007. TI Corruption Perceptions Index

van Asselt Marjolein BA, Rijkens-Klomp N (2002) A look on the mirror: reflection on participation in Integrated Assessment from a methodological perspective. Global Environmental Change 12: 167-184

van Edig X (2005) Measurement of Absolute Poverty and Indicators of Poverty Among Rural Households in Central Sulawesi, Indonesia. Diplomarbeit, Department of Geography, University of Göttingen, Göttingen

Wunder S (2008) Payments for environmental services and the poor: concepts and preliminary evidence. Environment and Development Economics 13 (3): 279-297

Young OR, Agrawal A, King LA, Sand PH, Underdal A, Wasson M (1999) Institutional Dimensions of Global Environmental Change - Science Plan. IHDP Report No. 9. Bonn, Germany: International Human Dimensions Programme on Global Environmental Change

Agricultural expansion in the Brazilian state of Mato Grosso; implications for C stocks and greenhouse gas emissions

Eleanor Milne[1,2]*, Carlos Eduardo P. Cerri[3], and João Luis Nunes Carvalho[3]

[1] The Macaulay Institute, Craigiebuckler, Aberdeen, AB158QH, UK
[2] Colorado State University (NREL), Fort Collins, CO 80523-1499 USA
[3] Center for Nuclear energy in Agriculture (CENA), The University of São Paulo, C.P. 96, C.E.P. 13.400-970, Piracicaba, SP, Brazil

*corresponding author: E. Milne, email: eleanor.milne@colostate.edu

Summary

The states of Rondonia and Mato Grosso in Brazil together make up the world's largest agricultural frontier. Between 2001 and 2004, deforestation, to provide land for highly mechanized agriculture in this area, reached unprecedented rates. The environmental consequences of this include increased greenhouse gas (GHG) emissions from above and below ground sources. In this area, where agriculture is a relatively new activity, land use and management practices change rapidly as farmers react to market pressures but also use trial and error to increase productivity/reduce losses. Farmers routinely make use of the latest technologies making agricultural expansion very different from historical examples. The fact that systems are in a state of flux has implications for sustainability and the ability of scientists to produce local and national GHG inventories and project future GHG emissions and carbon stock changes. Such projections are necessary to assess the full environmental impacts of such large scale native vegetation loss and to inform policy makers accordingly.

A pilot study of farmer interviews around the town of Sinop, Mato Grosso, found that land management methods have varied over time, with a recent trend emerging to move from monoculture systems to integrated crop/livestock systems. The interviews also revealed that management practices employed by the small (<500 ha) farms differed markedly from those employed by medium (> 500 ha) and large (> 10,000 ha) farms. Wider studies carried out in the neighbouring state of Rondonia have shown soil C stocks and GHG emissions to be highly dependent on agricultural management. We

T. Tscharntke et al. (eds.), *Tropical Rainforests and Agroforests under Global Change*, Environmental Science and Engineering, 447–460, DOI 10.1007/978-3-642-00493-3_21,

discuss the implications of rapidly changing management practices in the agricultural frontier region of Mato Grosso and Rondonia on past regional estimates of C stocks and GHG fluxes and future projections. The future outlook is discussed in the context of possibility of REDD being included in a successor to the Kyoto Protocol.

Keywords: Brazil, Mato Grosso, Rondonia, agricultural expansion, deforestation

1 Introduction

1.1 Deforestation in the Brazilian Amazon

Prior to the 1960s, lack of access limited deforestation in the Brazilian Amazon. By the 1970s construction of two federal highways (Belm-Brazlia and Cuiab- Porto Velho) and other smaller roads, increased access and deforestation began to happen on a large scale (Kirby et al. 2006). From the 1970s, the drivers and processes of deforestation in the Brazilian Amazon changed continually and are still changing today (Fearnside 2008). Historically, land was cleared for small scale agriculture or timber extraction before being consolidated into larger areas, mainly used for cattle ranching (Morton et al. 2006). Throughout the 1970s and early 1980s settlement in the Brazilian Amazon was encouraged by tax incentives and government subsidies (Fearnside 2005). Subsidies were officially discontinued in 1991, although in practice many persisted well beyond this (Fearnside 2008). Even with the removal of subsidies, migration into the area (and associated deforestation) continued, as cattle production and timber extraction remained economically profitable without incentives. During the 1990s deforestation was exacerbated by the recovery of the Brazilian Real following its devaluation. Cash was more readily available to buy land, and forested land was cheaper to purchase than land which had already been cleared.

In the 1990s, cattle ranching continued to be the main land use on deforested areas in the Brazilian Amazon, however the introduction of crops (particularly soybean) able to cope with local climate and soil conditions, fuelled an expansion of commercial cropland (Bickel and Maarten Dros 2003). Since the year 2000, the area under large scale mechanised agriculture has increased dramatically, pushing commercial crop production further north into the Amazon (D'Avila 2003; Sato 2003), although cattle ranching still remains the main land use following forest clearance. Morton et al. (2006), estimate that mechanised agriculture increased by 36,000 km^2 from 2001 - 2004. This drove deforestation either by using up cattle ranching land, which was then replaced by newly deforested land, or by land clearing specifically for crop production. From 2005 - 2007 deforestation to provide land directly for mechanised agriculture slowed as international prices for agricultural commodities dropped (soybean by 50%) (Fearnside 2008). However, 2007 then saw a subsequent rise in soybean prices and a surge in deforestation in the major soybean growing state, Mato Grosso.

In the next decade, crop production along the agricultural frontier in the Brazilian Amazon is set to expand massively, with an estimated 70% of future deforestation predicted to be carried out to provide land directly for crops by 2020 (Cerri et al. 2007). An understanding of the agricultural systems being implemented and their associated environmental impacts, particularly on GHG emissions, is therefore imperative.

1.2 Deforestation and agriculture in the state of Mato Grosso

The Brazilian states of Mato Grosso and Rondonia together make up the world's largest agricultural frontier. The drivers of deforestation in each state vary, with Rondonia having a high proportion of deforestation by small holder farmers (mainly for cattle production) and Mato Grosso a high proportion of deforestation by large scale commercial farmers (for ranching and mechanised crop production). In the period between 2001 - 2004, 87% of cropland expansion in the Brazilian Amazon occurred in Mato Grosso (Morton et al. 2006).

Mato Grosso is the third largest state in the Brazilian Amazon covering over 90 million hectares. The north of the state has a humid climate and originally comprised of 50 million ha of tropical forest, with 46% of the state falling inside the Amazon Biome. Some 37% of the total forested area of Mato Grosso had been cleared by 2007 (Fearnside 2008). The south has a climate with a distinct dry season and comprises mainly of savannah Cerrado'. In the 1970s and 80s private settlement projects brought people from the south of Brazil (where small farms were being consolidated and there was a labour surplus) to Mato Grosso. The original intention was for the settlers to grow rice for the domestic market, addressing food shortage problems in Brazil. However, many farmers switched to soybean production for the export market facilitated by the emergence of soybean varieties adapted to tropical conditions. Farming to supply the export market saw an increase in the size of farms with the number of farms larger than 10,000 ha rising from 643 in 1980 to 767 in 1996 (Bickel and Maarten Dros 2003). Today 90% of the soybean grown in the Brazilian Amazon is grown in Mato Grosso (Fearnside 2008).

The influence of export agriculture on deforestation and land clearing in Mato Grosso can be seen when the correlation between agricultural commodity prices and deforestation rates are considered. When the prices for soybean or rice have been high farmers have deforested to open up more land to increase production and when they have been low (as in 2006) deforestation rates have slowed. Similarly, farmers tend to switch crops as they become more profitable to produce. In 2007, international soybean prices reached record levels and there was an accompanied surge in the rates of deforestation, although any relationship between this surge and the price of soybean has been denied by the Brazilian Government at both the local and the national level

(Sant'Anna 2008). The consolidation of small farms into larger ones can also exacerbate deforestation as large farms tend to employ fewer people. Small holders who sell their land to agribusiness then have the money and the motivation to clear new patches of forest in order to start again (Figueiredo et al. 2006). Even if large farms deforest directly, they can still produce a surplus of workers, as a lot of workers are needed for the initial land clearing operation but only a few once the farms are established. Carvalho (1999) estimated that only one worker per 167-200 ha of soybean field would find employment in Mato Grosso. Large scale soy production probably employs fewer people per hectare than any other crop grown in Brazil.

At the forest frontier in Mato Grosso, there is a mix of small and large farms, with migrant farmers from diverse origins reacting to national and global market pressures, in addition to untested soils and climatic conditions (Figueiredo et al. 2006). This has led to a fluctuating agricultural situation. The range of crop rotations, inputs, off-takes and tillage practices have varied greatly and are still changing today. Detail of practices such as clearing methods, crops grown, crop rotations, tillage operations, inputs and off-takes are lacking. Farmers routinely make use of the latest technologies making agricultural expansion very different from historical examples, particularly in terms of the rate of deforestation, the size of farms on newly cleared land and the number of people employed on farms. The fact that systems are in a state of flux has implications for sustainability and the ability of scientists to produce local and national GHG inventories and project future GHG emissions and carbon stock changes. Such projections are necessary to assess the full environmental impacts of such large scale native vegetation loss and to inform policy makers accordingly. Without this information, the Brazilian government may be underestimating the environmental impacts of forest and Cerradão cleared to provide land for croplands, especially in terms of national reports of GHG emissions resulting from land use change.

2 A small pilot survey on cropping practices in the Sinop area, Mato Grosso

Over the past 6 years a group of scientists from The University of São Paulo, Brazil, and Colorado State University, USA, have been using the CENTURY Model (Figure 1), (Parton et al. 1987) and the GEFSOC modelling system, which includes CENTURY (Easter et al. 2007; Milne et al. 2007), to estimate changes in C stocks and GHG emissions resulting from land use change in the Brazilian Amazon (Cerri et al. 2004; Cerri et al. 2007). They estimated that projected deforestation across the entire Brazilian Amazon between 1990 and 2030 would result in a loss of 4200 Tg of C from the soil (0-30cm layer), a decline of 7%. However, they also estimated that in the areas where forest was ultimately replaced by no-till soybean production rather than pasture there could be an increase in soil C stocks in the short term (Cerri et al. 2007).

Century is a general ecosystem model which models the turnover of C, nitrogen (N) and phosphorus (P) in above and below ground pools of an ecosystem, but with emphasis on below ground dynamics (Figure 1). To date, efforts for the Brazilian Amazon have concentrated on modelling forest-to-pasture changes as this has been the dominant form of land use change. However, there is now a need to parameterize CENTURY and other ecosystem models for the forest to cropland changes for which (as stated above) little information is available.

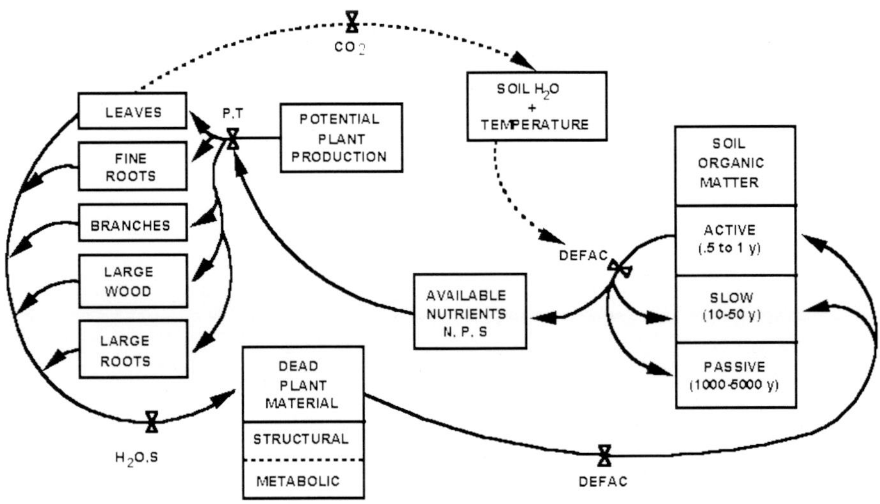

Fig. 1. Overview diagram of the Century Ecosystem Model (Parton et al. 1987)

With this in mind, during a visit to Sinop, Mato Grosso in December 2007, five farmer interviews were carried out to investigate cropping systems being used following deforestation and how this might vary with farm size. The overall purpose of the trip was to develop a future collaborative research program around parameterization of the Century General Ecosystem Model for cropping systems on newly cleared land in the Brazilian Amazon and was funded by a US National Science Foundation's Planning grant.

2.1 Study area and interviews

The study considered the area surrounding Sinop (11.85° S 55.46° W), a frontier city ∼ 400 miles north of Cuiab, the capital of the state of Mato Grosso, the largest soybean producing state in Brazil. Sinop lies on the edge of the agricultural expansion zone in the Southwest Amazon and has one of the highest grain productivity rates in the world. The city was established in 1974 during the time when soybean varieties were adapted for tropical conditions leading to agricultural expansion. However, the greatest growth in the city and clearance of the surrounding vegetation has occurred in the past 20 years, coinciding with a move to larger scale mechanized agriculture (primarily soybean) for the export market. Native vegetation in the area is Cerrdão meaning big Cerrado or big savannah'. It comprises of dense tree cover but lacks some of the key hardwood species that would formally classify it as tropical forest. It is characterized by trees to a height of 9 m and a density of 3,000 trees ha^{-1} (Carvalho et al. 2007).

To facilitate interviews, a representative from Embrapa (Brazil's agricultural and livestock agency) arranged visits to five farms in the Sinop area. Farms varied in size, from large (20,000 ha) to relatively small (200 ha). Farms were considered in terms of crop choice, crop rotations and cropping practices from the time of farm establishment (the mid 1980's onwards) through to anticipated activities for the future. At each farm farmers were interviewed to determine:

- Location
- Farm size
- The time the farm was established
- Current and past crops grown
- Past rotations (since farm establishment)
- Current rotations
- Tillage operations
- Inputs and timings of inputs (fertiliser, organic manure, herbicide)
- Crop residue returns
- Future intentions for crops/management practices
- Farmers opinion on areas in Mato Grosso most likely to change in the future

2.2 Findings

All five farms were initially set up to repeatedly grow a monoculture, generally soybean, with corn or millet as a second crop. All the farmers interviewed were responsible for the initial clearing of the land. Farms were set up between 8 and 24 years ago. Immediately following deforestation (the first 2-3 years), all farmers grew rice as this is a robust crop that grows well in the acid soil in the area. This practice prepares the soil for the subsequent more valuable crops. Management practices employed by the small (< 500 ha) farms differed markedly from those employed by medium (> 500 ha) and large ($> 10,000$ ha) farms (details are given below). This was an interesting finding as these differences are likely to have consequences for soil organic carbon (SOC) turnover, nitrogen cycling and GHG emissions from soils.

Fertiliser

Fertiliser inputs appear to have varied over the years as farmers have tried different systems and market prices of both fertiliser and crops have changed. Inputs were not necessarily associated with farm size. Only one farm (< 500 ha), added nitrogen (N) fertiliser to soybean. All farmers added N to rice at rates ranging from 34 - 70 kg ha^{-1}. Whether or not N was added to corn and the amount added depended on the farmer, the amount of rain and the value of corn on the market. None of the farmers added manure to any of the crops. Where the larger farms operated rotations involving livestock, manure was added by default, but in this provisional study farmers were unable to quantify how much. The variation in fertiliser application and the way this continues to fluctuate emphasizes the need for a larger study, as fertiliser use affects GHG emissions, SOC turnover and the sustainability of the soil.

Farm machinery

All of the farms were heavily mechanised, in contrast to the small scale slash-and-burn agriculture used 30 years ago. Large farms typically had a set of machines for every 700 ha, many of which used the latest technologies such as the application of Geographical Information Systems (GIS), remote sensing and tools associated with precision agriculture. Even the smaller farms had their own harvesting machines and tractors.

Tillage

All of the farms we visited of all sizes practiced no-tillage for soybean, with conventional till for the rice crop during the first 2-3 years after deforestation. The use of no-till is motivated primarily by cost, although all farmers were aware of the benefits of no-tillage to soil fertility and structure. Tillage increases decomposition rates, leading to a loss of soil organic matter (SOM) and therefore soil fertility and structural stability (Rowell, 1996). The breaking up of soil aggregates can lead to problems of soil compaction and root penetration having detrimental effects on crop productivity. The Century model allows the user to simulate tillage events on a monthly basis and slows or speeds up decomposition based on tillage related disturbance.

Crop rotations and integrated crop/livestock systems

It appeared that the smaller the farm the more simplistic the rotation used. Small and medium farms grew rice or soybean with a second crop (either corn or millet) and occasionally swapped the rice and soybean growing areas to avoid the build up of pests.

The larger farms appeared to be moving away, or to have moved away (in the last 5 years), from monoculture towards crop-livestock integration systems involving different crops and rotations. The owners of the larger farms that we interviewed informed us that almost all large farms in the area employed a similar rotation (see below and Figure 2) where the land was split into four areas, with three areas (comprising 60% of the total) rotated between constant corn, soybean-corn and rice-millet or soybean-millet and the remaining area (40% of the total) being an integrated livestock/cropping system where livestock grazed on crop residue or the whole crop (in the case of millet) and pasture planted with *Brachiaria* (a fast growing tropical grass species).

Rotation becoming popular among large farms around Sinop, Mato Grosso:

- 20% of the area is kept under corn (for cattle feed). This may vary from year to year depending on the price of corn.
- 20% ot the area is soybean with corn as a second crop.
- 20% of the area is soybean/millet or rice/millet on alternate years. Farmers may do more than one year rice or soybean depending on the price.
- 40% of the area integrated livestock and agriculture (soybean) system. They sometimes grow soybean/sorghum and allow the cows to go onto the field and eat the sorghum (the whole crop).

Every two years the farmer will rotate options 2, 3, and 4.

3 Implications for estimating C stock changes and GHG emissions; lessons from studies in Rondonia

Amazon deforestation is Brazil's largest source of GHGs emissions ($\sim$ 70%). Mato Grosso makes a significant contribution to Brazil's national GHG emissions as it has the highest annual deforestation and GHG emission rate (Nepstad et al. 2007). Although the amounts of CO_2 and other GHGs released during the forest clearing stage are by no means certain, estimates of emissions from land use following deforestation have even greater uncertainty associated with them (Morton et al. 2006). This is due to a lack of information on land management practices and the way these practices affect carbon stocks and associated GHG emissions in Amazonian soils.

One study which considered cropped systems on newly cleared areas of the Amazonian agricultural frontier was carried out by Carvalho et al. (2009). Carvalho and colleagues took SOC and GHG measurements on a large farm in Rondonia. Six fields were chosen, each of which had been deforested at different points in the past (1-6 years ago) and represented different stages in a land management system which was popular in the area. In addition, measurements were taken in an area of native vegetation, in this case Cerradão, to act as a reference. The land management system consisted of two years of rice under a conventional tillage (CT) system, followed by 1-3 years

Fig. 2. Cattle grazing on *Brachiaria brizantha* as part of an integrated livestock system at an Embrapa test site, Sinop, Mato Grosso, Brazil. This field is in a rotation with soybean and rice and will change use every 2 years. Teha natives Cerrdão vegetation can be seen in the background.

of soybean grown under a no-tillage system. Fields were limed after initial clearing and the first crop (rice) received N, P and K and were left fallow in the winter. Soybean was rotated either with sorghum or millet which received no N fertiliser, or maize which did receive N fertiliser.

Carvalho et al. (2009), evaluated the impacts of differeent crop management practices (CT and NT) which were introduced following the conversion of Amazon Cerrado into cropland. Soil C stocks, corrected for a mass of soil equivalent to the 0-30-cm layer under Cerrado, indicated that soils under NT generally had higher C storage compared to native Cerrado and CT soils. The annual C accumulation rate in the conversion of rice under CT into soybean under NT area was 0.38 Mg ha^{-1} $year^{-1}$. The authors also evaluated greenhouse gas emissions in the same study area in order to estimate soil C sequestration'. Recent definitions of soil C sequestration' for a specific agro-ecosystem in comparison with a reference ecosystem, state that sequestration should be considered as the result (for a given period of time and portion of space) of the net balance of all greenhouse gases, expressed in C-CO_2 equivalents or C-equivalents, computing all emission sources at the soil-plant-atmosphere interface. In this case although CO_2 emissions were not used in the C sequestration estimates to avoid double counting, Carvalho et al. presented CO_2 fluxes for both dry and wet seasons. In the wet season, CO_2 emissions were twice as high as in the dry season and the highest N_2O emissions occurred under the NT system. There were no CH_4 emissions to the atmosphere (negative fluxes) and there were no significant seasonal variations. When N_2O and CH_4 emissions in C-equivalent were subtracted (assuming that the measurements made on 4 days were representative of the whole year), the soil C sequestration rate of the conversion of rice under CT into soybean under NT was 0.23 Mg ha^{-1} $year^{-1}$. The authors conclude by stressing that although they found positive soil C sequestration rates, the results do not present data regarding the full C balance in soil management changes in the Amazon Cerrado.

Carvalho et al. (2009) found a decline in soil C stock of approximately 25% (compared to native vegetation) in the first year when rice was grown using CT. In the farmer interviews carried out around Sinop (reported in Section 2), all farmers grew rice immediately following deforestation (for at least 2 years) which will almost certainly be accompanied by a similarly large soil C loss. Many studies have shown an initial sharp decline in SOC following conversion from native vegetation to cropland followed by a slower steady decline until a new equilibrium is reached (Resck et al. 2000). The use of conventional tillage immediately following deforestation is carried out to ensure all remaining tree roots which were not removed by burning are grubbed up and either removed or decompose. The simultaneous break up of soil aggregates also leads to a loss of organic C from the soil (Six et al. 2002; Salton et al. 2005). The surveys around Sinop revealed that some farmers continued to grow rice using CT beyond the initial two year soil conditioning' period rather than switching to soybean. The reasons given were a perceived lack of stability in the international price of soybean and the inconvenience associated with switching (finding a buyer for the soybean, using different production methods, etc.). Across the frontier region of Mato Grosso, the number of farmers who continue to grow rice using CT is unknown but this obviously has implications for soil C stocks and long-term soil sustainability.

Rice is a less lucrative crop than soybean, therefore farmers in Mato Grosso tend to swap to soybean production after a couple of years when soil conditions are suitable (the pH has been increased by liming, and remaining tree roots have decomposed or have been picked up manually or with machines). Originally farmers were encouraged to move to the Brazilian Amazon in the 1980s to grow rice for the domestic market, however, as suitable soybean varieties became available, soybean became the dominant crop (Fearnside 2005). The surveys around Sinop found that the farmers interviewed grew soybean using no-tillage (NT) and that they all believed this to be standard practice in the area. Carvalho et al. (2009), found C accumulation rates of 0.38 Mg ha^{-1} yr^{-1} when moving from CT rice to NT soybean and even found soil C stocks in NT soybean to be similar to those under the native vegetation. Obviously these gains are small when set against the overall C losses from above ground vegetation following deforestation. They also found reduced CO_2 emissions from NT soybean compared to CT rice. So on first appraisal it appears that those farmers in Mato Grosso who choose to grow rice under CT may have higher C emissions. The picture is however complicated by another finding of Carvalho's that NT soybean had higher N_2O emissions than CT rice, a finding which is in line with other studies on tropical soils by Nobre (1994) and Six et al. (2002). As N_2O is a more potent GHG than CO_2, this could outweigh the apparent C accumulation found with the NT soybean system, depending on total emissions involved.

The findings discussed above both from Carvalho et al. (2009) and the farmer surveys have implications for past estimates of C stock changes in the Brazilian Amazon. Cerri et al. (2007) considered land use change scenarios where areas of forest were converted to pasture and then pasture to soybean production. At the time of writing, this was the dominant land use change, however today there is an increase in the area of forest and Cerradão being cleared and then put directly into agricultural production. A revision of the stock change estimates made by Cerri et al. (for the period 1990 - 2030) reflecting this new trend, would probably show a larger SOC stock loss in the first 3 years following deforestation, especially in the state of Mato Grosso. Cerri et al. (2007) did not have information on the variety of different agricultural practices across the agricultural frontier region and were therefore unable to use them in SOC stock change estimates. As Carvalho et al. (2009) showed, C stocks and GHG emissions vary for CT and NT systems. A scenario with less change to pasture and more to NT soybean could potentially reduce SOC losses but increase N_2O emissions.

Until details of the land management systems following deforestation, and the consequences that these have for C stocks and GHG emissions are understood, the true impact of land use change at the agricultural frontier in Mato Grosso on Brazil's GHG emissions will be difficult to ascertain. The preliminary interviews in the Sinop area (reported here) have shown a variety of land management practices which vary with farm size. They have also shown that new systems such as integrated livestock /cropping systems are emerging fast. A full survey across the entire area, in conjunction with measurements of GHG emissions and C stock changes, is needed in order to understand the environmental consequences of deforestation.

4 Future outlook

The UNFCCC and the wider scientific community are currently exploring ways to reward countries for reducing GHG emissions from deforestation and forest degradation - (REDD). Brazil is a prime candidate to benefit from REDD, as 70% of its national GHG emissions come from deforestation and it has a well established deforestation monitoring system (particularly in Mato Grosso (INPE 2007). However, the Brazilian government have been reluctant to endorse REDD for a number of reasons, not least the uncertainties surrounding emissions. In order for REDD to work, the real GHG cost of deforestation needs to be calculated and this should take into account GHG emissions that result from the land systems that replace the forest.

The farmers we interviewed in Mato Grosso and the workers at Embrapa all acknowledged that current production systems for both livestock and crops use land inefficiently and are, therefore, driving further deforestation. The potential for more efficient use of already cleared land is also recognised at the state level and was probably a factor which lead the governor of Mato Grosso (Brazil's largest soybean producer) to endorse an initiative put forward by a number of NGOs and endorsed by the Brazilian government National pact for valuing the Amazon forest and ending deforestation'. If deforestation is to be curbed across in the main deforesting states of the Brazilian Amazon, governments, NGOs and international organisations are going to have to work with ranchers and farmers in the area and this will only be possible if we have a clear understanding of the land management practices they employ.

Acknowledgements

We would like to thanks the USA National Science Foundation for providing a planning grant REF: OISE-0725310 to support this work. The work was also supported by a Defra grant (CEOSA 0803 REDD Phase I) which is endorsed by The Global Land Project

We would also like to thank EMPRAPA in Sinop for all their assistance, especially Flavio Jesus Wruck. The lead author would like to dedicate this chapter to her father, Mr Andrew Milne, September 1942 – February 2009.

References

Bickel U, Maarten Dros J (2003) The Impacts of Soybean Cultivation on Brazilian Ecosystems. Three Case Studies commissioned by the WWF Forest Conservation Initiative, Zurich, October 2003

Carvalho R (1999) A Amazônia rumo ao "Ciclo da soja". In: Amaznia Papers 2, Amigos da Terra, São Paulo, Setembro, 1999

Carvalho JLN, Cerri CEP, Cerri CC, Feigl BJ, Piccolo MC, Godinho VP, Herpin U (2007) Changes of chemical properties in an oxisol after clearing of native Cerrado vegetation for agricultural use in Vilhena, Rondonia State, Brazil. Soil Till. Res. 96: 95-102

Carvalho JLN, Cerri CEP, Feigl BJ, Godinho VP, Cerri CC (2009) Carbon sequestration in agricultural soils in the Cerrado region of the Brazilian Amazon. Soil Till. Res. 103: 342-349

Cerri CC, Cerri CEP, Davidson EA, Bernoux M, Feller C (2004) A ciência do solo e o sequestro de carbono. Soc. Bras. de Ci. Solo - Boletim Informativo 29: 29-37

Cerri CEP, Easter M, Paustian K, Killian K, Coleman K, Bernoux M, Falloon P, Powlson DS, Batjes NH, Milne E, Cerri C (2007) Predicted soil organic carbon stocks and changes in the Brazilian Amazon between 2000 and 2030. In: Milne E, Powlson DS, Cerri CEP (Eds.) Soil Carbon Stocks at Regional Scales. Agric. Ecosyst. Environ. 122: 58-72

D'Avila N (2003) Desmatamento na Amazônia: o novo nome da soja. Com ciência 10 October 2003 (online) http://www.comciencia.br/reportagens/agronegocio/14.shtml (Accessed 25/09/08)

Easter M, Paustian K, Killian K, Williams S, Feng T, Al Adamat R, Batjes NH, Bernoux M, Bhattacharyya T, Cerri CC, Cerri CEP, Coleman K, Falloon P, Feller C, Gicheru P, Kamoni P, Milne E, Pal DK, Powlson DS, Rawajfih Z, Sessay M, Wokabi S (2007) The GEFSOC soil carbon modelling system: a tool for conducting regional-scale soil carbon inventories and assessing the impacts of land use change on soil carbon. In: Milne E, Powlson DS, Cerri CEP (Eds.) Soil Carbon Stocks at Regional Scales. Agric. Ecosyst. Environ. 122: 13-25

Fearnside PM (2005) Deforestation in Brazilian Amazonia: History, rates, and consequences. Conservation Biology 19 (3): 680-688

Fearnside PM (2008) The roles and movements of actors in the deforestation of Brazilian Amazonia. Ecology and Society 13 (1) (online) http://www.ecologyandsociety.org/vol13/iss1/art23/

Figueiredo LD, Porro N, Pereira LS (2006) Associations in emergent communities at the Amazon forest frontier, Mato Grosso. London, UK, Instituto de Pesquisa Ambiental da Amazonia & IIED. 73 pp.

Instituto Nacional de Pesquisa Espacial - INPE (2007) Estimativas Annuais de Desmatamento. Projecto PRODES monitoramento da fl oresta Amaznica Brasileira por satlite, Available at http://www.obt.inpe.br/prodes

Kirby KR, Laurance WF, Albernaz KA, Schroth G, Fearnside PM, Bergen S, Venticinque EM, da Costa C (2006) The future of deforestation in the Brazilian Amazon. Futures 38: 432-453

Milne E, Al-Adamat R, Batjes N, Bernoux M, Bhattacharyya T, Cerri CC, Cerri CEP, Coleman K, Easter MJ, Falloon P, Feller C, Gicheru P, Kamoni P, Killian K, Pal DK, Paustian K, Powlson DS, Rawajfih Z, Sessay M, Williams S, Wokabi SM (2007) National and sub national assessments of soil organic carbon stocks and changes: the GEFSOC modelling system. In: Milne E, Powlson DS, Cerri CEP (Eds.) Soil Carbon Stocks at Regional Scales. Agric. Ecosyst. Environ. 122: 3-12

Morton DC, DeFries RS, Shimabukuro YE, Anderson LO, Arai E, Espirito-Santo F del Bon. , Freitas R Morisette J (2006) Cropland expansion changes deforestation dynamics in the southern Brazilian Amazon PNAS 103 (39): 14637-14641

Nepstad D, Soares-Filho B, Merry F, Moutinho P, Oliveira Rodrigues H, Bowman M, Schwartzman S, Almeida O, Rivero S (2007) The costs and benefits of reducing carbon emissions from deforestation and forest degradation in the Brazilian Amazon. Report from UNFCCC COP, 13th Session. 3-14 December 2007, Bali, Indonesia

Nobre AD (1994) Nitrous oxide emissions from tropical soils. Thesis (Ph.D) University of New Hampshire, New Hampshire, 103p.

Parton WJ, Schimel DS, Cole CV, Ojima DS (1987) Analysis of factors controlling soil organic matter levels in Great Plains grasslands. Soil Sci. Soc. Am. J. 51: 1173-1179

Resck DVS, VasconcelloS CA, Vilela L, Macedo MCM (2000) Impact of conversion of Brazilian Cerrados to cropland and pasture land on soil carbon pool and dynamics. In: Lal R, Kimble JM, Stewart BA (Eds.) Global climate change and tropical ecosystems. Adv. Soil Sci., Boca Raton, CRC Press. p.169-196

Rowell DL (1996) Soil Science Methods and Applications. Longman Group, U.K. pp 64-65

Salton JC, Mielniczuk J, Bayer C, Fabrício AC, Macedo MCM, Broch D, Boeni M, Conceição PC (2005) Matéria orgânica do solo na integração lavoura-pecuária em Mato Grosso do Sul. Dourados: Embrapa Agropecuria Oeste.

Sant'Anna L (2008) Queremos saber a serviço de quem o Inpe está mentindo. O Estado de São Paulo, 27 January 2008. (online) http://www.amazonia.org.br/noticias/noticia.cfm?id=260422

Sato S (2003) Novo perigo na Amazônia: plantio de soja. O Estadode São Paulo (São Paulo, Brazil) 4 July 2003

Six J, Feller C, Denef K, Ogle SM, Sá JCM, Albrecht A (2002) Soil carbon matter, biota and aggregation in temperate and tropical soils - Effects of no-tillage. Agronomie 22: 755-775

Contribution of agroforestry to biodiversity and livelihoods improvement in rural communities of Southern African regions

Kanungwe Felix Kalaba[1*], Paxie Chirwa[2], Stephen Syampungani[1], and Clifford Oluyede Ajayi[3]

[1] Copperbelt University, School of Natural Resources, P.O. Box 21692, Kitwe, Zambia
[2] Stellenbosch University, Department of Forest and Wood Science, Stellenbosch 7602, South Africa
[3] World Agroforestry Centre (ICRAF), P. O. Box 30798, Lilongwe 03, Malawi

*corresponding author: K.F. Kalaba, email: kanungwe@gmail.com or kanungwekalaba@yahoo.co.uk

Abstract

It has been widely documented that the traditional farming system of shifting cultivation contributes to huge annual losses of forest cover, altering the structure and distribution of species resulting in loss of biodiversity. On the other hand, formal institutional approaches to natural forest biodiversity conservation focused on protecting the tree species in parks and reserves while neglecting their conservation in farming systems. Improved agroforestry systems (AFS) such as improved fallows that mimic shifting cultivation and other AFS provide benefits that contribute to rural livelihoods, improved socio-economic status and ecosystem functioning of land use systems. Recently, there is an increasing recognition of the contribution of agroforestry to improve ecosystem services and livelihoods especially in rural areas. Compared with subsistence agriculture, AFS provides added benefit by generating cash income from the marketing of diverse products. In southern Africa, research that aims to addressed biodiversity and socio-economic issues includes domestication of diverse priority indigenous fruit tree species; and the evaluation of soil fertility replenishing Agroforestry technologies. This paper discusses the contribution of the natural forest resource and AFS to the improvement of the socio-economic livelihoods of smallholder farmers and the promotion of the conservation of biodiversity drawing on evidence from research conducted in southern Africa over the last two decades.

T. Tscharntke et al. (eds.), *Tropical Rainforests and Agroforests under Global Change*,
Environmental Science and Engineering, 461–476, DOI 10.1007/978-3-642-00493-3_22,

Keywords: Agroforestry systems, biodiversity, ecosystem, livelihoods, socio-economic, forest resource

1 Introduction

In Southern Africa and many developing countries, there is an inextricable link between the forest resource and the livelihoods of the rural communities. More than 80% of the rural population in sub-Saharan Africa is poor and traditionally relies on forests for most of their livelihoods including fuelwood and timber as well as other non timber forest products (Brigham et al, 1996; Schreckenberg *et al.*, 2006; Ngulube, 2000). Fuelwood provides the main source of the total household energy requirements in Southern Africa with the consumption varying from country to country namely: 85% in Mozambique (Brigham et al., 1996), 76% Zambia (Chidumayo, 1997), 91% Tanzania (SADC, 1993) and 14% in South Africa (Gander, 1994). Additionally, the rural dwellers also generate a wide range of non-timber products which include beeswax, honey, edible fruits, edible insects, wild vegetable, game meat, mushrooms, traditional medicines and fibres (Brigham et al., 1996). For example, Leakey *et al.* (2005) observed that harvesting of indigenous tree fruits from the wild can boost rural annual income by US $300-US $2000 per household. The use of wild foods such as fruits is observed throughout southern Africa: Malawi (Akinnifesi et al., 2006), Zambia (Chidumayo and Siwela, 1988), Zimbabwe (Campbell, 1987), and South Africa (Shackleton et al., 1998).

On the other hand, the widespread poverty in Southern African countries due to slow rates of economic growth (Kaimowitz, 2003) has resulted in deforestation and biodiversity loss due to overexploitation, conversion to farmland, slash and burn agriculture, charcoal production, bush fires and harvesting of wood (Akinnifesi et al., 2008; Chilufya and Tengnas, 1996; Hyde and Seve, 1993). While it could be argued that conversion to agriculture with improved inputs may have resulted in increased production; the converse is true that the economic restructuring programs in most of the developing countries resulted in prohibitively increased costs of these inputs. Consequently, the rural communities have reverted back to the use of the natural resource as a source of income albeit currently being used in an unsustainable manner as a result of increased commercialization. Hence, on the whole, the natural resource has borne the main brunt of both the agricultural revolution as well as the hard economic realities.

This paper seeks to highlight the important contribution that the forest resource is making to the livelihood of rural communities in some countries in southern Africa; pointing to the importance of maintaining biodiversity; and the contribution that agroforestry as a type of land use can make to the continued conservation and maintenance of agro-biodiversity. The paper also highlights some problems encountered in promoting agroforestry for sustainable livelihoods and maintenance of biodiversity.

2 The link of livelihood strategies of rural communities and biodiversity

Rural communities in southern Africa procure a wide variety of products from forest resources to meet their basic needs for food security, health and nutrition through collection of food, medicines, wood and pole (Shackleton and Shackleton, 2000). They are also important natural assets for rural households, providing both subsistence and market-oriented livelihood strategies. For example, in south-eastern Zimbabwe the average value of woodland goods collected was observed to be 30% of the average gross cash income per household per year (Campbell et al., 2002). The expanding commercialization of many woodland products also provides rural households with a range of market-oriented woodland livelihood opportunities. Serra and Zolho (2003) for instance, estimate that charcoal suppliers to Beira, Mozambique earn on average US$70-140 per month. The harvesting of woodland products is widely recognized as an integral component of the rural livelihoods throughout the developing world, offering goods for both household consumption and income generation (Kaimowitz, 2003).

2.1 Food security

The majority of rural communities in Southern Africa are engaged in agriculture for their livelihoods (Shackleton *et al.,* 2001). Additionally, they exploit natural resources as a means of supplementing their cash income (Brigham *et al.,* 1996). Many authors have highlighted the importance and use of woodland resource in spreading the risk associated with the availability of food over critical periods (Akinnifesi et al., 2008; Shackleton et al., 2005; Shackleton and Shackleton, 2004). Such periods include the beginning of the rainy season when food shortage is most acute as households have usually exhausted previous years harvest but the harvest for the new season has not commenced. The availability of NTFPs serves as an important gap-filler when food stocks are low (Chileshe, 2005) and also as a source of income. For example, the collection of indigenous fruits contributes between 5.5 and 6.5% to the total household income in the rural communities of Southern Africa (Akinnifesi et al., 2008). Mithoefer and Waibel (2003) estimated the returns to labour of collecting indigenous fruit tree products from communal forest areas to be equivalent to USD13.31 per day. Compared with a labour wage of less than one US dollar per day in Zimbabwe, this figure represents very high returns on investment (Akinnifesi et al., 2008). Indigenous fruit have been reported to contribute about 42% of the natural food basket in Southern Africa (Campbell et al., 1997). A survey conducted in Malawi, Zambia and Mozambique revealed that 6085% of the households lacked food during critical hunger period during the year and that these households confirmed the reduction in vulnerability by collecting fruits from woodlands (Akinnifesi et al., 2006).

Additionally, food security extends using natural woodlands for browse and fodder during drought periods (Chirwa *et al.*, 2008).

The most common edible insects are termites and caterpillars (Sileshi *et al.*, 2008). They are not only a source of food but also a source of income for the local communities surrounding the forest woodlands (Chidumayo and Mbata, 2002). According to Holden (1991), a one week collection of NTFPs can earn a person an equivalent of a months minimum wage in Zambia. Even in years of moderate abundance, edible caterpillars generate incomes of over US$60 per household that are comparable or even higher than incomes from sales of agricultural crops in the northern Zambia (Chidumayo and Mbata, 2002). In Malawi, Cunningham (1997) reported that people who participate in caterpillar collection earned an average of US$50 per person from the sale of caterpillars. Such earnings enable rural dwellers to buy goods and pay for basic services required by the household. Additionally, fruits, mushrooms, leafy vegetables, roots, bush meat and honey are important non-wood products from the woodlands.A detailed study of 36 farming households in one location in Malawi revealed that during a continuous period of 25 months a total of 37 different leafy vegetables, two root vegetables, 21 fruit and 23 mushrooms and 14 caterpillar species were collected (Lowore, 2006). The miombo ecoregion is also rich in edible mushrooms with 45 and 60 species reported in Zimbabwe and Malawi, respectively (Makonda and Gillah, 2007). Similar results were reported in Zambia (Chihongo, 1995). Honey from the woodlands of southern Africa is also an important food supplement to the rural communities; and is especially abundant under tree species such as *Isoberlinia angolensis, Julbernadia paniculata* and *Brachystegia spp.* This is especially important because of the flowering patterns of miombo species. For example, the flowering of *Brachystegia spp.* between October and December provides the farmers with food and also cash income needed to pay for agricultural inputs; while the flowering of *Julbernadia* and *Marquesia* spp. between May-June provide them with cash income without necessarily having to wait for income from the sale of agricultural crop (Jumbe *et al.*, undated).

2.2 Health and nutrition

It is estimated that more than 80% of the rural communities in sub-Saharan Africa depend on medicinal plants for most of their health needs and also for income generation (Garrity, 2004). Medicinal plants deserve special attention among the non wood products because of their importance in traditional healing and also their contribution to international trade (Syampungani *et al.*, 2008). The market in raw materials for medicinal or therapeutic plants and products of Southern Africa is readily available; between 5,000 and 10,000 tonnes are exported annually, and between 50,000 to 100,000 tonnes are consumed locally (Diederichs, 2006). The estimated market value of these products stands at US$150 million per year (Diedrichs,2006). The informal trade of medicinal plants and products in the region is dominated by four to five hundred thousand traditional healers that dispense medicines and herbal remedies to up to 100 million consumers (Diederichs, 2006). The range of plant health products has increased both within the southern African region and many parts of the world. Cunningham (1997) attributed the increase, especially in developing countries to increased urbanization and the inadequacy of conventional medicinal facilities. Local communities have exploited the leaves for treating several ailments, such as constipation, toothache, cold and cough, fever, pains, measles and Malaria (Syampungani *et al.,* 2008). There are a variety of species used in traditional healing including: *Albizia antunesiana, Brachystegia spiciformis, Rhus chirindensis, Julbernadia paniculata, Pseudolachnostylis maprouneifolia* (see Geldenhuys *et al.,* 2006). A recent study in Zambia has shown that some indigenous fruit trees are important sources of medicine for the rural people with almost two-thirds of the households using indigenous fruit trees for medicinal purposes (Kalaba *et al.,* 2009). Plant-based traditional remedies in many sub-Saharan African countries are becoming more frequently used for the treatment of HIV/AIDS related illnesses (Kayambazhinthu *et al.* 2003).

2.3 Energy and forest resource use and implications on biodiversity

Fuelwood is one of the major uses of miombo woodlands and it dominates the national energy budgets for most southern African countries (Coote *et al.* 1995; Campbell *et al.,* 2008) because it is the single most important energy source for cooking, heating and brick burning throughout Southern Africa (Geist, 1999). It accounts for high percentages of the total household energy requirements (Syampungani et al., 2009);85% in Mozambique (Brigham et al., 1996);91% in Tanzania (SADC,1993);76% in Zambia (Chidumayo, 1997) and 14% in South Africa (Gander,1994)Most of miombo woodland species are suitable to be used as fuelwoods. The preference for certain species and small dimensions (Chidumayo *et al.,* 1996) may have implications on the sustainability of supply and future availability of those particular species. Woodlands are also an important source for construction material such as poles and bark ropes. *Brachystegia* and *Bauhinia* species have been reported to be important due to their strong fibre that is also easy to peel. Rural dwellers make domestic implements used in households from the woodlands. This is because most *miombo* species used have certain attributes such as strength and resistance to splitting (Chidumayo *et al.,* 1996).

2.4 Impact of livelihoods strategies on biodiversity

The impacts of livelihoods strategies on the woodlands and individual species well-being varies. For example, the high levels of wood energy in the national energy budgets for most southern African countries make fuelwood consumption a major local and regional environmental issue (Chambwera,

2004). Neither natural nor artificial regeneration has been able to keep pace with the rate of harvesting (Syampungani et al., 2009). Fuelwood consumption together with slash and burn agriculture has accelerated forest degradation in the region (FAO, 2000). Biodiversity and nutrient losses have been cited as the major concern in the Southern African woodlands (Sileshi *et al.,* 2007). About 191 tree species are endangered while a number of animal species and small plants are threatened due to forest conversion (FAO, 2000). Conversion of woodlands due to agriculture and charcoal production deplete terrestrial carbon by drastically reducing carbon density as well as soil organic carbon (Sileshi *et al.,* 2007).

Harvesting of bark of trees for various products such as medicines, rope fibre and for making beehives can be highly destructive and lead to increased tree mortality (Chidumayo *et al.,* 1996). Beekeeping has always been considered detrimental to forestry in Miombo from time immemorial, because of the large number of trees used in hive construction, and the indiscriminate burning that was sometimes caused by honey-hunters. However, a number of methods for reducing the negative impact of bark harvesting have been proposed and tested for obtaining bark from woody material that has already been cut for other purposes and also substitution such as use of leaves for medicine (Shackleton & Clarke, 2007); improved harvesting methods (Geldenhuys *et al.,* 2006) that prevent ring barking and reduce fungal infection. Unsustainable methods for harvesting edible caterpillars have contributed to deforestation in southern Africa woodlands. Local extinction of some species due to the loss of their natural habitat and host plants, and eradication of some considered pests has been a major constraint to their sustainable utilization and harvesting (Chidumayo and Mbata, 2002; Munthali and Mughogho, 1992).

3 Agroforestry technologies and biodiversity management

The services provided by agroforestry practices to rural livelihoods and conservation of biodiversity have attracted wide attention among agroforestry and conservation scientists (Mcneely and Schroth 2006). Agroforestry technologies (AF) focus on the role of trees on farms and agricultural landscapes to meet economic, social and ecological needs (Garrity, 2006). Traditional agroforestry practices have a huge potential in supporting biodiversity conservation.

The use of agroforestry technologies mitigate biodiversity loss and provide opportunities for improving diversification and range of livelihood options for rural households (Akinnifesi *et al.* 2008).

In southern Africa, farmers have from time immemorial maintained and included trees in their landscapes. Traditionally farmers grow crops under scattered trees of different species (Akinnifesi *et al.* 2007). Some of the agroforestry technologies that are being implemented by the rural communities in southern Africa include: improved fallows, rotational woodlots and indigenous fruit trees in the parklands system.

3.1 Rotational woodlots

Rotational woodlot is an agroforestry practice whose primary goal is to increase fuelwood production. Woodlots are stands of trees planted on farms, communal land or degraded lands to provide products and services. Trees grown in rotational woodlots or scattered on crop land provide large quantities of fuel wood. Woodlots are one of the agroforestry options with the capacity to arrest deforestation and shortage of wood fuel energy in southern Africa (Akinnifesi et al., 2008). The establishment of woodlots reduces the pressure on indigenous forest by alternatively providing both wood and non-wood products to the rural communities; and so maintain the biodiversity in the natural forests. Woodlots contribute significantly to the reduction of deforestation and conserving biodiversity. The trees have the potential to increase the soil fertility and improve soil structure through nutrients contributed via the decomposition of biomass or leaf residues. The fast-growing trees provide products services such as provision of vegetative cover to reduce soil erosion.

3.2 Improved fallows

It has been widely documented that the traditional farming system of shifting cultivation contributes to huge annual losses of forest cover, altering the structure and distribution of species resulting in loss of biodiversity (Chidumayo and Mbata, 2002; Chidumayo *et al.*, 1996). On the other hand, formal institutional approaches to natural forest biodiversity conservation focused on protecting the tree species in parks and reserves while neglecting their conservation in farming systems. "Improved fallow" involves the planting of fast growing nitrogen-fixing tree species for one to two years followed by two years of cropping (Sanchez, 1999). The practice builds on the knowledge that nitrogen is the most limiting macro nutrient in the soil, but it is highly abundant in the atmosphere. Improved fallows mimic shifting cultivation but they are an improvement as they replenish the soil fertility system in a shorter period thereby contributing to the rural livelihoods and ecosystem functioning of land use systems. This can be attributed to the careful choices of species, management of tree density and accompanying silvicultural practices that distinguish improved fallows from natural fallows. Several agroforestry technologies have subsequently been developed with the aim of addressing the soil fertility improvement including alley cropping, improved fallows, coppicing fallows (see Matthews *et al.* 1992 in Akinnifesi *et al.* 2008) and using *Leucaena leucocephalla and Gliricidia sepium* to increase the yield of associated crops (Sileshi *et al.,* 2005; Matthews *et al.* 1992 in Akinnifesi *et al.* 2008). Various tree species have been utilized in improved fallows to improve yields especially by mixing species with compatible and complementary rooting or shoot-growth patterns in fallow systems and thereby diversifying the system and maximizing growth and resource utilization above and below ground (see Akinnifesi *et al.* 2008).

3.3 Indigenous fruit trees parklands system

The inclusion of indigenous fruit trees (IFTs) on agricultural land in southern Africa has been highlighted by various authors (Ngulube *et al.* 2006; Kalaba *et al.* 2008) In Zambia, Kalaba *et al.* (2009) reported that rural households intentionally retain fruit trees on their fields, by leaving trees standing in agricultural land. In Malawi, Ngulube *et al.*, (2006) highlighted the prevalence of cultural-religious restrictions governing the use and exploitation of indigenous fruit trees. For example, during woodland clearing prior to cultivation or settlement, important fruit trees such as *Parinari curatellifolia*, *Strychnos cocculoides* and *Uapaca kirkiana*, are customarily left uncut and scattered around homesteads or crop fields. Packham (1993) has reported similar cases for Tanzania, Zambia and Zimbabwe where *Parinari curatellifolia* and *Uapaca kirkiana* are left deliberately in cultivated fields. The integration of IFTs in agricultural production systems has been reported to reduce the risks inherent to monocultures of staple food crops, such as susceptibility to pests and diseases, soil nutrient depletion (Hughes and Haq, 2003). This also improves the landscape mosaic which ultimately reduces the risks of monocrops while increasing agro biodiversity in the landscape.

4 Socio-economic conditions that affect the adoption of agroforestry

Rural communities in southern Africa are faced with high poverty levels. In Zambia, Kalaba *et al.* (2009) revealed that over 90% of rural households experience regular hunger periods during the rainy season between November and April. Similar findings have been reported for Malawi, Zambia and Mozambique (Akinnifesi *et al.*, 2004). This implies that most households suffer from food insecurity, offering enough evidence of the high prevalence of rural poverty. Rural households are characterized by low literacy and lack inadequate skills and training, such as production and marketing skills. Given the profitability of agroforestry technologies (Franzel et al., 2002; Ajayi et al., 2007) and the impact that they have on households and the environment (Ajayi et al., 2004; Kwesiga et al., 2005), efforts are being made to scale up the adoption of the technology and enhance its acceptability among many more potential farmers who could benefit from the technology. Results of studies conducted in the southern African region show that farmers do appreciate agroforestry and its potential linkage to food security and household welfare indicators, but they face some challenges to the widespread uptake of agroforestry including land constraints, property rights, availability of seeds, knowledge-intensive nature of the technology. Farmer acceptability and improved adoption of the technology will be influenced by the extent to which efforts are taken to meet these challenges (Ajayi, 2007). The process of adoption of agroforestry technologies is more complicated than those for annual crops and modern agricultural development packages based on chemical inputs (Mercer 2004; Scherr and Müller 1991) because of the multi-components and the multi-years through which testing, modification and "adoption" of agroforestry takes place (Ajayi et al., 2003). A synthesis of the studies on the adoption of agroforestry in Zambia (Ajayi et al., 2003) revealed that the adoption of agroforestry is not a direct relationship based on the technological advantages of an agroforestry practice alone, but is influenced by several factors. The broad category of the factors are technology-specific (e.g. soil type, management regime), household-specific factors (e.g. farmer perceptions, resource endowment, household size), policy and institutions context within which agroforestry technologies is disseminated (input and output prices, land tenure and property rights), and geo-spatial such as tree species performance across bio-physical conditions, location of village (Ajayi et al., 2007). One way to enhance the adoption of agroforestry technologies is to target them to their biophysical and social niches, facilitate appropriate policy and institutional context for the dissemination of the technologies, understanding the broader context and dynamics of the adoption process (Ajayi et al., 2007).

Given the strong influence of the policy and institutional context within which agroforestry technologies are disseminated to potential users, it is important that efforts to scale up agroforestry should complement farmer training at the farm level with active engagement of policy makers and shapers (advocates) to facilitate policy incentives and regulations that are conducive to and encourage smallholder farmers to adopt agroforestry technologies (Ajayi et al., 2007).

4.1 Adoption of agroforestry

Farmers show their appreciation of diversity by retaining and managing trees on their farmlands. It is worth noting that farmers will only be involved in conservation of biodiversity if there is a perceived benefit. The World Agroforestry Centre (ICRAF) reports that over 480, 000 smallholder farmers in southern Africa are practicing agroforestry (ICRAF 2007). The successive adoption rate has been attributed to the adoption criteria which has been participatory in nature. Akinnifesi *et al.* (2008) reported an increase in demand in the adoption of agroforestry by farmers. The impact of agroforestry adoption on livelihoods of farmers in Malawi, Mozambique and Zambia includes increase in crop yields, increase in income, increased savings resulting in change of wealth and soil improvement. Qualitative assessments of the impact of agroforestry adoption on livelihoods of farmers in Malawi, Mozambique and Zambia are presented in Table 1.

Table 1. Qualitative assessments of impact of agroforestry adoption on livelihoods of farmers in Malawi, Mozambique and Zambia

Impact indicator	Malawi (n=31)	Zambia (n=184)	Mozambique (n=57)	Regional range
		% of respondents		
1. Increase in area under agroforestry	55	87	65	83-100
2. Yield increase (> quarter to triple)	70	90	71	83-100
3. Significant food security (> 2 months of hunger reduction)	94	84	54	66-100
4. Increase in income	58	68	53	33-83
5. Firewood availability	90	nd*	59	nd*
6. Increased savings	87	94	71	nd*
7. Change in wealth	77	84	77	77-100
8. Strong reduction in *Striga spp*	90	93	88	71-100
9. Soil improvement	84	82	59	71-100
10. Other benefits	65	nd*	24	nd*

nd* Not determined
Adapted from Akinnifesi *et al.* 2008

There are a number of factors that influence the adoption of species. Among them are household or community preferences, land tenure and inheritance rights and the availability of germplasm.

Germplasm accessibility is a critical factor that affects the adoption of agroforestry technologies. In the absence of the germplasm, rural people are left with no option but to abandon the technologies despite their superiority which can be established scientifically. For example, in Malawi, the number of farmers using *Gliricidia*-maize intercropping is relatively low as compared to those using *Tephrosia spp., Sesbania sesban and Cajanus cajan* (Akinnifesi *et al.* 2008). They cited the lack of availability of germplasm for *Gliricidia* as a factor retarding its adoption whereas the seeds of Tephrosia spp are easily accessible to farmers at minimal costs. Land tenure and inheritance rights also significantly affect adoption of AF technologies. Tree based agroforestry technologies are more negatively affected by land and tree tenure arrangements (Akinnifesi *et al.* 2008). According to Ajayi (2007), the extent to which land tenure affects adoption of agroforestry technologies varies by geographical location, type of culture and whether the technologies is tree-based or annual shrub based.

4.2 Policy and governance

The types of institutions and legislation and sectors governing the management of natural resources in Southern Africa include government ministries, para-state agencies, international conventions and religious or faith-based institutions (Oduol *et al.*, 2008). The institutions have been changing according to the changing state and administrative frameworks from colonial to post-colonial times (see Kowero, 2003). These changes have undermined the traditional institutions and organisations responsible for the management of woodland resources (Matose and Wily, 1996) and therefore have resulted in the exclusion of the local communities in managing the resources. Additionally, there is lack of clear and appropriate policies that support the development of important agroforestry products such as non wood forest products with economic potential (Oduol *et al.*, 2008). The available policies and laws governing forest exploitation are restrictive in nature, through control for protection (Kayambazinthu *et al.*, 2003). They put emphasis on non-consumptive utilization of protected resource (Munthali and Mughogo, 1992). Policies conducive to the promotion of agroforestry are lacking in Southern Africa (Syampungani *et al.*, 2008). It is therefore important that policies that promote agroforestry are put in place.

5 Conclusion

The natural forest resource continues to play a major role in improving the livelihood of rural communities in Southern Africa, and this it does, because of the rich biodiversity in forests. Thus, natural forests are able to provide for energy, food and nutrition and health. However, the current levels of deforestation which cause land degradation, soil nutrient depletion, loss of natural habitats and therefore change in structure and composition of the natural woodlands pose a threat to the contribution of the southern woodlands to rural livelihoods. Improved agroforestry systems have the potential to contribute to the maintenance of biodiversity in natural systems due to the reduction in overreliance of rural communities on natural forest resources; as they are able to maintain their production systems through improved AFS. AFS have inadvertently resulted in improved agro-biodiversity because of the multiple components involved; and through mimicking traditional systems in some cases, the so called new AF technologies have easily been adopted. However, there is still a strong need to promote AFS through increased dissemination of germplasm and advocacy to policy makers.

References

Ajayi OC, Akinnifesi FK, Sileshi G, Chakeredza G (2008) Comparative Evaluation of Labor Use and Profitability of Renewable Soil Fertility Replenishment Technologies in Southern Africa. Proceedings of the Tropentag Conference Competition for Resources in a Changing World: New Drive for Rural Development, October 7-9, University of Hohenheim, Germany. http://www.tropentag.de/abstracts/full/708.pdf

Ajayi OC, Place F, Kwesiga F, Mafongoya P (2007) Impacts of Improved Tree Fallow Technology in Zambia. In: Waibel H, Zilberman D (eds.) International Research on Natural Resource Management: Advances in Impact Assessment. CABI Wallingford, UK and Science Council/CGIAR, Rome pp.147-168 ISBN: 976-1-84593-283-1

Ajayi OC, Franzel S, Kuntashula E, Kwesiga F (2003) Adoption of improved fallow soil fertility management practices in Zambia: synthesis and emerging issues. Agroforestry systems 59 (3): 317-326

Ajayi OC , Akinnifesi FK, Gudeta S, Chakeredza S (2007) Adoption of Renewable Soil Fertility Replenishment Technologies in Southern African Region: Lessons Learnt and the Way Forward. Natural Resource Forum 31 (4): 306-317

Akinnifesi FK, Sileshi G, Ajayi OC, Chirwa PW, Mngomba S, Chakeredza S, Nyoka BI (2008) Domestication and conservation of indigenous Miombo fruit trees for improving rural livelihoods in Southern Africa. Tropical Conservancy 72-74

Akinnifesi FK, Sileshi G, Ajayi OC, Chirwa PW, Kwesiga F, Harawa R (2008) Contributions of agroforestry research and development to livelihood of smallholder farmers in Southern Africa: 2. Fruit, medicinal, fuelwood and fodder tree systems. Agricultural Journal 3 (1): 76-88

Akinnifesi FK, Kwesiga FR, Mhango J, Chilanga T, Mkonda A, Kadu CAC, Kadzere I, Mithofer D, Saka JDK, Sileshi G, Ramadhani T, Dhliwayo P (2006) Towards the developmentof Miombo fruit trees as commercial tree crops in southern Africa. Forests, Trees and Livelihoods 16: 103121

Brigham T, Chihongo A, Chidumayo E (1996) Trade in woodland products from the miombo region. In: Campbell B (ed.), The Miombo in Transition: Woodland and Welfare in Africa. Centre for International Forestry Research (CIFOR), Bogor,pp. 136-174

Campbell BM (1987) The use of wild fruits in Zimbabwe. Economic Botany 41 :375-385

Campbell B, Jeffrey S, Kozanayi W, Luckert M (2002) Household Livelihoods in Semi-Arid Regions: Options + Constraints. Jakarta, Indonesia, Center for International Forestry Research (CIFOR)

Chambwera M (2004) Economic Analysis of Urban Fuelwood Demand: The case of Harare in Zimbabwe. PhD Thesis, Wageningen University, The Netherlands

Chidumayo EN, Gambiza J, Grundy I (1996) Managing miombo woodlands. In: Campbell BM (ed.), The Miombo in Transition: Woodlands Welfare in Africa. Centre for International Forestry Research, Bogor, pp.175-193

Chidumayo EN (1997) Miombo ecology and management: an introduction. Stockholm Environment Institute, Stockholm, Sweden

Chidumayo EN, Mbata KJ (2002) Shifting cultivation, edible caterpillars and livelihoods in the Kopa area of northern Zambia. Forests, Trees and Livelihoods 12: 175-193

Chihongo AW (1995) Tanzanian perspective towards non wood products from Africa. A resource paper prepared for the African Academy of Sciences round table discussion on non wood forest products. Pretoria, South Africa, 21-23 November, 1995

Chikamai B, Tchatat M (undated) Forest Management For Non-wood Products and Services in Africa. A report prepared for the project lessons on Sustainable Forest Management in Africa. AFORNET/FAO/KSLA. http://www.ksla.se/sv/retrieve_file.asp. Date visited: 19/07/2008

Chileshe RA (2005) Land tenure and rural livelihoods in Zambia: A case studies of Kamena and St. Joseph. PhD Thesis. Faculty of Arts, University of The Western Cape, South Africa

Chilufya H, Tengnäs B (1996) Agroforestry extension manual for northern Zambia. Regional Soil Conservation Unit (RSU), Technical Handbook Series 11, p.124

Chirwa TS, Mafongoya PL, Chintu R (2003) Mixed planted-fallows using coppicing and non-coppicing tree species for degraded Acrisols in eastern Zambia. Agroforest. Systems. 59: 243-251

Chirwa PW, Akinnifesi FK, Sileshi G, Syampungani S, Kalaba FK, Ajayi OC (2008) Opportunity for conserving and utilizing agrobiodiversity through agroforestry in Southern Africa. Biodiversity 9 (1&2) 45-48

Cunningham AB (1997) An Africa-wide overview of medicinal plant harvesting, conservation and healthcare. Medicinal plants for forest conservation and healthcare. Non wood forest products 11, FAO, Rome, Italy

Diederichs N (ed., 2006) Commercialising Medicinal Plants: A Southern African Guide. Sun Press, Stellenbosch

Franzel S, . Phiri D, Kwesiga F (2002) Assessing the adoption potential of improved fallows in eastern Zambia. In: Franzel S, Scherr SJ (eds.) Trees on the Farm: Assessing the Adoption Potential of Agroforestry Practices in Africa. CAB International, Wallingford, UK, pp 37-64

Food and Agriculture Organization (FAO, 2000a) Global Forest Resources Assessment Available at
http://www.fao.org/docrep/004/Y1997E/y1997elr.htm bm63. Last accessed: July 22, 2008

Food and Agriculture Organization (FAO, 2000) The elimination of food insecurity in the Horn of Africa, A strategy for concerted government and UN agency action. FAO, Rome.
Available at http://www.fao.org/docrep/003/x8406e/X8406e00.HTM

Gander M (1994) Domestic Energy used by Farm workers Living on Farmland in Natal and Transvaal. Report to the Department of Mineral and Energy Affairs, South Africa

Garrity DP (2004) Agroforestry and achievement of the Millennium Development Goals. Agroforestry Systems 61: 5-17

Geist JH (1999) Global assessment of deforestation related to tobacco farming. Tobacco Control 8: 18-28

Geldenhuys CJ, Syampungani S, Meke G, Vermeulen WJ (2006) Response of different species on bark harvesting for traditional medicine in South Africa. In: Beste JJ, Seydack AWH, Vorster T, van der Merwe IJ, Dzivhani S (eds.) Multiple use management of natural forests and woodlands: Policy refinement and scientific progress. Natural Forests and Savanna woodlands Symposium IV Port Elizabeth, South Africa, 15-18 May 2006

Holden S (1991) Edible caterpillars - A potential agroforestry resource? Food Insects Newsletter 4 (3): 3-4

Hyde WF, Seve JE (1993) The economic role of wood products in tropical deforestation: The severe example of Malawi. Forest Ecology and Management 57: 283-300

Hughes A, Haq N (2003) Promotion of indigenous fruit trees through improved processing and marketing in Asia. International Forestry Review 5 (2 special issue: NTFPs revisited): 176-181

Jumbe BL, Bwalya SM, Husselman M (Undated) Contribution of dry forests to rural livelihoods and the national economy in Zambia. http://www.cifor.cgiar.org/miombo/doc. Dated visited: 19/09/2008

Kalaba FK, Chirwa PW, Prozesky H, Ham C (2009) The role of indigenous fruit trees in rural livelihoods: the case of communities around the Mwekera area, Copperbelt province, Zambia. In: Jaenicke H, Ganry J, Hoschle-Zeledon I, Kahane R (eds.) Underutilized plants for food, Nutrition, Income and Sustainable Development. Proceedings of International Symposium held in Arusha, Tanzania 3-7 March, 2008. Acta Horticulturae 806. International Society for Horticultural Science. Leuven, Belgium 739pp

Kaimowitz D (2003) Not by bread alone: forests and rural livelihoods in sub-Saharan Africa. In: Oksanen T, Pajari T, Tuomasjukka T (eds.) Forests in Poverty Reduction Strategies: Capturing the Potential. EFI: joensuu 65-86

Kayambazinthu D, Matose F, Kajembe G, Nemarundwe N (2003) Institutional arrangements governing natural resource management of the miombo woodland. In: Kowero G, Campbell BM, Sumaila UR (eds.) Policies and Governance Structures in Woodlands of Southern Africa. CIFOR, Bogor, Indonesia, p 45-64

Kowero G (2003) The challenges to natural forest management in sub-Saharan Africa rural development: experiences from the miombo woodlands of southern Africa. In: Kowero G, Campbell BM, Sumaila UR (eds.) Policies and Governance Structures in Woodlands of Southern Africa. CIFOR, Bogor, Indonesia, p 1-8

Kwesiga F, Franzel S, Mafongoya P, Ajayi OC, Phiri D, Katanga R, Kuntashula E and Chirwa T (2005) Successes in African Agriculture: Case Study of Improved Fallows in Eastern Zambia. Environment and Production Technology Division (EPTD) Discussion Paper 130, IFPRI, Washington DC

Leakey RRB, Tchoundjeu Z, Schreckenberg K, Shackleton SE, Shackleton CM (2005) Agroforestry Tree Products (AFTPs): Targeting poverty reduction and enhanced livelihoods. International Journal of Agricultural Sustainability 3: 1-23

Lowore J (2006) Miombo woodlands and rural livelihoods in Malawi. International Forestry Research (CIFOR), Bogor, Indonesia

Makonda FBS, Gillah PR (2007) How to balance variable wood and non wood products and uses in miombo woodlands.
http://www.metla.fi/hanke/8512/esitelmat-tanzania-2007.
Date visited 17/08/08

Mithöfer D, Waibel H (2003) Income and labour productivity of collection and use of indigenous fruit tree products in Zimbabwe. Agroforestry Systems 59: 295305

Muntali SM, Mughogho DEC (1992) Economic incentives for conservation: bee-keeping and saturinid caterpillar utilization by rural communities. Biodiversity Conservation 1: 143-54

Ngulube MR, Mwabumba L, Chirwa PW (2006) Socio-economic aspects of Miombo woodlands in Malawi: The case of Non-Timber Forest Products. Natural Forests and Woodland Savanna Symposium IV, Port Elizabeth, South Africa. 15-18 May 2006

Ngulube MR (2000) The utilization and marketing of non-timber forest products from the Miombo woodlands of Malawi: a case study. In: Ngulube RM, Mwabumba L, Chirwa PW (eds.) Community-based management of Miombo woodlands in Malawi: proceedings of a national workshop, Mangochi, Malawi, 2729 September 1999. Forestry Research Institute of Malawi, Zomba. p 4469

Oduol PA, Ajayi OC, Matakala P, Akinnifesi FK (2008) The role of Institutional Arrangements and Policy on the Conservation, Utilization and Commercialization of Indigenous Fruits in Southern Africa. In: Akinnifesi FK, Leakey RRB, Ajayi OC, Sileshi G, Tchoundjeu Z, Matakala P, Kwesiga FR (eds.) Indigenous Fruit Trees in the Tropics: Domestication, Utilization and Commercialization. CABI/ICRAF, Wallingford. pp 310-321.

Packham J (1993) The value of indigenous fruit- bearing trees in miombo woodland areas of South Central Africa. Rural Development Forest Network Paper, Overseas Development Institute, London

Pattanayak SK, Sills EO (2001) Do tropical forests provide natural insurance? The microeconomics of non-timber forest product collection in the Brazilian Amazon. Land Economics 77 (4): 595-612

Peham APK (1996) NWFP development in Malawi. In: FAO (Ed.) Commonwealth Science Council: Non-Wood forest products: A regional expert consultation for English-speaking African Countries, 17-22, October 1993, Arusha, Tanzania

Schreckenberg K, Awono A, Degrande A, Mbosso C, Ndoye O, Tchoundjeu Z (2006) Domesticating indigenous fruit trees as a cntribution to poverty reduction. Forests, Trees and Livelihoods 16: 35-51

Serra A, Zolho R (2003) Inqurito Sobre a Produo e Consumo de Combustvel Lenhoso na Cidade da Beira. SAfMA Internal Report

Shackleton SE, Dzerefos CM, Shackleton CM, Mathabela FR (1998) Use and trading of wild herbs in the central lowveld savanna region, South Africa. Economic Botany 52 (3): 251-259

Shackleton CM, Shackleton SE (2000) Direct use values used in woodlands. In: Owen DL (ed.) The Southern African Forestry Handbook. Pretoria, South African Forestry Institute. pp 635-41

Shackleton CM, Shackleton SE, Cousins B (2001) The role of land-based strategies in rural livelihoods: the contirbution of arable production, animal husbandry and natural resource harvesting in communal areas in South Africa. Development Southern Africa 18 (5): 582-604

Sileshi G, Akinnifesi FK, Ajayi OC, Chakeredza S, Kaonga M, Matakala PW (2007) Contributions of agroforestry to ecosystem services in the miombo ecoregion of eastern and southern Africa. African Journal of Environmental Science and Technology 1 (4): 68-80

Southern African Development Community (SADC) (1993) Energy Statistics Year Book for 1991. Energy Sector, TAU, Luanda, Angola

Syampungani S, Chirwa PW, Akinnifesi FK, Sileshi G, Ajayi OC (2009) The miombo woodlands at the cross roads: potential threats, sustainable livelihoods, policy gaps and challenges. Natural Resources FORUM 33: 150-159

Human ecological dimensions in sustainable utilization and conservation of tropical mountain rain forests under global change in southern Ecuador

Perdita Pohle[1]*, Andrés Gerique[1], Martina Park[1], and María Fernanda López Sandoval[2]

[1] Institute of Geography, Friedrich-Alexander-University Erlangen-Nürnberg, Kochstrasse 4/4, 91054 Erlangen, Germany
[2] Escuela de Geografía, Facultad de Ciencia Humana, Pontificia Universidad Católica del Ecuador (PUCE), Quito

*corresponding author: P. Pohle, email: ppohle@geographie.uni-erlangen.de

Summary

Profound knowledge of region-specific human ecological parameters is crucial for the sustainable utilization and conservation of tropical mountain rain forests in southern Ecuador, a region with heterogenic ethnic, socio-cultural and socio-economic structures. In order to satisfy the objectives of forest conservation on the one hand and the utilization claims of the local population on the other, an integrated concept of nature conservation and sustainable land use development is being sought (e.g. Ellenberg 1993). Within the human ecological research project of the German Research Foundation (DFG) presented here, four research topics have been explored in detail in indigenous Shuar and Saraguro as well as local Mestizo communities of southern Ecuador:

Research topic 1 is concerned with the use of wild plants and local agrobiodiversity. In the research area land use is focused on cattle ranching, which poses the main threat to forests and to biodiversity. Based on an ethnobotanical survey conducted in Shuar, Saraguro and Mestizo communities, the use of non-timber forest products (NTFPs) and the cultivation of plant species in demand in home gardens have been identified as promising options for increasing household incomes.

Research topic 2 deals with the local people's perception and evaluation of the natural environment, environmental stress/risk factors and conservation measures. Although deforestation is an ongoing process within the research area, Saraguro and Mestizo farmers clearly prefer to live in areas with forest

T. Tscharntke et al. (eds.), *Tropical Rainforests and Agroforests under Global Change*,
Environmental Science and Engineering, 477–509, DOI 10.1007/978-3-642-00493-3_23,

and highly value the economic functions of the forests, e.g. as an agricultural reserve which can be inherited by their children. Conservation measures have long since been established in the research area, but local people are not aware and have not been informed about resource use regulations and restrictions.

Research topic 3 investigates livelihood strategies of local communities which to varying degrees depend on natural resources. Whereas the Shuar's livelihood strategies to a large extent depend on subsistence agriculture (shifting cultivation) combined with fishing, hunting, and gathering of wild plants, the Saraguros and Mestizos are mainly engaged in agro-pastural activities that combine market economy (cattle ranching) and subsistence economy (crop production and horticulture).

Research topic 4 is concerned with the determination of the political and administrative use agreements including land tenure systems. North of Podocarpus National Park current land use and land tenure conflicts are founded primarily on the colonization process starting in the 1950s, and are severely dependent on state policies on land adjudication, increase of accessibility, and national or international concern for nature conservation and environmental protection.

Despite the remoteness of many communities living in the tropical rainforests of southern Ecuador, the local people have never been completely isolated from global processes of ecological, political, cultural and economic changes. Recently, the research area was declared as Biosphere Reserve Podocarpus – El Cóndor. Since biosphere reserves are strongly rooted in cultural contexts, in southern Ecuador it thus can be the vehicle for protecting tropical mountain ecosystems and developing sustainable forms of land use at the same time.

Keywords: Human ecological dimensions, tropical mountain forests, sustainable utilization, conservation, Ecuador, indigenous people, agrobiodiversity, ethnobotany, non-timber forest products (NTFPs), local perception of the environment, local perception of conservation measures, livelihood strategies, household incomes, land tenure conflicts, global change, conservation strategies, cultural diversity, human ecology, ethno-ecology

1 Introduction: conceptual framework, research area and ethnic groups

The forests of the Andean Amazon constitute one of the most important hotspots of biodiversity worldwide (Jørgensen & Ulloa Ulloa 1994, Myers et al. 2000, Barthlott et al. 2007). However, this region contains some of the world's most rapidly changing landscapes due to deforestation (FAO 2007), and also faces the threat of stress from climate change (Malhi et al. 2008). Both problems are directly linked, as deforestation is considered to be responsible for approximately 20% of the annual global carbon dioxide emissions

(UNFCCC 2008). Most land conversion can be attributed to new settlers or colonists (Pichón 1996, Gerold & Lanfer 2001, Tourrand et al. 2008) but indigenous peoples as well are turning to more intensive land uses and are assimilating themselves into local and regional market economies (Sierra 1999, Putsche 2000, Rudel et al. 2002, Gray et al. 2008). According to the Fourth Assessment Report of the IPCC (2007) the reduction and prevention of deforestation represents the mitigation option of carbon dioxide emissions with the largest and most immediate carbon stock impact in the short term. On its part, the Stern review (2006: 217) considers this option to be the most economical to stop global warming. Thus, to counteract climate change, projects that address the drivers of deforestation should be developed. Moreover, to understand how local people use forest resources is of utmost importance to develop sustainable productive alternatives that reduce deforestation while encouraging local development and poverty alleviation. In order to be successfully developed and implemented, these alternatives should take local and ethnic particularities into account.

In the agricultural frontier zone of southern Ecuador, a region of heterogenic ethnic, socio-cultural and socio-economic structures, profound knowledge of human ecological dimensions is crucial for the sustainable utilization and conservation of tropical mountain rain forests. Human ecological dimensions may encompass various aspects of the interplay of individuals or social groups with their natural environment (Weichhart 2007). In biodiversity-rich places local people usually have a detailed ecological knowledge, e.g. of species, ecosystems, ecological relationships and historical or recent changes to them (Warren et al. 1995, Alcorn 1999). At the local level, utilitarian and socio-cultural values as well as perceptions and beliefs are the driving forces behind use, management and conservation of natural resources. Economic and political factors also influence people's decision-making. Under current pressures of deforestation, fragmentation and species extinction, there is a need to thoroughly study the issues of environmental knowledge and perception, rural livelihood strategies, land use conflicts and land tenure systems. The analysis and evaluation of these four research topics or human ecological dimensions is essential for the sustainable management of a megadiverse mountain ecosystem and has thus been the focus of research within the human ecological research project of the DFG Research Unit FOR 816: "Biodiversity and sustainable management of a megadiverse mountain ecosystem in South Ecuador".

The conceptual framework of the human ecological research project is given in Figure 1. The overall problems in the research area are deforestation (Mosandl et al. 2008), loss of biodiversity (Koopowitz et al. 1994 cit. in Mosandl et al. 2008: 38), land degradation (Harden 1996, Göttlicher et al. 2009), land use conflicts (Pohle 2004: 20) and rural poverty (INEC 2003) – shown in their sometimes explosive character. At the centre of the research are the local people – either individuals or social groups - in their interaction with the natural environment. The main goal of the research project is to identify

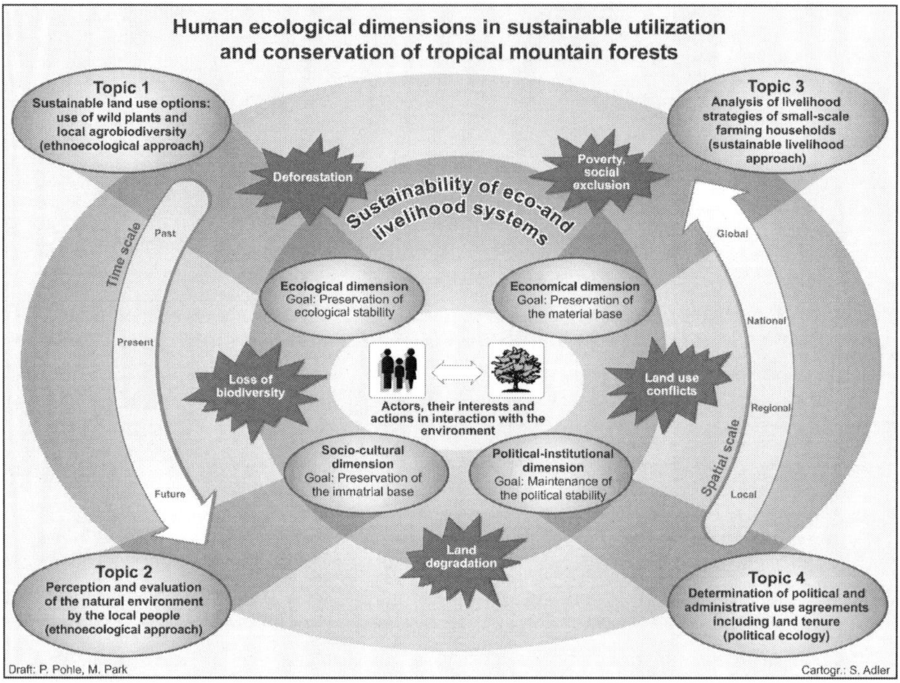

Fig. 1. Conceptual framework of the research project.

development strategies for achieving the sustainability of eco- and livelihood systems (e.g. Farrington et al. 1999). The project chose to use the human-environment related approaches of geographical development research (Scholz 2004), including the concepts of human ecology, ethno-ecology, political ecology and the sustainable livelihood approach. These integrative approaches are applied to explore interrelations and interactions between humans and societies with their natural environment. As conceptual and methodological frameworks they can considerably contribute to the solution of the conflict "protection versus utilization" of tropical mountain forests. The ethnoecological approach (e.g. Münzel 1987, Posey & Balée 1989, Warren et al. 1995, Müller-Böker 1999, Nazarea 1999) is used with particular focus on the use of non-timber forest products (NTFPs) and local agrobiodiversity, as well as the perception and evaluation of the natural environment by local ethnic groups (Fig. 1). The sustainable livelihood approach (Chambers & Conway 1992, Chambers 1995, Bohle 2001, Derichs & Rauch 2001, Krüger 2003) is used to identify ethno-specific rural livelihood strategies and their dependence on natural resources. The concept of political ecology as an actor-oriented multi-layer approach is used to identify conflicts over utilization and agreements on resource use and land tenure systems (Blaikie & Brookfield 1987, Bryant & Bailey 1997, Krings 2008). All three concepts (ethnoecology, sustainable livelihood approach, political ecology) are part of human ecology (geographical human ecology: e.g. Steiner & Wiesner 1984, Meusburger & Schwan 2003, Serbser 2004, Nentwig 2005).

The research area comprises the northern and eastern surroundings of the Podocarpus National Park within the Biosphere Reserve Podocarpus – El Cóndor (Fig. 2). Given their location between the Andean highlands and the lowlands of the Amazon, the tropical mountain rainforests of the eastern Andean slopes in southern Ecuador have an extraordinarily rich biodiversity. Podocarpus National Park and its surroundings are especially noteworthy for their species diversity and are a hotspot of vascular plant diversity worldwide (Myers et al. 2000, Barthlott et al. 2007). The tropical mountain rainforests of southern Ecuador play an important role as habitat for flora and fauna, and are of great relevance for the preservation of genetic resources (Die Erde 2001, Beck et al. 2008). At the same time, humans have lived here and sustained themselves for centuries. However, during the past five decades, these mountain rainforest ecosystems have come under enormous pressure due to the expansion of agricultural land - especially pastures (Mosandl et al. 2008, Göttlicher et al. 2009), the extraction of timber (Wunder 1996), the mining of minerals (Confeniae 2007, Mining Watch Canada 2007) the tapping of water resources and other forms of human intervention (e.g. road construction, power supply lines, Beck et al. 2008). According to the FAO report (2007), the annual deforestation rate of 1.7% for Ecuador is the highest of all South American countries. Within the research area (catchment area of the Tambo Blanco) a deforestation rate of 1.16% (1976-1987) and 0.86% (1987-2001) was estimated by Tutillo (2009) using satellite images and aerial photograph analysis.

The population figures in the research area vary significantly between the two provinces of Loja in the west and Zamora Chinchipe in the east (Pohle 2008). Loja province, representing an ancient cultural landscape, has a significantly larger total population with 404,835 inhabitants according to the 2001 census, and a significantly higher population density with 36.8 inhabitants/ km^2. The province of Zamora Chinchipe, being an area of recent agricultural colonization, has a total population of 76,601 inhabitants and a population density of only 7.3 inhabitants/ km^2. Further, the two provinces differ in ethnic composition. In each case, Mestizos represent the major population group at 92.8% in Loja, and 83.2% in Zamora Chinchipe province (2001), but Zamora Chinchipe has a significantly larger proportion of indigenous inhabitants (12.2%) than Loja (3.0%). In Loja province the resident indigenous communities are for the most part Saraguros, while in Zamora Chinchipe the dominant indigenous communities are the Shuar (Fig. 2).

The ***Shuar*** area of settlement extends from the lower levels of the tropical mountain rainforest (approx. 1,400 m a.s.l.) down to the Amazonian lowland (Oriente) in the region bordering Peru (Fig. 2). The Shuar, Amazonian Indians, belong to the Jívaro linguistic group. They are typical forest dwellers who practice shifting cultivation in a subsistence economy (Fig. 3). They also

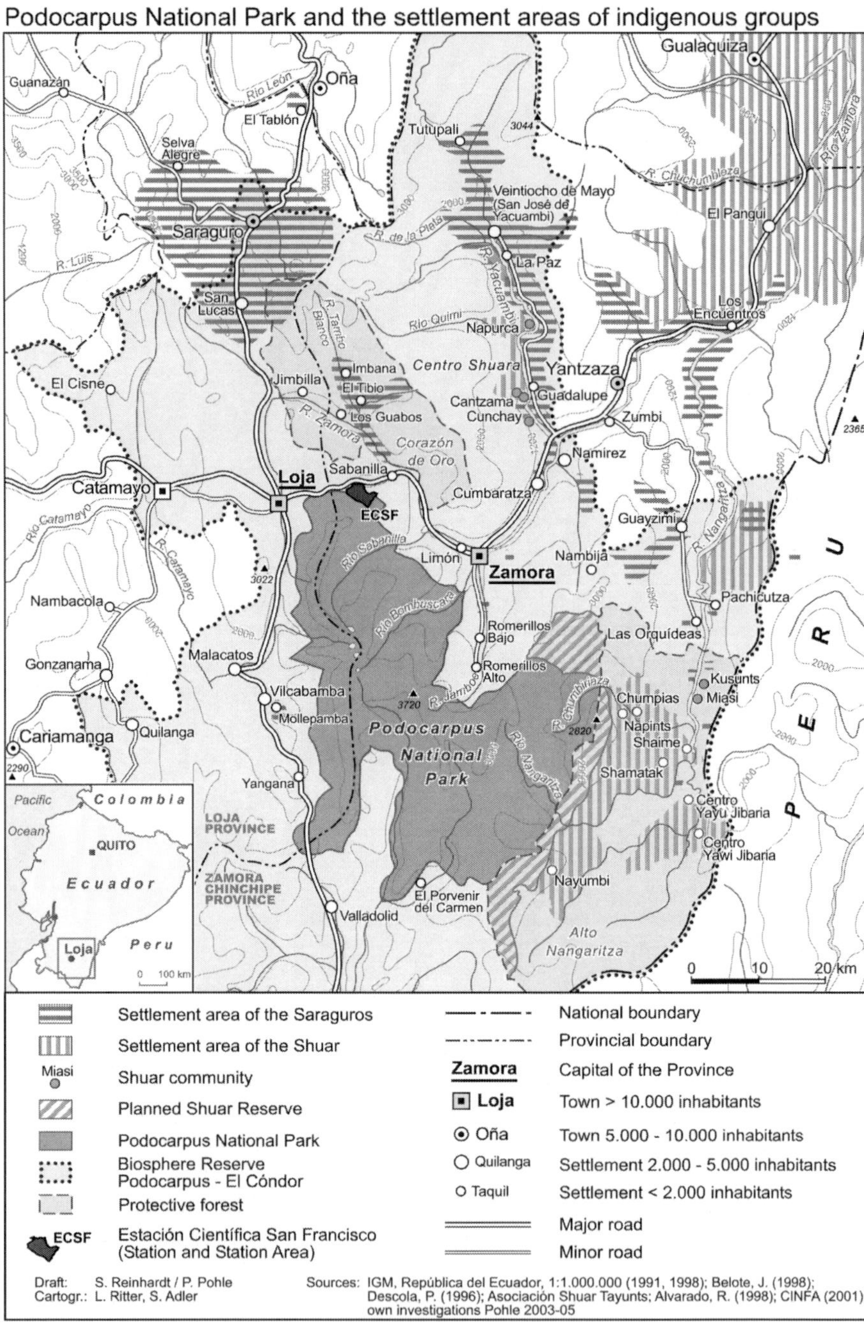

Fig. 2. The Podocarpus National Park and the settlement areas of indigenous groups.

fish, hunt and gather forest products. During the last decades some Shuar have begun to raise cattle and some are also engaged in timber extraction.

Fig. 3. Shuar women from Shaime taking a rest in their forest garden. Photo by A. Gerique.

The ***Saraguros*** are Quichua-speaking highland Indians who live as agro-pastoralists, in the most part in the temperate mid-altitudes of the Andes (Sierra) between 1,700 and 2,800 m a.s.l. (Fig. 4). As early as the 19th century the Saraguros kept cattle to supplement their traditional "system of mixed cultivation", featuring maize, beans, potatoes and other tubers (Gräf 1990). Now, cattle ranching has developed as the main branch of their economy.

The ***Mestizos***, a term generally used to indicate people of mixed Spanish and indigenous descent, are a very heterogeneous group who either live in towns, rural communities or scattered farms (*fincas*). In the area north of Podocarpus National Park they arrived from the 1960s onwards, encouraged by the national land reform of 1964, to log timber and to practice cattle farming and agriculture. As colonizers they converted large areas of tropical mountain rainforests into almost treeless pastures (Fig. 5).

Fig. 4. Saraguro woman with her husband and children from El Tibio. Photo by P. Pohle.

2 Aims and methods of the human ecological research in southern Ecuador

Within the DFG Research Unit FOR 402 and FOR 816, beginning in 2004 and continuing until now, ethnoecological, agro-geographical, household-economical and political ecological research was undertaken in sample communities of indigenous Shuar and Saraguros as well as local Mestizos settling around Podocarpus National Park. The overall goal was to document the traditional ecological knowledge among the three ethnic groups, to analyse current forms of land use including the cultivation of forest and home gardens, and to evaluate ethno-specific life-support strategies as well as strategies for natural resource management. On the basis of these overriding objectives of inquiry, the aims of the presented human ecological research project may be summarized as follows (cf. Fig 1):

Topic 1: to identify sustainable land use options, especially regarding the use of wild plants (non-timber forest products, NTFPs) and local agrobiodiversity;

Fig. 5. Mestizos on their pasture in the Río Zamora valley. Photo by E. Tapia.

Topic 2: to make the ethno-specific knowledge, perception and evaluation of the natural environment, environmental stress/risk factors and conservation measures transparent;

Topic 3: to document and analyse ethno-specific livelihood strategies of rural farming households and their impact on natural resources;

Topic 4: to determine the political and administrative use agreements including land tenure systems.

Topic 1: To document traditional ecological knowledge and to identify sustainable land use options, an ethnobotanical survey was carried out between 2004 and 2007 in sample communities of the Shuar (Shaime, Chumpias, Napints), the Saraguros (El Tibio, El Cristal) and the Mestizos (Los Guabos, 12 fincas along the Loja – Zamora road). Various techniques of unstructured and semi-structured inquiry, such as participant observation techniques, artifact and plant use interviews, checklist interviews and group interviews (Alexiades 1996) were used. The ethnobotanical inventory technique was the main procedure used to collect ethnobotanical information. An inventory of wild and cultivated plants was compiled including botanical names as well as indigenous and Spanish names, with plants classified according to use categories such as medicine, food, fodder, construction material etc. (Gerique & Veintimilla 2008). A total of 68 male and female informants ranging from 14 to 75 years of age were interviewed. In addition, 13 semi-structured expert interviews were conducted.

Topic 2: To reveal the local people's perception and evaluation of the natural environment, environmental stress/risk factors, and conservation measures, the following investigations have been undertaken in 2008 and 2009 in the Saraguro community of El Tibio and the Mestizo community of Los Guabos:

- Gathering of qualitative data with the help of contrasting photographs concerning the perception and evaluation of different cultural landscapes;
- Based on a standardised questionnaire, assessment of environmental stress/ risk factors like land degradation (landslides), deforestation, loss of biodiversity etc. perceived by the rural population (awareness, reaction, risk-avoiding strategies);
- Based on a half-standardised questionnaire, investigation of the local peoples' attitudes towards nature conservation and conservation measures.

Since the perception survey is still under way, the findings are for the time being based on a limited number of interviews: 2 group discussions as a pre-test, and 22 individual interviews, each directed towards the perception of different cultural landscapes, environmental stress/risk factors and conservation measures.

Topic 3: In accordance with the concept of sustainable livelihoods (Chambers & Conway 1992), a household survey was conducted from September to November 2008 in six rural communities, including the Saraguro community of El Tibio and the Mestizo community of Los Guabos.[1] In these two communities a complete inventory of households was undertaken comprising data from 48 households and 240 permanently present household members. The household survey included 161 mainly standardized questions directed towards the five capitals or resources human, social, natural, physical, and financial (Fig. 6). Also included was general information about household composition and biographical data of each household member. All data were entered into a SPSS database. Analysis and interpretation of the data is being undertaken.

Topic 4: Data and information concerning the colonization process, land use conflicts and lend tenure systems were obtained through a combination of secondary data research with empirical field studies undertaken in 2008 and 2009. Historical data on land tenure and forest policy were gathered from literature, and in public and private archives. Statistical data on land tenure were obtained from the rural land register of the Cantón Zamora and the agrarian census of 1954, 1974, and 2001. To get in-depth information about land use and land tenure conflicts, 11 thematically focused interviews were conducted with representatives of governmental (Ministry of Environment, Municipality

[1] Four communities are situated in the Yacuambi Valley north of Zamora and will not be discussed in this paper. In total, 148 households comprising 731 persons participated in the survey.

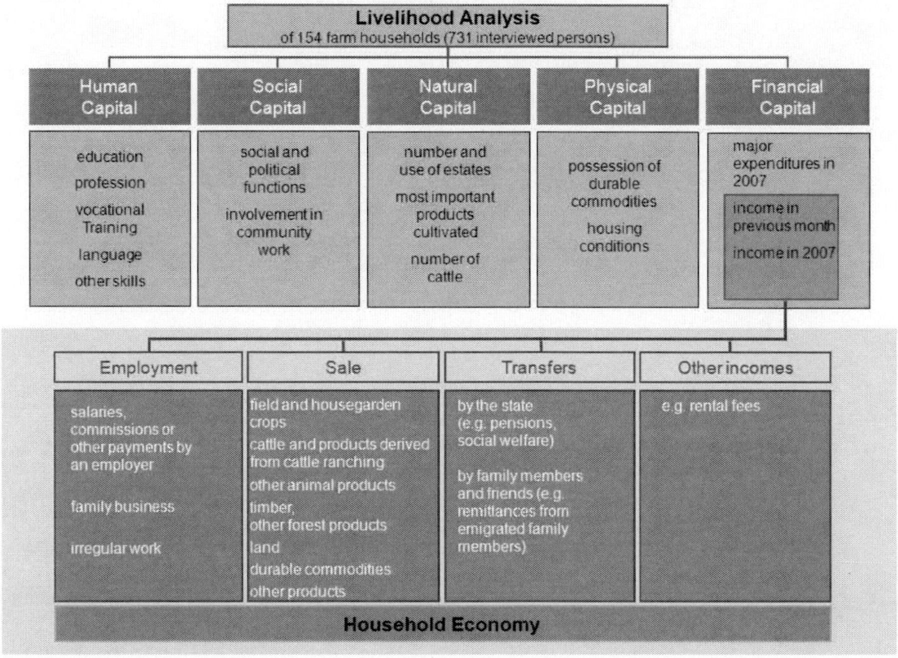

Fig. 6. Framework of the livelihood analysis conducted in six rural communities in southern Ecuador. Survey by M. Park.

of Loja and Zamora, Instituto Nacional de Reforma Agraria y Colonización - IERAC) and non-governmental (Nature and Culture International) institutions, village administrators and respected elders of the villages.

Further to the methods applied for the specific research topics mentioned above, various open interviews and group discussions with villagers were undertaken, along with land use mapping and multi-temporal analysis of satellite images and aerial photographs.

3 The use of wild plants and local agrobiodiversity

North of Podocarpus National Park land use is focused on cattle ranching. This type of land use constitutes the main threat to the forests and to biodiversity. Therefore, alternative activities for securing rural livelihoods are needed in order to reduce the pressure on the forests. Two promising options for increasing household incomes might be the use of non-timber forest products (NTFPs) and the production of plant species in demand in home gardens.

The ethnobotanical inventory undertaken in Shuar, Saraguro and Mestizo communities indicates differences in plant knowledge and plant use among the three ethnic groups which are closely related to their specific ways of life. In Figure 7 the total number of useful plant species and the total number of plant uses recorded among the three ethnic groups is given. The uses of plants are presented according to use categories like medicine, food, construction etc., and according to the plant categories cultivated, ruderal and forest plants.

As typical rainforest dwellers, the Shuar have the most comprehensive knowledge of plants and their uses (Fig. 7). Their vast plant knowledge is for example reflected in the highest number of useful plant species (314) and the highest number of plant uses (489) recorded among the three ethnic groups.[2] The Shuar use most of the plants as medicine (106) and dietary supplement (100). Many plants are also used as construction materials (66). The high number in the category "fodder" (44) results from wild plant species eaten by game, as the Shuar consider such plants to be fodder. Most of the plants used by the Shuar are collected in the forest. 43.5% (213) of the uses recorded are of forest plants. Like other Amazonian ethnic groups (Cerón 1991, Moya 2000, Yépez et al. 2005), the Shuar base their livelihoods on subsistence slash and burn agriculture besides gathering, hunting and fishing. Thus, the forest is of great economic relevance to them and provides them with medicinal and edible plants, construction items and other materials that cover almost all their basic needs.

Unlike the Shuar, the Saraguros and Mestizos make very limited use of forest resources. Only about 7% of the plant uses recorded among the Saraguros and Mestizos are of forest plants (Fig. 7). As agro-pastoralists they have converted most of the primary forest into pastures, fields and home gardens, leaving forest remains only along mountain ridges or in river ravines. The forest basically supplies the Saraguros and Mestizos with timber for their own use or to sell occasionally outside the community. However, the Saraguros and Mestizos have considerable knowledge of cultivated plants, mainly used for food and medicine (Fig. 7). They also have profound knowledge of fodder species and plant species typical for living fences. Corn and bean provision is guaranteed by small crop fields *(chacras)* in ecologically favourable locations. Fruits, vegetables, spices, medicinal and ornamental plants are cultivated in home gardens *(huertas)*, which constitute regional reservoirs of agrobiodiversity. For the Mestizos, the ornamental use of plants is significant (66). Ornamental plants are cultivated for the decoration of houses, gardens and chapels. Some are even sold outside the community like Azucena blanca *(Lilium candidum)*.

Previous agrogeographical investigations of home gardens *(huertas)* of the Shuar and Saraguros (Pohle & Gerique 2008) have clearly shown their relevance as places of great agrobiodiversity and refuges of genetic resources. Furthermore, they contribute significantly to securing and diversifying food supplies. Within the home gardens studied, the majority of plants are of nu-

[2] Differences in data compared to previous publications reflect the collection of new data and a different categorization of plant uses.

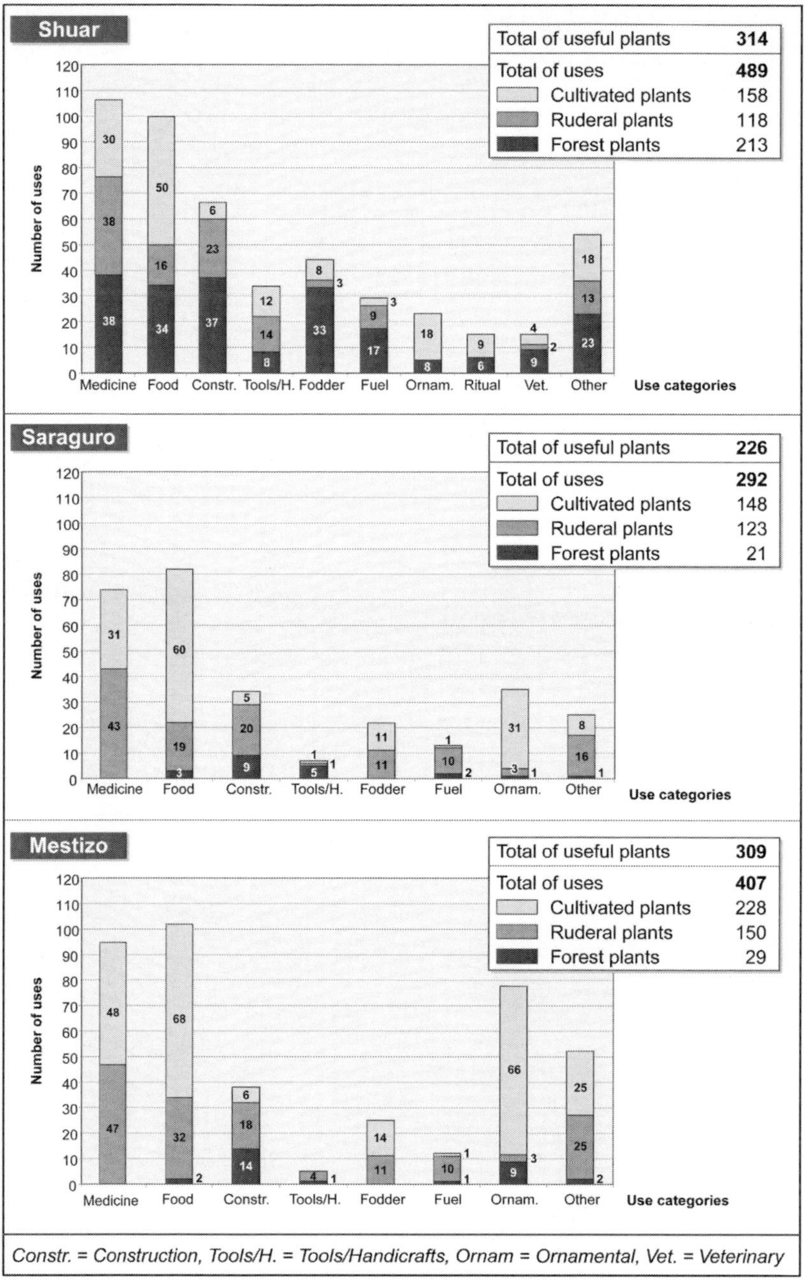

Fig. 7. Useful plants of the Shuar (Shaime, Chumpias, Napints), the Saraguros (El Tibio, El Cristal) and the Mestizos (Los Guabos, Sabanilla). Note: one plant species can be found in more than one use category. Ethnobotanical survey by A. Gerique.

tritional value, followed by medicinal and ornamental plants. The most important cultivated products are plantains, tubers and various types of vegetables and fruit. Given their relatively dense and tall stands of trees, the multi-tiered arrangement of plants, and the great diversity of species, the gardens of the three ethnic groups can be seen as an optimal form of exploitation in the region of tropical rainforests.

As the ethnobotanical and agrogeographical surveys indicate, there is a potential for the production of plant species in demand in home gardens. Two options which may improve local livelihoods and decrease the pressure on forests should be considered:

- the sustainable commercialization of non-timber forest products (NTFPs), and
- the sustainable production of plant species and plant products of a high demand in regional markets[3].

Since the use of NTFPs for income generation is in contest, due to the risks of destructive harvesting practices and because of the low densities of many NTFP species in natural forests (Cunningham 2001, Van Deijk 1999), special attention was paid to NTFPs that could be domesticated or produced in a sustainable way, e.g. in home gardens. Based on the ethnobotanical research, an inventory of NTFPs and horticulture products that grow in the region under study and that could be of commercial interest was elaborated. The inventory includes wild species like *Fuchsia canescens* (edible fruits), *Trianea* sp. (edible fruits), *Solanum caripense* (edible fruits), and a number of Bromeliaceae and Orchidaceae (ornamental plants) that are being used by the local people and that could be considered for domestication. Other species (e.g. *Tibouchina oroensis,* ornamental) are not being used but could be considered for domestication as well. Table 1 shows further plant species that might be of potential interest and that are being cultivated in the region. Since the Mestizos of Los Guabos and the Saraguros of El Tibio already have experience in market oriented gardening – some of them sell ornamental plants (e.g. *Lilium candidum, Zantesdechia aethipica*) or ingredients for the famous Horchata tea in the markets of Loja – the cultivation of plant species in demand in regional markets could generate additional income opportunities without requiring high investments.

[3] A market analysis of NTFPs, horticulture products, and products derived from cattle ranching was conducted from March – June 2009 in the market places of Loja. Analysis of data takes place at the moment.

Table 1. Potential plant species for market-oriented production in home gardens.

Family	Species	Use (plant parts used)
AMARANTHACEAE	*Alternanthera porrigens*	Medicine, tea (leaves)
BUDDLEJACEAE	*Buddleja americana*	Medicine (leaves)
CANNACEAE	*Canna indica*	Food (leaves, tubers)
JUGLANDACEAE	*Juglans neotropica*	Food (seeds), medicine (leaves)
MIMOSACEAE	*Inga* spp.	Food (fruits)
MUSACEAE	*Musa* spp.	Food (fruits)
PASSIFLORACEAE	*Passiflora* spp.	Food, juices (fruits)
PIPERACEAE	*Piper aduncum*	Teas (leaves)
ROSACEAE	*Fragaria* spp.	Food (fruits)
ROSACEAE	*Rubus* spp.	Food (fruits)
SAPOTACEAE	*Pouteria lucuma*	Food (fruits)
SOLANACEAE	*Physalis* cf. *Peruviana*	Food (fruits)
SOLANACEAE	*Trianea* sp.	Food (fruits)

4 Local people's perception and evaluation of the natural environment, environmental stress/risk factors and conservation measures

Although the Shuar have not yet been integrated into the perception survey, from the literature (e.g. Münzel 1977, 1987) it is evident that the Shuar as traditional forest dwellers have a special relationship to the forest. The Shuar not only make a livelihood from the forest (cf. chapter 5), but the forest has a deep cultural and spiritual meaning for them and provides them with their cultural identity. The Shuar have an expert's feel for the forest, being well acquainted with the biological diversity of their circumscribed living space; they can distinguish between an astonishingly large number of plant and animal species, and possess what, from a Western point of view, is an extensive knowledge of the ecosystem. Their detailed knowledge of plants, for instance, is reflected in a precise plant terminology in the Shuar language (Van den Eynden et al. 2004). But the Shuar not only know a large number of forest plants, they also know how to use them (chapter 3) - until now in a sustainable way (Pohle & Gerique 2008). While the Shuar's forest management at present low population levels can be evaluated as preserving plant biodiversity, the sustainability of the Saraguros' and Mestizos' use of the environment has yet to be rated. Market-oriented cattle raising in particular has led to the rapid increase of pastures at the expense of forest in recent decades.

To find out how Saraguros and Mestizos perceive and evaluate different cultural landscapes, two photographs taken north of the Podocarpus National Park were shown to them, one with a totally deforested landscape and another still with forests (Fig. 8 & 9). They were then asked: *Where would you prefer to live?* The Saraguro and Mestizo farmers both clearly prefer to live in areas with forests. "I will not stay in an area that looks like a desert", was the comment of one farmer. The Saraguro and Mestizo farmers highly value economic functions of the forests: forests are agricultural reserves (potential land for grass), forests can be inherited by the children, forests are good to provide work, forests are a safety net, forests provide people with construction wood, fuel, and to a limited extent with food and medicinal plants – were the forest benefits mentioned by the interviewees. Further to the economical functions of the forests, the landscape with forest was also highly valued as a clean and tranquil place, characterised by a healthy environment with fresh water supply and pure air. The photograph without forest was described as exhausted and empty land. "If you have fertilizer, you can use it", was one farmer's statement. However, gender and age related differences could be observed as well. While the household heads and the men actively engaged in agro-pastoralism, in all interviews pointed out the benefits from the forests, some women and old people preferred the open landscape. Among young women forests were even perceived as fearful and dangerous places.

Fig. 8. Photo comparison to evaluate farmers' perception of two cultural landscapes, one totally deforested and one still with forests (Figures 8 & 9). Photo by P. Pohle.

Fig. 9. Photo comparison to evaluate farmers' perception of two cultural landscapes, one totally deforested and one still with forests (Figures 8 & 9). Photo by P. Pohle.

When people were asked about their estimation of whether forest land is increasing or decreasing, astonishingly almost all interviewees of Los Guabos answered: "it remains the same". In contradiction to this incorrect perception, half of them could enumerate disappeared native tree species due to deforestation. Romerillo (*Prumnopitys montana*), Cedro (*Cedrela odorata*), Laurel (*Aniba cf. hostmanniana*), and Nogal (*Juglans neotropica*) were the most commonly named disappeared species. When people were asked if they consider afforestation important, only a few interviewees stated that it is not important because there is still enough forest, but the majority estimated afforestation as an important task to protect nature, as a measure against species extinction, and to maintain fresh water supply and pure air. Almost all interviewees had experience with planting trees, especially along the edges of their pastures and to a lesser degree in their home gardens. As the most commonly planted species Pino (*Pinus patula*) and Cipre (*Cupressus macrocarpa*) were named; only a few had also planted native tree species like Romerillo (*Prumnopitys montana*) and Cedro (*Cedrela odorata*).

Environmental stress/risk factors as perceived by the rural population form - beside other parameters - the basis for the need for sustainable development. Every year the Saraguros of El Tibio have to struggle with flooding and river erosion; a large landslide is endangering the settlement due to backward erosion. According to the perception survey the local people are very much aware of environmental risks such as erosion and landslides. All people consider too much rain as a stressor and almost all number erosion and landslides as environmental risk. Only the moving away from endangered places could be observed as a traditional risk-avoiding strategy. Besides, remnants of forests preserved in river ravines can be taken as a form of biological erosion control. Severe environmental problems reported by all interviewees included plant diseases (*lancha*), animal pests and the lack of timber.

The research area has a long history of conservation efforts: in 1982 the Podocarpus National Park was declared, in 2002 the Bosque Protector Corazón de Oro and in 2007 the Biosphere Reserve Podocarpus – El Cóndor (Fig 2). However, concerning the perception of conservation measures, the survey indicates that only a minority of local inhabitants know about conservation areas. The majority do not even know of the Podocarpus National Park and, although they are living within it, they do not know of the Bosque Protector Corazón de Oro. All of the interviewees claimed to have no idea of what the specific conservation areas are for, where the borders are, and what resource use regulations and restrictions exist. Moreover, at the moment there are considerable jurisdiction problems between national institutions in the land legalization process within the Bosque Protector Corazón de Oro that in future will make local people very sceptical about conservation measures (cf. chapter 6).

Although forests are highly valued economically by the local Saraguro and Mestizo farmers, deforestation is an ongoing process in the area. In colonization areas the main reason for forest clearing is the arrival of new settlers, the founding of new households and the subsequent conversion of forests into agricultural land. In the vicinity of the two investigated villages of El Tibio (Saraguro) and Los Guabos (Mestizo) the colonization process, however, is mainly completed. In these areas the clearing of forests seems to be a household-based decision which takes place under the following circumstances:

- if land is divided after inheritance, the clearing of new lots often follows;
- if soil and pasture gradually lose their fertility, forest is cleared to gain new pastures;
- if the farmer has the strong desire to improve livelihood, this is mainly realised by means of extensive cattle ranching and consequently in forest clearing.

The latter might be the main reason for deforestation in the area. In this respect the distribution of land among the farmers and landownership has to be investigated more closely to find out which household type is contributing most to the deforestation process (chapter 5).

5 Analysis of livelihood strategies of small-scale farming households

Livelihood strategies of local communities still depend largely on natural resources which show clear signs of over-use in the area (Fig. 8). Some of the key questions are:

- Which livelihood strategies are typical for the different ethnic groups?
- To what extent do households' livelihood strategies depend on natural resources?
- How does use and over-use of natural resources correspond with economic welfare or poverty (measured in terms of monetary income and dependency on social transfer by the state) of households and communities?

The Shuar are traditionally engaged in a number of livelihood activities that typically include fishing, hunting, gathering of wild plants and animal products, keeping domesticated animals, and cultivating forest gardens with starchy roots, tubers and plantains (Fig. 3). Their traditional livelihood system was based to a large extent on subsistence activities that enabled them to live almost autonomously and made little or no use of market integration. Until today, some Shuar try to conserve this traditional way of life. However, it cannot be ignored, that in recent times a number of Shuar households is increasingly involved in markets. The selling of timber, fruits and vegetables (sometimes to intermediary vendors), cattle raising and the production of handicrafts provide Shuar with monetary income. Employment can still be considered as an exception, but is sometimes the case for those Shuar working for governmental institutions or within Shuar political organizations. Despite this partial market integration, until today, the Shuar's livelihood system still depends to a large extent on natural resources.

The Saraguros and Mestizos, in contrast to the Shuar, have long been involved in the market system. They are engaged in agro-pastoral activities that combine market economy (cattle ranching) and subsistence economy (crop production and horticulture). Whereas corn and beans are cropped in shifting fields (*chacras*), vegetables, fruits, spices and other useful plants are cultivated in permanent home gardens (*huertas*). The main products drawn from cattle raising are neither meat nor milk, but cheese, which is called *quesillo* or *cuajada* by Ecuadorians. It is a white fresh unsalted cheese that is produced by adding lab-ferment to fresh milk and separating the precipitating proteins from the whey. Production usually takes place in the pastures, since cheese is easier to transport and less perishable than fresh milk.

Data from the livelihood analysis show that the sale of mainly agricultural products is one of the most important sources of income for households in the studied Saraguro and Mestizo communities: an average of 52.4% of the total income is obtained from sales (Tab. 2). Among the agricultural products sold, cheese plays an especially important role. In those households selling cheese, an average of 30.9% of the total household income in 2007 was derived from cheese vending. Of 47 households 11 stated to have obtained even more than half of last year's income from cheese. Households not able or willing to maintain their livelihood by selling agricultural products depend to a great degree on payments for informal work (day labour). On average, 52.3% of the households' income is generated from irregular work in the form of day labour. Formal employment, in contrast, plays an inferior role in the livelihood systems of the communities studied. Only two out of 47 households interviewed receive regular income from formal employment. Illegal logging seems to be of importance as well, although data are difficult to obtain from household interviews. Migration has gained importance during the past decades and must be considered as another livelihood strategy (Pohle 2008). It is directed either towards the towns of Ecuador, mainly the city of Loja and the capital Quito, or is directed abroad, to foreign countries like the U.S.A. and Spain. Out-migration is more frequently practised among the Mestizos.

Table 2. Average household incomes for selected income categories in 2007 in the communities of Los Guabos and El Tibio.

	All households (N = 47)		Households that received income from this category		
	Average income 2007 [US $]	Average percentage of total income [%]	N	Average income 2007 [US $]	Average percentage of total income [%]
Total income	4773	–	47	–	–
Income from paid work	1875	37.3	28	3039	60.5
Income from irregular work	765	**27.8**	25	1438	**52.3**
Income from sale[1]	1842	37.4	35	2578	52.4
Income from sale of cheese	377	**15.1**	23	770	**30.9**
Income from transfers[2]	938	22.7	36	1277	30.9

[1] E.g. sale of field and housegarden crops, cattle and products derived from cattle ranching, timber and other forest products, commodities, land etc.
[2] Transfers by the state (= social aid, *"Bono de pobreza"*) and remittances from emigrated family or community members.

Household types are heterogeneous in the Saraguro and Mestizo communities studied and vary by, for example, household size, age and composition, owned land and cattle, as well as income structure in general. Since data were only obtained recently, the statistical investigation is still under way. Informal conversations during field work indicate that there is quite a number of landless households and households with extremely small holdings (less than one ha) especially among the Mestizos. Their household income mainly stems from day labour on the pastures and fields of the bigger landowners (Tab. 2).

It was difficult to empirically substantiate this finding, since only a minority of the inhabitants of the study area hold official land titles, and questions directed towards land ownership are delicate, especially if asked as part of a questionnaire, as was done for the livelihood analysis. Concerning the question of which household type participates most in the deforestation process, these yet to be quantified landless and poorest households are probably no relevant actors in this respect. Rather, the decision to clear forest is apparently undertaken by the small number of more privileged landowners who also have the equipment and can afford to hire workers. Their decision is often based on a strong desire to improve their livelihood by integrating more into the market economy and this is realised especially by means of extensive cattle ranching. Concerning the "poverty-forest-debate", these findings would contradict the frequently held belief that poverty increases deforestation, at least on a household level (cf. also Wunder 1996).

6 Determination of the political and administrative use agreements including land tenure systems

In the research area numerous land use conflicts exist, such as forest conservation versus forest exploitation and pasture extension. Knowledge of actual and historical land tenure systems gives insight into land use and land right conflicts and shows the explosive nature of the land issues in times of rapid economic, social and cultural change. To understand the mechanisms of deforestation and forest degradation, knowledge of land tenure and access rights to both land and resources and the associated regulations are an essential precondition (Ostrom & Schlager 1996, Wunder 1996, Cunningham 2001).

Land tenure conflicts are based on irregular and unequal land tenancy, which is the result of historical and political developments. The Spanish conquerors established the foremost restrictions to land access to indigenous people and introduced "formal" or "legal" property rights, based on written law. A first element for conflictive land tenure, especially in the Andes (Sierra), was the asymmetric distribution of land between big landowners and indigenous people who were kept in a state of dependency (Barsky 1984, Pohle 2008). A second element was the disparity between traditional customary (ancestral) community based land tenure and legal land property rights (Viteri 2007). Starting in the 1950s up to the 1970s, land redistribution and colonization policies introduced through modern laws of colonization and agrarian reform resulted in numerous tenure conflicts dealing with expropriation of used land and adjudication of colonized areas. Both land redistribution and colonization policies promoted intensive processes of agricultural frontier expansion and therefore, deforestation (Brown et al. 1992, Sierra 1996), which affected especially the Andean foothills and the lowlands of the Pacific coast and the Amazon area. In the past two decades, the modern policy of environmental conservation and the establishment of a system of protected areas added further constraints to land regulation processes.

North of Podocarpus National Park current land use and land tenure conflicts are founded primarily on the colonization processes starting in the 1950s, and are severely dependent on state policies on land adjudication, increase of accessibility, and national or international concern for nature conservation and environmental protection (Fig. 10). Furthermore, local city governments (Loja, Zamora) have strong interest in watershed management and water supply, adding a new element to land tenure conflicts. The conflicts over land tenure (as an effect of the land occupation processes described above), resulted in unsustainable land use forms. Both timber extraction and pasture expansion at the beginning of the colonization process in the past century were basic requirements to prove possession of land in order to get property titles from the state (Barsky 1984). Although these requirements were eliminated in the law of Agrarian Development (1994), they were the agents for deforestation and the clearing up of most of the forest cover of the entire Amazonian region, including the eastern Andean slopes (Barsky et al. 1982, Gerold & Lanfer 2001).

One consequence of the varying land tenure regimes in the region is the insecurity of access to land and resources. Data of the agrarian census of 2001 report less than 30% of farms having property titles in the province of Zamora Chinchipe, the lowest proportion at the provincial level in Ecuador (INEC 2002). Field information gathered in 2008 through semi-structured interviews showed that informal buy-sell contracts of possessor rights *(compra de derechos posesorios)* have been a widely accepted form of land market transactions. These contracts have neither been inscribed in the office of property registration nor been registered in the cadastre.

Population pressure, land scarcity and resource use restrictions in conservation zones have recently increased the interest of farmers in entitlement of their land. Currently, interest in regularization of land property is severely influenced by three factors:

1. As a strategy for "reactivating the economy and alleviating poverty" the improvement of land tenancy security and the enhancement of the dynamism of rural property markets were the main issues for the neo-liberal politics in the 1990s (Francescutti 2000: 1).
2. Nowadays, several forms of social aid introduced by the current government (e.g. bonus for housing, formal leasing from the National Development Bank) demand land titles in order to qualify for them.
3. With the declaration of protective areas within the National System of Protected Areas (SNAP) formal limitations for frontier expansion and timber extraction were introduced and therefore, clear limits of property and borders of conservation areas demanded in order to assure conservation practices.

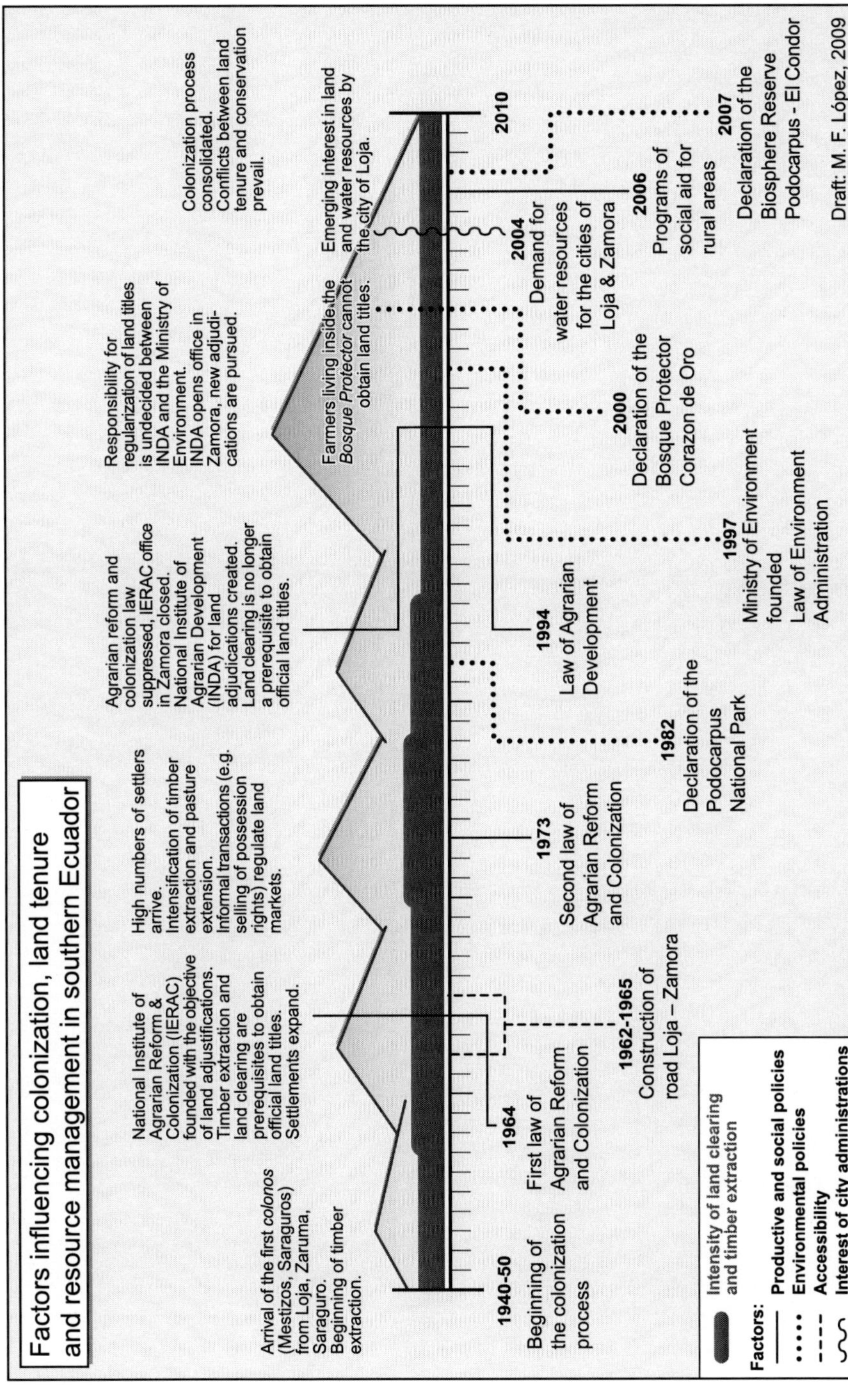

Fig. 10. Factors influencing colonization, land tenure and resource management in southern Ecuador.

Around 14 conservation areas including protective forests exist within the Biosphere Reserve Podocarpus – El Cóndor (NCI-MAE 2006). People who have settled the region for more than 50 years and live inside these areas or in the buffer zones either do not know about it or do not have enough information about its implications for land entitlement or legal access to forest and water resources. In the research area a special conflict exists between the local people and the Bosque Protector Corazón de Oro. At the time of declaration in 2002, the Bosque Protector encompassed more than 15 settlements; 30% of the area was already converted into pastures and another 30% of the forest cover was interspersed with pastures (CINFA 2006). Since the establishment of the Bosque Protector, tenure of land which has been in use for decades (sometimes a half century) cannot be passed to the next generation and farmers cannot get legal land titles for formal leasing. This is due to jurisdiction problems for land entitlement between the National Institution of Agrarian Development (INDA) and the Ministry of Environment. While the latter is responsible for entitlement in protected areas and national forest patrimony, entitlement in rural areas not being under any category of conservation, is in legal jurisdiction of INDA. In both cases procedures are complicated, time-consuming and expensive (Burbano 2008). Semi-structured interviews on the local perception of conservation measures carried out in the villages of El Tibio and Los Guabos in 2008 and 2009 by M.F. Lopez and P. Pohle showed a remarkable lack of information and participation that led to uncertainty and distrust among the local people. These unsolved problems have resulted in people in the study area being highly sceptical towards conservation measures: for them, conservation means restrictions for resource use and land entitlement.

7 Human ecological dimensions under global change: conservation strategies of biological and cultural diversity

Despite the remoteness of many communities living in the tropical rainforests of Ecuador, the local population has never been completely isolated from global processes of ecological, political, cultural and economic change. As an example of external influence in earlier centuries, the cultural transformation in the fields of religion or language by the Spaniards can be mentioned. In more recent times the global improvement in medical care has resulted in an increase in population. In the past few years, people in even distant villages have been connected to roadways and have become integrated in global information and communication systems via telephone, radio and the internet. Currently, the following effects of globalization are of special relevance in the rural areas of southern Ecuador:

1. The confrontation of traditional, non-industrial societies with the western concept of nature conservation and global protection of biodiversity and climate (Pohle 2004, Sælemyr 2004, Pohle & Gerique 2006).
2. The exploitation of minerals by international companies (Confeniae 2007, Mining Watch Canada 2007).
3. The integration of local people into the worldwide labour market by economically induced out-migration (Caritas Española et al. 2004, Pohle 2008: 34f.).
4. The vulnerability of local populations to the fluctuations of the world economy, e.g. in the field of labour migration and tourism (Süddeutsche Zeitung 2008).

The experience in international nature conservation during the past 2-3 decades (Ellenberg 1993, Ghimire & Pimbert 1997, Müller-Böker et al. 2002) has shown that resource management, if it is to be sustainable, must serve the goals of both nature conservation and the use claims of the local population. The strategy is one of **"protection by use"** instead of "protection from use", a concept that has emerged throughout the tropics under the philosophy "use it or lose it" (Janzen 1992, 1994). In line with this concept is the integrated concept of conservation and development exemplified by Unesco's Biosphere Reserve (Unesco 1984). Since September 2007 the research area belongs to the Biosphere Reserve Podocarpus – El Cóndor. Programmatically, Biosphere Reserves are strongly rooted in cultural contexts and traditional ways of life (Bridgewater 2002). With regard to the area under study, in the buffer and transition zone of Podocarpus National Park measures to be taken should take into account ethno-specific environmental knowledge and know-how. A Biosphere Reserve in southern Ecuador could thus be the vehicle for protecting tropical mountain ecosystems and developing sustainable forms of land use at the same time.

The Shuar's traditionally practiced plant biodiversity management, which has been described as clearly sustainable at present low population levels (Pohle & Gerique 2008), will only be preserved by the following preconditions: the complete legalization of their territorial claims, a comprehensive protection of their territories and support for the Shuar's traditional way of life and cultural identity. To improve livelihood in an economic sense, additional sources of financial income are essential. In line with this, promotion of ecotourism, support of traditional handicrafts, and the cultivation of useful plants for a regional market are potential options requiring discussion.

For the Saraguro and Mestizo households alternative incomes should be found since cattle ranching poses the main threat to the forests in southern Ecuador. Decades of global efforts to conserve biodiversity have shown that people are more likely to incorporate new sources of income as complements to their existing activities than as substitutes for them (Ferraro & Kiss 2002). Moreover, in the research area cattle ranching fulfils multiple objectives within the farmers' livelihood strategies that are very difficult to substitute: as well as providing households with regular income, cattle awards a prestigious social status; cattle also represents a way of accumulating wealth as a private insurance, which is especially important in regions with weak loan and pension systems. Therefore, only a partial substitution of pasture land appears to be realistic. As one promising approach, the cultivation of useful plants for a regional market could be discussed. In this respect the cultivation of medicinal herbs, fruits, vegetables and ornamental flowers in home gardens may be considered. Another option could be the use of non-timber forest products like the production of honey, teas, liquors and conserves (Añazco et al. 2005, Ordoñez & Lalama 2006). However, in order to stem the further loss of forests and biodiversity, it may be necessary to convince local people that scrub and wasteland (*matorral*) should be replanted with native tree species, preferably with useful native trees in demand (Mosandl & Günter 2008, Stimm et al. 2008, Knoke et al. 2009). The introduction of silvipastoral and agroforestry systems should also be taken into account (v. Walter et al. 2008). Additionally, the improvement of the pasture economy as well as the veterinary service is indispensable. Finally, the promotion of off-farm employment opportunities as well as payments for environmental services to protect the watershed area of Loja would benefit the local farmers.

The chaotic land property rights regime in southern Ecuador represents a further chance, if sustainable activities in exchange for land property rights contracts could be arranged. Local inhabitants would get legal land titles only after a partial renunciation of cattle raising. These measures could be accompanied by technical assistance, capital input to establish alternative production systems and monitoring. Finally, a system of payments for preserved forests that works as a private pension insurance fund is conceivable; land owners would get payments for preserving and enlarging their forest land only after retirement. Pension payments would be in accordance with the total number of years of forest possession, thus allowing the purchase and sale of land and making forest possession and its preservation attractive. Assurances to cover forest fires and similar disasters could complete the system. After a series of financial crisis and bankruptcies the Ecuadorian State as well as the traditional banking sector count with little credibility among Ecuadorians. Thus, all these activities would be managed by recognized social investing institutions (e.g. Oikocredit); their small loan programmes could help to start the implementation of these innovations.

And what could the carbon markets do? To date, afforestation and reforestation project activities under the Clean Development Mechanism (CDM) of the Kyoto Protocol have shown very limited potential. As of October 2008, only one afforestation project had been registered in the CDM executive board worldwide (SAF 2008, Settelmayer & Schlamadinger 2008). Nevertheless, a more promising possibility could be the voluntary carbon market (Gillenwater et al. 2007: 87, Speckman 2008: 8). Moreover, the expected post-Kyoto agreement will in all probability include REDD (Reducing emissions from deforestation and degradation) projects that could provide the seed money for such initiatives (Hamilton et al. 2008: 61, Speckman 2008).

Acknowledgements
The authors thank Eduardo Tapia (Loja), Tatjana Ramon (Loja), Darío Veintimilla (Loja), Paúl Castro (Loja), and the Reinaldo Espinosa Herbarium of the Universidad Nacional de Loja (UNL) for contributing to the results presented here. We are grateful to the German Research Foundation (DFG) for funding this project within the research units FOR 402 and FOR 816.

References

Alcorn J (1999) Indigenous Resource Management Systems. In: Posey DA: Cultural and Spiritual values of biodiversity. A complementary contribution to the global biodiversity assessment: 203-206. London

Alexiades MN (1996) Collecting ethnobotanical data: An introduction to basic concepts and techniques. In: Alexiades MN (ed.) Selected Guidelines for Ethnobotanical Research: A Field Manual. The New York Botanical Garden, New York, 53-94

Añazco M, Lojn L, Yaguache R (2005) Productos forestales no madederos en el Ecuador. DFC/FAO/Ministerio de Ambiente, Gobierno de los Países Bajos. Quito

Barsky O (1984 reprinted 1988) La Reforma Agraria Ecuatoriana. Quito.

Barsky O, Díaz-Bonilla E, Furche C, Mizrahi R (1982) Politicas Agrarias, Colonización y Desarrollo Rural en Ecuador. Quito

Barthlott W, Hostert A, Kier G, Küper W, Kreft H, Mutke J, Rafiqpoor D, Sommer H (2007) Geographic Patterns of Vascular Plant Diversity at Continental to Global Scales. Erdkunde 61 (4): 305-315

Beck E, Bendix J, Kottke I, Makeschin F, Mosandl R (eds.) (2008) Gradients in a Tropical Mountain Ecosystem of Ecuador. Ecological Studies 198. Berlin

Beck E, Hartig K, Roos K, Preuing M, Nebel M (2008) Permanent Removal of the Forest: Construction of Roads and Power Supply Lines. In: Beck E, Bendix J, Kottke I, Makeschin F, Mosandl R (eds.) Gradients in a Tropical Mountain Ecosystem of Ecuador. Ecological Studies 198: 361-370. Berlin

Berkes F (1999) Sacred Ecology: Traditional Ecological Knowledge and Resource management. Philadelphia.

Bohle HG (2001) Neue Ansätze der geographischen Risikoforschung. Die Erde, 132: 119-140

Bridgewater PB (2002) Biosphere reserves: special places for people and nature. Environmental Science & Policy 5 (1): 9-12

Brown LA, Sierra R, Southgate D, Lobao L (1992) Complementary perspectives as a means of understanding regional change: Frontier Settlement in the Ecuador Amazon. Environment and Planning 24 (7): 939-961

Bryant RL, Bailey S (1997) Third World political ecology. London

Burbano M (2008) La conservación y el desarrollo en zonas rurales habitadas: implicaciones entre el Bosque Protector Corazón de Oro y la comunidad de El Tibio, Zamora. Thesis, PUCE. Quito

Cerón CE (1991) Manejo florístico shuar-achuar (Jíbaro) del ecosistema amazónico del Ecuador. Hombre y Ambiente. Año V, enero-febrero Nr. 17: 7-32

Chambers R (1995) Poverty and Livelihoods: Whose Reality Counts? IDS-Discussion Papers, 347. Brighton

Chambers R, Conway G (1992) Sustainable rural livelihoods: practical concepts for the 21st century. IDS-Discussion Papers 296. Brighton

CINFA (2006) Informe técnico. Estado de conservación de reas Naturales Protegidas y Bosques Protectores de Loja y Zamora. UNL, Loja

Colding J, Folke C (1997) The relations among threatened species, their protection, and taboos. Conservation Ecology [online] 1 (1): 6

Confeniae (2007) Las Nacionalidades indígenas continan las luchas contra las compañías mineras y petroleras que han invadido nuestros territorios. http://www.confeniae.org/es/comunicados/07073_mafia_mineras.html

Cunningham AB (2001) Applied Ethnobotany. People, Wild Plant Use and Conservation. People and Plants Conservation Manuals. London

Derichs A, Rauch T (2001) LRE und der "Sustainable Rural Livelihoods" Ansatz. Gemeinsamkeiten, Unterschiede, Komplementaritäten. Entwicklungsethnologie, 9 (2): 60-78.

Die Erde (2001) Themenheft "Tropische Wald-Ökosysteme". Die Erde 132 (1)

Ellenberg L (1993) Naturschutz und Technische Zusammenarbeit. Geographische Rundschau 45 (5): 290-300

Farrington J, Carney D, Ashley C (1999) Sustainable livelihoods in practice: Early applications of concepts in rural areas. Overseas Development Institute (ODI) - Natural Resource Perspectives, 42

FAO (2007) State of the World's Forests 2007. Food and Agriculture Organisation of the United Nations. Rome

Ferraro PJ, Kiss A (2002) Direct Payments to Conserve Biodiversity. Science 298: 1718-1719

Francescutti D (2002) Regularización de la tenencia de tierras, evolución, costos beneficios y lecciones. El caso de Ecuador. FAO, Roma

Fujisaka S, Escobar G, Veneklaas EJ (2000) Weedy fields and forests: interactions between land use and the composition of plant communities in the Peruvian Amazon. Agriculture, Ecosystems and Environment 78: 175-186

Gerique A, Veintimilla D (2008) Useful plants and weeds occurring in Shuar, Saraguro and Mestizo communities. Checklist Reserva Biológica San Francisco (Prov. Zamora-Chinchipe, S. Ecuador). Ecotropical Monographs 4: 237-256.

Gerold G, Lanfer N (2001) Agrarkolonisation und Bodennutzungsprobleme im Oriente Ecuadors. Erdkunde 55: 362-378

Ghimire KB, Pimbert MP (eds.) (1997) Social Change and Conservation: An Overview of Issues and Concepts. London

Gillenwater M, Broekhoff D, Trexler M, Hyman J, Fowler R (2007) Policing the voluntary carbon market. Nature Reports Climate Change, 6: 85-87 www.nature.com/reports/climatechange

Göttlicher D, Obregón A, Homeier J, Rollenbeck R, Nauss T, Bendix J (2009) Land-cover classification in the Andes of southern Ecuador using Landsat ETM+ data as a basis for SVAT modelling. International Journal of Remote Sensing 30: 1867-1886

Gräf M (1990) Endogener und gelenkter Kulturwandel in ausgewählten indianischen Gemeinden des Hochlandes von Ecuador. München

Gray CL, Bilsborrow RE, Bremmer JL, Lu F (2008) Indigenous Land Use in the Ecuadorian Amazon: A Cross-Cultural and Multilevel Analysis. Human Ecology 36: 97-109

Hamilton K, Sjardin M, Marcello T, Xu G (2008) El Estado de los Mercados Voluntarios de Carbono 2008. Ecosystem Marketplace and New Carbon Finance. London and Washington DC (Translation from the English Version)

Harden, CP (1996) Interrelationships between land abandonment and land degradation: a case from the Ecuadorian Andes. Mountain Research and Development 16: 274-280

INEC (Instituto Nacional de Estadísticas y Censos) (2002) III Censo Agropecuario Resumen Nacional. Quito

INEC (Instituto Nacional de Estadística y Censos) (2003) VI Censo de Población y Vivienda 2001 http://www.inec.gov.ec/web/guest/descargas/basedatos/cen_nac/ cen_pob_nac_2001 Verified 23.05.08

IPCC (2007) Climate Change 2007. International Panel on Climate Change. Fourth IPCC Assessment Report. Geneva

Janzen DH (1992) A south-north perspective on science in the management, use and economic development of biodiversity. In: Sandlund OT, Hindar K, Brown AHD (eds.): Conservation of Biodiversity for Sustainable Development: 27-52. Oslo

Janzen DH (1994) Wildland biodiversity management in the tropics: where are we now and where are we going? Vida Silvestre Neotropical 3: 3-15

Jørgensen PM, Ulloa Ulloa C (1994) Seed plants of the high Andes of Ecuador a checklist. Aarhus (AAU) University Reports 34: 1-443

Koopowitz H, Thornhill AD, Andersen M (1994) A general stochastic model for the prediction of biodiversity losses based on habitat conversion. Conservation Biology 8: 425-438

Knoke T, Calvas B, Aguirre N, Roman-Cuesta RM, S Günter, Stimm B, Weber M, Mosandl R (2009) Can tropical farmers reconcile subsistence demands with forest conservation? Frontiers in Ecology and the Environment, Front Ecol Environ 2009; doi:10.1890/080131

Krings T (2008) Politische Ökologie. Grundlagen und Arbeitsfelder eines geographischen Ansatzes der Mensch-Umwelt-Forschung. Geographische Rundschau 12: 4-9

Krüger F (2003) Handlungsorientierte Entwicklungsforschung: Trends, Perspektiven, Defizite. Petermanns Geographische Mitteilungen, 147 (1): 6-15

Malhi Y, Timmons Roberts J, Betts RA, Killeen TJ, Li W, Nobre CA (2008) Climate Change, Deforestation and the Fate of the Amazon. Science 319: 169-172

Meusburger P, Schwan T ed. (2003) Humanökologie. Ansätze zur Überwindung der Natur-Kultur-Dichotomie. Erdkundliches Wissen 135. Wiesbaden

Mining Watch Canada (2007) Victim or Villain? Canadian Mining Investment in South-East Ecuador Exacerbates Divisions, Conflicts. http://www.miningwatch.ca/updir/Ecuador_Analysis_final.pdf

Mosandl R, Günter S, Stimm B, Weber M (2008) Ecuador suffers the Highest Deforestation Rate in South America. In: Beck E, Bendix J, Kottke I, Makeschin F, Mosandl R (eds.) Gradients in a Tropical Mountain Ecosystem of Ecuador. Ecological Studies 198: 37-47, Berlin

Mosandl R, Günter S (2008) Sustainable management of tropical mountain forests in Ecuador. In: Gradstein SR, Homeier J, Gansert D (eds.) Biodiversity and Ecology Series (2): The tropical mountain forest. Göttingen, 179-195

Moya A (2000) Ethnos. Atlas etnogrfico del Ecuador. Quito

Müller-Böker U (1999) The Chitawan Tharus in Southern Nepal. An Ethnoecological Approach. Nepal Research Centre Publications, 21. Stuttgart

Müller-Böker U, Kollmair M, Soliva R (2002) Der Naturschutz in Nepal im gesellschaftlichen Kontext. Asiatische Studien LV 3 (2001) Bern: 725-775

Münzel M (1977) Jívaro-Indianer in Südamerika. Roter Faden zur Ausstellung, 4. Museum für Völkerkunde, Frankfurt/ M.

Münzel M (1987) Kulturökologie, Ethnoökologie und Ethnodesarrollo im Amazonasgebiet. Entwicklungsperspektiven 29. Kassel

Myers N, Mittermeier RA, Mittermeier CG, da Fonseca GAB, Kent J (2000) Biodiversity hotspots for conservation priorities. Nature 403: 853-858

Nazarea VD ed. (1999) Ethnoecology, situated knowledge/located lives. The University of Arizona Press, Arizona

NCI-MAE (2006) Reserva de Biósfera Podocarpus- El Cóndor. Resmen de la propuesta para la declaratoria de Reserva de Biosfera dirigida a la UNESCO. Loja

Ordoñez O, Lalama K (2006) Experiencias del Manejo Apícola en Uritusinga. PROBONA, Fundación Ecológica Arco Iris, Samiri-ProGeA. Loja

Ostrom E & Schlager E (1996) The formation of property rights. In: Hanna S, Folke C & Maler KG (ed.) Rights to nature, 126-156. Washington

Pichón FJ (1996) The Forest Conversion Process: A Discussion of the Sustainability of Predominant Land Uses Associated with Frontier Expansion in the Amazon. Agriculture and Human Values 13 (1): 32-51

Pohle P (2004) Erhaltung von Biodiversität in den Anden Südecuadors. Geographische Rundschau 56 (3): 14-21

Pohle P (2008) The People Settled Around Podocarpus National Park. In: Beck E, Bendix J, Kottke I, Makeschin F, Mosandl R (eds.) Gradients in a Tropical Mountain Ecosystem of Ecuador. Ecological Studies 198: 25-36. Berlin

Pohle P, Gerique A (2006) Traditional ecological knowledge and biodiversity management in the Andes of southern Ecuador. Geographica Helvetica 4: 275-285

Pohle P, Gerique A (2008) Sustainable and Non-Sustainable Use of Natural Resources by Indigenous and Local Communities. In: Beck E, Bendix J,

Kottke I, Makeschin F, Mosandl R (eds.) Gradients in a Tropical Mountain Ecosystem of Ecuador. Ecological Studies 198: 331-345. Berlin

Posey DA (1985) Indigenous management of tropical forest ecosystems: the case of the Kayapó indians of the Brazialian Amazon. Agroforestry Systems 3: 139-158

Posey DA, Balée W (eds.) (1989) Resource Management in Amazonia: Indigenous and Folk Strategies. Advances in Economic Botany, 7. New York

Putsche L (2000) A Reassessment of Resource Depletion, Market Dependency, and Culture Change on a Shipibo Reserve in the Peruvian Amazon. Human Ecology 28 (1): 131-140

Rudel TK, Bates D, Machinguiashi R (2002) Ecologically Noble Amerindians? Cattle Ranching and Cash Cropping among Shuar and Colonists in Ecuador. Latin American Research Review 37 (1): 144-159

Sælemyr ST (2004) People, park and plant use: Perception and use of Andean "nature" in the southern Ecuadorian Andes. Nors Geogrfisk Tidsskrift-Norwegian Journal of Geography 58: 194-203. Oslo

SAF (Society of American Foresters) (2008) Forest Offset Projects in a Carbon Trading System. A Position Statement of the Society of American Foresters. http://www.safnet.org/policyandpress/psst/offset_projections_expires12-8-20

Scholz F (2004) Geographische Entwicklungsforschung: Methoden und Theorien. Stuttgart.

Settelmyer S, Schlamadinger B (2008) A carbon market solution? Global Carbon 2008. A Carbon Expo Special Report from Environmental Finance and Carbon Finance. 12-14 London

Sierra R (1996) La Deforestación en el Noroccidente del Ecuador 1983-1993, Quito

Sierra R (1999) Traditional resource-use systems and tropical deforestation in a multi-ethnic region in North-West Ecuador. Environmental Conservation 26 (2): 136-145

Speckman J (2008) REDD under way. Global Carbon 2008. A Carbon Expo Special Report from Environmental Finance and Carbon Finance: 6-11. London

Stern N (2006) Stern Review on the Economics of Climate Change. London

Stimm B, Beck E, Günter S, Aguirre N, Cueva E, Mosandl R, Weber M (2008) Reforestation of Abandoned Pastures: Seed Ecology of Native Species and Production of Indigenous Plant Material. In: Beck E, Bendix J, Kottke I, Makeschin F, Mosandl R (eds.) Gradients in a Tropical Mountain Ecosystem of Ecuador. Ecological Studies 198: 417-429. Berlin

Süddeutsche Zeitung (2008) Spanien will arbeitslose Einwanderer loswerden. Sueddeutsche.de 18.07.08 http://www.sueddeutsche.de/politik/14/448747/text/

Toledo VM, Ortiz B, Medellín-Morales S (1994) Biodiversity islands in a sea of pasturelands: indigenous resource management in the humid tropics of Mexico. Etnoecologica 2 (3): 37-50

Tourrand JF, Morales Grosskopf CH, Wood CH (2008) Amazonian cattle-ranching: Towards a new social-environmental agreement. In: Tiessen H, Steward JWB (eds.): Applying Ecological Knowledge to Land-Use Decisions. Scope. Inter-American Research Institute for Global Change Research. So José dos Campos, Brazil

Tutillo Vallejo A (2009) Die Nutzung der natürlichen Ressourcen bei den Saraguro und Mestizen im Wassereinzugsgebiet des Tambo Blanco in Südecuador. PhD thesis unpublished

UNESCO (ed.) (1984) Action plan for biosphere reserves. Nature and Resources 20/4

UNFCCC (2008) Fact Sheet: Reducing emissions from deforestation in developing countries: approaches to stimulate action. United Nations Framework Convention on Climate Change.

Van den Eynden V, Cueva E, Cabrera O (2004) Of "climbing peanuts" and "dog's testicles", Mestizo and Shuar plant nomenclature in Ecuador. Journal of Ethnobiology 24 (2): 279-306

Viteri F (2007) Reforma Agraria en el Ecuador. MEF, Quito

Von Walter F, Barkmann J, Olschewski R (2008) Ex-ante analysis of an agroforestry management option in Southern Ecuador – The Tara example. EURECO-GfÖ Proceedings: 527

Warren DM, Slikkerveer LJ, Brokensha G (1995) The cultural dimension of development: Indigenous knowledge systems, London

Weichhart P (2007) Humanökologie. In: Gebhardt H, Glaser R, Radtke U (2007): Geographie. Physische Geographie und Humangeographie: 941-958. Heidelberg

Wunder S (1996) Deforestation and the uses of wood in the Ecuadorian Andes. Mountain Research and Development 16 (4): 367-382

Yépez P, de la Torre S, Cerón CE, Palacios W (eds.) (2005) Al inicio del Sendero: Estudios Etnobotnicos Secoya. Editorial Arboleda. Quito

Linkages between poverty and sustainable agricultural and rural development in the uplands of Southeast Asia

Manfred Zeller*, Tina Beuchelt, Isabel Fischer, and Franz Heidhues

Department of Agricultural Economics and Social Sciences in the Tropics and Subtropics, Section Rural Development Theory and Policy, Universität Hohenheim, (490a), D-70593 Stuttgart, Germany

*corresponding author: M. Zeller, email: manfred.zeller@uni-hohenheim.de

Summary

Most of the upland areas of Southeast Asia are characterized by insufficient infrastructure, low productivity in smallholder crop and animal production, mounting environmental problems such as soil and forest degradation and loss of biodiversity, increasing population pressure, and widespread poverty, particular in rural areas. While some upland areas in Southeast Asia have been experiencing considerable progress during the past twenty years, others have stagnated or even declined economically, socially and environmentally. This paper focuses on the disadvantaged upland areas and discusses sustainable development in the upland areas of Southeast Asia, notably Cambodia, Lao PDR, Thailand, Vietnam, and Indonesia. We distinguish three explanatory approaches for land use change and agricultural and rural development. Apart from the market approach and the population approach, we suggest to focus more on governance issues as a major driving force of land use change. The governance approach appears particularly relevant for upland areas which are often politically and institutionally marginalized. The paper concludes with implications for rural and agricultural development policies.

Keywords: Deforestation, development policies, governance approach, land use change, market approach, population approach

T. Tscharntke et al. (eds.), *Tropical Rainforests and Agroforests under Global Change*, Environmental Science and Engineering, 511–527, DOI 10.1007/978-3-642-00493-3_24,

1 Explanatory approaches to sustainable development linking poverty reduction with economic growth and the environment

It is generally agreed that "sustainable development" as defined by the Brundtland Report (WCED 1987:43), i.e., "development that meets the needs of the present without compromising the ability of future generations to meet their own needs", rests on three pillars: biodiversity and resource conservation; economic development; and poverty reduction.

Many countries in Southeast Asia have an exceptional biodiversity with a very high concentration of endemic species (Myers et al. 2000). This biodiversity is severely threatened by human activities, because the species are concentrated in areas where rural poverty is widespread (Tonneijck et al. 2006, Snel 2004). Therefore, biodiversity conservation, economic development and poverty reduction should be addressed simultaneously (Adams et al. 2004, Coxhead 2003). However, in practice many projects aiming at the integration of conservation and equitable development tend to bring unsatisfactory results. For upland areas where sustainable development is intimately linked to forest use and conservation the hypothesis that income generation from forest products encourages sustainable use and conservation of forests, while at the same time alleviating poverty, is not confirmed by evidence; benefits to both, conservation and poverty alleviation have not yet been convincingly documented (Fisher 2000, Gilmour et al. 2004).

A number of more recent studies explored the underlying factors of agriculturally driven land use change, and its interaction with deforestation, biodiversity loss, poverty, inequality, and economic growth (see for example, Kaimowitz and Angelsen 1998, Tonneijck et al. 2006, Pham 2005). Modelling the relationships between changing natural environments to agricultural production systems and biodiversity, Tonneijck et al. (2006) found out that the relation between biodiversity conservation and agriculture depends on the livelihood and income improvement strategy people embark on in rural areas. Of the three main strategies, intensification of agriculture, diversification of rural income sources, and expansion of agricultural production, the expansion of land to increase agricultural production would result in the greatest loss of biodiversity. They estimate that up to 80 percent of species diversity would be lost due to full conversion because even in low-input production systems species diversity is below 20 percent. Therefore, the best strategy to conserve biodiversity is to intensify agricultural production on a limited area, leaving the remainder untouched. Srivastava et al. (1998) emphasize that agricultural intensification is even possible while conserving and enhancing biodiversity in that same area; this requires proper agricultural practices, a supportive policy environment and institutional development.

Earlier studies also conclude that intensification of agriculture and rural income diversification through enhanced market access, creation of employment in rural areas, and technical progress in smallholder agriculture can stabilize forest frontiers by enabling smallholder farmers and rural households to earn sustainable incomes. For example, Deininger and Minten (1999), Shively (1991), Zeller et al. (2000) as reviewed by Maertens et al. (2006) find that irrigation development with subsequent increases in the level and stability of crop yields significantly reduces forest clearing.

There are also a number of studies that show that improved access to agricultural markets and technology as well as roads have an accelerating effect on forest clearing (Maertens et al. 2006). Indeed, the literature on the effect of improved technology on deforestation and agricultural land expansion is – according to Kaimowitz and Angelsen (1998) – divided into two approaches with quite opposite policy conclusions. The "population approach" based on subsistence models identifies population growth, hunger and poverty linked to local conditions of lacking market access and low levels of technology as the main drivers of agricultural expansion into upland and forest areas. Expansion into these areas leads to deforestation and soil degradation while productivity levels remain low. According to this approach, investment in human capital and technological progress through research and appropriate pro-poor technologies would result in higher agricultural productivity, and thus induce farmers to crop less land for meeting subsistence needs. However, this land-saving effect of agricultural technology has been much larger in lowland compared to upland areas as technological progress, in particular for rice, has been much faster for irrigated lowland areas compared to upland areas in Southeast Asia (Heidhues and Rerkasem 2006, von Uexkull 1998).

The "market-based approach" (Kaimowitz and Angelsen 1998), on the other hand, emphasizes the role of access to markets, institutions and technology for enhancing the profitability of agriculture as the main driver for agricultural expansion. Favourable local conditions provide smallholder farmers access to inputs and markets. While agricultural production is increased and poverty rates are falling, the environmental degradation may increase. Therefore, the pro-poor policies related to human capital, infrastructure, access to markets and institutions can lead only to sustainable development when they are coupled with policies protecting the environment and providing payments for ecological services.

These two approaches, however, do not adequately capture the governance issue linked to large-scale logging by national and multinational firms, followed by the expansion of plantations such as oil palm in Indonesia and rubber in Lao PDR. We therefore distinguish a third explanatory approach which we term the "governance approach". Here, we recognize institutional and power factors as well as individual or collective greed, corruption and policy failure playing the pivotal role in the conversion of forested land for plantation agriculture. Not smallholders are the key actors, but large-scale corporations colluding with national or local government. This type of agricultural expansion can result in increased marginalization of indigenous ethnic groups, possibly also worsening their poverty situation if they do not benefit as wage workers or outgrowers from plantation agriculture. Additionally, a large-scale destruction of natural habitats occurs. Adequate policy response would be to fight corruption, to strengthen transparent decentralized, community-based systems, to induce national reforms in order to improve governance, to strengthen the non-governmental organization (NGO) / civic sector, and to give a political voice to the poor and marginalized groups.

Depending on which of the three explanatory approaches best describes the underlying causes of land use change and related changes in socio-economic and environmental development, the preferred policy mix also will differ. In our review of case studies, we seek to identify which of the three approaches explains best the observed development. In our view, the three approaches can - depending on the conditions of a particular region - be useful in explaining the linkages between agricultural development, environment and poverty (Table 1).

2 Case studies of linkages between poverty, environment and economic growth in upland agriculture of Southeast Asia

Our hypothesis in reviewing the following case studies is that policies for sustainable development are doomed to fail if they are not properly addressing the underlying drivers of agricultural and rural development and related land use change. As the economic, social and institutional conditions and natural characteristics vary widely, so must the policies. In section 2.1, we first review findings from meta-analyses on land use change that support this view. Second, the diverse underlying drivers can be broadly attributed to the three explanatory approaches as shown in section 1, each requiring different policies.

2.1 Upland Southeast Asia: an overview

The upland areas of Southeast Asia are ecologically, economically and socio-culturally heterogeneous and hence affected by various pressures. Once richly endowed, Southeast Asian countries have in varying degrees undergone natural resource degradation and depletion, which is caused by the interplay of population and state-promoted economic growth (Heidhues and Rerkasem 2006, ADB 2000, FAO 2007). About one third of the population (80 million out of 200 million people) in the Mekong region -including Cambodia, Lao PDR, Vietnam, Myanmar and Thailand- is poor (Sunderlin 2004). According to Dauvergne (1999, cited in ADB 2000), environmental degradation tends to mainly affect the poor who live in remote areas. They frequently suffer from polluted and unsafe water, inadequate sanitation, erosion and flooding. The livelihoods of the rural poor often depend on forest resources to meet their subsistence needs. Even though there has been very little research on the poverty alleviation potential of community-based forestry management

Table 1. Major explanatory approaches for agricultural expansion in Southeast Asia

Type of approach	Underlying drivers	Actors	Local conditions	Result	Policy responses
Population approach	Poverty, food insecurity	Small-holders	Low level of technology, lack of political voice and market access	Deforestation, soil degradation, expansion of agric area at low levels of productivity	Agricultural research, investment in human capital, pro-poor technology policies
Market-based approach	Increased market access and improved infra-structure/technology raise profitability in agriculture	Small-holders	Smallholder farmers obtain access to inputs and markets	Agricultural intensification, falling poverty rates, but environmental degradation like in population approach if not coupled with policy interventions to protect environment	Pro-poor policies related to human capital, infra-structure, access to markets and institutions, but coupled with policies protecting the environment (e.g. community-based forestry, protection of parks, agricultural extension), and providing payments for ecological services
Governance approach	Institutional factors and poor governance lead to the marginali-zation of certain social/ethnic groups and their traditional land rights, and favour external investors and migrants	Large-scale companies	Lack of decentra-lized decision-making and local voice/ control,corruption, collusion, and strategic alliances by the powerful	Increased marginalization of indigenous ethnic groups, possibly rising poverty levels if these groups do not benefit as wage workers or outgrowers from plantation agriculture. Large-scale destruction of natural habitats	Fight corruption, strengthen transparent decentralized, community-based systems, national reforms improving governance, strengthening of NGO/- civic sector, and political voice of the poor/marginalized

Source: Own compilation

(CBFM) in Southeast Asia so far, it is seen as a way to improve the poor's livelihoods - if supported by policy change (Sunderlin 2004).

Fox and Vogler (2005) summarize the results of eight case studies from Thailand, Yunnan (China), Vietnam, Cambodia and Lao PDR. Research results revealed that land cover has been stable and swidden cultivation has remained the dominant land-use practice during the last 50 years even though the countries tried to control swidden cultivation through different policies, e.g. banning shifting cultivation, declaring forest reserves or implementing resettlements. Swidden cultivation is performed mainly by poor smallholders in Southeast Asia. Here, the population approach considering poverty as a causal factor of land use change and agricultural expansion largely explains the development.

Geist and Lambin (2001, cited in Fox and Vogler 2005) conducted a meta-analysis of 152 sub-national cases of tropical deforestation, 55 of these were from Asia. A major result of their study was that the causes and drivers of land cover change cannot be reduced to a single variable or to even a few variables. Geist and Lambin (2001) as well as Fox and Vogler (2005) show that most cases of agricultural expansion into forests were driven by multiple causes, with economic factors playing a major role, followed by policy and institutional, technological, socio-political or cultural, and demographic factors (Geist and Lambin 2001).

According to FAO (2007), most Southeast Asian countries have updated their forestry policies in the past 15 years, creating a legal basis to implement sustainable forest management. However, national statistics on income, employment and production of the forestry sector only focus on the formal sector, while data on the informal sector is still lacking. People in the informal sector are usually poor and therefore lack the necessary resources to practice sustainable forest management. They often do not have land and/or forest rights that make the collection of wood and other forest products "illegal". Moreover, large-scale illegal logging, which is (amongst others) enabled through corrupt government officials and high ranking members of the military, leads to a continuous decline of primary forests. While destroying the livelihoods of the rural poor, the benefits of the illegal transactions are shared between logging companies and government officials[1] This situation suggests that the governance approach provides the most useful base for proper policy design.

2.2 Cambodia and Lao PDR

Although Cambodia has twice the number of people than Lao PDR, the poverty characteristics are similar. In the Human Development Report (HDR) 2007/2008 Cambodia and Lao PDR are ranked 131 and 130 respectively, out

[1] More details on "illegal logging" are provided by EIA and Telapak (2008) and FAO (2007). WWF (2008) also analyses the role of Southeast Asia and China related to illegal wood exports to the European market.

of 177 countries (UNDP 2007). Like in other Southeast Asian countries, the majority of the people in Cambodia and Lao PDR are living in rural areas and depend on agriculture and forestry for their livelihood.

Cambodia has one of the world's highest deforestation rates, its primary rainforest cover decreased from over 70 percent in 1970 to 3.1 percent in 2006. Sokh and Lida (2001) as well as Sunderlin (2004) state that the development of CBFM in Cambodia is affected by various obstacles, including unclear and insecure tenure rights, lack of land use planning and benefit sharing arrangements. Extension for CBFM is fragmented and limited in scale and many ongoing projects lack proper evaluation or monitoring. Hence, the support of foreign donors and NGOs is considered crucial due to weak government financial resources, institutional and personal capacity.

In Lao PDR, CBFM mainly concentrates on production forests and benefit sharing arrangements for village's access to a portion of timber wealth. There are also some reforestation efforts included. Unlike in Cambodia, central authorities have a strong role in promoting and administering community forestry, with support from a limited number of international organizations and NGOs (Braeutigam 2003). In both countries, Braeutigam (2003) identified the lack of capacity in government services as one of the main obstacles for a successful implementation of the national community forestry program.

In Lao PDR, land use planning and land allocation programs are closely tied to CBFM. While the goal of these programs is to provide tenure security for rural households, to reduce shifting cultivation, and to conserve forest resources these programs prove not to be successful for disadvantaged groups and even had negative impacts on rural communities through reducing the available agricultural areas (Braeutigam 2003). Also Lestrelin et al. (2005), Fox (2000, cited in Lestrelin 2005) and Chamberlain (2002) identify the role of government policies like the relocation and the implementation of the Land-Forest Allocation program as one of the major causes of poverty, depriving people of their land and customary land use practices.

In addition, illegal logging in Cambodia and Lao PDR is a worrying issue. A report from the Hmong National Development, Inc. (HND 2008) states that in Lao PDR, illegal logging is operated by Vietnamese military-owned companies which are cooperating with Lao's military officials. They mention systemic corruption which facilitates the illegal trade and that contributes to many deaths among the Hmong ethnicity due to enforced displacement by the Lao and Vietnamese military. After harvesting, the lion's share of the illegal timber and logs are transported to Vietnam and made into furniture for foreign markets.

Table 2 summarizes the major problems regarding sustainable agricultural and rural development, policy measures and its effects in Lao PDR and Cambodia. The case studies from Lao PDR and Cambodia provide evidence for the "governance approach". Extensive policy failure, due to corruption and weak government financial resources, institutional and personal capacity led to large-scale deforestation. Consequently, the ethnic minorities have to bear the burden to make their livelihoods in even more degraded areas. These worsened preconditions lead to further unsustainable development, which then can be explained by the "population approach".

Table 2. Policy measures and its effects in Lao PDR and Cambodia.

Case	Major problems	Policy measures	Effects on Economy	Effects on Ecology	Effects on Poverty
Lao PDR, Cambodia	• Land degeneration • Enlargement of protected areas at expense of smallholders • Lack of tenure rights • Illegal logging • Corruption	• National resettlement and land reallocation policies • Community-based forest and land management • Enlargement of protected areas	• Agricultural production area ⇓ • Forest area ⇓ • (illegal) timber export ⇑	• Degeneration of agricultural & forest area ⇑	• Increased

Source: Own compilation

2.3 Vietnam

The "doi moi" reform process, which was launched in 1986, led to a dynamic economic development. In the HDR 2007/2008 Vietnam is ranked 105 out of 177 countries. This development reflects the country's successful reforms to reduce poverty. However, while the deltas documented rapid improvements, the central highlands and northern uplands have experienced a much slower economic growth. Especially the mountainous areas of northern Vietnam are underdeveloped and the poverty rate in this region was still 68 percent in 2002, which is the highest in Vietnam (World Bank 2003). The area is mainly inhabited by "socially and politically marginalized ethnic minorities" (Pandey et al. 2006:2). Deforestation, soil erosion, and loss of biodiversity are apparent signs of poverty and environmental degradation, which are both, interrelated and widespread (ADB 2002, Gomiero et al. 2000, Sunderlin and Thu Ba 2005, all cited in Pham 2005). Consequently, Vietnam's government is promoting policies to advance forest protection, sedentarization of shifting cultivators, assignment of land-use rights to farmers, and the provision of economic assistance to poor farmers. In addition to a reduced logging quota and a log export ban, the Vietnamese government expanded timber plantations to supply raw material. Nevertheless, Vietnam has a track record for importing illegal timber, first from Cambodia in the late 1990s, then from Indonesia in 2003, and now from Lao PDR (EIA and Telapak 2008).

Pham (2005) analyzes land use changes in Son La Province of northern Vietnam, a mountainous region with severe poverty and environmental problems, and their impacts on agricultural and economic growth, the environment, and in particular on forest loss and degradation, and poverty. Pham combines commune-level data for the entire Son La Province with remote-sensing data in a geo-referenced information system. According to Pham and Zeller (2006), agricultural growth in general, and expansion of crop production and the introduction of high-yielding rice and maize varieties in particular, contributed significantly to the enhancement of food security. However, agricultural expansion and intensification was undertaken on fragile hillsides, often as a result of encroachment into previously forested areas. This development led to massive forest losses during the 1980s and early 1990s. In response, the government of Vietnam initiated a major reforestation program during the 1990s. The environmental degradation through forest loss, soil degradation, and biodiversity loss certainly will contribute to future losses in agricultural productivity in Son La Province. The results by Pham (2005) suggest a likely "downward spiral" situation: agricultural growth – environmental deterioration – decline of agricultural productivity and food insecurity.

Müller and Zeller (2002) conducted research in the southeastern part of Vietnam's Dak Lak province, which has shown an increase in forest cover with currently 52 percent of forest, compared to 33 percent in the overall country (General Statistical Office 2001, cited in Müller and Zeller 2002). During the last decade, traditionally practiced shifting cultivation almost entirely disappeared, while agricultural production became locally more concentrated with potential environmental benefits for preserving the integrity of ecosystems and endangered species populations. The presented data reveals that forest regeneration at the expense of agricultural area predominantly occurs near ethnic villages, which are usually closer to forested areas and further away from all-year roads and political centers. The increase in forest cover over the last decade despite the observed population growth does not correspond to the widely stated positive correlation between higher population density and lower forest cover. In summary, Müller and Zeller (2002) showed that a combination of the right policy instruments and investments in infrastructure (e.g. improved access to roads, markets, and services) can facilitate agricultural intensification without adverse consequences for forest resources. However, the presented results are based on two purposely selected districts and are not representative of the entire province or country. In other areas of the central highlands, deforestation continued due to expansion of cash crops, e.g. coffee and pepper plantations. These plantations were also set up by external investors, collaborating with local political elites, infringing on customary land use rights of indigenous ethnic groups. The major problems for sustainable agricultural and rural development, policy measures and their effects for Vietnam are listed in Table 3.

Past land use changes in the northern highlands can be best explained by the population approach (at least during the 1990s) and later – with improvements in market access – by the market-based approach. Smallholders are here the major agents of land use change and agricultural expansion. In the central highlands of Vietnam, the agricultural expansion is caused by a multitude of causal factors that can find their basis in all three explanatory approaches.

Table 3. Policy measures and its effects in Vietnam.

Case	Major problems	Policy measures	Effects on		
			Economy	Ecology	Poverty
Vietnam	• Poverty, food insecurity • Shifting cultivation, agriculture with low productivity • Deforestation	• Reforestation • Massive infrastructure investments (roads, irrigation, etc.) • Agric. intensivication & commercialization • Dissemination of high-yielding varieties	• Agricultural productivity ⇑ • Smallholder income ⇑ • Export of wooden furniture ⇑, but no economic benefit for ethnic minorities	• Soil degeneration ⇑ • Forest cover ⇑ (=plantation forest, not primary forest) • Biodiversity ⇓	• Poverty rates ⇓ • Food insecurity ⇓

Source: Own compilation

2.4 Indonesia

In the HDR 2007/2008, Indonesia is ranked 107 out of 177 countries. The data for the case studies from Indonesia, based on Birner et al. (2002, 2006), Maertens et al. (2006), Schwarze and Zeller (2005), and van Rheenen et al. (2003), were collected in villages near the Lore Lindu National Park in Central Sulawesi, Indonesia. The park is located in an ecological and socio-cultural divers region and was, due to its rich biodiversity and high endemism, declared a World Heritage Site by UNESCO (Birner and Mappatoba, 2002). The national park's administration, in collaboration with different NGOs, played a pioneering role in promoting negotiated community agreements on conservation, which strive for dealing with major hazards of the national park.

Following Birner et al. (2006), biological diversity is still rapidly declining and encroachment of protected areas for agricultural production continues to be a major problem. Consequently, conservation organizations demand an expansion of protected areas, as well as a better enforcement of regulations in already existing protected areas. Birner et al. (2006) identified three major issues concerning the encroachment of protected areas: (1) population density in the area, (2) the availability of suitable land inside the park, and (3) customary rights. Another important issue, which was previously addressed by Chomitz and Grey (1996) and Cropper et al. (2001), is avoiding the creation of pull factors, e.g connecting protected areas and parks by roads. Birner et al. (2006:12) conclude that "strengthening law enforcement without at the same time reducing the need for encroachment created by poverty will not be a viable policy option".

Birner and Mappatoba (2002) highlight the potential of CBFM agreements for improving nature conversation and rural development in the region, as they are negotiated at the local level und thus take the specific ecological, socio-economic and cultural conditions into account. Due to their voluntary character, they can reduce conflicts and problems of state regulations. However, the overall success of the implementation is influenced by the ideals and opinions of both, the facilitating organizations and the village leaders responsible for the implementation on the local level. Hence, the "problem of unequal power relations and conflicts of interest within the villages may well jeopardize the deliberation process. The role of intermediaries, or representatives, should, therefore, be considered carefully" (Birner and Mappatoba 2002:13).

Focusing on constraints, Birner and Mappatoba (2002:25-26) found that "80 per cent of the respondents mentioned at least one problem, which the national park causes for them or their community". About half of the respondents addressed future land scarcity for their children due to park protection. Furthermore, land with traditional property rights was located inside the park and while the agreement allowed them to collect certain non-timber forest products, it did not allow them to use their traditional land for agricultural purposes. Finally, the loss of income provided by rattan collection was listed as the third most important problem.

Similar results are presented by Schwarze et al. (2007), who analyse the importance of forest products, especially for the rural poor. It was found that poor households depend most on income from forest products. Consequently, “any improvements in law enforcement concerning the collection of forest products within the national park will hit the poorest households hardest” (Schwarze et al. 2007: 221). The study presents different policy options to reduce the collection of forest products and, at the same time, improve the livelihood of the poorest households, including e.g. better access to primary education as well as the construction and improvement of irrigated rice-fields for the poorest households in order to improve the nutrition status of the family and to gain additional income.

Rosyadi et al. (2005) examine the challenges of implementing devolution policies at the local level in Banyumas district, Central Java Province, analyzing a reforestation project which allows farmers to cultivate crops on government-owned forest land during the first years after establishing a forest plantation. As in previous case studies, the boosting role of NGOs (e.g. in mobilizing village communities and creating political capital) and international donor organizations could be demonstrated. The organizations openly criticized corruption, collusion, and nepotism which finally helped to approve the villagers’ CBFM plan (Rosyadi et al. 2005). Summarizing, Rosyadi et al. (2005:25) show that devolution policies have the potential to decrease deforestation “by reducing the incentives for villagers to participate in illegal logging networks and by empowering those villagers who are concerned about the environmental damage caused by illegal logging”.

In contrast, Casson and Obidzinski (2002) and McCarthy (2000b, both cited in Rosyadi et al. 2005) present examples from Kalimantan and South Aceh respectively, where illegal logging is still common among local authorities as it guarantees substantial contributions to the government budgets. According to Larsen (2002:1), “illegal logging alone has destroyed 10 million hectares of Indonesia’s rich forests”, and it is still widespread (cf. WWF 2008). A report of Nellemann et al. (2007), published by the United Nations Environmental Programme (UNEP), also highlights that large-scale threats to Indonesia’s pristine forest areas are illegal logging done by timber companies as well as the spread of oil palm plantations and processing facilities. Nellemann et al. (2007) state that community-based monitoring and control of forested and protected areas is less effective if deforestation is driven by external companies instead of smallholders. As the largest-scale threat to Indonesian forests comes from domestic and foreign timber and oil palm companies, Nellemann et al. (2007) state: “Among the most promising and important Indonesian government initiatives is the further development, support and training of the 'SPORC'rapid response ranger units. However, it is essential that these units and selected parks have the necessary paramilitary training, equipment and mandate to prevent illegal loggers from operating inside protected areas” (Nellemann et al. 2007: 43).

Table 4 summarizes again the major problems for sustainable development, the policy measures taken and their effects on the economy, environment and poverty in Indonesia.

Table 4. Policy measures and its effects in Indonesia.

Case	Major problems	Policy measures	Effects on		
			Economy	Ecology	Poverty
Indonesia	• Food insecurity & poverty • Violation of customary land rights • Investments in roads / transmigration near protected areas • External investors, illegal logging firms or plantations • Corruption	• Agricultural development programs & agreements with local communities • Devolution policies (local level) • Community conservation agreements • Strengthening voice of community organizations • Strengthening law enforcement in protected areas	• Agricultural production area ⇑ on new forest plantations	• Illegal logging ⇓ (only in protected areas)	• Poorest people depend most on forest products ⇒ have to go to other villages for rattan collection

Source: Own compilation

The Indonesian case studies show that the governance approach - by strengthening community organizations or state-controlled ranger units controlling agricultural expansion and logging - can make important contributions. Also, as corruption is still widespread in many parts of Indonesia, institutional reforms and specific policies based on the governance approach would be most effective in combining resource and biodiversity conservation, economic development and poverty reduction.

3 Conclusion

In this paper, we distinguish three explanatory approaches to land use change and agricultural and rural development. The population approach considers population pressure linked to poverty and food insecurity as the major impetus for land use change. The market-based approach places the emphasis in the role of markets and land-saving technologies in boosting agricultural incomes and reducing poverty. The governance approach focuses on political and institutional factors that are to guide the allocation of land and control the expansion of land use driven mainly by external investors.

Overall, economic growth and agricultural development benefited smallholders in many countries. However, these economic gains can be short-lived if looming environmental problems are not addressed more effectively in the future. For example, the agricultural expansion on hillsides in northern Vietnam comes at a massive degradation of soils. A number of countries suffers from weak governance, providing possibilities for illegal logging and land acquisition at the expense of the environment, poor smallholders and ethnic minorities.

In the extremely diverse cultural, socio-economic and agro-ecological conditions prevailing in the uplands of Southeast Asia reforms and policies need to be designed to properly take into account and address the major underlying factors of agricultural and rural development. They may have to be geared primarily to deal with population pressure, poverty, and food insecurity; they may have to focus on market access and mechanisms; or they need to address governance failures. This implies that there will not be a blue-print recipe for rural development policy that adequately takes into account the relationships between social justice, economic growth and environmental sustainability.

A successful policy strategy is likely to combine specific policy measures from the three explanatory approaches, and addresses governance and market access issues as well as the nexus between population, poverty and food security. We argue that a combination of policy instruments that derive their justification from all three explanatory approaches will prove most successful in the long-run.

Acknowledgements
This paper was supported under the scope of the special research programs SFB 564 in Thailand and Vietnam, funded by the Deutsche Forschungsgemeinschaft (DFG). We also thank for the constructive comments obtained from an anonymous review.

References

ADB (2000) Environments in transition: Cambodia, Lao PDR, Thailand, Viet Nam. Programs Department (West). Asian Development Bank, Manila

Adams WM, Aveling R, Brocktington D, Dickson B, Elliott J, Hutton J, Roe D, Vira B, Wolmer W (2004) Biodiversity conservation and the eradication of poverty. Science 306: 1146-1149

Birner R, Mappatoba M (2002) Community agreements on conservation in central Sulawesi: A Coase solution to externalities or a case of empowered deliberative democracy? STORMA discussion paper series, No. 3. Georg-August-Universität, Göttingen

Birner R, Maertens M, Zeller M (2006) Need, greed or customary rights - which factors explain the encroachment of protected areas? Empirical evidence from a protected area in Sulawesi, Indonesia. Contributed paper presented at the 26th Conference of the International Association of Agricultural Economists (IAAE) in Brisbane, Queensland, Australia, August 12-18, 2006

Braeutigam D (2003) Community based forest management in Cambodia and Laos: Frame conditions, selected examples and implications. Working Paper 03. Consultancy Report. Phnom Phen: MRC-GTZ Cooperation Programme, Agricultural, Irrigation and Forestry Programme; Watershed Management Component

Coxhead I (2003) Development and the environment in Asia. Asian-Pacific Economic Literature 17 (1): 22-54

Environmental Investigation Agency (EIA) and Telapak (2008) Borderlines. Vietnam's booming furniture industry and timber smuggling in the Mekong region. http://www.eia-international.org/files/reports160-1.pdf (accessed August 2008)

Fisher RJ (2000) Poverty Alleviation and Forests: Experiences from Asia. Paper prepared for Workshop "Forest ecospaces, biodiversity and environmental security", 5 October 2000, Amman, Jordan. Pre-congress workshop. IUCN World Conservation Congress 2000

Food and Agriculture Organization of the United Nations (FAO) (2007) State of the World's Forests 2007. FAO, Rome

Fox J, Vogler JB (2005) Land-use and land-cover change in Montane Mainland Southeast Asia. Environmental Management 36 (3): 394-403

Gilmour D, Malla Y, Nurse M (2004) Linkages between community forestry and poverty. Regional Community Forestry Training Center for Asia and the Pacific (RECOFTC), Bangkok

Heidhues F, Rerkasem B (2006) IRRI's Upland Rice Research. External Assessment, http://www.sciencecouncil.cgiar.org/meetings/meeting/SC5/Item_12_IRRI_Upland_Rice_Review.pdf (accessed February 2007)

Hmong National Development (HND) (2008) (PressZoom) - Washington, D.C., August 28, 2008. Link: http://presszoom.com/story_145827.html (accessed August 2008)

Kaimowitz D, Angelsen A (1998) Economic models of deforestation: A review. Centre for International Forestry Research (CIFOR), Bogor

Larsen J (2002) High price of illegal logging. http://www.peopleandplanet.net/doc.php?id=1715 (accessed September 2008)

Lestrelin G, Giordano M, Keohavong B (2005) When "conservation" leads to land degradation: Lessons from Ban Lak Sip, Laos. Research Report 91. International Water Management Institute (IWMI), Colombo

Maertens M, Zeller M, Birner R (2006) Sustainable agricultural intensification in forest frontier areas. Agricultural Economics 34 (2), pp. 197-206

Myers N, Mittermeier RA, Mittermeier CG, da Fonseca GAB, Kent J (2000) Biodiversity hotspots for conservation priorities. Nature 403: 853-858

Müller D, Zeller M (2002) Land use dynamics in the Central Highlands of Vietnam: A spatial model combining village survey data with satellite imagery data. Agricultural Economics 27 (3): 333-354

Nellemann, C., Miles, L., Kaltenborn, B.P., Virtue, M. and H. Ahlenius (eds.) (2007) The last stand of the orangutan - State of emergency: Illegal logging, fire and palm oil in Indonesia's national parks. United Nations Environment Programme. GRID-Arendal, Norway

Pandey S, Khiem NT, Waibel H, Thien TC (2006) Upland rice, household food security, and commercialization of upland agriculture in Vietnam. International Rice Research Institute, Los Banos

Pham, M C (2005) Land-use change in the Northwestern uplands of Vietnam. Dissertation. Georg-August Universität Göttingen, Göttingen

Pham M C, Zeller M (2006) Bringing space into land-use change models: A case study in the northwestern uplands of Vietnam. Selected paper presented at the International Symposium "Towards Sustainable Livelihoods in Mountainous Regions". Chiang Mai, Thailand: Chiang Mai University, University of Hohenheim, and World Agroforestry Centre, March 7-9, 2006

Rosyadi S, Birner R, Zeller M (2005) Creating political capital to promote devolution in the forestry sector - A case study of the forest communities in Banyumas district, Central Java, Indonesia. Forest Policy and Economics 7: 213-226

Schwarze S, Zeller M (2005) Income diversification of rural households in Central Sulawesi, Indonesia. Quarterly Journal of International Agriculture 44 (1): 61-73

Schwarze S, Schnippers B, Weber R, Faust H, Wardhono A, Zeller M, Kreisel W (2007) Forest products and household incomes: Evidence from rural household living in the rainforest margins of Central Sulawesi. In: Tscharntke T, Leuschner C, Zeller M, Guhardja E, Bidin A (eds) (2007) The stability of tropical rainforest margins, linking ecological, economic and social constraints of land use and conservation.Springer Verlag, Berlin, 209-224

Snel M (2004) Poverty-conservation mapping applications. Discussion paper presented at IUCN World Conservation Congress, November 17-25 2004, Bangkok, Thailand http://www.povertymap.net/publications/doc/iucn_2004/report.cfm (accessed June 2008)

Srivastava J, Smith N, Forro DA (1998) Toward a Strategy for Mainstreaming Biodiversity in Agricultural Development. In: Lutz E (ed), Agriculture and the environment: Perspectives on sustainable rural development. A World Bank Symposium. The World Bank, Washington DC, 9-21

Sunderlin WD (2004) Community forestry and poverty alleviation in Cambodia, Lao-PDR, and Vietnam: An agenda for research. Regional Consultation Workshop for ADB-RETA 6115, 1-2 September 2004

Tonneijck F, Hengsdijk H, Bindraban PS (2006) Natural resource use by agricultural systems: linking biodiversity to poverty. Plant Research International B.V., Wageningen

United Nations Development Programme (UNDP) (2007) "2007/2008 Human development index rankings". http://hdr.undp.org/en/statistics/ (accessed April 2008)

United Nations Development Programme (UNDP) (2007a) "HDR 2007/2008 - Country data sheet Viet Nam". http://hdrstats.undp.org/countries/data_sheets/cty_ds_VNM.html (accessed January 2008)

Van Rheenen T, Elbel C, Schwarze S, Nuryartono N, Zeller M, Sanim B (2003) Encroachments in primary forests: Are they really driven by despair? In: Gerold G, Fremerey M, Guhardja E (eds) (2003) Land use, nature conservation and the stability of rainforest margins in Southeast Asia. Springer Verlag, Berlin, 199-214

Von Uexkull HR (1998) Constraints to agricultural production and food security in Asia: challenges and opportunities. In: Johnston, AE and JK Syers (1998) Nutrient management for sustainable crop production in Asia: Proceedings of an international conference held in Bali, Indonesia, 9-12 December 1996. Cambridge, UK: CAB International

World Commission on Environment and Development (WCED) (1987) Our Common Future. Oxford University Press, Oxford

World Bank (2003) Vietnam Development Report 2004 - Poverty. The International Bank for Reconstruction and Development, Washington DC

World Wide Fund For Nature (WWF) (2008) Illegal wood for the European market. An analysis of the EU import and export of illegal wood and related products. http://assets.panda.org/downloads/illegal_wood_for_the_european_market_july_2008.pdf (accessed August 2008)

Index of keywords

adaptation 352
adaptive capacity 352
agricultural expansion 448
agricultural intensification 16
agrobiodiversity 478
agroecosystem 97
agroforestry 16, 177
Agroforestry systems 462
agroforestry systems 74
amphibians 16
Andes 240
ants 16
arthropods 16

bees 16
Biodiversity 240
biodiversity 16, 462
biosphere reserve 414
birds 16
boundary demarcation 414
Brazil 448
bryophytes 16
buffer zone 414
butterflies 16
butterfly conservation 74

cacao 16
cacao agroforestry 310
carbon sequestration 432
Castanopsis acuminatissima 392
Central Sulawesi 192, 328, 414
climate change 240
cocoa 16
Cocoa agroforests 116
cocoa production 142
common property resources 142
community resource management 432
community structure 16
Conopomorpha cramerella 16
conservation 478
conservation strategies 478
cultural diversity 478
customary community 142

decomposition 16
Deforestation 511
deforestation 213, 448
development policies 511
diversification 161
dung beetles 16

economic development 213
ecosystem 462
Ecuador 478
ENSO 328
ethno-ecology 478
ethnobotany 478
extreme weather events 240

farming activities 177
focus groups 432
forest butterflies 74
forest conversion 74
forest distance 16
forest resource 462
fungal disease 16

gender empowerment 177

T. Tscharntke et al. (eds.), *Tropical Rainforests and Agroforests under Global Change*, Environmental Science and Engineering, 529–531, DOI 10.1007/978-3-642-00493-3,

Gender roles 177
global change 478
GlobCover 270
governance approach 511

habitat fragmentation 97
herbivores 16
herbivory 16
herbs 16
herbs diversity 378
household incomes 478
Human ecological dimensions 478
human ecology 478
hydraulic architecture 392
hydraulic conductivity 392
Hydrological Modelling 328
Hymenoptera 16

income dynamics 161
indigenous people 478
Indonesia 97, 161, 192, 294, 310, 352, 378, 414
innovations 352
insects 16
institutions 213, 432
intensification 116
interception 310

land access 142
land rights 213
land tenure conflicts 478
land title 142
land use change 97, 511
land-use change 16
Landsat 270
landscape ecology 16
law enforcement 414
lianas 16
livelihood strategies 478
livelihoods 462
local knowledge 352
local perception of conservation measures 478
local perception of the environment 478

mammals 16
market approach 511
Mato Grosso 448
migration 213
mitigation 352
Muridae 16

natural hazards 352
non-timber forest products (NTFPs) 478
nutrients 310
Nymphalidae 16

Panama 294
parasitoids 16
Payments for environmental services 432
Philippines 294
plants 16
plot structure 116
pollination 16, 97
population approach 511
poverty 161
poverty assessment 192
poverty indicators 192
predation 16
premontane rainforest 392
protected area management 414

range size 74
rats 16
rattan palms 16
reptiles 16
resilience 352
River discharge development 328
Rondonia 448

sap flux 294
secondary forests 74
Seville Strategy 414
shade management 116
shade trees 16
social capital 352
social vulnerability 352
socio-economic 462
soil analysis 116
South Ecuador 240
Southeast Asia 270
species range shift 240
species richness 378
spiders 16
Sulawesi 378, 392

sustainable utilization 478

tenure security 142
Theobroma cacao 16
timberline 240
time robustness of targeting tools 192
tree height 392
tree plantations 74
tree size 294
tree species 294
trees 16
trophic interactions 16
tropical biodiversity 74
Tropical Deforestation 328
Tropical deforestation 270
tropical forest cover 270
tropical landscape 97
tropical mountain forest 378
tropical mountain forests 478

vertebrates 16
vessel anatomy 392

xylem sap flux 392

yield determinants 116
yields 116